Advances in Intelligent and Soft Computing

148

Editor-in-Chief

Prof. Janusz Kacprzyk
Systems Research Institute
Polish Academy of Sciences
ul. Newelska 6
01-447 Warsaw
Poland
E-mail: kacprzyk@ibspan.waw.pl

For further volumes:
http://www.springer.com/series/4240

David Jin and Sally Lin (Eds.)

Advances in Electronic Commerce, Web Application and Communication

Volume 1

Editors
David Jin
Wuhan Section of ISER Association
Wuhan
China

Sally Lin
Wuhan Section of ISER Association
Wuhan
China

ISSN 1867-5662 e-ISSN 1867-5670
ISBN 978-3-642-28654-4 e-ISBN 978-3-642-28655-1
DOI 10.1007/978-3-642-28655-1
Springer Heidelberg New York Dordrecht London

Library of Congress Control Number: 2012932758

© Springer-Verlag Berlin Heidelberg 2012
This work is subject to copyright. All rights are reserved by the Publisher, whether the whole or part of the material is concerned, specifically the rights of translation, reprinting, reuse of illustrations, recitation, broadcasting, reproduction on microfilms or in any other physical way, and transmission or information storage and retrieval, electronic adaptation, computer software, or by similar or dissimilar methodology now known or hereafter developed. Exempted from this legal reservation are brief excerpts in connection with reviews or scholarly analysis or material supplied specifically for the purpose of being entered and executed on a computer system, for exclusive use by the purchaser of the work. Duplication of this publication or parts thereof is permitted only under the provisions of the Copyright Law of the Publisher's location, in its current version, and permission for use must always be obtained from Springer. Permissions for use may be obtained through RightsLink at the Copyright Clearance Center. Violations are liable to prosecution under the respective Copyright Law.
The use of general descriptive names, registered names, trademarks, service marks, etc. in this publication does not imply, even in the absence of a specific statement, that such names are exempt from the relevant protective laws and regulations and therefore free for general use.
While the advice and information in this book are believed to be true and accurate at the date of publication, neither the authors nor the editors nor the publisher can accept any legal responsibility for any errors or omissions that may be made. The publisher makes no warranty, express or implied, with respect to the material contained herein.

Printed on acid-free paper

Springer is part of Springer Science+Business Media (www.springer.com)

Preface

In the proceeding of ECWAC2012, you can learn much more knowledge about Electronic Commerce, Web Application and Communication all around the world. The main role of the proceeding is to be used as an exchange pillar for researchers who are working in the mentioned field. In order to meet high standard of Springer, the organization committee has made their efforts to do the following things. Firstly, poor quality paper has been refused after reviewing course by anonymous referee experts. Secondly, periodically review meetings have been held around the reviewers about five times for exchanging reviewing suggestions. Finally, the conference organization had several preliminary sessions before the conference. Through efforts of different people and departments, the conference will be successful and fruitful.

During the organization course, we have got help from different people, different departments, different institutions. Here, we would like to show our first sincere thanks to publishers of Springer, AISC series for their kind and enthusiastic help and best support for our conference.

In a word, it is the different team efforts that they make our conference be successful on March 17–18, 2012, Wuhan, China. We hope that all of participants can give us good suggestions to improve our working efficiency and service in the future. And we also hope to get your supporting all the way. Next year, In 2013, we look forward to seeing all of you at ECWAC2013.

January 2012 ECWAC2012 Committee

Committee

Honor Chairs

Prof. Chen Bin	Beijing Normal University, China
Prof. Hu Chen	Peking University, China
Chunhua Tan	Beijing Normal University, China
Helen Zhang	University of Munich, China

Program Committee Chairs

Xiong Huang	International Science & Education Researcher Association, China
LiDing	International Science & Education Researcher Association, China
Zhihua Xu	International Science & Education Researcher Association, China

Organizing Chair

ZongMing Tu	Beijing Gireida Education Co. Ltd, China
Jijun Wang	Beijing Spon Technology Research Institution, China
Quanxiang	Beijing Prophet Science and Education Research Center, China

Publication Chair

Song Lin	International Science & Education Researcher Association, China
Xionghuang	International Science & Education Researcher Association, China

International Committees

Sally Wang	Beijing Normal University, China
LiLi	Dongguan University of Technology, China
BingXiao	Anhui University, China

Z.L. Wang	Wuhan University, China
Moon Seho	Hoseo University, Korea
Kongel Arearak	Suranaree University of Technology, Thailand
Zhihua Xu	International Science & Education Researcher Association, China

Co-sponsored by

International Science & Education Researcher Association, China
VIP Information Conference Center, China
Beijing Gireda Research Center, China

Reviewers of ECWAC2012

Z.P. Lv	Huazhong University of Science and Technology
Q. Huang	Huazhong University of Science and Technology
Helen Li	Yangtze University
Sara He	Wuhan Textile University
Jack Ma	Wuhan Textile University
George Liu	Huaxia College Wuhan Polytechnic University
Hanley Wang	Wuchang University of Technology
Diana Yu	Huazhong University of Science and Technology
Anna Tian	Wuchang University of Technology
Fitch Chen	Zhongshan University
David Bai	Nanjing University of Technology
Y. Li	South China Normal University
Harry Song	Guangzhou Univeristy
Lida Cai	Jinan University
Kelly Huang	Jinan University
Zelle Guo	Guangzhou Medical College
Gelen Huang	Guangzhou University
David Miao	Tongji University
Charles Wei	Nanjing University of Technology
Carl Wu	Jiangsu University of Science and Technology
Senon Gao	Jiangsu University of Science and Technology
X.H. Zhan	Nanjing University of Aeronautics
Tab Li	Dalian University of Technology (City College)
J.G. Cao	Beijing University of Science and Technology
Gabriel Liu	Southwest University
Garry Li	Zhengzhou University
Aaron Ma	North China Electric Power University
Torry Yu	Shenyang Polytechnic University
Navy Hu	Qingdao University of Science and Technology
Jacob Shen	Hebei University of Engineering

Contents

Analysis of College Sports Consumption and the Sunshine Sports in China

Yan-qing Chen, Jian-min Qi, and Ling-juan Shi

Physical Department, Hebei University, Baoding 071002, China

Abstract. Sunshine sports consumption is not only an ordinary material consumption but also mainly spiritual consumption which was a product of the development of social productive forces to a certain stage. The sunshine sports industry's services include the provision of gym, the organization of fitness activities, fitness consultation and so on. Owing to the constant and rapid development of China's economy and the enlarged size of higher education enrollment, in the context of universal concern of health, development of sunshine sports industry seems increasingly important. This paper will also investigate impacts of its development constraints, lead people to consume rationally, and raise the level of sports consumption.

Keywords: sports consumption, sunshine sports, college students, development.

1 Introduction

In virtue of the constant and rapid development of China's economy and the enlarged size of higher education enrolment, people involve in sports activities more and more common today, and also become a main force of sports consumption which has its own characteristics and problems[1]. The appropriate sports consumption is not only beneficial to physical and mental health of people, but also helpful to the growth of consumer demands. However, the inappropriate sports consumption not only increases the financial burden on parents, but also is not conducive to the healthy development of people. Therefore, the study of sports consumption and sports industry has important theoretical and practical significance in reasonable guides to sports consumption of people[2].

2 Current Situations of Sunshine Sports Consumption of College Students in China

According to characteristics of physical sports consumption of university students, several factors to be analyzed, such as the capacity of sports consumption of college students, capacity of sports consumption of different aged university students and so on.

D. Jin and S. Lin (Eds.): Advances in ECWAC, Vol. 1, AISC 148, pp. 1–5.
springerlink.com © Springer-Verlag Berlin Heidelberg 2012

2.1 Sunshine Sports Consumption Capacity of College Students in China

Questionnaire is designed based on research purposes, 1,000 copies sent out, 967 copies recycled with a recovery rate of 96.7%. Among the recycled ones, 950 are valid with an effective rate of 95.0%.In table 1[3], it's clear to see that the average consumption ability for most college students is below 500 RMB, accounting for 70.7% of the total. The higher the sports consumption is, the smaller the number of people is. According to the synthetic statistics calculation of average number, the sports physical consumption power of university students in China is about 398.43RMB per year. Sports consumption is greatly stimulated along with rising consumption level and improving economic level in China. Since there are a large number of college students, it is apparent that university sports consumption is a huge market.

Table 1. Distribution of Sports Consumption Capacity

Amount of sports consumption	Number of people	Percentage
Below 400	454	47.8
401 --500	218	22.9
501 -- 600	186	19.5
Above 600	92	9.8
Total number of people	950	100

2.2 Types of Sports Consumption of College Students

According to the external manifestations of sports consumption, sports consumption can be divided into material sports consumption[4], participatory sports consumption, and spectator sports consumption. The material sports consumption refers to personal consumption for sports apparel, shoes, hats, sports equipment, sports books and magazines etc; participatory sports consumption refers to the fee charged for sports contributions; spectator sports consumption means to purchase sports tickets and watch sports television. All these patterns are shown in Table 2.

It is obvious that the material consumption takes the dominant place, which is followed by participatory sports consumption. college students tend to buy sportswear, equipment, newspapers and magazines, because they need to pay for their sports hobbies, while college students spend more time studying. The least is spectator sports consumption, accounting for about 14.3%.

Table 2. Types of Sports Consumption of College Students in China

Types of sports consumption	Number of people	Percentage
Material sports consumption	620	65.3
Participatory sports consumption	194	20.4
Spectator sports consumption	136	14.3
Total number of people	950	100

3 It Is in Line with the Direction of Market Economic Development

The physical fitness meets the need of market economy reforms, market economy reforms require to speed up the process of sports industry. The basic meaning of the sports industry should be to maximize the development of sports undertakings operated by industry, to give full play sports on their economic functions, to expand sports and enhance self-development space. Sports industry is in large part to emphasize that the conversion of the mechanisms and the adjustments of the mode.

Seen from abroad, in the last 1990s, sports industry has been among the top 10 backbone industries of GDP in western developed countries. In contrast, sports industry in China is still on the initial stage of development, which is with enough space to be improved. Sports consumption per capita in Europe and America is 300 to 500 US dollars and no more then 100 RMB in China, over 30 times gap produced. Sports industry contributes about 11% to the national economy in America and in China merely 0.7%, with over 15 times gap produced.

Seen from China, in 2009, 10 industrial adjustment and industrial reinvigoration layout plans were formulated in succession and in 2010, 'Guidance on Accelerating the Development of Sports Industry' was issued by General Office of the State Council, which directed the industrial reinvigoration to the sports industry on behalf of the soft strength in national economic development. with the approval of 'National Sports Industry 'in china. With the industrial adjustment and people's living standard improving in china, sports industry possesses a huge market for social requirement and has the basic capacity to build up the national economy.

4 Sunshine Sports Consumption on the Effects of Sport Industry Development in China

The key to developing sports industry is scientific and technological research, development and innovation. Good human resources and humanistic environment act as the important foundation of scientific and technological progress.

Sport consumption[5], including game attendance, can be affected by many factors. The Psychological Continuum Model suggests a four-phase hierarchy that explains the psychological connection to a sport or sport organization. This process begins with the initial awareness of the sport or organization, which leads to attraction, then develops into attachment, and finally allegiance to that sport or organization. Mass media influences the default level of awareness, which reiterates the importance of media relations and performances on developing effective strategies to foster consumer awareness.

Sports consumption is essentially a cultural phenomenon. The success of the management of fitness club inspires people's enthusiasm for participation in sports activities, driving the sports consumption by the roots. Take 1.3 billion population of China as the base, each one percentage point increase, fitness consumer will bring the value of the sports industry and the benefits are considerable.

National and government policy-oriented have a direct relationship with upgrade of the level of sports consumption[6]. Since 1995 the state promulgated the "National Fitness Program" to the State Sports General Administration issued "National Health Science and Technology Action Plan" in 2008, government has upgraded mass sports to a new height, and actively promoted the construction of information service application system of national fitness.

Approved by the State Council of China, the day each 8th, August has been regarded as "National Fitness Day". The establishment of the National Fitness Day is to adapt to the needs of the public sports consumption, improve the needs of National Fitness Campaign start; is the demand of further developing the comprehensive functions of sport and social effects, enriching sports cultural life and promoting the comprehensive development of people.

5 Conclusions and Recommendations

From the above analysis, it can be seen that although sports consumption of college students in China has made rapid progress, but it is still at a low level and at an initial embryonic stage. Compared with some developed countries, there is also a large gap.

5.1 Conclusions

(a) Consumer awareness of sports consumption is not fully established among university students. The overall level of consumption is low, and economic ability to pay is limited.

(b) Constrains of sports consumption in college students in proper sequence are poor economic conditions, lack of time, too small sports stadiums, no interest in sports, only a few consumer types, poor quality of sports venues and others.

(c) Sports consumption of students is diverse in nature. Therefore, the characteristics of college students should be paid more attention, and sports items which are suitable to college students should be increased to enhance participation in sports.

5.2 Recommendations

(a) Attach importance to sports interest of college students, establish a life-time sports values, encourage them participate in sports activities actively and help them develop a healthy lifestyle. Awareness of physical concepts and cultivation of sports consumption should be enhanced in order to enable college students to understand that sports consumption is the major forms of consumption to meet the health needs and accept that "spending money on health" is an important part to improve the quality of life.

(b) Reinforce the construction of sports facilities as far as possible, and solve the existing problems of insufficient venues. Sports facilities are sites for sports consumption, whose quantity and quality are main restraints to sports consumption for college students in China.

(c) Teachers should teach students all kinds of basic sports skills and techniques in order to arouse their interests in sports, and also adopt various forms of teaching tools such as watching the competitions to arouse the level of sports consumption of college students.

Acknowledgements. This paper is supported by the Social Science Foundation of Hebei Province (HB11TY013).

References

1. Hu, C.-W., Lu, F., Guo, W.-G.: City Leisure Sports consumption Survey and Countermeasures of the Development. Chengdu Institute of Physical Education (5), 9–11 (2009)
2. Yan, B.: Probe for development strategy of China's sports economy. Sports Culture Guide (4), 65–68 (2010)
3. Dai, J.-H., Luo, M., Gu, J.-Q.: Sports Consumer Behavior Analysis of Urban Residents in Jiangsu Province. Shanghai Institute of Physical Education (1), 18–23 (2003)
4. Zhen, J.: School Sport Consumer Behavior Research. Journal of Physical Education (6), 24–125 (2000)
5. Fu, Y.-N., Han, F.-Y.: Sports Physical Consumption Quality Analysis of Guangzhou University. Guangzhou Institute of Physical Education (2), 92–94 (2003)
6. Wang, L., Wu, W.: Analysis of Market Behavior and Performance of Sport Goods Industry in China under Industrial Organization Theory. Journal of Tianjin University of Sport 23(6), 469–471 (2009)

The Research on the Sports Consumption and the Development of Physical Fitness Industry

Yan-xia Zhang and Long Wang

Physical Department, Hebei University, Baoding 071002, China

Abstract. Sports consumption, as a product of the development of social productive forces to a certain stage, is a personal consumption behavior aiming to pursue fitness and entertainment after satisfying the existence consumption. Sports industry is the third industry of consumer services. Therefore, the interdependence of fitness and sports industry must exist. If we take as an assumption that a healthy, positive and wholesome lifestyle is the goal of individuals in society, then clearly sport should be able to contribute to achieving that goal. That implies both the consumption of and participation in sporting activities. This paper will also investigate impacts of its development constraints, lead people to consume rationally, and raise the level of sports consumption.

Keywords: sports industry, physical fitness, economy, development.

1 Introduction

The goals of fitness and competitive sport is different. The ultimate goal of fitness is to improve the nation's physical and mental qualities, reduce disease, and prolong lives. The target system needs to be supported in stages in order to achieve this ultimate goal[1]: the first step is to increase awareness of fitness; the second step is to improve the scientific nature of fitness; the last step is to improve the organization of fitness.

2 The Sports Industry Will Help the Realization of the Fitness

2.1 The Development of Sports Industry Will Increase Awareness of Fitness

With the development of sports industry, particularly in the form of the sports community, the fitness can be achieved at home. Family life and physical and mental health can get a good sense of unity and the physical fitness is from a single exercise to the comprehensive physical and mental health.

2.2 The Development of Sports Industry Will Provide the External Conditions for Fitness

Fitness has a nature of public goods consumption, it can not only rely on individuals to provide. The issuance of sports lottery is an important industry of sports industry, it

D. Jin and S. Lin (Eds.): Advances in ECWAC, Vol. 1, AISC 148, pp. 7–11.
springerlink.com © Springer-Verlag Berlin Heidelberg 2012

provides strong financial support for the realization of the national fitness plan. According to statistics[2], since 1997, national and local governments have invested 2.5 billion RMB to built 1182 sports venues, 1089 category sports venues in more than 300 cities in china.

2.3 The Development of the Sports Industry Will Provide a Special Fitness Organizations

Unlike individual fitness training, physical fitness is a planned health consumption activities, it broke through the exercise of individual dispersion and disorder, which must be a special organization to be guaranteed.

Fitness club is an important body sports industry, which has a strict form of organization and usually a membership-based management model, which allows people to natural health spending according to plan. The state have allocated 24 million RMB for creation of fitness clubs which provided a strong guarantee to conduct fitness activities.

2.4 The Development of the Sports Industry Will Provide Scientific Guidance

The final evaluation criteria for fitness is not only to participate in, but also a substantial increase in the physical and mental state. Sports industry is not only able to attract the mass organizations, but also able to guide their scientific fitness activities to solving their various problems in physical and mental health, which is very important.

3 The Fitness Will Help the Development of the Sports Industry

Sports industry needs to be supported by popular consumer demand. Sports industry is the third industry. Person's time is divided into two parts: the working time and leisure time. Only more choices of leisure time, sports consumption will have the support of time in order to develop the sports industry. China's urbanization level is low, people have less leisure time, sports consumption will to overcome this obstacle in order to become a mass consumer[3].

3.1 Fitness Will Promote the Development of Popular Consumer Demand and Market Potential

China's Engel coefficient declined significantly in moderate consumption stage, and the spending of spirit consumer increased[4]. As fitness itself, including moral, intellectual, physical health and overall development, which can not only improve the national physical fitness, but also has its unique role in terms of improving the quality of the national culture.

Fitness is a human capital investment because it can improve people's health status, reinforce their ability to work, extend life expectancy and working hours, thereby increasing people's income. On the other hand people's exercise and fitness activities are carried out in free time after work, this time can be used for rest, Fitness replace the general sense of the rest which is more conducive to the elimination of fatigue and physical recovery.

In summary, fitness will become a consumer trends in new economic conditions, and the same time it has accumulated a lot of purchasing power for the sports industry.

3.2 Fitness Will Provide a Broad Market for the Sports Industry

The public sport has played a significant role in the development of sports industry from foreign experience. Fitness has a great potential market which has been covered 1.2 billion people in china. People have participated in sports activities with unprecedented enthusiasm for the correct guidance of national fitness program. The new pattern of the sports industry has been formed which promoted the development of the sports industry.

3.3 Fitness Will Provide New Opportunities for the Sports Industry

The development of fitness clubs need to break the traditional forms which will introduce scientific methods into general fitness sports venues. In addition, fitness will be given new content in the knowledge economy era, the development of sports industry can also rely on fitness to drive.

4 Major Constraints of College Students' Sports Consumption

It is bound to help guide students' sports consumption and foster awareness of sports consumption by studying its specific constraints and identifying its crux. The survey results in Table 1 reveal that correct values of the health of sports consumption, regular economic income, more abundant leisure time, proper environment and places for consumption, and enough choices of consumer items are main factors influencing and constraining sports consumption behavior[5].

Table 1. Major constraints of sports consumption of college students in China

Constraints factor	Percentage
Fewer financial resources	63.52
Little spare time	18.25
Small sports venues	15.78
No interest in sports	2.45

4.1 Consumption Values

Consumption values are the primary constraint which influences college students' sports consumption. Survey results suggest that college students not interested in sports consumption is only 2.45%, indicating that values of sports consumption for fitness establishes fundamentally among the students, but on the other hand, the health physical education should also be deepened further among them.

4.2 Economic Situation

Economic situation is the main constraint which limits students' sports consumption. It can be seen in the above tables that 63.52% of the students still consider that current paying ability greatly restricts their sports consumption behavior. Obviously, it is the most direct and important impact on students sports consumption. Most students play the role of consumers in society, but don't have an independent economic resource, and most of them rely on parents for monthly living expenses for living, learning and entertainment etc. Under the premise of limited economic capacity, they render it is justifiable to spend on living and learning, and it is followed by recreation and leisure consumption. However, the former proportion of consumer spending is second to the later one because its high prices keep many students away.

4.3 Physical Environment

Physical environment is an important constraint of college students' sports consumption. Sports consumption values of students are subject to multiple effects. It is subject to not only their learning process, the acceptance of sports and health education, but also their living environment and social environment. Once the atmosphere for sports consumption is formed, taking a little time to relax can be accepted even if study is busy. In addition, these objective factors such as the lack of sports venues, limited consumption items to choose etc. also greatly restrict sports consumption to some extent.

5 Conclusion and Suggestion

The physical fitness is not only provide fast-growing economy with huge demand stimulation, which is propitious to the development of local technology, but also arouse the enthusiasm of economic prospects and stimulate consumption which helps to promote the formation and development of the sports industry chain[6].

Sports industry is one of the important material bases to sports economic development and social progress, which has an important stimulating role on sports economy and even national economic growth, and is the emergence of modern human economy in the form of a new industrial economy. It can be seen that, China's sports industry is still in its infancy, there is much room for development.

The government should overall plan the excavation, exploitation and utilization of management mode of the fitness club[7]. Based on Chinese current situation, the government should to promote industrial interaction of sporting industry, sports tourism Industry, leisure sports industry, and the standard of sports consumption and so on, and the same time to support cyclic development of sports industry chain.

The sports consumption plays an important role to the physical fitness and sports industry. However, China's sports market mechanism is in the transitional stage of reform, the ability of regulating can not fully achieve the interaction of physical fitness and sports industry, which must be supported by the Government's industrial policy and macro-control.

Acknowledgements. This paper is supported by the project of livelihood research of Hebei Province (201101142) and by the project of Department of Education of Hebei Province (S090603).

References

1. Tian, H., Zhou, H.: Leisure, Recreational Sport, and Developing Trend in China. China Sports Science 26(4), 23–26 (2006)
2. Zeng, G.: Think about the development of China's sports industry. Guangming Daily, September 17 (2008)
3. Bao, M.: Basic theoretical issues of the sports industry. Sports Scientific Research 26(4), 22–29 (2005)
4. Wang, L., Wu, W.: Analysis of Market Behavior and Performance of Sport Goods Industry in China under Industrial Organization Theory. Journal of Tianjin University of Sport 23(6), 469–471 (2009)
5. Zhou, W.: Effects of Sports Events on the Economic Development of the Host City. Business Times 36 (2008)
6. Yan, B.: Probe for development strategy of China's sports economy. Sports Culture Guide (4), 65–68 (2010)
7. Li, D.: Current Situations and Prospects of Sports Industry in China. Journal of Physical Education 4(3), 11–15 (2005)

L_2 Performance Index Robust Control of Arc Furnace Electrode Regulator System

XiaoHe Liu, Nan Gao, and Yuan Gao

School of Automation, Beijing Information and Technology University,
Beijing100192, China

Abstract. Electric arc furnace electrode regulator system is a strong coupling and strong non-linear system. Nonlinear robust control problem lies in how to solve the HJI inequality, but there is no universal method to obtain the solution of HJI inequality. In this paper, it is given a new nonlinear robust controller design method which based on the exact state feedback linearization, linear robust control theory and L_2 interference suppression performance theory. Then it has made some simulations to prove the tracking control law and robust control law.

Keywords: feedback linearization, L_2 performance index, disturbance, simulation.

1 Introduction

Arc furnace electrode regulator system is the heat generated by the electrode and metal ore smelting furnace. It is made of graphite electrode, using as three-phase electrode with metal burden from the arc of high temperature between smelting metal. Because of the electric arc furnace electrode regulator system of its own three-phase coupled complex characteristics, the traditional PID control and fuzzy control is difficult to achieve good control effect. In this paper, the robust adaptive control is by far the best option.

2 Description of Electric Arc Furnace Electrode Regulator System

Because the main circuit time constant of electric arc furnace is far less than that of electric arc furnace adjusting electrode regulation system in the process of time. So when do not care about the arc voltage and current transient waveform, the main circuit can be considered as nonlinear components that the length of electric arc is mapped to the current of electric arc, such as the following formula

$$I = h(L) \tag{1}$$

The block diagram of electric arc furnace regulator system is shown in Fig.1.

D. Jin and S. Lin (Eds.): Advances in ECWAC, Vol. 1, AISC 148, pp. 13–21.
springerlink.com © Springer-Verlag Berlin Heidelberg 2012

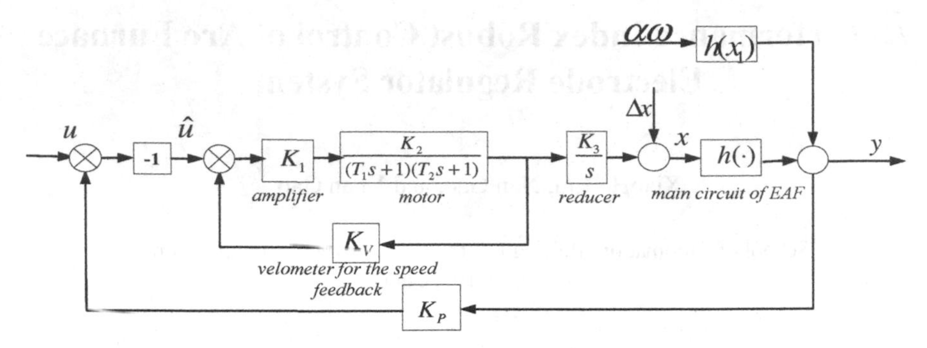

Fig. 1. Block diagram of simple-phase arc furnace electrode regulation system

Where I is RMS of arc current, and L is the arc length.

Then $\frac{d^3x}{dt^3} = \frac{-(T_1+T_2)}{T_1T_2}\frac{d^2x}{dt^2} + \frac{-(1+K_1K_2K_V)}{T_1T_2}\frac{dx}{dt} + \frac{-K_1K_2K_3}{T_1T_2}\{u - K_p\left[h(x_1+\Delta x)+\alpha\omega h(x_1)\right]\}$

Let $x_1 = x$, $x_2 = \dot{x}$, $x_3 = \ddot{x}$,

Then $\frac{dx_1}{dt} = x_2$, $\frac{dx_2}{dt} = x_3$, $\frac{dx_3}{dt} = a_1x_3 + a_2x_2 + a_3[u - K_p\left[h(x_1+\Delta x)+\alpha\omega h(x_1)\right]]$

Where $\alpha_1 = \frac{-(T_1+T_2)}{T_1T_2}$, $\alpha_2 = \frac{-(1+K_1K_2K_V)}{T_1T_2}$, $\alpha_3 = \frac{-K_1K_2K_3}{T_1T_2}$.

3 Electric Arc Furnace of Electrode Regulation System Feedback Linearization Tracking Control

For the following single-input single-output system

$$\begin{cases} \dot{x} = f(x) + g(x)u \\ y = h(x) \end{cases} \tag{2}$$

Where x is n-dimensional vector; $f(x)$ and $g(x)$ are n-dimensional smooth vector functions; u and y are scalar quantities; $h(x)$ is scalar function. When the relative degree of the system (2) is the same as the order of this system, the system can be described as exact feedback linearization system. The state feedback equation of the system is

$$u = \frac{1}{L_g L_f^{\rho-1} h(x)}[-L_f^{\rho} h(x) + v] \tag{3}$$

The input and output are simplified as the following form

$$y^{(\rho)} = v \tag{4}$$

Where ρ is the degree of the system; v is feedback control signal.

When the disturbance of system is 0, the system is described as followed

$$\dot{x} = Ax - BK_p h(x_1) + Bu$$
$$y = h(x_1)$$

For system (2), finding its Lie derivative, relative degree of the system is 3, and it is the same as the degree of the system. So the electric arc furnace electrode regulator system can be described as exact feedback linearization system.

Let $z_1 = h(x_1)$, $z_2 = L_f h = \frac{\partial h}{\partial x_1} x_2$, $z_3 = L_f^2 h = \frac{\partial^2 h}{\partial x_1^2} x_2^2 + \frac{\partial h}{\partial x_1} x_3$, then the state feedback equation of the system is

$$u = \frac{1}{L_g L_f^2 h}(-L_f^3 h + v) = \frac{1}{a_3}[a_3 K_p h(x_1) - a_2 x_2 - a_1 x_3 + v] \quad (5)$$

Then the original nonlinear system can be transformed into a linear system, and the linear system is (6) form

$$\begin{cases} \dot{z}_1 = z_2 \\ \dot{z}_2 = z_3 \\ \dot{z}_3 = v_1 \end{cases} \quad (6)$$
$$y = z_1$$

The above equation can be written as $\begin{cases} \dot{z} = Az + Bv \\ y = Cz \end{cases}$,

where $A = \begin{bmatrix} 0 & 1 & 0 \\ 0 & 0 & 1 \\ 0 & 0 & 0 \end{bmatrix}, B = \begin{bmatrix} 0 \\ 0 \\ 1 \end{bmatrix}, C = [1 \ \ 0 \ \ 0]$.

Because the system requirements of the input signal having a derivative exist, not directly with the Laplace domain as input signal. So with a three order system step response to order the output of the idea of what the ideal as output y_d. Purpose is to choose an appropriate feedback control signal, making the actual output can asymptotically track reference output. Choose feedback control signal for:

$$v_1 = y_d^{(3)}(t) - \sum_{i=1}^{3} c_{i-1}(z_i - y_d^{(i-1)}) \quad (7)$$

Where, c_0, c_1, c_2 are backlog constants. Define the actual output and the reference for the error between the output for: $e(t) = y(t) - y_d(t)$. Then

$$e^{(i-1)}(t) = y^{(i-1)}(t) - y_d^{(i-1)}(t) = z_{(i)}(t) - y_d^{(i-1)}(t)$$

So the feedback signal can be written as $v_1 = y_d^{(3)}(t) - \sum_{i=1}^{3} c_{i-1} e^{(i-1)}(t)$,

then $\dot{z}_3 = y_d^{(3)}(t) - \sum_{i=1}^{3} c_{i-1} e^{(i-1)}(t)$,and $\dot{z}_3 = y^{(3)}(t)$, so

$$\dddot{e} + c_2 \ddot{e} + c_1 \dot{e} + c_0 e = 0 \quad (8)$$

Because $c_0, \cdots, c_{r-1}$ are arbitrary setting, so type (8) corresponding to the characteristics of the differential equation root can any configuration. It can set $c_0, \cdots, c_{r-1}$, making all features are located in the plane of the complex root left half open, namely $\mathrm{Re}(s_i) < 0$, even located in $\mathrm{Re}(s_i) < \sigma_0$. So it can make error to faster than the rate of the output of the attenuation $e^{-\sigma_0 t}$, and system can asymptotically track reference output.

4 Based on the L_2 Performance of the Control Law

For the following system

$$\begin{cases} \dot{x} = f(x) + g_1(x)w + g_2(x)u \\ y = h(x) \end{cases} \quad (9)$$

Where $x \in R^n$, $u \in R^m$, $w \in R^p$ and $y \in R^s$ are state vector, control vector, vector and adjust output vector interference; $f(x)$, $g_1(x)$, $g_2(x)$ and $h(x)$ are smooth vector functions, and meet $f(0) = 0$, $h(0) = 0$.

For system (9) of the nonlinear robust control problem is to find a small enough $\gamma^* > 0$ and a control strategy $u = u^*(x)$ to make it all $\forall \gamma > \gamma^*$:

$$\int_0^T (\|y\|^2 + \|u\|^2) dt \le \gamma^2 \int_0^T \|w\|^2 \, dt \quad \forall T \ge 0 \quad (10)$$

And at that time $w = 0$ closed-loop system is asymptotically stable.

System (9) of the nonlinear robust control problem of the penalty functions to take the following output: $z(t) = \begin{bmatrix} h(x(t)) \\ 0 \end{bmatrix} + \begin{bmatrix} 0 \\ u(t) \end{bmatrix}$.

For convenience, in the research system (9) of the feedback linearization when hypothesis $x \in R^n$, $w \in R^2$, $u \in R$, $y \in R$. Definition, $g_1(x) = [g_{11}(x) \quad g_{12}(x)]$, $w = [w_1, w_2]^T$.

Hypothesis from control u to output y , the relationship ρ is n . From interference w_1 and w_2 to the output y, the relationship are μ_1 and μ_2. And using μ to represent the minimum value between μ_1 and μ_2, obviously $\mu \le n$.

Using differential geometry, structure the following form of coordinate transformation and the nonlinear feedback:

$$z = KT(x), u = \alpha(x) + \beta(x)u \tag{11}$$

The nonlinear affine system

$$\begin{cases} \dot{x} = f(x) + g_2(x)u \\ y = h(x) \end{cases} \tag{12}$$

Conversion of linear system as follows:

$$\begin{cases} \dot{z} = Az + B_2 \upsilon_2 \\ y_z = Cz \end{cases} \tag{13}$$

Where

$$z = \begin{bmatrix} z_1 \\ \vdots \\ z_\mu \\ z_{\mu+1} \\ z_{\mu+2} \\ \vdots \\ z_n \end{bmatrix} = KT(x) = \begin{bmatrix} k_1 h(x) \\ \vdots \\ k_\mu L_f^{\mu-1} h(x) \\ k_{\mu+1} L_f^{\mu} h(x) \\ k_{\mu+2} L_f^{\mu+1} h(x) \\ \vdots \\ k_n L_f^{n-1} h(x) \end{bmatrix}$$
$$\alpha(x) = L_f^n h(x) \qquad \beta(x) = L_{g_2} L_f^{n-1} h(x)$$

Where, $L_f^r h(x)$ representatives of the first order Lie along the function derivatives $f(x)$; $0 \le r \le n$; $K = diag[k_1, k_2, \ldots k_n]$ are pending diagonal constant matrix. The geometric meaning can be understood as a vector in mapping ϕ from space x to space z length compression ratio. That is a factor (A, B_2, C) by Brunovsky standard.

In the coordinate transformation and the feedback system (11), the function of system (9) can be written for

$$\begin{cases} \dot{z} = Az + B_2 \upsilon_2 + \dfrac{\partial KT(x)}{\partial x} g_1(x) w \\ y_z = Cz \end{cases} \tag{14}$$

Assume

$$\bar{w} = \frac{\partial KT(x)}{\partial x} g_1(x) w = \begin{bmatrix} 0 & 0 \\ \vdots & \vdots \\ 0 & 0 \\ k_\mu L_{g_{11}} L_f^{\mu-1} h(x) & k_\mu L_{g_{12}} L_f^{\mu-1} h(x) \\ k_{\mu+1} L_{g_{11}} L_f^{\mu} h(x) & k_{\mu+1} L_{g_{12}} L_f^{\mu} h(x) \\ \vdots & \vdots \\ k_n L_{g_{11}} L_f^{n-1} h(x) & k_n L_{g_{12}} L_f^{n-1} h(x) \end{bmatrix} \begin{bmatrix} w_1 \\ w_2 \end{bmatrix} \tag{15}$$

So the system (14) can be written for

$$\begin{cases} \dot{z} = Az + B_1 w + B_2 \upsilon_2 \\ y_z = Cz \end{cases} \tag{16}$$

Where $B_1 = \begin{bmatrix} 0_{(\mu-1)\times(\mu-1)} & 0_{(\mu-1)\times(n-\mu+1)} \\ 0_{(n-\mu+1)\times(\mu-1)} & I_{(n-\mu+1)\times(n-\mu+1)} \end{bmatrix}$.

The following linear $H\infty$ control theory can be used to design robust control law of the system. The solvability condition of the system's robust control is that if and only if the Riccati inequality

$$A^T P + PA + \frac{1}{\gamma^2} PB_1 B_1^T P - PB_2 B_2^T P + C^T C < 0 \tag{17}$$

There is a non-negative solution P^*. Now the optimal control strategy υ_2^* is :

$$\upsilon_2^* = -B_2^T P^* z \tag{18}$$

The worst possible interference $\overline{w}^*$ is: $\overline{w}^* = \frac{1}{\gamma^2} B_1^T P^* z$

From the (11) and (18) can know that the control law u^* of the x coordinate is:

$$u^* = -\beta^{-1}(x)\left[\alpha(x) + B_2^T P^* KT(x)\right] \tag{19}$$

Where u^* is that the nonlinear robust control law of the original system which expressed by the (9) . The following from differential game will do brief explanation.

The so-called robust control problem is for $\forall T > 0$, structure u^*, to make control law performance index

$$J(\upsilon, \overline{w}) = \int_0^T (\|y\|^2 + \|u\|^2 - \gamma^2 \|w\|^2) dt \tag{20}$$

As far as possible severe interference minimum, this problem under the equation can be equivalent described as (9) of the constraints, the countermeasures to solve differential problem:

$$\min_{\upsilon \in L_{2e}} \max_{w \in L_{2e}^2} J(\upsilon, w) \le 0 \tag{21}$$

For the same type (16) description of the linear system, its robust control problem is tectonic control law to make performance indicators

$$\overline{J}(\upsilon_2, \overline{w}) = \int_0^T (\|y_z\|^2 + \|\upsilon_2\|^2 - \gamma^2 \|w\|^2) dt \tag{22}$$

As far as possible severe interference in the minimum, namely

$$\min_{\upsilon\in L_{2e}} \max_{\bar{w}\in L_{2e}^{n-\mu+1}} \bar{J}(\upsilon_2,\bar{w}) \le 0 \tag{23}$$

Add (11) into (22), and considering (23), and get

$$\int_0^T (\|CKT(x)\|^2 + \|\alpha(x)+\beta(x)u^*\|^2)dt \le \gamma^2 \int_0^T (\left\|\frac{\partial KT(x)}{\partial x} g_1(x)w\right\|^2)dt \le \gamma^2 c_0^2 \int_0^T \|w\|^2 dt \tag{24}$$

Where $c_0^2 = \sup \tau(x)$; $\tau(x)$ is the matrix maximum Eigen value of function $g_1^T(x)\frac{\partial^T KT(x)}{\partial x}\frac{\partial KT(x)}{\partial x} g_1(x)$.

Inequality (24) is the solution of robust control of the system

$$\begin{cases} \dot{x} = f(x) + g_1(x)w + g_2(x)u \\ \bar{y} = \begin{pmatrix} CKT(x) \\ \alpha(x)+\beta(x)u \end{pmatrix} \end{cases} \tag{25}$$

and the corresponding closed-loop system from the gain L_2 (w to $\bar{y}$) do not over γc_0.

According to the above account, type (19) that the control law is the optimal control strategy for system (9) of the performance index.

$$\min_{\upsilon\in L_{2e}} \max_{w\in L_{2e}^2} \hat{J}(u,w) = \min_{\upsilon\in L_{2e}} \max_{w\in L_{2e}^2} \int_0^T (\|y\|^2 - \gamma^2 \|w\|^2)dt \le 0$$

In the type of system (25) is a vector. Therefore, control law (19) for system (9) has robust.

5 System Simulations

The parameter values of the electric arc furnace electrode regulation system in Figure 1 are as follows:

$K_1 = 30.49$, $K_2 = 2.45$, $K_3 = 0.038$, $T_1 + T_2 = 0.172$, $T_1T_2 = 0.000425$, $K_V = 0.05$, $K_P = 1$ .The transfer function of the model third-order system is $\frac{2000}{s^3 + 15s^2 + 120s + 2000}$. Choose $c_0 = 19000$, $c_1 = 2900$, $c_2 = 45$.

So feedback linearization tracking control is $v_1 = y_d^{(3)}(t) - \sum_{i=1}^{3} c_{i-1}(z_i - y_d^{(i-1)})$.

According to (17), to solve Riccati inequality get a nonnegative solution:

$$P^* = \begin{bmatrix} 2.5982 & 2.6013 & 2.1574 \\ 2.6013 & 3.4641 & 2.8228 \\ 2.1574 & 2.8228 & 4.0418 \end{bmatrix}.$$

According to the type (18) get linear robust control law is:

$$v_2 = -B_2^T P^* z = -2.1574 z_1 - 2.8228 z_2 - 4.0418 z_3$$

The simulated results are as follows:

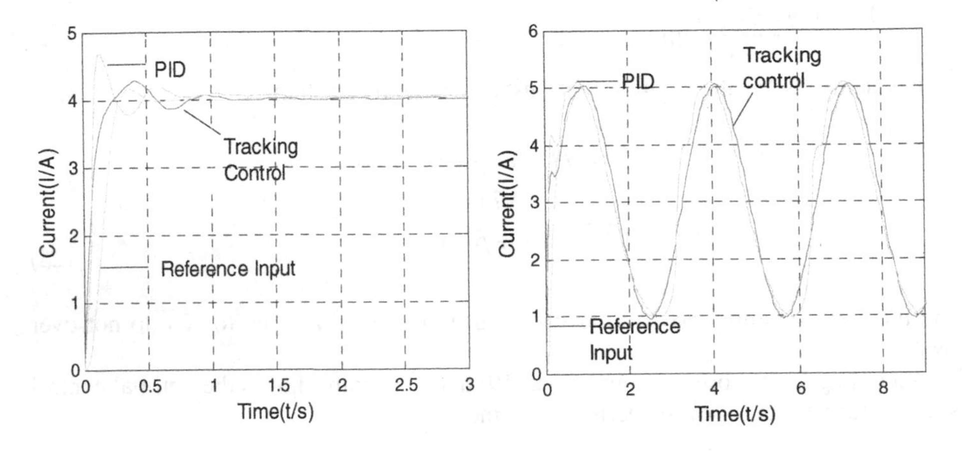

Fig. 2. Simulated result of Step response

Fig. 3. Simulated result of Sine -wave response

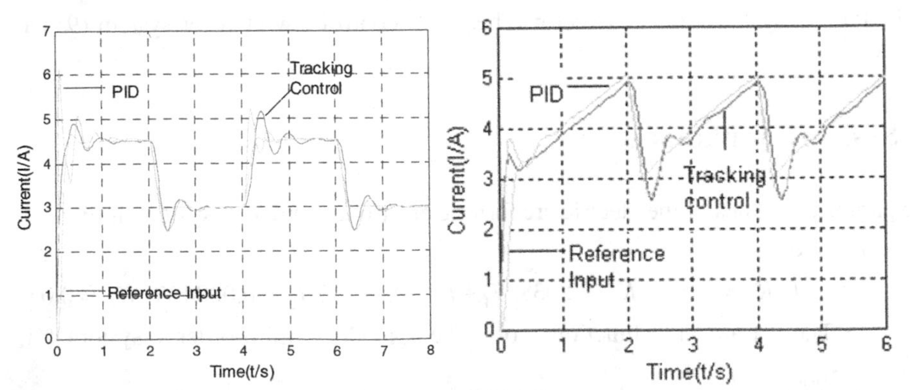

Fig. 4. Simulated result of Square-wave response

Fig. 5. Simulated result of Triangle-wave response

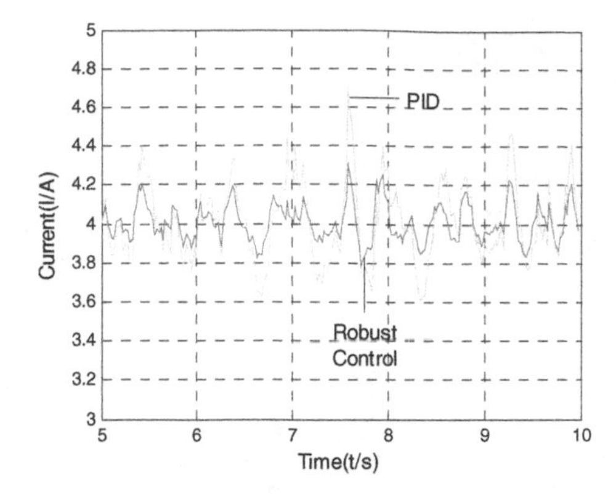

Fig. 6. Simulated result of Step response by adding White noise

6 Conclusion

This paper described based on the L_2 performance of the robust adaptive control method can be used in electric arc furnace. Through electrode regulation system of single-phase electric arc furnace electrode regulation system simulation results can be seen that adjust process is short, the small overshoots, tracking the effect is good. The simulation results of the same kind of nonlinear system for robust control, and has strong theoretical meaning and reference value.

References

1. Kuai, R., Liu, X.-H.: Electric arc furnace of electrode regulation system fuzzy-pid control study. Machine and Hydraulic 36(7) (July 2008)
2. Zhang, X.-Z., Liu, X.-H.: Robust model reference adaptive control of Arc furnace electrode regulation system. Automation and Instrumentation (6) (2005)
3. Zhu, Y.-H., Jiang, C.-S., Li, L.-Y.: A class of nonlinear uncertain system robust adaptive control. Data Acquisition and Processing 18(3) (September 2003)
4. Hassan, H.A., Rao, M.P.R.V.: Novel e-Modification Robust Adaptive Control.Scheme Using Non-Quadratic Lyapunov Functions for Higher Order Systems. Serbia & Montenegro, Belgrade, November 22–24 (2005)
5. Zhang, Y.-H., Liu, X.-H.: Based on the feedback linearization of electric arc furnace control electrode regulation system. Machine and Hydraulic 36(7), 264–267 (2008)
6. Liu, X.-H., Guan, P., Liu, L.-H.: Adaptive control theory and application. Science Publishing Company (January 2010)

A Foreground Segmentation Scheme

Shilin Zhang, Heping Li, and Shuwu Zhang

High Technology and Innovation Center, Institute of Automation,
Chinese Academy of Sciences, Beijing, China
zhangshilin@126.com

Abstract. Speech signal, video caption text and video frame images are all key factors for a person to understand the video content. Through above observation, we bring forward a scheme which integrating continuous speech recognition, video caption text recognition and object recognition. The video is firstly segmented into a number of shots by shot detection. Then the caption text recognition and speech recognition are carried out and the results are treated as two paragraphs of text. Only the noun words are kept. The words are further depicted as a graph. The graph vertices stand for the words and the edges denote the semantic relation between two neighboring words. In the last step, we apply the dense sub graph finding method to mine the video semantic meaning. Experiments show that our video semantic mining method is efficient.

Keywords: Video Semantic Mining, Auto Speech Recognition, Information Fusion, Object Recognition.

1 Introduction

There has been an explosive growth of available videos on the internet. Good examples are the video sharing website "youtube", hosting billions of videos with thousands of uploads each minute. The advent of such collections has sparked research on video semantic mining within these collections to implement a system processing videos without human intervention; see Snoek and Worring for a recent overview [1]. One active line of research is on using only the visual contents for concept-based management tasks. But in this paper, we intend to integrate three aspects of information, namely the visual content, the speech signal and the video caption text. The Bag-of-Words method [2-3] has proven to be the most efficient strategy as a generic classification scheme for individual concepts. This is proven by their top performance in various major benchmarks over the past few years such as the TRECVID high-level feature extraction task [4] and the Pascal [5]. In these benchmarks, concept detectors are able to detect classes such as chair, cat, car, boat, building, meeting, and sports with varying degrees of success [4-5].But just the visual contents are not enough for video content understanding and management. Multi-modal information fusion based video semantic mining is the frontier research direction, as is reported in [7].

Section 2 presents the speech, object and caption text recognition procedures. Section 3 shows the text processing and the graph construction. Section 4 depicts dense sub graph finding procedure and the video semantic mining method. Experiments on

D. Jin and S. Lin (Eds.): Advances in ECWAC, Vol. 1, AISC 148, pp. 23–27.
springerlink.com © Springer-Verlag Berlin Heidelberg 2012

real world videos are shown in Section 5. We end the paper with a conclusion and some prospective future work in Section 6.

2 Video Content Analysis

2.1 Optimization Problem

Through above analysis, the video is represented by three high level semantic features, namely the Text_{ASR} , Text_{VOCR} and Text_{OBJECT} . They can be seen as three sub spaces of the video high level feature space. The three folds of texts can be processed by part of speech and stemming, and only the noun words are kept to form three bags of words. So the video feature space can be seen as a tensor which constitutes three vectors. We can formally denote it as:

$$\Psi \in \Re^{\mathrm{Word}_{OBJECT},\ \mathrm{Word}_{ASR},\ \mathrm{Word}_{VOCR}} \quad (1)$$

In the optimization procedure, we check every word in every modal. If a word is far away from the other two modal's words in semantic relation measure, in other words, if there are no corresponding words in the other two modals, so this word is isolated and will be dropped off. Though the above procedure is simple and comprehensible, the computation complexity is terrible. So in the next paragraph we transform the above optimization problem to a sub graph finding procedure.

2.3 Dense Sub Graph Finding

Three communities represented by three color circles stand for the speech recognition words, caption text recognition words and object recognition words respectively. If a word is recognized wrong, then its semantic relation with other words should be isolated and the relation measure must be far away from the other modal's words. By dense sub graph finding procedure, we can find the most compact clique in the graph which forms a semantic clustering. The isolated vertex can be dropped by the methods because its semantic is not near to any other words. The densest sub graph problem which is used to find a sub graph with the maximum average degree without any constraint of the sub graph can be solved in polynomial time. Algorithms from Goldberg and Lawler find the densest sub graph in polynomial time using a series of s-t min cuts problems. The min cut is defined by two sets of disjoint vertices (A and B), one containing the source (s) and one containing the sink (t) such that the sum of the edges from A to B is minimized. The in-degree of the source must be zero, and the out-degree of the sink must also be zero.

The following describes Goldberg's algorithm to calculate the densest sub graph in a graph. Goldberg's algorithm sets up a network flow graph by creating a source and a sink to add to an existing graph (G). Every undirected edge with weight w in the original graph is replaced by two directed edges with weight w. An edge from the source to every vertex in G is added with weight $m' = \Sigma_{e\epsilon E(G)}$ w(e), where E(G) is the set of all edges in G and w(e) is the weight of edge e. An edge from every vertex in G to the sink is added with weight $m' + 2g - d_i$, where g is a guess of the density of the densest sub graph and d_i is the sum of all weights of edges in G adjacent to vertex i. The algorithm maintains an upper bound (u) and lower bound (l), initially set to zero and m' respectively. It performs a binary search while $l\text{-}u \geq 1/(n*n\text{-}1)$, where n is the number of vertices in the graph. In each iteration, g is set to (u+l)/2 and the network flow graph are reconstructed based on the current g. Then, the min cut is calculated. If the source is the only vertex on the source side of the cut (A), then u = g, otherwise l = g. When the algorithm terminates, the densest sub graph is equivalent to the vertices on the source side of the cut minus the source, with density g if G is unweighted. If G is weighted then g is not actually a guess of the density of G but merely a variable. The density must be calculated when the algorithm terminates and finds the set of vertices in the densest sub graph.

If the density of the densest sub graph is D and all edges in the residual graph with weight less that α are set to zero, then all sub graphs with density $D + \alpha$ will be computed. The following outlines the algorithm to find all densest sub graphs and "almost" dense sub graphs:

```
Step1:
    l ← 0; u ← m';
    while u - l ≥ 1/(n*(n-1)):
           g ← (u + l) / 2;
      Construct the flow network N as described
    above;
           Find min-cut {A, B};
           If A = = {s} then u ← g;
         Else l ← g;
Step 2:
   Construct the residual graph R of N where edges <
     are set to   zero
    Compute a graph of the strongly connected compo-
    nents of R, called SCC
    Remove T and its predecessors from SCC
     Remove S and its successors from SCC, called
     them SS
   All densest subgraphs ← closure on remaining SCC
  combined with (SS - {s})
```

The dense sub graph contains the most important and representative words and they are more believable. The bag of words found by the dense sub graph can be used to classify videos or annotate a video.

3 Experiment

We evaluate the method on TRECVID dataset and Chinese CCTV programs. The TRECVID dataset comprises 100 hours of documentary video with ~40, 000 key frames. The dataset is equally divided into development and test sets, with each containing approximately 50 hours and around 20,000 key frames. In our experiments, we use the 30 semantic concepts listed in Figure 6 used in the TRECVID valuation. We used the ground truth annotations as baseline from MediaMill with a lexicon of 101 semantic concepts but we used only the most interesting 30 ones. Performance was then evaluated on the test set. We used the ground truth annotations as baseline from MediaMill with a lexicon of 101 semantic concepts but we used only the most interesting 30 ones. Performance was then evaluated on the test set.

In Table 1, we illustrated some multi-modal recognition fusion results. The first column denotes the key frame of a video shot. The second column lists the annotation of our system.

Table 1. Video shot annotation result

Shot image	annotation	Shot image	annotation
	Tree Environment Child Grass Spring		Swim Woman Track Champion Record
	Train Railway Hill Success Plan	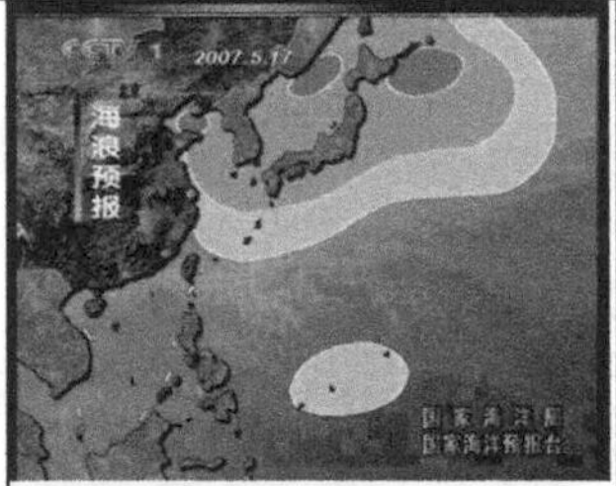	Climate Weather Sea Wind Shore Forecast
	Cloud Sea Weather Program		Summer Tree Dew Grass

4 Conclusion

We have presented a semantic mining method which exploits the three aspects of modals in videos. By a graph based late fusion method, the text, audio and visual information can be unified to annotate a video. The experiments on TRECVID and CCTV news programs showed that the graph based fusion can outperform other fusion method, such as baseline method, with statistically significant improvements. Several issues remain open. Our method depends on the accuracy of caption text recognition, speech recognition and object recognition. So the achievements in the above three aspects will boost out method. Secondly an effective semantic similarity and correlation computation method will make our method more accuracy.

References

1. Snoek, C.G.M., Worring, M.: Concept-based video retrieval. Trends Inf. Retriev. 4(2), 215 (2009)
2. Csurka, G., Dance, C., Fan, L., Willamowski, J., Bray, C.: Visual categorization with bags of keypoints. In: Proc. ECCV Int. Workshop Statistical Learning in Computer Vision, Prague, Czech Republic (2004)
3. Sivic, J., Zisserman, A.: Video Google: A text retrieval approach to object matching in videos. In: Proc. IEEE Int. Conf. Computer Vision (2003)
4. Smeaton, A., Over, P., Kraaij, W.: Evaluation campaigns and TRECVID. In: Proc. ACM SIGMM Int. Workshop Multimedia Information Retrieval (2006)
5. Winn, J.: The PASCAL Visual Object Classes Challeng 2010 (VOC 2010) Development Kit. Tech. Rep., University of Leeds (2010)
6. Zhang, X., Liu, Y., Liang, C., Xu, C.: A visualized communication system using cross-media semantic association. In: Proceedings of the 17th International Conference on Advances in Multimedia Modelling, Taipei, Taiwan, January 05-07 (2011)
7. Hanbury, A., Müller, H.: Automated component-level evaluation, present and future. In: Proceedings of the 2010 International Conference on Multilingual and Multimodal Information Access Evaluation: Cross-Language Evaluation Forum, Padua, Italy, September 20-23 (2010)
8. Lee, H., Yu, J., Im, Y., et al.: A unified scheme of shot boundary detection and anchor shot detection in news video story parsing. Multimedia Tools and Applications, 1127 (2011)
9. Open Source Toolkit For Speech Recognition, `http://cmusphinx.sourceforge.net/`
10. Rleon, M., Mallo, S., Gasull, A.: A tree structured-based caption text detection approach. In: Proceedings of the Fifth IASTED International Conference on Visualization, Imaging, and Image Processing, p. 220 (2005)

Research on Fuzzy Comprehensive Evaluation Based on Rough Set Theory

XiangHui Li[1] and KeNing Da[2]

[1] Economic College of Shenyang University, Shenyang, China, 110044
[2] Management College of Shenyang Architecture University, Shenyang, China, 110168

Abstract. In order to overcome the subjective characteristic of the existing evaluation method, we use a fuzzy comprehensive evaluation method based on rough set theory. Through attribute reduction, it can determine the weight coefficient of indexes according to attribute significance. The weight coefficient will be obtained based on attribute significant calculating in information system. The real case indicates that our presented method can solve the conflict of multi-objective and fuzzy, further, it has good feasibility in practice.

Keywords: evaluation method, human capital, rough set theory(RST).

1 Introduction

RST is presented by Poland mathematician Pawlak in 1982. It is a data analysis tool for fuzzy and uncertain knowledge. After 1990, scholars from many countries take more concern on rough set theory. Now, it becomes one of the most active research area of information science. By attribute reducing, the fuzzy comprehensive evaluation method based on rough set can effectively get weight coefficient through attribute significance. Thus, it can overcome shortcoming of weight determine subjectivity.

In this paper, we take index of human capital force for example and use RST to determine fuzzy comprehensive weight coefficient. In further, the presented method can effectively deal with un-accurate, un-consistent, un-complete information and reveal potential rule.

2 Steps of Fuzzy Comprehensive Evaluation Based on RST

Fuzzy comprehensive evaluation base on rough set obeys the following steps:

Step 1. Construction of decision information list
In decision list $S=<U,A>$, U denotes domain, also means the set of evaluated small towns, and A denotes index system constituted by all evaluation indexes.
Step 2. Attributes reduction
According to attributes reduction and the concept of core, we calculate the undistinguishable relationship to obtain a reduction in decision list, and then get the main index system of core competitiveness for small towns.
Step 3. Calculation of attribute importance

D. Jin and S. Lin (Eds.): Advances in ECWAC, Vol. 1, AISC 148, pp. 29–34.
springerlink.com © Springer-Verlag Berlin Heidelberg 2012

In line with the importance of each evaluation index and learning from AHP, we calculate the weights of indexes on the bottom level, and then turn to the indexes on the higher level. In this way, we will get the importance of indexes on every level.
Step 4. Calculation of each index weight
Step 5. Calculation of index evaluation coefficient on all levels
Step 6. Entire evaluation
By the weights and values of all attribute, we make a comprehensive decision to obtain the rules of describe information, and evaluate each object at last.

3 Human Capital Force Evaluation Index

We evaluate human capital force for different areas and divide it into 2 secondary indexes and 17 third grade indexes shown in table 1.

Table 1. Human capital force evaluation index

Population structure	Non-agricultural population and agricultural population proportion(%)
	Annual growth rate difference between non-agricultural population and agricultural population
	Resident proportion for the person whose age less than 15 years old (opposite)
	Resident proportion for the person whose age more than 65 years old (opposite)
	Proportion of actual work population in total population
	Difference between foreign population and out population
Talent and education state	Proportion of university degree or above population in total population
	Proportion of secondary education population in total population
	Proportion of high school diploma population in total population
	Proportion of junior high school education population in total population
	Proportion of illiterate population in more than 15 years old population (opposite)
	Consumption expenditure on cultural and educational entertainment products and services (yuan)
	Proportion of talent population in total population (%)
	Talent density index
	Average payment of cities and towns work unit (yuan)
	Growth rate of total staffs (%)
	Wage growth rate of total staffs (%)

4 Attribute Reduction of Human Capital Force Bases on RST

4.1 Reduction of Population Structure Index

Take seven towns A, B, C, D, E, F, G for example, we implement discretization for continuous numerical of population structure, which shown in table 2. C1-C6 denote six third grade indexes of population structure (in table 1).

Table 2. Evaluation information system of population structure

Town	NO.	C1	C2	C3	C4	C5	C6
A	1	3	3	3	2	3	1
B	2	2	2	2	3	2	1
C	3	1	2	2	2	1	3
D	4	1	2	1	1	1	3
E	5	1	2	1	1	2	3
F	6	1	2	2	1	2	3
G	7	2	1	2	3	2	2

Decision System $S=<U,C>$, in which $U=\{1,2,3,4,5,6,7\}$
$C=\{C1,C2,C3,C4,C5,C6\}$
$U/ind(C)=\{\{1\},\{2\},\{3\},\{4\},\{5\},\{6\},\{7\}\}$. By undistinguishable relationship, we can get $U/ind(C_3,C_5,C_6)=\{\{1\},\{2\},\{3\},\{4\},\{5\},\{6\},\{7\}\}$
Thus, $\{C_3,C_5,C_6\}$ is one of the reductions, but it is not the only one.
Calculate importance of each attribute:
$U/ind(C-\{C_3\})=\{\{1\},\{2\},\{3,4\},\{5,6\},\{7\}\}$
$U/ind(C-\{C_5\})=\{\{1\},\{2\},\{3,6\},\{4,5\},\{7\}\}$
$U/ind(C-\{C_6\})=\{\{1\},\{2,6,7\},\{3\},\{4\},\{5\}\}$
Let, $P=U/ind(C)$, $Q=U/ind(C-\{C_3\})$, $S=U/ind(C-\{C_5\})$,
$T=U/ind(C-\{C_6\})$
$POS_Q(P)=\{1,2,7\}$, $POS_S(P)=\{1,2,7\}$, $POS_T(P)=\{1,3,4,5\}$
Each attribute's importance is as follows:

$$\mu_{C_3}=\frac{|pos_P(P)|-|pos_Q(P)|}{|pos_P(P)|}=\frac{4}{7}$$

$$\mu_{C_5}=\frac{|pos_P(P)|-|pos_S(P)|}{|pos_P(P)|}=\frac{4}{7}$$

$$\mu_{C_6}=\frac{|pos_P(P)|-|pos_T(P)|}{|pos_P(P)|}=\frac{3}{7}$$

The weight coefficients for $\{C_3, C_5, C_6\}$ are as follows:

$$\omega_{C_3} = \frac{\mu_{C_3}}{\mu_{C_3} + \mu_{C_5} + \mu_{C_6}} = 4/11, \ \omega_{C_5} = \frac{\mu_{C_5}}{\mu_{C_3} + \mu_{C_5} + \mu_{C_6}} = 4/11, \ \omega_{C_6} = \frac{\mu_{C_6}}{\mu_{C_3} + \mu_{C_5} + \mu_{C_6}} = 3/11.$$

In order to avoid the dimension affection, we use discrete index coefficient for evaluation. The evaluation results are as follows:

$$F_1 = A_1 * \omega = \begin{bmatrix} 3 & 2 & 2 & 1 & 1 & 2 & 2 \\ 3 & 2 & 1 & 1 & 2 & 2 & 2 \\ 1 & 1 & 3 & 3 & 3 & 3 & 2 \end{bmatrix}' * \begin{bmatrix} 4/11 \\ 4/11 \\ 3/11 \end{bmatrix}$$

$$= [\ 2.45 \quad 1.72 \quad 1.90 \quad 1.54 \quad 1.90 \quad 2.27 \quad 2]'$$

Thus, we get the evaluation coefficient of 7 towns shown in table 3.

Table 3. Evaluation coefficient1

Town	A	B	C	D	E	F	G
Coefficient	2.45	1.72	1.90	1.54	1.90	2.27	2

4.2 Reduction of Talent and Education State Index

Through discretization for continuous numerical of talent and education state, we get information system which shown in table 4. C1-C11 denote 11 third grade indexes of talent and education state (in table 1).

Table 4. Evaluation information system of talent and education state

Town	NO.	C1	C2	C3	C4	C5	C6	C7	C8	C9	C10	C11
A	1	3	3	3	3	3	3	2	1	1	1	2
B	2	1	1	2	3	2	3	2	1	1	3	1
C	3	1	1	1	3	2	2	2	2	1	2	1
D	4	1	2	1	2	1	1	3	3	1	3	2
E	5	1	1	1	3	1	2	2	1	2	1	2
F	6	1	1	1	1	1	2	1	1	1	1	2
G	7	1	2	1	3	3	3	3	2	3	1	3

Decision System $S = \langle U, C \rangle$, in which $U = \{1,2,3,4,5,6,7\}$
$B = \{C_1, C_2, C_3, C_4, C_5, C_6, C_7, C_8, C_9, C_{10}, C_{11}\}$

$U / ind(B)=\{\{1\},\{2\},\{3\},\{4\},\{5\},\{6\},\{7\}\}$. By undistinguishable relationship, we can get $U / ind(C_2,C_8,C_9,C_{11})=\{\{1\},\{2\},\{3\},\{4\},\{5\},\{6\},\{7\}\}$
Thus, (C_2,C_8,C_9,C_{11}) is one of the reductions, but it is not the only one.
Calculate importance of each attribute:

$$U / ind(B-\{C_2\})=\{\{1,6\},\{2\},\{3\},\{4\},\{5\},\{7\}\}$$

$$U / ind(B-\{C_8\})=\{\{1\},,\{2,3\},\{4\},\{5\},\{6\},\{7\}\}$$

$$U / ind(B-\{C_9\})=\{\{1\},\{2\},\{3\},\{4\},\{5,6\},\{7\}\}$$

$$U / ind(B-\{C_{11}\})=\{\{1\},\{2,6\},\{3\},\{4\},\{5\},\{7\}\}$$

Let $P=U / ind(B)$, $Q=U / ind(B-\{C_2\})$, $S=U / ind(B-\{C_8\})$, $T=U / ind(B-\{C_9\})$, $M=U / ind(B-\{C_{11}\})$

$POS_Q(P)=\{2,3,4,5,7\}$, $POS_S(P)=\{1,4,5,6,7\}$, $POS_T(P)=\{1,2,3,4,7\}$, $POS_M(P)=\{1,3,4,5,7\}$

Each attribute's importance is as follows:

$$\mu_{C_2}=\frac{|pos_P(P)|-|pos_Q(P)|}{|pos_P(P)|}=\frac{2}{7}\quad \mu_{C_8}=\frac{|pos_P(P)|-|pos_S(P)|}{|pos_P(P)|}=\frac{2}{7}\quad \mu_{C_9}=\frac{|pos_P(P)|-|pos_T(P)|}{|pos_P(P)|}=\frac{2}{7}$$

$$\mu_{C_{11}}=\frac{|pos_P(P)|-|pos_M(P)|}{|pos_M(P)|}=\frac{2}{7}$$

The weight coefficients for (C_2,C_8,C_9,C_{11}) are as follows:

$$\omega_{C_2}=\frac{\mu_{C_2}}{\mu_{C_2}+\mu_{C_8}+\mu_{C_9}+\mu_{C_{11}}}=1/4\quad \omega_{C_8}=\frac{\mu_{C_8}}{\mu_{C_2}+\mu_{C_8}+\mu_{C_9}+\mu_{C_{11}}}=1/4\quad \omega_{C_9}=\frac{\mu_{C_9}}{\mu_{C_2}+\mu_{C_8}+\mu_{C_9}+\mu_{C_{11}}}=1/4$$

$$\omega_{C_{11}}=\frac{\mu_{C_{11}}}{\mu_{C_2}+\mu_{C_8}+\mu_{C_9}+\mu_{C_{11}}}=1/4$$

In order to avoid the dimension affection, we use discrete index coefficient for evaluation. The evaluation results are as follws:

$$F_2=A_2*\omega=\begin{bmatrix}3&1&1&2&1&1&2\\1&1&2&3&1&1&2\\1&1&1&1&2&1&3\\2&1&1&2&2&2&3\end{bmatrix}'*\begin{bmatrix}0.25\\0.25\\0.25\\0.25\end{bmatrix}$$

$$=[\,1.75\quad 1\quad 1.25\quad 2\quad 1.5\quad 1.25\quad 2.5\,]'$$

Thus, we get the evaluation coefficient of 7 towns shown in table 5.

Table 5. Evaluation coefficient 2

Town	A	B	C	D	E	F	G
Coefficient	1.75	1	1.25	2	1.50	1.25	2.5

5 Summaries

Since evaluation index are at different levels in multi-index evaluation system, they have different effecting on evaluation, even though they are at the same level, every attribute may have difference in significance. Therefore, it is important to determine the weight coefficient in fuzzy comprehensive evaluation, which will affect the evaluation result directly. By the analysis in this paper, we use fuzzy comprehensive evaluation based on RST to reduce attributes of core competency for human capital force indexes. The method can resolve the conflict problem of fuzzy and multi-object, it also overcomes the subjective in weight determination during evaluation.

Acknowledgements. The authors would like to thank Humanities and Social Sciences Planning Fund of Ministry of Education (no. 10YJA790097), Social Sciences Planning Fund of Liaoning Province (no. L10BJL033), science and technology project of shenyang (no.F11-263-5-19)for the financial support of this research. We thank the anonymous reviewers for their careful review and valuable suggestions on the manuscript.

References

1. Almor, T., Hashai, N.: The competitive advantage and strategic configuration of knowledge-intensive, small- and medium-sized multinationals: a modified resource-based view. Journal of International Management 10, 479–500 (2004)
2. Zhao, H.B.: The competitiveness of the industry of-a theory and reviewed in this paper. The Contemporary Financial 12 (2004)
3. Zhou, H.S., Wu, Y.M., Lu, W.C.: City competitiveness evaluation index and methods. The Economy 12 (2003)

Research of Engine Department Team Based on Engine Room Resource Management

DeZhi Jiang

Qingdao Ocean Shipping Mariners College, Jiangxistr. 84,Qingdao, China
Jiangdzh@coscoqmc.com.cn

Abstract. In order to study engine department team of engine room resource management, introduced the related research results of team type, elaborated manifestation of the engine department team, proposed how to cultivate the engine department team spirit. According to "engine room resource management" the application environment, proposed carries on the team work through engine simulator to the crew's training. The training result showes that it plays an important role to improve the team spirit of the crews.

Keywords: engine room resource, engine department team, team spirit.

1 Introduction

The engine room resource management is the compulsion training content which the STCW Manila amendment increases newly, it belongs to the management science category. It is a management science concrete branch and the application, it involves the scope is really broad, the actual content is also quite complex. And human resources management is core of engine room resource management. Because, in the engine room resource management system the human is a main body, engine room each work must depend upon the human to complete. The engine room management is the man management to a great extent. The engine room resource management's goal is that engine department personnel guaranteed that ships' safe navigation, reduces and avoids the latent human error accident, when has in the emergency case or engine room current management. Meanwhile, Plays each person's in team work role, carries out the related work operation sequence methodically. Therefore, the construction of engine department team is an important part in the engine room resource management.

2 Team Types

The team may according to the different way to carry on sets up. Team's type also has many kinds. The west manages the scholar also often to make concrete to team's research in the specific team type. The early time are not many in view of the team itself type's research. In recent years, the scholar only then starts to pay attention to the team the type. For example, Sundstrom et al. divide into the staff officer team, the producer services team, the project development team, the motion negotiations team.

D. Jin and S. Lin (Eds.): Advances in ECWAC, Vol. 1, AISC 148, pp. 35–40.
springerlink.com © Springer-Verlag Berlin Heidelberg 2012

And Suan and the Diane viewpoint most has the representation. They had summarized the team research in the literature, then they has divided four kind of team types (table 1): work team, parallel team, project team and management team.

Table 1. Team type and team characteristic

Team type	Team characteristic
work team	Long-term stable member. for example producer services team.
parallel team	The trans-departmental personnel compose, the unofficial organization unit. for example decision-making team.
project team	Has the time period, for example new product development team.
management team	Coordinated and instruction company overall level business through the judgment, helps the company to enhance the achievement or the competitive power, for example high-level.

3 Engine Department Team

In the reality team's manifestation are always many and varied. When we constructs the team, We pondered that from three aspect's thoughts.

3.1 Problems and Solutions

Some abroad experts divide team into the policy-based team, the problem solving team and the creative team. Facing these questions known and solution known, we must construct the policy-based team. Facing these questions known and response options unknown, we must construct the problem solving team. Questions and response options for the unknown, we need to play the role of creative team change.

When the organization confronted with problem to try to construct the team, we must pay attention to each team type the demand are different. The policy-based team's main demand is each person needs to know how to do. The members of problem-solving team need to know where the problem lies and have been granted limited autonomy or freedom to solve the problem. Creative team possible need complete freedom.

The work and tasks of engine room are varied. The problems facing the engine room are also ever changing. Sometimes is some routine work. Its problems and solutions are known. At this time, engine department team is a policy-based team. Sometimes is the failure processing. Its problems known and response options unknown. At this time, engine department team is a problem-solving team. Therefore, problems and solutions of different combinations, there is a different team types. Question known, the solution from unknown evolves known, team evolved into the policy-based team from the problem-solving team.

3.2 Member and Task

The team members constituted team's objective entity. Team members are coming together by the task, so they have a collaboration. The internal members of team have the specific role. When we constructs the team, the member mainly manifests in three latitudes: team cycle, member stability and member operating time disposition. Static team was composed of stable member. It also consistent with the expectations of the future and a common participation. On-the-job time of dynamic team member is often shorter. The join and leave of members depends on task. Members may also participate in different tasks within the team and other. The tasks undertaken by the team usually has two kinds: convention duty and non-convention duty. The convention duty is refers to the people can complete task prospective according to the usual rule. But the non-convention duty is arises suddenly, cannot define beforehand, require a variety of knowledge on techniques in order to better solve the task. According to the above thought and organization's special details, we may set up the different type team.

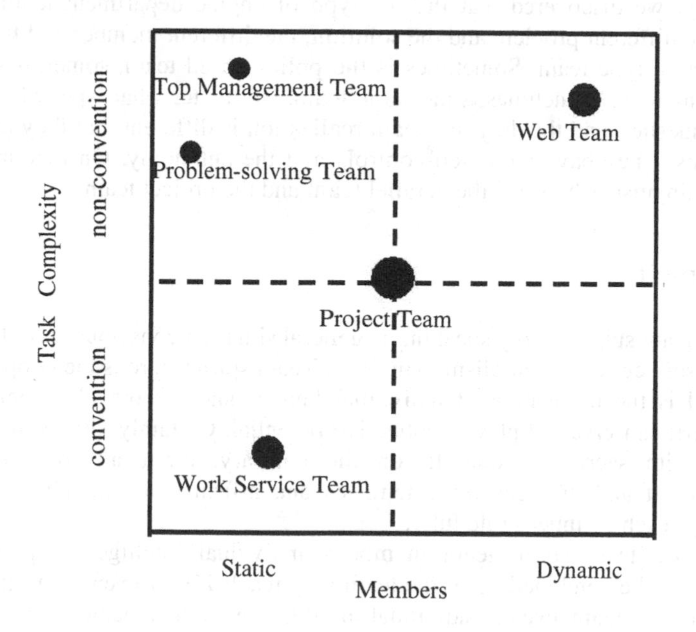

As for the member latitude, the engine department members have the specific role, such as chief engineer, second engineer, third engineer, forth engineer and motorman on duty, etc. In certain time, the engine department member very little changes. They participate in the team work stably. The member often has the approximate skill. They cooperate with one another, complete the work of engine room together, therefore, the team might regard as the static team. As for the duty latitude, some are the convention tasks, for example the engine room duty, by now, might regard as is the work service team; Some are the non-convention tasks, for example emergency treatment of various faults suddenly of engine room, by now, might regard as is the problem-solving team. Therefore, different combinations of member and task, there is a different team types.

3.3 Authority and Autonomy

When we constructs the team in the organization, authorization concern and the management pattern possibly are different to different types team. The parallel team which composes from the different department engaged in generally part-time task, and team's existence period is limited. Therefore this kind of team often has the low autonomy. The project team often needs to solve a concrete duty in the limited period. At this time, it needs the independent management and autonomous. The work team has a stabler member and the fixed duty, therefore it need more self-controls and autonomy.

The member and the task of engine department are stabler, conforms to characteristic of the work team, therefore it has more self-controls and the autonomy.

In summary, we discovered that the the type of engine department team is not changeless, the different problem and the solution, the different member and the duty, have the different type team. Sometimes is the policy-based team, sometimes is the problem-solving team, sometimes is the static team. No matter what type, when faces the different question and the duty, the team realization is different, but they are both the work teams. They have more self-controls and the autonomy. On this spot, the work team distinguishes between the parallel team and the project team.

4 Team Spirit

The so-called team spirit, simply speaking is general situation consciousness, the team spirit and the service spirit centralism manifests. Team spirit's core is the cooperation, what reflected is the unification of individual benefit and the overall benefit. The good team spirit can give full play to collective potential. Certainly, the team spirit is not premise with sacrificing oneself, on the contrary, the team spirit respects individual interest and the achievement, raises and affirms each member's special skill, thus plays each member's role fully.

The team has team spirit, team member's individual intelligence quotient is possibly 100, but the combined IQ of the team may reach 150 or even higher;; in turn a lack of team spirit team, even if individual intelligence quotient achieves 120, Team grouped together only 60 to 70 IQ. The key elements of this situation occurs is the cultural composition of the team, which is called team spirit.

4.1 Team Spirit Reflected on the Ship

When we constructs the team in the organization, authorization concern and the management pattern possibly are different to different types team. The parallel team which composes from the different department engaged in generally part-time task, and team's existe.

The ocean ship are far away from the land most of time, completion of each job relies on co-operation by the crew. If you only have the “Let every fox take care of his own tail” attitude is not enough, especially in special circumstances, each one goes his own way, it will not only cause the entire ship's task not to be able to complete, but also cause present the dangerous situation and the dangerous situation. The good team spirit manifests at least four aspects in the ship:

1) The good team spirit may against accident's occurrence, be beneficial in the safety. Accident's occurrence has various factor, human's factor contained a high percentage, everybody cooperates, each other reminder, the accident certainly large scale will reduce.
2) The good team spirit is helpful to munication between the crew, realizes be on time of ship. The production cost for efficiency is not an empty slogan, we should all work together to be able to achieve significant results.
3) The good team spirit may promote the development of crew individual in enterprise. Each person possibly meets like this or such question in the work, if communicates frequently with the periphery person, promptly will hoarkaed some contradictions, will solve the related problem, will also have the promotion and the help to individual enterprise, once had the development opportunity also to be able the very good assurance.
4) The good team spirit may improve the personality, consummates individual quality. Each person has each strong point and the shortcoming respectively in collective. Only then integrates this team, we will looked for the best in others, simultaneously can also see in the comparison own insufficiency.

4.2 Cultivate Team Spirit

The good team spirit may the harmonious ships atmosphere, eliminate the negative influence which each pressure brings. Makes the good team spirit on the ships, its particularity requests our each people to undertake the responsibility, all hands on deck, unity is strength.

1) construct with a atmosphere of confidence in each other
2) Setting up effective machinery for crew communication
3) Reinforcement learning of professional knowledge and professional dedication
4) a leading role of ship managers

5 Team Training Based on Simulator

What engine room resource management emphasis is management skills of marine engineer in the team and the team work, the correspondence and the communication, the leadership and policy-making, the raise and maintain of scene consciousness. These skill are uneasily-realized through some operation and management of equipments. Therefore, a person's management level not easy direct quantification, only in this process of completing the task it is able to manifest. Full mission engine room simulator is best platform of completing these tasks. That is, engine room simulator is only a carrier, through completing the different task on the simulator, plays the role of each person in team work, guarantees ships' safe navigation.

The content is various in engine room resource management. It is to complete the training through different task. These tasks may divide into the current management and the emergency response.

In the current management, completes engine room S/B using the team, navigation duty of on-duty group, the contact of engine room and bridge, the simulation deduces of PSC.

In the emergency response, we may establish the different breakdown in the normal navigation, for example the main engine sudden stop, the emergency reverse, the main engine system failure, the power plant breakdown, the entire ship lose the electricity, the ships collision. So that inspection team situational awareness, decision-making capacity, communication and coordination ability, the ability of contingency deployment.

6 Conclusion

Based on the analysis of team type, this research proposed the manifestation of engine department team and training of team on the engine room simulator. Its key point is that completing the different task on this platform, enables the engine personnel to use each kind of mechanical power equipment, the safety equipment effectively, plays each person's role in team work, thus completes the related work. The training result indicated that it has the influential role to raises crew's team consciousness as well as the team spirit.

Ackowledgement. This work is supported by COSCO(Projece No:2011-1-H-008).

References

1. Jiang, H., Sun, L.: How to construct the team: Team type and construction thought. Shanghai: Shanghai Econmic Review, 87–91 (May 2007)
2. Sundstrom, E., Demeuse, K.P., Futrell, D.: Work team: application and effectiveness. America: American Psychologist 45, 120–133 (1990)
3. Hetherington, C., Flina, R., Mearns, K.: Safety in shipping: The human element. Journal of Safety Research 37(4), 401–411 (2006)
4. Hu, F.: On the Factors Influencing the Work Eff iciency of the Global Virtual Team and Possible Solutions. Kunming: Journal of Yunnan Nationalities University 22(6), 105–108 (2005)
5. Liang, E.-S.: Human factor and the engine room resource management. China Water Transport 9(9), 52–53 (2009)
6. Young, C.: The STCW Convention. IMO Technical Support Program (1995)
7. Jiang, D., Zhao, X.: Engine Room Resource Management Using Engine Room Simulator. Shanghai: Navigation of China 34(1), 22–25 (2011)
8. Quick, T.L.: Successful Team Building. American Management Association, New York (1992)

Trajectory Optimization and Analysis for Near Spacecraft's Unpowered Gliding Phase

Yongyuan Li[1], Yi Jiang[1], and Chunping Huang[2]

[1] School of Mechatronical Engineering, Beijing Institute of Technology, Beijing, China
lyy6912@yeah.net

[2] R&D Centre, China Academy of Launch Vehicle Technology, Beijing, China

Abstract. According to the disadvantages of indirect method for trajectory optimization, that initial value the for conjugate variables are highly sensitive and can not obtain the global optimal solution, this paper introduced hybrid genetic algorithm to search initial value of conjugate variables, solved equations to obtain the track of Near Spacecraft at unpowered gliding phase. In the optimization process, the author takes the constraints of heating rate, dynamic pressure and overload into consideration, and optimizes the trajectory of under different initial factors. Numerical simulations confirm the applicability of the presented algorithm. Considering practical applications, some typical spacecraft gliding trajectory are optimized. The discussion of the numerical simulations gives some valuable opinions in practical applications.

Keywords: Near Spacecraft, hybrid genetic algorithm, Trajectory Optimization.

1 Introduction

The pathways which reentry vehicle returns to the atmosphere from the exoatmosphere are multitudinous. According to reentry process of reentry vehicle whether be effected by aerodynamic lift can be divided into zero-lift reentry (also known as ballistic reentry) and lifting reentry. The lifting reentry is further divided into ballistic - lift-style reentry (i.e. reentry of small lift-drag ratio, L/D≤0.3~0.5) and lift-type reentry (i.e. reentry of large lift-drag ratio, L/D>1.0~1.5). Among them, the lift-style reentry vehicle can change its re-entry trajectory through the use of vehicle control technique to adjust the force aircraft.

The numerical methods for Optimal Control Problems are divided into two categories: indirect method and direct method. In the indirect method, the first-order necessary conditions for optimal variation method that can be obtained from the optimal control problem are derived. These necessary conditions are uesed to form a Hamiltonian Boundary Value Problem (HBVP), and then we can obtain the optimal trajectory through the numerical method to slove the optimal problem. The main advantages of indirect method are its solution accuracy and it can ensure that the solution to meet the first-order optimality conditions. The shortcomings of indirect method be embodied in that the process of derive the optimal solution are more complex and cumbersome; the domain of convergence is very small for solving two point boundary value; The unknown boundary conditions initial estimates of high

D. Jin and S. Lin (Eds.): Advances in ECWAC, Vol. 1, AISC 148, pp. 41–48.
springerlink.com © Springer-Verlag Berlin Heidelberg 2012

precision; and many indirect method for solving problems need to estimate the initial value of coordination variables, these variables without physical meaning, and so to increase of initial estimates difficult. In the direct method, the continuous-time optimal control problem is converted to a non-linear programming (NLP) problem solver. Direct method including direct shooting method that only discrete control variables and point collocation method that discrete both control variables and state variables. Advantages of direct method is: do not need to derive the first order optimal conditions, the convergence domain of the indirect method is more broad, less demanding on the accuracy of initial estimates; drawback is no guarantee of non-linear programming problem solution is optimal.

Currently, two methods are used in the trajectory optimization problem widely. Zhao Jisong and oter authors are used indirect method for Design and optimization of hypersonic skip-glide trajectory, Zhou Hao and other authors are used indirect method for Time-shortest Trajectory Optimization for Hypersonic Vehicle. Wu Delong and other authors based on the maximum principle for optimal air-assisted orbit conducted in-depth research, and details the results of their research in the literature [3]. Kangbing Nan an other authors convert trajectory optimization problems into nonlinear programming problem by using direct single shooting method, and solve the problem of hypersonic glide trajectory optimization.In order to overcome the shortcomings of the indirect method, this paper introduces the hybrid genetic algorithm for searching initial value of conjugate variable, and solves the unpowered gliding trajectory optimization problem for near spacecraft.

2 Hybrid Genetic Algorithm

In this paper, gradient method is combined with genetic algorithm, Using the local search capabilities of gradient method to offer more effective search direction, so that the algorithm has better flexibility and easier to parallelize the algorithm.

Hybrid genetic algorithm main steps:

(a) Initialization, a uniform distribution initial population is randomly generated; (b) Mating, by two-two matching principles, the best individual from the group contains new individuals and the male parent is chooses to the next generation; (c) Mutation, each individual of population is mutated as probability of Pm; (d) Local search, using the gradient method is repeated local optimization operation; (e) Termination, if termination conditions are met, then the algorithm terminates, otherwise go to step (b).

Gradient method main steps as follow:

(a) $\varepsilon > 0$ is selected as the termination. $X^{(0)}$ is selected as initial point, let $k = 0$; (b) calculate $\Delta f(X^{(k)})$, if $\|\Delta f(X^{(k)})\| < \varepsilon$, Iteration termination, to take approximate optimal solution $X^* = X^{(k)}$, else let $S^{(k)} = -\Delta f(X^{(k)})$, Starting from $X^{(k)}$ for one-dimensional search along $S^{(k)}$, get λ_k, make $\min_{\lambda > 0} f(X^{(k)} + \lambda S^{(k)}) = f(X^{(k)} + \lambda_k S^{(k)})$;(c) let $X^{(k+1)} = X^{(k)} + \lambda_k S^{(k)}$, $k+1 \to k$, return to Step(b).

3 Mathematical Model

Considering the earth as a non-rotating ball, and supposing the vehicle no sliding, we can obtain simplified plane motion equations for unpowered near spacecraft.

$$\begin{cases} dr / dt = V \sin\gamma \\ d\beta_e / dt = V \cos\gamma / r \\ dV / dt = -\rho S_{ref} C_D V^2 / r - g \sin\gamma \\ d\gamma / dt = \rho S_{ref} C_L V^2 / (2mV) - (rg - V^2)\cos\gamma / (rV) \end{cases} \tag{1}$$

Which r, β_e, V and γ were geocentric distance, range angle, velocity(relative to the ground) and velocity path angle respectively, as shown in Fig. 1. ρ, g m and S_{ref} were atmospheric density, Gravitational acceleration, mass of vehicle, reference area of vehicle respectively.

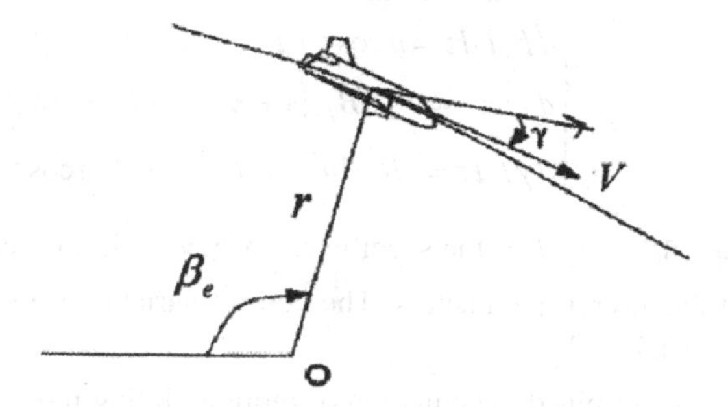

Fig. 1. Parameter schematic diagram of plane reentry

To study the problem of convenience, convert the equations of motion as dimensionless and reflect the characteristics of ballistic parameters. Define the dimensionless height z , dimensionless speed u and dimensionless time τ .

$$z = r / R_0, u = V / V_c, \tau = t / \sqrt{R_0 / g_0} \tag{2}$$

Which, R_0 , g_0 , respectively, the mean radius of the earth and gravitational acceleration of the sea level acceleration, $V_c = \sqrt{g_0 R_0}$. In hypersonic flight conditions, aerodynamic drag coefficient approximately meet parabolic drag polar relationship, and when the Mach number is larger, the zero-lift drag coefficient C_{D0} and induced drag factor K can be approximated a constant.

$$C_D = C_{D0} + K \cdot C_L^2 \tag{3}$$

The lift coefficient C_L^* and drag coefficient C_D^* when the lift-drag ratio achieve maximum can be determined by

$$C_L^* = \sqrt{C_{D0} / K}, \quad C_D^* = 2C_{D0} \tag{4}$$

Thus the maximum lift-drag ratio is expressed as

$$E^* = C_L^* / C_D^* = 1 / \sqrt{2KC_{D0}} \tag{5}$$

Definition of generalization lift coefficient

$$\lambda = C_L / C_L^* \tag{6}$$

So lift coefficient and drag coefficient can be expressed as function of λ, that

$$C_L = \lambda C_L^*, \quad C_D = 0.5C_L^* \left(1 + \lambda^2\right) / E_L^* \tag{7}$$

Put (2) and (7) into the equations of motion (1), we obtain the dimensionless parameters with characteristic plane reentry equation.

$$\begin{cases} dz / d\tau = u \sin \gamma \\ d\beta_e / d\tau = u \cos \gamma / z \\ du / d\tau = -0.5\overline{B}\rho\left(1 + \lambda^2\right) u^2 / E^* - \sin \gamma / z^2 \\ d\gamma / d\tau = \left[\overline{B}\rho\lambda u^2 - (1/z^2 - u^2/z)\cos \gamma\right] / u \end{cases} \tag{8}$$

Among them, $\overline{B} = 0.5R_0 S_{ref} C_L^* / m$, the specific re-entry vehicles, to reflect the physical characteristics of the aircraft parameters. The control variable of reentry trajectory is generalized lift coefficient $\lambda(t)$.

For equation (8) determined the equation of motion gliding re-entry vehicle for the maximum gliding distance of the optimal control problem, hoping to find a control law to allow it to by the equation (8) describes the system from the initial state to state by the termination provisions, performance indicators and make the minimum.

$$\beta_e = \beta_{e0}, z = z_0, u = u_0, \gamma = \gamma_0 \tag{9}$$

$$\beta_e|_{\tau=\tau_f} = \beta_{ef}, z(\beta_{ef}) = z_f, u(\beta_{ef}) = u_f, \gamma(\beta_{ef}) = \gamma_f \tag{10}$$

$$J(\lambda) = -\beta_e(\tau_f) \tag{11}$$

The above problem is a time-varying system, the end value of performance index, control constrained optimal control problem, and Hamiltonian can be constructed as follows:

$$H = \lambda_z \cdot u \sin \lambda + \lambda_{\beta_e} \cdot \frac{u \cos \lambda}{z} + \lambda_u \cdot \left(-\frac{\overline{B}\rho\left(1 + \lambda^2\right)u^2}{2E^*} - \frac{\sin \gamma}{z^2} \right) \\ + \lambda_\gamma \cdot \left(\frac{1}{u}\left(\overline{B}\rho\lambda u^2 - \left(\frac{1}{z^2} - \frac{u^2}{z} \right) \cos \gamma \right) \right) \tag{12}$$

Where, λ_z, λ_{β_e}, λ_u and λ_γ are undetermined Lagrange multipliers. From (12) yields the Association state equation (with function) is

$$\begin{cases} \frac{d\lambda_z}{d\tau} = -\lambda_u \cdot \left(\frac{\sin\gamma}{z^3}\right) - \lambda_\gamma \cdot \left(\frac{2}{uz^3} - \frac{u}{z^2}\right) \\ \frac{d\lambda_u}{d\tau} = -\lambda_z \cdot \sin\gamma - \lambda_u \cdot \left(-\frac{\overline{B}\rho(1+\lambda^2)u}{E^*}\right) - \lambda_\gamma \cdot \left(\overline{B}\rho\lambda + \frac{\cos\gamma}{z}\right) \\ \frac{d\lambda_\gamma}{d\tau} = -\lambda_z \cdot u\cos\gamma - \lambda_u \cdot \left(\frac{-\cos\gamma}{z^2}\right) - \lambda_\gamma \cdot \frac{1}{u}\left(\frac{1}{z^2} - \frac{u^2}{z}\right)\sin\gamma \end{cases} \tag{13}$$

4 Optimization Problems Described

One of the advantages of gliding re-entry is increased range than normal re-entry, when applied to long-range attack weapon system, is very conducive to penetration. It can be made to optimize the aircraft's maximum range of objectives; namely

$$\max : \beta_e = \beta_e(\tau_f) \tag{14}$$

As the spacecraft thermal protection systems, structure and control system requirements, hypersonic aircraft flight path need to meet heat flux $\dot{Q}_{\max}$, total dynamic pressure $q_{\max}$ and overload constraint $n_{\max}$. That is

$$\dot{Q} = \left(\frac{C_1}{R_d}\right)\left(\frac{\rho}{\rho 0}\right)^{0.5}\left(\frac{V}{V_c}\right)^{3.15} \le \dot{Q}_{\max}, n = \frac{\sqrt{L^2+D^2}}{mg_0} \le n_{\max}, q = \frac{1}{2}\rho V^2 \le q_{\max} \tag{15}$$

Which, $\dot{Q}_{\max}$, $n_{\max}$ and $q_{\max}$ for the corresponding constraint value. We also need take into account the height and speed of the terminal constraints, i.e.

$$u(\tau_f) = u_f, h(\tau_f) = h_f \tag{16}$$

Constraint handling using penalty function method, the physical meaning of different constraints into the same order of magnitude, of feasible solutions by the penalty into the unconstrained problem, penalty function of the specific type of entry table

$$p_1(x) = \begin{cases} 0 & V(t_f) \ge V_f \\ \left[\frac{(V_f - V(t_f))}{1000}\right]^2 & V(t_f) < V_f \end{cases}, p_2(x) = \begin{cases} 0, & \dot{Q}_{\max} < \overline{\dot{Q}}_{\max} \\ \left[\frac{(\dot{Q}_{\max} - \overline{\dot{Q}}_{\max})}{1000}\right]^2 & \dot{Q}_{\max} \ge \overline{\dot{Q}}_{\max} \end{cases} \tag{17}$$

$$p_3(x) = \begin{cases} 0 & n_{\max} < \overline{n}_{\max} \\ \left[\frac{(n_{\max} - \overline{n}_{\max})}{1000}\right]^2 & n_{\max} \ge \overline{n}_{\max} \end{cases}, p_4(x) = \begin{cases} 0, & q_{\max} < \overline{q}_{\max} \\ \left[\frac{(q_{\max} - \overline{q}_{\max})}{1000}\right]^2 & q_{\max} \ge \overline{q}_{\max} \end{cases}$$

Which $\dot{\bar{Q}}_{max}$, $\bar{n}_{max}$ and $\bar{q}_{max}$ is the maximum heat flux, the maximum total overload, the maximum dynamic pressure respectively in the optimization calculating process.

The maximum range is as one of the main performance indicators of hypersonic vehicle. This paper mainly analysis the impact of initial parameters under the conditions that not consider the case of path constraints.

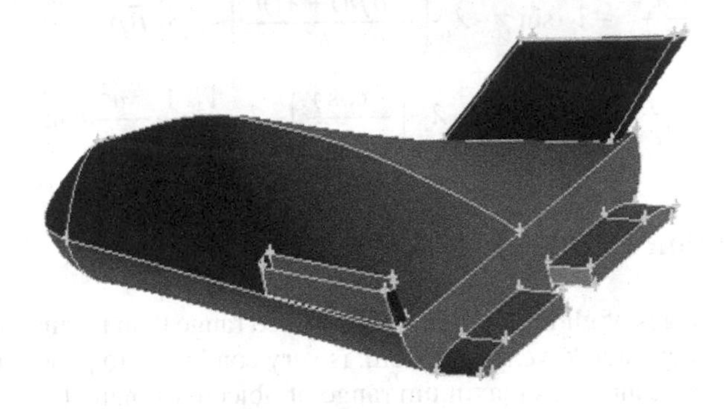

Fig. 3. The shape of vehicle

The basic parameters are as follows: Quality m =257kg, Reference area S_{ref} =0.0857m^2, Maximum lift-drag ratio E^* =1.978. Other desired trajectory parameters: the mean radius of the earth takes R_0 =6378000, gravitational acceleration g uses the following model to calculate: $g = g_0 R_0^2 / r^2$, Atmospheric density fitting using the formula $\rho = 1.9917e^{(-h/6452.56073)}$ 。

Consider the following three initial factors: (a) different initial inclination of velocity: (b) different initial velocity: (c) different initial velocity. The specific values shown in Table 1.

Table 1. Three different initial conditions

	γ_0(deg)	$V_0(km/s)$	$H_0(km)$
(a)	-5/-3/-1/1/3/5	5.5	30
(b)	3	5.0/5.5/6.0	30
(c)	3	5.5	30/50/70/90

Set the genetic algorithm population size 100, crossover probability 0.8, mutation probability 0.2, and the maximum evolution generation 100. Evolution about 50 generations, a variety of initial conditions has been optimized convergence. The simulation results are shown in Figure 4.

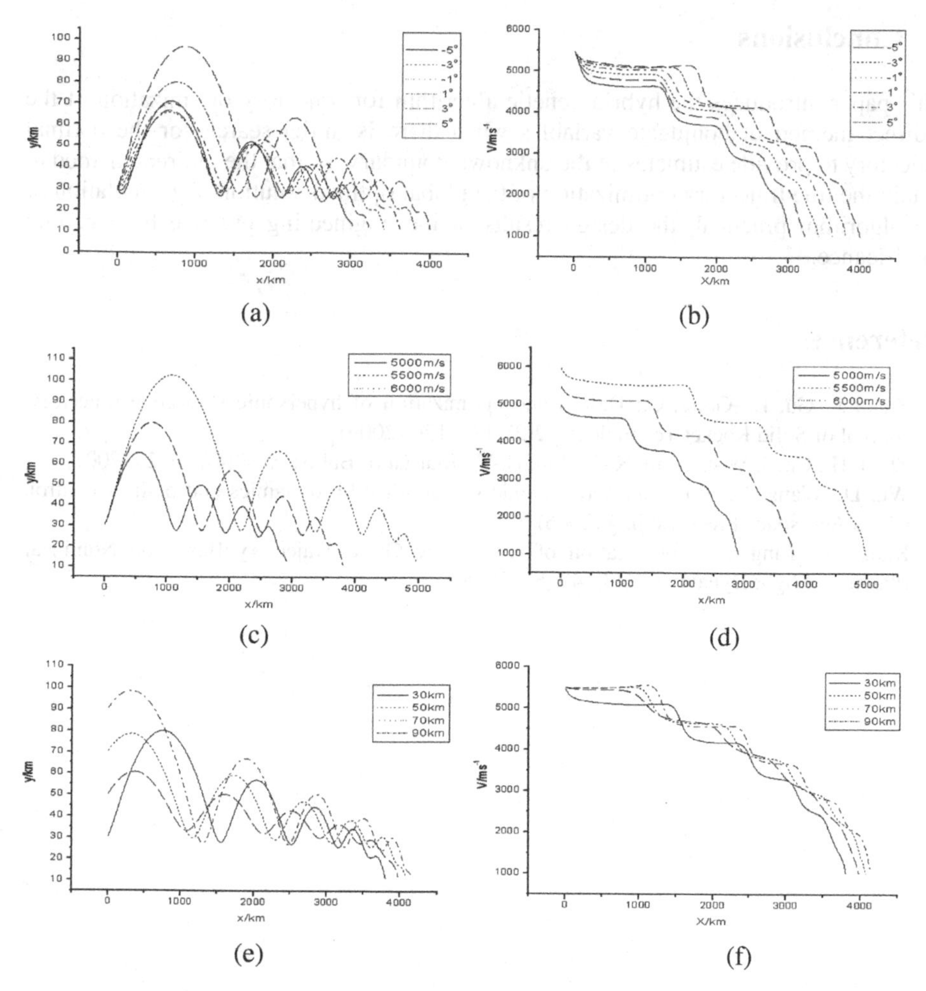

Fig. 4. Optimization results under different initial conditions

In Fig.4, (a) and (b) are the height curves and velocity curves change with range when the inclination of velocity take different value. The free range is increases when the inclination of velocity is increase, which, when $\gamma_0 = 5^o$ the maximum range for 3999km, but when $\gamma_0 = -5^o$ the minimum of 3029km. (c) and (d) are the height curves and velocity curves change with range when the velocity take different value. The free range is increases when the velocity is increase. Increasing trend is clear, that is a great influence on the velocity. (e) and (f) are the height curves and velocity curves change with range when the height take different value. The free range is increases when the height is increase. Increasing trend is not very clear, that height has little effect.

5 Conclusions

This paper introduces the hybrid genetic algorithm for trajectory optimization of the indirect method of conjugate variables when there is initial search for the optimal trajectory to provide estimates of the unknown boundary, so that the indirect method to obtain the true trajectory optimization of the global optimal solution. By simulation of the algorithm practical, the design results of the engineering practice have certain significance.

References

1. Zhao, J., Gu, L., Gong, C.: Design and optimization of hypersonic skip-glide trajectory. Journal of Solid Rocket Technology 32(2), 123–126 (2009)
2. Zhou, H., Chen, W.-C., Yin, X.-L., Liu, H.-T.: Journal of Ballistics 19(4), 26–29 (2007)
3. Wu, D., Wang, X.: Auxiliary aerodynamic spacecraft orbit dynamics and optimal control. China Aerospace Press, Beijing (2006)
4. Kang, B., Tang, S.: Optimization of Hypersonic Glide Trajectory Based on Nonlinear Planning. Flight Dynamics 26(3), 49–52 (2008)

New Stability of Markovian Jump Delayed Systems with Partially Unknown Transition Probabilities

Yanfang Zuo[1], Lianglin Xiong[2,*], and Junhui Wang[2]

[1] Element Courses Department Kunming Metallurgy, Kunming, 650031, P.R. China
[2] School of Mathematics and Computer Science,
Yunnan University of Nationalities, Kunming, 650031, P.R. China
lianglin_5318@126.com

Abstract. This paper addresses the problem of the delay-dependent stability for Markovian jump time-varying delayed systems with partial information on transition probability. Combined the new constructed Lyapunov functional with the introduced free matrices, and using the analysis technique of inequalities, the delay-dependent stability conditions are derived in form of LMIs. A numerical example is given to show the validity and potential of the developed results.

Keywords: Delay-dependent stability, Markovian jump, Time-varying delay, Partly unknown transition probabilities, Linear matrix inequality (LMI).

1 Introduction

In the past few decades, Markovian jump systems (MJS) have been widely studied due to the fact that many dynamical systems subject to random abrupt variations can be modeled by MJSs such as manufacturing systems, networked control systems, fault-tolerant control systems, etc. There are a lot of useful results have been presented in the literature, such as [1]-[4], and the references therein. Typically, MJS are described by a set of classical differential (or difference) equations and a Markov stochastic process(or Markov chain) governing the jumps among them. Since the transition probabilities in the jumping process determine the system behavior to a large extent, and because the ideal knowledge on the transition probabilities are definitely expected to simplify the system analysis and design, many analysis and synthesis results have been reported, assuming the complete knowledge of the transition probabilities[5]-[10].

However, the likelihood of obtaining such available knowledge is actually questionable, and the cost is probably expensive. A typical example can be found in NCS, where the packet dropouts and channel delays are well-known to be modeled by Markov Chains with the usual assumption that all the transition probabilities are completely accessible [11]-[13]. However, in almost all types of communication networks, either the variation of delays or the packet dropouts can be vague and

* Corresponding author.

D. Jin and S. Lin (Eds.): Advances in ECWAC, Vol. 1, AISC 148, pp. 49–57.
springerlink.com © Springer-Verlag Berlin Heidelberg 2012

random in different running periods of networks-all or part of the elements in the desired transition probabilities matrix are hard or costly to obtain. The same problems may arise in other practical systems with jumps. Therefore, rather than having a large complexity to measure or estimate all the transition probabilities, it is significant and necessary, from control perspectives, to further study more general jump systems with partly unknown transition probabilities. Recently, many results on the Markovian jump systems with partly unknown transition probabilities are obtained [14]-[19]. Most of these papers haven't considered the effect of delay in the stability or stabilization conditions.

In this paper, the delay dependent stability and stabilization problems of Markovian jump linear system with partly unknown transition probabilities are investigated. The number of matrix inequalities conditions obtained in this paper is less than the existing results to some extent, due to the proposed lemmas in the proof of our theorem of this paper. Finally, a numerical example is provided to illustrate the validity of our results.

2 Problem Statement and Preliminaries

Consider the following delay systems with Markovian jump parameters:

$$\begin{cases} \dot{x}(t) = A(r_t)x(t) + B(r_t)x(t-\tau(t)) \\ x(t_0+\theta) = \varphi(\theta), \forall \theta \in [-\tau, 0] \end{cases} \tag{1}$$

where $x(t) \in R^n$ is the state vector , $\tau(t) > 0$ is time-varying delay which satisfies $0 \leq \tau(t) \leq \tau, \dot{\tau}(t) \leq \tau_d < 1$, and $\varphi(\theta)$ is the initial condition function. $\{r_t\}, t \geq 0$ is a right-continuous Markov process on the probability space taking values in a finite state space, $\wp = \{1, 2, \cdots, N\}$ with generator $\Pi = (\lambda_{i,j}), i, j \in \wp$ given by

$$Pr\{r_{t+\Delta} = j | r_t = i\} = \begin{cases} \lambda_{ij}\Delta + o(\Delta), j \neq i \\ 1 + \lambda_{ij}\Delta + o(\Delta), j = i \end{cases},$$

where $\Delta > 0, \lim_{\Delta \to 0} \frac{o(\Delta)}{\Delta} = 0$, $\lambda_{ij} \geq 0$ for $j \neq i$ is the transition rate from mode i at time t to mode j at time $t+\Delta, \lambda_{ii} = -\sum_{j=i}^{N} \lambda_{ij}$. $A(r_t)$, $B(r_t)$, $D(r_t)$ are known matrix functions of the Markov process.

Since the transition probability depends on the transition rates for the continuous-time MJSs, the transition rates of the jumping process are considered to be partly accessible in this paper. For instance, the transition rate matrix Π with $\mathbb{N}$ operation modes may be expressed as

$$\begin{pmatrix} \lambda_{11} & ? & \lambda_{13} & \cdots & ? \\ ? & ? & ? & ? & \lambda_{2N} \\ \vdots & \vdots & \vdots & \ddots & \vdots \\ ? & \lambda_{N2} & \lambda_{N3} & \cdots & \lambda_{NN} \end{pmatrix} \tag{2}$$

where? represents the unknown transition rate.

For notational clarity, $\forall i \in \wp$, the set U^i denotes with $U_k^i \triangleq \{ j : \lambda_{ij}$ is known for $j \in \wp\}$; $U_{uk}^i \triangleq \{ j : \lambda_{ij}$ is unknown for $j \in \wp\}$. Moreover, if $U_k^i \neq 0$, it is further described as

$$U_k^i = \{ k_1^i, k_2^i, \cdots, k_m^i \} \tag{3}$$

where m is a non-negation integer with $1 \leq m \leq N$ and $k_j^i \in Z^+, 1 \leq k_j^i \leq N\,(j = 1, \cdots, m)$ represent the *jth* known element of the set U_k^i in the *ith* row of the transition rate matrix Π.

Remark 1: It is worthwhile to note that if $U_k^i = 0$, $U^i = U_{uk}^i$ which means that any in - formation between the *ith* mode and the other $N-1$ modes is not accessible. Then MJSs with N modes can be regarded as ones with $N-1$ modes. It is clear that when $U_{uk}^i = 0$, $U^i = U_k^i$ the system (1) becomes the usual assumption case.

For the sake of simplicity, the solution $x(t, \varphi(\theta), r_0)$ with $r_0 \in \wp$ is denoted by $x(t)$. It is known from [21] that $\{x(t), t\}$ is a Markov process with an initial state $\{\varphi(0),\ r_0\}$, and its weak infinitesimal generator, acting on function V is defined in [22], as following

$$EV(x(t), t, i) = \lim_{\Delta \to 0^+} \frac{1}{\Delta} \Big[\varepsilon\big(V(x(t+\Delta), t+\Delta, r_t + \Delta) \big| x(t), r_t = i \big) - V(x(t), t, i) \Big] \tag{4}$$

Before presenting the main results, we first state the following lemmas which will be used in the proof of our main result.

Lemma1.([23]) For a positive matrix $Q > 0$, any matrices $F_1, F_2, F_3, F_4, F_5, F_6$ and scalar $h \geq 0$, the following inequality holds:

$$\int_{t-h} \dot{y}^T(s) Q \dot{y}(s) ds \leq -\xi^T(t) \tilde{F} \xi(t) + h \xi^T(t) F^T Q^{-1} F \xi(t)$$

where

$$F = (F_1 \quad F_2 \quad F_3 \quad F_4 \quad F_5 \quad F_6),$$

$$\xi^T(t) = \left[y^T(t) \quad y^T(t-h) \quad \left(\int_{t-h} y(s) ds \right)^T \quad \dot{y}^T(t) \quad \dot{y}^T(t-h) \quad g^T(y(t-h)) \right],$$

$$\tilde{F} = \begin{pmatrix} 0 & 0 & F_1^T & 0 & 0 & 0 \\ * & 0 & F_2^T & 0 & 0 & 0 \\ * & * & F_3^T + F_3 & F_4 & F_5 & F_6 \\ * & * & * & 0 & 0 & 0 \\ * & * & * & * & 0 & 0 \\ * & * & * & * & * & 0 \end{pmatrix},$$

3 Main Results

In this section, a stochastic stability criterion for MJSs is firstly given based on the free-connection weighting matrix method as follows, and then the feedback controllers are designed according to the obtained stability condition.

Theorem 3.1: The system (1) with a partly unknown transition rate matrix (2) is stochastically stable if there exist matrices $P_i, Q_i, R_1 > 0, R_2 > 0, R_3 > 0$, any matrices $W_i = W_i^T, M_i = M_i^T, F_1, F_2, F_3$ with appropriate dimensions satisfying the following linear matrix inequalities:

$$\varphi_i = \begin{pmatrix} \varphi_{i11} & \varphi_{i12} & \varphi_{i13} & \tau F_1^T & \tau A_i^T \\ * & \varphi_{i22} & F_3 & \tau F_2^T & 0 \\ * & * & \varphi_{i33} & \tau F_3^T & \tau B_i^T \\ * & * & * & -\tau R_2 & 0 \\ * & * & * & * & -\tau R_2 \end{pmatrix} < 0, \tag{5}$$

$$\sum_{j \in U_k^i} \lambda_{ij} \left(Q_j - M_i \right) \le R_3, \tag{6}$$

$$P_j - W_i \le 0, j \in U_{uk}^i, j \ne i, \tag{7}$$

$$Q_j - W_i \le 0, j \in U_{uk}^i, j \ne i, \tag{8}$$

$$P_j - W_i \ge 0, j \in U_{uk}^i, j = i, \tag{9}$$

$$Q_j - W_i \ge 0, j \in U_{uk}^i, j = i, \tag{10}$$

where

$$\varphi_{i11} = P_i A_i + A_i^T P_i + \sum_{j \in U_k^i} \lambda_{ij} \left(P_j - W_i \right) + Q_i + \tau R_3 + R_1 - F_1 - F_1^T,$$

$$\varphi_{i12} = -F_2 + F_1^T, \varphi_{i13} = P_i B_i - F_3, \varphi_{i22} = F_2 + F_2^T - R_1, \varphi_{i33} = -(1 - \tau_d) Q_i.$$

Proff. Construct a stochastic Lyapunov functional candidate as

$$\begin{aligned} V(x_t,t,r_t) &= x^T(t)P(r_t)x(t) + \int_{t-\tau(t)}^{t} x^T(s)Q(r_t)x(s)ds + \int_{t-\tau}^{t} x^T(s)R_1x(s)ds \\ &+ \int_{-\tau}^{0}\int_{t+\theta}^{t} \dot{x}^T(s)R_2\dot{x}(s)dsd\theta + \int_{-\tau}^{0}\int_{t+\theta}^{t} x^T(s)R_3x(s)dsd\theta \end{aligned} \tag{11}$$

where $P(r_t), Q(r_t), R_1>0, R_2>0, R_3>0, r_t \in \wp$ are all matrices with appropriate dimensions to be determined. Then, for given $r_t = i \in \wp, P(r_t)=P_i, Q(r_t)=Q_i$ and the weak infinitesimal operator E of the stochastic process $x(t)$ along the evolution of $V(x_t,t,i)$ is given as

$$\begin{aligned} EV(x_t,t,i) &\le x^T(t)\left(P_iA_i + A_i^TP_i\right)x(t) + 2x^T(t)P_iB_ix(t-\tau) + x^T(t)Q_ix(t) + x^T(t)R_1x(t) \\ &+ \tau\dot{x}^T(t)R_2\dot{x}(t) - (1-\tau_d)x^T(t-\tau(t))Q_ix(t-\tau(t)) - x^T(t-\tau)R_1x(t-\tau) \\ &- \int_{t-\tau}^{t} \dot{x}^T(s)R_2\dot{x}(s)ds + \tau x^T(t)R_3x(t) - \int_{t-\tau(t)}^{t} x^T(s)R_3x(s)ds \\ &+ \int_{t-\tau(t)}^{t} x^T(s)\left(\sum_{j=1}^{N}\lambda_{ij}Q_j\right)x(s)ds + x^T(t)\sum_{j=1}^{N}\lambda_{ij}P_jx(t) \end{aligned} \tag{12}$$

Following a similar line as in proof of Lemma 1 in [20], it is trivial to obtain the following inequality

$$\begin{aligned} \int_{t-\tau}^{t} \dot{x}^T(s)R_2\dot{x}(s)ds &\le -2\left(\int_{t-\tau}^{t}\dot{x}(s)ds\right)^T F\xi(t) + \tau\xi^T(t)F^TR_2^{-1}F\xi(t) \\ &= -\xi^T(t)\widetilde{F}\xi(t) + \tau\xi^T(t)F^TQ^{-1}F\xi(t), \end{aligned} \tag{13}$$

where $F = \begin{pmatrix} F_1 & F_2 & F_3 \end{pmatrix}$, $\widetilde{F} = \begin{pmatrix} -F_1 - F_1^T & -F_2 + F_1^T & -F_3 \\ * & F_2 + F_2^T & F_3 \\ * & * & 0 \end{pmatrix}$.

Taking into account the situation that the information of transition probabilities is not accessible completely, the following zero equations hold for arbitrary matrices $W_i = W_i^T$ and $M_i = M_i^T$,due to $\sum_{j=1}^{N}\lambda_{ij} = 0$.

$$-x^T(t)\sum_{j=1}^{N}\lambda_{ij}W_ix(t) = 0, \forall i \in \wp, \tag{14}$$

$$-\int_{t-\tau(t)}^{t} x^T(s)\left(\sum_{j=1}^{N}\lambda_{ij}M_i\right)x(s)ds = 0, \forall i \in \wp. \tag{15}$$

Defined $\widetilde{G}=\begin{pmatrix}A_i & 0 & B_i\end{pmatrix}$, substituting (13) into (12), and adding the left side of (14) and (15) into $EV(x_t,t,i)$ results in

$$EV(x_t,t,i)\le x^T(t)\Big(P_iA_i+A_i^TP_i+\sum_{j\in U_k}\lambda_{ij}(P_j-W_i)+Q_i+\tau R_3+R_1\Big)x(t)$$
$$+2x^T(t)P_iB_ix(t-\tau(t))-(1-\tau_d)x^T(t-\tau(t))Q_ix(t-\tau(t))-x^T(t-\tau)R_1x(t-\tau)$$
$$-\int_{t-\tau(t)}^{t}x^T(s)R_3x(s)ds+\tau\xi^T(t)\widetilde{G}^TR_2\widetilde{G}\xi(t)+\tau\xi^T(t)F^TQ^{-1}F\xi(t)-\xi^T(t)\widetilde{F}\xi(t)$$
$$+x^T(t)\sum_{j\in U_{uk}^i}\lambda_{ij}(P_j-W_i)x(t)+\int_{t-\tau(t)}^{t}x^T(s)\Big(\sum_{j=1}^{N}\lambda_{ij}(Q_j-M_i)\Big)x(s)ds \tag{16}$$

Note that $\lambda_{ii}=-\sum_{j=i}^{N}\lambda_{ij}$, and $\lambda_{ij}\ge 0$ for all $j\ne i$, namely $\lambda_{ii}<0$ for all $i\in\wp$. Therefore it follows from Schur's complement and easy computation that, if $i\in U_k^i$ inequalities (19)-(22) imply that $EV(x_t,t,i)<0$. On the other hand, for the same reason, if $i\in U_{uk}^i$, inequalities(19)-(24) also imply inequality (17) holds. Therefore $\varepsilon\left\{\int_0^\infty\|x(t)\|^2dt\,|x_0,r_0\right\}<\infty$, which means that system (1) is stochastically stable according to the definition in [4]. This proof is completed.

Remark 2: Similar to [18], in order to obtain the less conservative stability criterion of MJSs with partial information on transition probabilities, the free-connection weighting matrices are introduced by making use of the relationship of the transition rates among various subsystems, i.e. $\sum_{j=1}^{N}\lambda_{ij}=0$ for all $i\in\wp$, which overcomes the conservativeness of using the fixed connection weighting matrices. Moreover, the result obtained in this theorem is delay-dependent stability condition. Using the technique of free weighting matrices would reduce the conservativeness.

When $\tau(t)=\tau$, the system is time-invariant delay systems. Similar to the proof of Theorem 1, the stability condition of this system is obtained as following corollary.

Corollary 1: The system (1) with a partly unknown transition rate matrix (2) and $\tau(t)=\tau$ is stochastically stable if there exist matrices $P_i,Q_i,R_2>0,R_3>0$, any matrices $W_i=W_i^T,M_i=M_i^T,F_1,F_2$ with appropriate dimensions satisfying the following linear matrix inequalities

$$\varphi_i=\begin{pmatrix}\varphi_{i11} & \varphi_{i12} & \tau F_1^T & \tau A_i^T\\ * & \varphi_{i22} & \tau F_2^T & \tau B_i^T\\ * & * & -\tau R_2 & 0\\ * & * & * & -\tau R_2\end{pmatrix}<0, \tag{17}$$

$$\sum_{j\in U_k^i} \lambda_{ij}\left(Q_j - M_i\right) \le R_3,$$

$$P_j - W_i \le 0, j \in U_{uk}^i, j \ne i,$$

$$Q_j - W_i \le 0, j \in U_{uk}^i, j \ne i,$$

$$P_j - W_i \ge 0, j \in U_{uk}^i, j = i,$$

$$Q_j - W_i \ge 0, j \in U_{uk}^i, j = i,$$

$$\varphi_{i11} = P_i A_i + A_i^T P_i + \sum_{j\in U_k^i} \lambda_{ij}\left(P_j - W_i\right) + Q_i + \tau R_3 - F_1 - F_1^T,$$

$$\varphi_{i12} = P_i B_i - F_2 + F_1^T, \varphi_{i33} = -Q_i + F_2 + F_2^T.$$

4 Example

In this section, one example is given to show the effectiveness and the merits of the free-connection weighting matrix method and combined matrix inequalities approach.

Example 1. Consider the MJLS (1) with four operation modes, which state matrices are as following:

$$A_1 = \begin{bmatrix} -4.15 & 0.75 \\ 0.15 & -1.5 \end{bmatrix}, A_2 = \begin{bmatrix} -2.15 & 0.49 \\ 0.15 & -2.1 \end{bmatrix},$$

$$A_3 = \begin{bmatrix} -1.3 & 0.15 \\ 0.15 & -2.8 \end{bmatrix}, A_4 = \begin{bmatrix} -1.9 & 0.34 \\ 0.15 & -1.65 \end{bmatrix},$$

$$B_1 = \begin{bmatrix} -2.2 & 0.12 \\ 0.24 & -0.25 \end{bmatrix}, B_2 = \begin{bmatrix} -0.45 & -0.96 \\ 0.47 & -1.57 \end{bmatrix},$$

$$B_3 = \begin{bmatrix} 0.58 & -1.68 \\ -0.13 & 0.96 \end{bmatrix}, B_4 = \begin{bmatrix} -1.67 & -1.5 \\ 1.39 & 1.23 \end{bmatrix}.$$

The partly transition rate matrix Π is considered as

$$\Pi = \begin{bmatrix} -1.3 & 0.2 & ? & ? \\ ? & ? & 0.3 & 0.3 \\ 0.6 & ? & -1.5 & ? \\ 0.4 & ? & ? & ? \end{bmatrix} \quad (25)$$

Using the Theorem 1, the time-varying delay are $\tau(t)=0.0342*\sin(t), \tau(t)=0.0742*\sin(t)$, and $0.1742*\sin(t)$ respectively, then the upper bound of the delay τ can been obtained as large as 0.327, 0.2682 and 0.1443 respectively. It is clear to see that as the τ_d increasing and the upper bound of the delay τ decreasing, and that is true of practice. It also means that MJS (1) whose state matrices described as above is stable. Ho -wever, according to the approach of Theorem 3 in [18], we can not find the feasible solutions which contain time delay to verify the stability of the system. Therefore, this example shows that the stability criterion in this paper gives much less conserve-ative delay dependent stability conditions as Lemma 1 is introduced.

5 Conclusion

The delay-dependent stability analysis of the Markovian jump time-varying delay system is studied in this paper. Firstly, Combined with some novel Lyapunov functionals and the analysis technique of inequalities, the delay-dependent stability conditions are presented. Finally, one example is provided to show the effectiveness of our results ob -tained in this paper.

Acknowledgments. This work was supported by the science study fund of Yunnan Province Office of Education (2010Y430).

References

1. Boukas, E.K.: Stochastic switching systems: Analysis and design. Birkhauser, Basel (2005)
2. Costa, O.L.V., Fragoso, M.D., Marques, R.P.: Discretetime Markovian jump linear systems. Springer, London (2005)
3. Liberzon, D.: Switching in systems and control. Birkhauser, Berlin (2003)
4. Lin, H., Antsaklis, P.J.: Stability and stabilizability of switched linear systems: a survey of recent results. IEEE Transactions on Automatic Control 54, 308–322 (2009)
5. Li, H.Y., Chen, B., Zhou, Q., Qian, W.Y.: Robust stability for uncertain delayed fuzzy Hopfield neural networks with Markovian jumping parameters. IEEE Transactions on Systems, Man and Cybernetics, Part B (Cybernetics) 39, 94–102 (2009)
6. Lou, X.Y., Cui, B.T.: Stochastic exponential stability for Markovian jumping BAM neural networks with time-varying delays. IEEE Transactions on Systems, Man and Cybernetics, Part B (Cybernetics) 37, 713–719 (2007)
7. Zhang, H.G., Wang, Y.C.: Stability analysis of Markovian jumping stochastic Cohen Grossberg neural networks with mixed time delays. IEEE Transactions on Neural Networks 19, 366–370 (2008)
8. Chen, W.H., Guan, Z.H., Lu, X.M.: Delay-dependent output feedback stabilization of Markovian jump systems with time-delay. IEE Proceedings-Control Theory and Applications 151, 561–566 (2004)
9. Shu, Z., Lam, J., Xu, S.Y.: Robust stabilization of Markovian delay systems with delay-dependent exponential estimates. Automatica 42, 2001–2008 (2006)

10. Wang, G.L., Zhang, Q.L., Sreeram, V.: Design of reducedorder $H\ \infty$ filtering for Markovian jump systems with modedependent time delays. Signal Processing 89, 187–196 (2009)
11. Krtolica, R., Ozguner, U., Chan, H., Winkelman, Goktas, J.H., Liubakka, M.: Stability of linear feedback systems with random communication delays. International Journal of Control 59, 925–953 (1994)
12. Seiler, P., Sengupta, R.: An H approach to networked control. IEEE Transactions on Automatic Control 50, 356–364 (2005)
13. Zhang, L., Shi, Y., Chen, T., Huang, B.: A new method for stabilization of networked control systems with random delays. IEEE Transactions on Automatic Control 50, 1177–1181 (2005)
14. Zhang, L.X., Boukas, E.-K.: Stability and stabilization of Markovian jump linear systems with partly unknown transition probabilities. Automatica 45, 463–468 (2009)
15. Zhang, L.X., Boukas, E.-K., James, L.: Analysis and synthesis of Markov jump linear systems with time-varying delays and partially known transition probabilities. IEEE Transactions on Automatic Control 53, 1177–1181 (2008)
16. Zhang, L.X., Boukas, E.-K., Baron, L.: Fault detection for discrete-time Markov jump linear systems with partially known transition probabilities. In: Proceedings of the 47th IEEE Conference on Decision and Control Cancun, Mexico, December 9-11, pp. 1054–1059 (2008)
17. Zhang, L.X., Boukas, E.-K.: Mode-dependent H filtering for discrete-time Markovian jump linear systems with partly unknown transition probabilities. Automatica 45, 1462–1467 (2009)
18. Zhang, Y., He, Y., Wu, M., Zhang, J.: Stabilization for Markovian jump systems with partial information on transition probability based on free-connection weighting matrices. Automatica 47, 79–84 (2011)
19. Luan, X., Liu, F., Shi, P.: Finite-time filtering for non-linear stochastic systems with partially known transition jump rates. IET Control Theory Applications 4, 735–745 (2010)
20. Park, J.H., Kwon, O.M., Lee, S.M.: LMI optimization on stability for delay neutral networks of neutral type. Applied Mathematics and Computation 196, 236–244 (2008)
21. Xu, S.Y., Mao, X.R.: Delay-dependent H control and filitering for uncertain Markovian jump system with time-varying delay. IEEE Transactions on Circuits and Systems Part I: Regular Papers 54, 2070–2077 (2007)
22. Skorohod, A.V.: Asymptotic methods in the theory of stochastic differential equation. American Mathematical Society, Providence, RI (1989)

Analysis on Audience's Attitudes toward and Behaviors Caused by Product Placement

Yue Yang[1], Xianliang Wang[2], and Hu Liu[1]

[1] School of Economic& Management, Communication University of China, Beijing, China
[2] Northeastern University at Qinhuangdao, Qinhuangdao, China
yyqhd@163.com, vickywxl@163.com, liuhu@cuc.edu.cn

Abstract. There is no systematic framework has been used to investigate the attitudes toward and behaviors caused by product placement. We modified the consumer socialization framework to examine placement-related attitudes and behaviors. Findings also reveal differences in both attitudes and behaviors across a range of factors, including gender, age and education level.

Keywords: product placement, movies & TV programs, empirical research, attitude & behavior.

1 Introduction

Product placement is favored by advertisers and media recently, which has developed into a key component of China's media industry, especially appearing most often in the movies and TV programs. It is very helpful for improving the effectiveness of this kind of adverting to consider the relationship between product placement and audience experience in the advertising operation.

The recent studies on product placement and audience experience mainly focus on the following four dimensions to analyze the trends on the audiences' attitude to product placement and their consumption-related behaviors.

Cultural Background of the Audience. The audiences of different ethnicity, different nations, and different educations are different in the degrees for accepting the product placement in movies and TV programs (Federico de Gregorio, 2010). Martin Eisend (2009) indicated that product placement acceptability of ethically charged/controversial products is generalizable over different cultures, but not product placement acceptability of neutral products. The higher the education of audience is, the more negative their attitude tends to be (Federico de Gregorio, 2010; Yin Xiaolin, 2009).

Age and Gender of the Audience. The audience of different age and gender has different attitude toward product placement in movies and TV programs. But results of those researches are somewhat not the same. Some researchers argue that the non-students groups have more positive attitude to product placement than the student groups (Yongjun Sung, 2009). And some researches show that the young people are more inclined to accept product placement (Federico de Gregorio, 2010; Cristina, 2010, and Yin Xiaolin, 2009). Some researchers conclude that the female has more positive attitude to product placement than the male (Federico de Gregori, 2010). While, some

D. Jin and S. Lin (Eds.): Advances in ECWAC, Vol. 1, AISC 148, pp. 59–65.
springerlink.com © Springer-Verlag Berlin Heidelberg 2012

researchers think that the gender could only in some time affect the audience's acceptance degree of this kind of advertising (Cristina, 2010 and Yin Xiaolin, 2009). It is worth suggesting that Zhao Qian (2010) and Yin Xiaolin (2009) have surveyed the same area in China, but got different conclusions. Zhao Qian concludes that there is no significant difference between the male and the female audiences in degrees of accepting product placement.

Product Category. Not all kinds of products can be accepted to be placed into movies and TV programs by audiences. Ignacio Redond (2008) shows that the most suitable product categories for product placement in movies and TV programs include alcoholic drinks, alcohol-free drinks, groceris, cultural offerings, and personal care product. But Yongjun Sung (2009) finds different results, which reveal that tobacco, firearm, gambling, politically orientated and alcohol brands or products are the most difficult to accept for product placement, and it is easiest to accept car, camera, electronics, sunglasses, charity, and heath care product for product placement.

Form of Product Placement. Although the classification of product placement is slightly different, scholars have similar findings. It is that blunt and highlighting type of product placement is more likely to remember for audience, but is not in favor of audience to accept the ad's content, so it is more difficult to induce the audience to change attitudes and behavior by this type of placement. But the cleverly integrated into the plot and subtle style product placement is more positive, especially in this form of product placement repeated gentle.(Pamela Miles Homer, 2009; Eva Van Reijmersdal, 2009; Peter Neijens, 2009; Moonhee Yang, 2007; Nie yanmei, 2008 and Dai tingting, 2010).

While there has been a growing stream of research focusing on product placement, there has also been a dearth in the use of systematic, conceptual groundings to help guide such investigations. There has also been very limited analysis of the origins and/or influencing social factors of these attitudes and how they interact with consumers' demographic characteristics.

2 Modified Conceptual Model of Consumer Socialization and Hypotheses

Essentially, the original consumer socialization framework (Moschis and Churchill, 1978) provided a means of analyzing the influences on and sources of how people learn to perform their roles as consumers in society. Moschis and Churchill （1978） though that stratum and gender were important social structural variables, while family, peer, media and schools were main impact factors. Their research indicated that the family played the biggest role during the consumer socialization, the effect of peer and media were secondly significant, but school were not significant from the statistical point of view. Besides, back in the 1950s, a lot of scholars studied on the impact of mass media, and they considered the correlation between consumer socialization and viewing time and individual age of movie and TV program. So, adapted from Moschis and Churchill (1978), Fig.1 models the general CS framework, and also includes the specific variables used in the current study. The three core components of antecedents, socialization processes, socialization agents, and outcomes are common to CS studies.

However, while these are central to CS, the framework is flexible as to the specific variables that can be investigated within each of these components. In our study, we focus on movie and TV program as the socialization agents during the consumer socialization process, and systematically analyze the attitude towards and behaviors caused by product placement and traditional advertising.

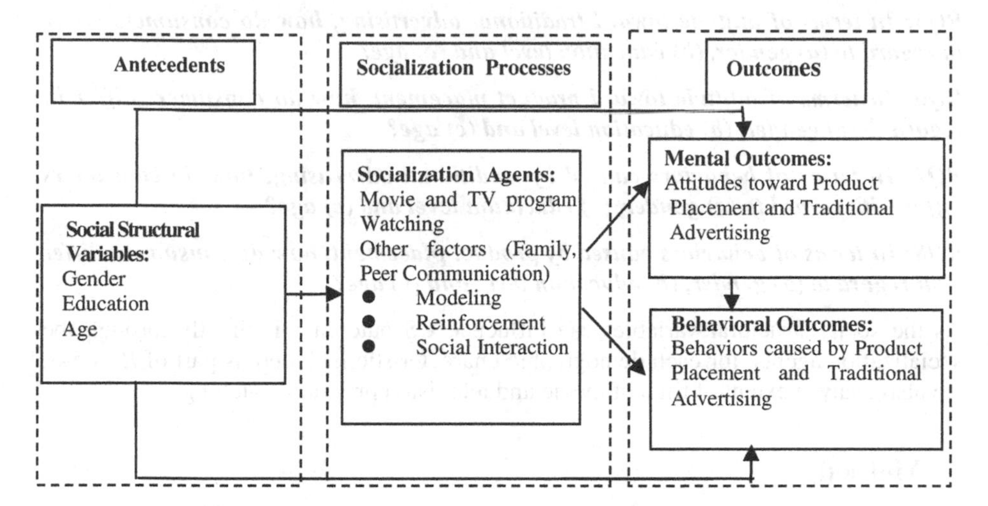

Fig. 1. A modified conceptual model of CS (consumer socialization) about product placement and traditional advertising

The media exposure has been consistently documented as a significant influencer of their consumer attitudes and behaviors among consumers according to CS. Because our study is focused on product placements in movie and TV program, we operationalize media exposure as time of movie and television program watching in China. And the comparison of influence degree is made between traditional advertising and product placement.

RQ1: Time of movie and television program watching will be positively related with traditional advertising attitudes.

RQ2: Time of movie and television program watching will be positively related with product placement attitudes.

RQ3: Time of movie and television program watching will be positively related with behaviors caused by traditional advertising.

RQ4: Time of movie and television program watching will be positively related with behaviors caused by product placement.

It should be noted that although CS outcomes can be influenced by social structural variables directly, as well as indirectly via the socialization agents (see Fig 1), the social learning mechanisms described earlier apply only to the processes by which the socialization agents impact the outcomes.

The framework does not provide a theory-oriented means of explaining how the social structural variables directly influence CS outcomes. Given that the CS framework does not propose a theoretical basis for how social structural variables directly influence CS outcomes, and in light of the research review, we propose the following interrelated research questions:

RQ5: In terms of attitude toward traditional advertising, how do consumers differ in regard to (a) gender, (b) education level and (c) age?

RQ6: In terms of attitude toward product placement, how do consumers differ in regard to (a) gender, (b) education level and (c) age?

RQ7: In terms of behaviors caused by traditional advertising, how do consumers differ with regard to (a) gender, (b) education level and (c) age?

RQ8: In terms of behaviors caused by product placement, how do consumers differ with regard to (a) gender, (b) education level and (c) age?

As the social structural variables can influence CS outcomes indirectly through the socialization agents, for each demographic characteristic included as part of *RQs 5-8*, we also analyze extent of time of movie and television program watching.

3 Method

3.1 Sample

The current study object was people under 45 years old. Trough online questionnaires, the answers were obtained by random respondents. The final sample size was 533. Demographic variables (gender, age, educational level) and average time watching movies and TV programs per week was assessed by a single question. Two questions measured the product placement attitudes toward by Likert 5 scale. Two dichotomous scale questions measured the behaviors in response to traditional advertising and product placement. Among the 553 samples, 58%were males and 42% were females. Approximately 51.4% of the samples were ages 19~24, followed by age 25~34(36.6%). 76.7% of the samples hadn't received the higher education. The data of demographic variable was in line with China's basic national conditions. Additional demographic characteristics of the respondents are shown in Table.1.

3.2 Variables and Measures

TIME (Time of Movie and TV program Watching). This variable was operationalized as the degree of media exposure. Respondents were asked to indicate how much time they spend watching TV and movie in a week.

A1 (Attitude toward Traditional Advertising). This variable was operationalized to assess attitudes toward traditional advertising by Likert 5 scale in general.

A2 (Attitude toward Product Placement). This variable was operationalized to assess attitudes toward product placement by Likert 5 scale in general.

B1 (Behavior caused by Traditional Advertising). This variable was operationalized to obtain the consumer behaviors caused by traditional advertising.

B2 (Behavior caused by Product Placement). This variable was operationalized to obtain the consumer behaviors caused by product placement.

Table 1. Demographic Profile of the Sample

		Frequency	Percentage
GEN	Male	309	58.0
	Female	224	42.0
	Total	533	
AGE	≤18	6	1.1
	19~24	274	51.4
	25~34	195	36.6
	35~44	58	10.9
	Total	533	
EDU	Yes	124	
	No	409	76.7
	Total	533	

GEN, AGE and EDU was operationalized to represent the gender and age of the respondent and whether he(or she) receives a higher education respectively.

Table 2. Descriptive Statistics and Correlation Coefficients Matrix

	TIME	A1	A2	B1	B2	Mean/SD
TIME	1.00					3.35/1.460
A1	.153**	1.00				2.65/.975
A2	.194**	.157*	1.00			3.11/1.00
B1	.181**	.159**	.176**	1.00		1.53/.500
B2	.017	.095*	.079	.567**	1.00	1.32/.467

**. Correlation is significant at the 0.01 level (2-tailed)
*. Correlation is significant at the 0.05 level (2-tailed)

4 Results of Research Questions

4.1 Results of RQs1~4

The relationships between TIME and A1, A2, B1, B2 were examined using correlation analyses. As results show in Table 2, there are positive correlations between the TIME and A1(r=.153), A2(r=.194), B1(r=.181). The results support *RQs1~3* and which is statistically significant at the 0.01 level. The relationship between TIME and B2 is positive but it is not statistically significant. So *RQ4* can't be supported.

4.2 Results of RQs5~8

In order to analyze the influences of social structural variables to CS outcomes and answer *RQs5~8*, the ANOVA was conducted for GEN, AGE and EDU.

EDU. The lower the respondents' education level, the lower their A1 mean score (p=.005) and the higher their A2 mean scores (p=.315). While, it indicated that the lower the respondents' education level, the higher their mean score of B1 (p=.054) and B2 (p=.003). That means respondent with higher education level are more willing to accept product placement, but they will not buy the products because of these ads.

GEN. As shown in Table3, the results suggested that female respondents hold more negative A1 (M=2.56) than male counterparts (M = 2.71) (p=.072). For A2, the influence was positive to female (M=3.29) and negative to male (M=2.98) (p=.000). It also indicated that different gender had different degree toward traditional advertising and product placement. To product placement, female had more positive attitude. Further analysis suggests that there had no significant difference between male and female to B1 (p=.165), B2 (p=.296). The behavior caused by traditional advertising (M=1.53) was more than by product placement (M=1.32).

Table 3. Summary of ANOVA(analysis of variance)

		GEN		AGE				EDU	
		Male	Female	≤18	19~24	25~34	35~44	YES	NO
A1	mean	2.71	2.56	2.00	2.56	2.95	2.10	2.86	2.58
	SD	.956	.996	.000	.859	1.039	.975	.810	1.012
	F(sig)	3.25*(.072)		14.890**(.000)				8.00**(.005)	
A2	mean	2.98	3.29	3.50	3.01	3.21	3.26	3.19	3.09
	SD	1.021	0.944	.548	.983	1.026	.983	.852	1.040
	F(sig)	12.825**(.000)		2.362*(.070)				1.011(.315)	
B1	mean	1.50	1.56	1.50	1.44	1.69	1.40	1.45	1.55
	SD	0.501	0.497	.548	.497	.463	.493	.500	.498
	F(sig)	1.931(.165)		12.041**(.000)				3.716*(.054)	
B2	mean	1.30	1.34	1.50	1.33	1.32	1.26	1.21	1.35
	SD	0.459	0.476	.548	.471	.467	.442	.409	.478
	F(sig)	1.092(.296)		.662(.576)				8.999**(.003)	

**F is significant at the 0.05 level *F is significant at the 0.10 level

AGE. The results of the ANOVA indicated a significant age difference for A1, A2. But the mean A2 of every age group was higher than 3, it indicated that they had positive attitude to product placement. Less than 18 years old group was the highest (M=3.50) and followed by 35~44. There is no obvious trend with age change. B1 had significant difference in age group. Respondents 25 to 34 years old showed the highest mean score (M=1.69). There are no significant differences between age group to B2. It indicated that they had more positive influence by traditional advertising (M=1.53) than product placement (M=1.32).

5 Discussion

The results of the current study show that consumer socialization contributes to product placement research by providing a flexible yet theory-grounded framework. Besides, attitudes toward product placement as well as placement-related behaviors may be assessed and better understood by comparing the traditional advertising and product placement. The flexibility of the framework is particularly attractive in that it allows for an expansive yet theoretically grounded set of influencing factors to be examined in relation to behaviors and attitude.

Our findings that both attitude toward and behaviors caused by product placement demonstrate differences among a range of demographic characteristics has practical implications for placement agents and marketers more broadly. Our results suggest that advertisers would do well to match up their placement efforts based on a particular target audience of movie and TV programs. Furthermore, our results provide some indications that advertisers would do well to focus their placement efforts on movies popular among less than 18 years old, female consumers in China. These audiences exhibit higher levels of receptiveness to product placement as a practice, and marketers can also follow these indications. However, the above should not be interpreted by advertisers as a blank check to use product placement at their pleasure or to mean that audiences will be willing to tolerate any kind of placement.

We noted previously that our data indicates some level of effectiveness on the part of product placement in convincing audiences to take consumption-related actions after viewing placements. While this is good news for marketers, but our study does not address the specific characteristics of a placement that may enhance the likelihood of consumption-related behaviors.

References

1. de Gregorio, F., Sung, Y.: Understanding Attitudes Toward and Behaviors in Response to Product Placement. Journal of Advertising (39), 83–96 (2010)
2. Sung, Y., de Gregorio, F., Jung, J.-H.: Non-student consumer attitudes towards product placement. International Journal of Advertising (28), 257–285 (2009)
3. Redondo, I., Holbrook, M.B.: Illustrating a systematic approach to selecting motion pictures for product placement and tieins. International Journal of Advertising (27), 691–714 (2008)
4. Eisend, M.: A Cross-Cultural Generalizability Study of Consumers' Acceptance of Product Placements in Movies. Journal of Current Issues and Research in Advertising (31), 15–25 (2009)
5. Cristina, Raluca, Delia: Product placement in Romanian movies produced after 1989. Journal of Media Research (8), 46–73 (2010)
6. van Reijmersdal, E.: Brand Placement Prominence: Good for Memory Bad for Attitudes. Journal of Advertising Research (6), 151–153 (2009)
7. Yang, M., Roskos-Ewoldsen, D.R.: The Effcetivencess of Brand Placements in the Movies: Levels of Placements, Explicit and Implicit Memory, and Brand-Choice Behavior. Journal of Communication (57), 469–489 (2007)
8. Yin, X.: Empirical research on the effect of product placement in films. China's Collective Economy (1) (2009)
9. Moschis, G.P., Churchill Jr., G.A.: Consumer Socialization: Theoretical and Empirical Analysis. Journal of Marketing Research (1978)

Considerations on Strengthening Enterprise Financial Control

Jie Li, XuYing Jiang, and JiChun Chen

Dept. of Air Force Finance, Xuzhou Air Force College, Xuzhou, China
jiangziya@163.com

Abstract. Financial control is an important component of enterprise financial management. How to strengthen enterprise financial control and improve operation and management level have already become important factors that constraint enterprise's development. In the market economy, the enterprise should take standardized operation, risk prevention, benefit improvement and sustainable development as the criterion and take pursuance of enterprise maximum profits as the goal to establish comprehensive and multi-level financial control system. This paper gives a deep research on how to strengthen financial control for Chinese enterprises under the conditions of market economy by combining main problems existing in current enterprise financial control.

Keywords: Enterprise, Financial control, Financial management.

1 Introduction

The enterprise strengthens the supervision and control over financial activities by establishing effective financial control system, which is an important link in perfecting enterprise organization management, promoting the identification of enterprise organizations for strategic objectives and its implementation, overcoming objective adverse selection preference and improving financial resource allocation and use efficiency. Therefore, establishing efficient and reasonable financial control system directly concerns activeness, creativity and responsibility of stakeholders of the enterprise, and concerns orderliness and efficiency of allocation and operation of financial resources and other various economic resources, so that it is a strategic problem relating to the survival and existence of enterprises.

2 Main Problems Existing in Current Enterprise Financial Control

2.1 Enterprise Financial Management System Is Imperfect

At present, property and material management system for Chinese enterprises is still imperfect, which has certain management loopholes and lacks various kinds of efficient and effective supervision system, so that the phenomena [1] that great loss of property and material and appropriating public property for private use due to improper storage of card and material easily occurs. Almost no scientific feasibility study is conducted

D. Jin and S. Lin (Eds.): Advances in ECWAC, Vol. 1, AISC 148, pp. 67–71.
springerlink.com © Springer-Verlag Berlin Heidelberg 2012

during investment of projects, thereby leading to blind investment, repeated construction, idle capital and unsmooth operation. In addition, control for enterprise accounting is still inadequate. Financial supervision and control themselves are built on mutual restraint. If persons in enterprise who should be mutually restrained join together and cheat jointly, it will make rules and regulations highly ineffective.

2.2 Enterprise Financial Data and Information Are Inaccurate

There are phenomena that financial data and information are not timely and accurately obtained during financial management of enterprises. Partial financial statements are not true and accurate enough so that it is difficult to reflect financial conditions of the enterprise through the financial statement obtained coupled with setting of enterprise financial accounting system and accounting subject failing to be implemented strictly according to standards. At the same time, methods for compiling and consolidating financial statement are unscientific and unreasonable, which also leads to inaccuracy of overall asset information of the enterprise, so that the overall financial condition of the enterprise is untrue, thereby easily results in deviation of important decisions of the enterprise.

2.3 Enterprise Budget Management Is Imperfect

At present, budget management for most enterprises has three main problems: the first is there is inconsistency between departmental budget set by the finance and comprehensive budget of the enterprise both in compilation time and compilation scope, so that two times of budget are required. The second is no corresponding tracking, analysis and evaluation systems are established for the implementation of the budget. It is difficult to accurately and specifically analyze and evaluate various budgets of departments and also correctly and completely evaluate utilization benefit of funds. Only basic and overall analysis and evaluation can be carried out at the moment. The third is there are still phenomena that contents of budget are untrue, constraint force of budget is not strong and budget fails to be implemented, etc during the implementation of budget.

2.4 Incentive Mechanism of Financial Control Is Ineffective

During the reform of property rights of enterprises, modes of financial control are various and reward and punishment coexist. Seeing from the type of reward, it mainly includes yearly salary system, bonus share, option share, share award for asset increment, and employee shareholding. These measures play certain incentive roles for standardizing enterprise financial management and strengthening enterprise financial control. Cases that lead to serious loss of state-owned assets due to embezzlement or violation of laws and disciplines will be correspondingly punished administratively and are prosecuted for criminal responsibility for serious ones. However, because there is great difference for reward and punishment policies in various regions and different enterprises in the same region are also different, degree of functioning of this incentive and constraint means also has great difference in financial control, which restricts the activeness of manager to some extent [2].

3 Countermeasures for Strengthening Enterprise Financial Control

3.1 Strengthening Enterprise Internal Control Management

All from chairman to general manager, then to divisional manager and finally to employee, or from factory director to workshop manager and then to each staff are represent by a series of principal-agent relationships. Effective monitoring for agents shall be ensured under the principal-agent relationships of each level. Therefore, level-to-level multi-path internal control system combining self-control with external constraint and combining internal supervision with external auditing must be established by enterprises. The first is to firmly grasp key point, key man, key link, fund control and key post for effective internal control. Expected objectives can be reached through various precise and effective policies and procedures. Smooth implementation normally requires proper duty separation, reasonable approval procedure, precise evidence record and independent achievement inspection. The second is to establish and improve accounting approval system, uniform accounting policy and strict hierarchical authorization, stipulate authorities and responsibilities of different levels, implement uniform fund allocation and transfer, and to fully play the monitoring function of financial settlement center to restrain the operation through systems and procedures. The third is financial control has different levels under existing enterprise system. The internal audit represents the supervision of board of directors for operating activities of subsidiary companies and functional departments, the external audit represents the supervision of the nation and all shareholders and board of supervisors represents the supervision of shareholders and investors [3]. Therefore, enterprises should establish relatively independent audit institution to objectively and fairly evaluate asset structure, business performance and cash status. In a word, strengthening enterprise internal control management is the process giving play to subjective initiative of personnel and improving various enterprise systems, and also the process pursuing rule by men on the basis of law.

3.2 Comprehensively Controlling by Utilizing Network Finance

The network finance is developed following the development of network enterprises and electronic commerce. With the increasing complexity of financial control activities, modern enterprises urgently need to implement dynamic financial control and management and also need to strengthen the control of shareholders for enterprises due to the existence of behavior of "insider-control". The popular network at present is to be contributive to the control for complex activities and also to be contributive to the control of shareholders for enterprises. Enterprises can unify logistics of economic activity of functional departments with the information flow through network to make management departments keep informed of objective information, thereby making financial control scientifically penetrate into each level of organization management and into overall process and each operation link of production, and vigorously combining financial control of each level, each process and each link in order to improve effectiveness of fund utilization of enterprises. Shareholders can be timely informed of operation of enterprises through the network, which is easy for shareholders to participate in decision-making of production and operation, thereby avoiding the behavior that agent damages interests of shareholder.

3.3 Implementing Comprehensive Budget Management System

Comprehensive budget management refers to vigorously combining enterprise internal conditions with external environment, introducing market competition into enterprise and clearly defining responsibilities, authorities and benefits of each subject of liabilities; introducing dynamic mechanism and vigorously combining responsibilities, authorities and benefits of each subject of liabilities to form comprehensive budget, thereby forming interest balance mechanism within the enterprise and realizing scientification, standardization and elaboration of enterprise decisions. A series of judgment and analysis concerning capital structure, capital organization mode, capital investment opportunity and capital operation means, etc are carried out through the comprehensive budget management on the basis of adapting to enterprise competition and promoting enterprises' medium-term and long-term development. Resource allocation is further optimized by changing from operating matters to be concerned by leaders to identifying the gap through participating in market competition by all staff in order to achieve the goal of improving economic profits of enterprise [4]. In conclusion, implementing comprehensive budget management is a powerful weapon to realize strategic target and also an important tool for enterprises to implement financial control.

3.4 Perfecting Enterprise Incentive Mechanism

In modern enterprises, constructing one incentive and restraint mechanism of manager behavior with consistent interests between the owner and the manager is of critical importance due to the gradual separation of ownership and management right. The manager shall be effectively motivated to prevent the manager from pursuing individual goals by taking advantage of his position within the managing scope and at the expense of interests of the investor, which is an important measure guaranteeing capital gains of the investor. The first, money is an important incentive means of enterprise. Money often has more value than the money itself. When money is used as reward and punishment means of financial control, the enterprise should provide managerial staff and employee who exceed standard score with corresponding award money and impose a fine for those who fail to reach the standard score based on the result of comprehensive evaluation for departments. The second, honor is an important part of enterprise incentive mechanism. The enterprise can periodically organize various competitions and excellence evaluation. Managerial staff and employee who have good performance and obtain good results will be given honors of verbal or written praise, commendation and group outstanding employee, etc, and also provided with corresponding remunerations, whereas relevant personnel with poor performance will be criticized in a circulated notice , etc.

4 Conclusions

Following three main conclusions can be reached by studying enterprise financial control: the first, enterprise financial internal control is guarantee of the survival and development of an enterprise. Financial activities of enterprise are controlled and supervised through standardized control means to ensure asset safety, operation efficiency and realization of strategic target of development of enterprise; the second, scientific budget target is the direction of strategic development of enterprise. Budget is the bridge between decision and implementation. Only scientific and reasonable budget target can make budget operable and implementation results corresponding with strategic demand of enterprise development; the third, perfect incentive mechanism is the basis of vitality of enterprise. Effective incentive mechanism can fully mobilize the initiative of all quarters, give free rein to advantages in various respects and make the company full of vitality.

References

1. Wang, Y.: Analysis of Problems Existing in Enterprise Financial Control and Countermeasures. Chinese Economy and Business. 287, 253–254 (2010)
2. Zhang, Y.: Discussions on Important Links of Enterprise Financial Control. New Accounting 291, 33–36 (2011)
3. Zhu, Y.: Enterprise Financial Control and Management. Market Modernization 288, 182 (2010)
4. Li, X.: How to Strengthen Enterprise Financial Control. Modern Accounting 278, 28–31 (2009)

Considerations on Strengthening State-Owned Asset Supervision and Administration

Jie Li, XuYing Jiang, and JiChun Chen

Dept. of Air Force Finance, Xuzhou Air Force College, Xuzhou, China
jiangziya@163.com

Abstract. State enterprises are the pillar of our national economy and they are facing the biggest difficulty and greatest challenge since this century, operating profit drops drastically and phenomenon of loss of state-owned assets is still serious under the superposed influence of global financial crisis and self structural imbalance. Under such circumstances, problems of state-owned asset supervision and administration have aroused great concern. This paper describes main problems existing in state-owned asset supervision and administration and analyzes causes of problems in detail. On this basis, countermeasures are proposed to strengthen state-owned asset supervision and administration.

Keywords: State-owned asset, Supervision and administration, Property right, Loss.

1 Preface

How to strengthen state-owned asset supervision and administration and ensure the value maintenance and value increment of state-owned assets are problems to be studied and solved urgently. With entry into the new century, both the nation and governments at various levels released many policies concerning the administration of state-owned assets, which enables the administration of state-owned assets to be placed on a preliminary normalized track, and achieved some achievements. However, it still has or is facing many problems that cannot be ignored. Among them, the main problem is that problem of loss of state-owned assets caused by poor supervision and administration is still serious. Therefore, the problem of state-owned asset supervision and administration has become one of the most attractive research fields in recent years.

2 Main Problems Existing in State-Owned Asset Supervision and Administration

2.1 There Is Phenomenon of Defaulted Supervision and Administration on Property Right of State-Owned Assets

With the development of market economy, partial state-owned enterprises implement mixed ownership coupled with the adjustment of the pattern of our national economy. For such enterprises, it is difficult to determine boundary and nature of property right

D. Jin and S. Lin (Eds.): Advances in ECWAC, Vol. 1, AISC 148, pp. 73–78.
springerlink.com © Springer-Verlag Berlin Heidelberg 2012

after merging many times, so that some enterprises still follow traditional mode that defining nature of property right first and then applying state-owned property right for administration. This not only violates legitimate rights and interests of other investors but also makes responsibility system of state-owned property right management difficult to be effectively implemented, which easily leads to defaulted supervision and administration on the property right of state-owned assets [1].

2.2 Problem of Loss of State-Owned Assets Is Still Serious

Firstly, loss of state-owned assets during management: unclear responsibilities, rights, and interests of management main body resulted from ambiguous property right of state-owned assets lead to blind decision-making, repeated construction, wasting of resources and idling of assets, thereby causing loss of state-owned assets. The first, management of physical assets has loopholes. The management of physical assets is not standardized, there are many unlisted assets and the phenomena that account doesn't agrees with other accounts and account doesn't agree with physical inventory, etc are serious. The second, asset disposal is relatively arbitrary. When assets require selling, breakage or discarding, no disposal procedures of review, assessment and technical evaluation, etc are implemented and they are handled without authorization, which causes asset management out of control.

Secondly, loss of state-owned assets during investment and management: some units invest abroad blindly, lacking the feasibility study, and the investment management is far lagging behind, which lead to wrong investment and low benefit from invested projects, or even failing to recover capitals invested. This is also the cause resulting in serious loss of state-owned assets.

Thirdly, loss of state-owned assets during restructuring of enterprise: when transferring or disposing state-owned assets, some enterprises violate regulations. The first, transfer to non-state-owned units or individuals without compensation or at the price clearly lower than the market price; the second, there is lack of effective supervision and administration for bankruptcy of enterprises; the third, in enterprise stock asset reorganization, value of state-owned assets are not evaluated or undervalued. All of these cause the loss of state-owned assets to some degree.

2.3 Supervision and Constraint on Enterprises by the Nation Is Weak

The principal-agent link of state-owned capital in our country is too long and levels are too many. Many enterprises expect to control large-scale assets and numerous enterprises with less capital through multilayer stock-holding to achieve development by leaps and bounds. This caused rapid expansion of capital link and dramatic growth of group size [2]. But results have never looked so good. This multilevel principal-agent relationship makes the supervision and constraint on enterprise operators by the nation decrease gradually, weakens the control on these enterprises by the nation and easily leads to wrong decision.

3 Cause Analysis of Problems Existing in State-Owned Asset Supervision and Administration

3.1 Laws & Regulations and Systems Are Imperfect

Some existing laws and regulations have lagged behind requirements of market economy development in some aspects. Standardization and operability are weak for some of them, so that no standardized and effective mechanism is formed. Imperfectness of system will cause arbitrary implementation, which makes investors easier to take thc advantage of political loopholes and results in loss of state-owned assets.

3.2 Management System Is Unscientific

The current state-owned asset management system of our country is gradually reformed and developed from traditional economic system after the reform and opening up, which has experienced hard theoretical and practical exploration and played an essential role. However, many problems still exist in protecting legitimate rights and interests of state-owned assets, etc for this system and phenomena that unclear functions of administrative departments, mutual shuffle and irresponsibility, etc still exist.

3.3 Function of Organization Is Insufficient

After the establishment of State-owned Assets Supervision and Administration Commission of the State Council (SASAC), although the nation set up state-owned asset management organizations in local governments at various levels and implemented state uniform ownership and classified management, the current situations of highly centralized ownership and widely distributed state-owned assets inevitably will form the gap in the supervision and management and functions of the organization failing to be fully displayed. This is an important factor resulting in the loss of state-owned assets.

3.4 Inspection and Punishment Are Inadequate

Supervision and Administration problems of state-owned assets haven't been well solved and one of the important reasons is the lack of strict law and discipline enforcement. Liability ascertainment is still inadequate for the management personnel who cause losses to the nation due to neglect of duty, which provides an opportunity for lawbreakers or speculators in law and discipline enforcement.

4 Countermeasures for Strengthening State-Owned Asset Supervision and Administration

4.1 Perfecting Laws and Regulations of State-Owned Asset Supervision and Administration

State-owned asset supervision and administration is an important work. Therefore, establishing one system of state-owned asset supervision and administration laws and regulations with internal relations, tight structure, concise level and harmonious content is required. The first is to adhere to national basic economic system and make state-owned asset supervision and administration laws and regulations always mutually adapt to the construction and development with Chinese characteristics; the second is to scientifically define responsibilities of state-owned asset supervision and administration organization and continuously improve state-owned asset supervision and administration laws and systems; the third is to establish and improve accessory rules and regulations concerning supervision and administration of state-owned assets according to requirements of existing laws and relevant administrative regulations; the fourth is to strengthen the construction of relevant systems to achieve the unification of rights, responsibilities and duties.

4.2 Standardizing Transaction Behavior of Property Rights of State-Owned Assets

The first is to improve enterprise property right transaction system. China should work out and promulgate relevant enterprise property right transaction laws and regulations as soon as possible so that the enterprise property right transaction behavior can have laws and regulations to abide by. The second is to perfect information transmission mechanism of enterprise property right transaction and play functions of government, market, enterprise and media. The third is to establish uniform and open property right market network, fully utilize professionalism of intermediary organization and adopt the operation mode of mechanism linkage and information sharing. The fourth is to standardize form of the transaction and severely limit the avoidance of market open and fair competition by adopting any flexible means.

4.3 Improving State-Owned Asset Supervision and Administration System

The fist is to improve state-owned asset supervision and administration organization. The state-owned asset supervision and administration organization hasn't been completely established seeing from three levels of central authority, province and districts & cities. Therefore, it is required to adjust and optimize functional departments to form standardized and order coordination system, and strengthen the dynamic integration between external supervision and enterprise internal supervision to form more effective supervision mechanism, thereby creating efficient internal organization operation system [3].

The second is to perfect the supervision and administration and operation of state-owned assets. At present, the mode of "state-owned asset supervision and administration organization—state-owned asset operation organization—enterprise" for state-owned asset supervision and administration and operation hasn't really formed. The middle level is the weak link and the phenomenon of "defaulted" main body of asset operation is prominent. In order to solve this problem, it is required to actively advance the reform of property rights system and the transformation of modern enterprise system of state-owned asset operation main body and state-invested enterprise and combines the state-owned asset supervision and administration with its operation.

The third is to implement matrix state-owned asset supervision and administration. In order to solve problems existing in state-owned asset disposal, it is required to intensify the asset disposal supervision and establish "matrix" asset disposal supervision and administration mechanism, for which each performs its own functions, cooperates closely and is mutually interrelated. The "matrix" supervision and administration mode is one intersecting and bidirectional management mode, that is, the vertical management taking state-owned asset management specific organization, state-owned asset management intermediary organization and state-owned asset use management unit as mainline, and the lateral management composed of asset use department, financial department and auditing department, etc of the enterprise are formed [4].

4.4 Implementing Responsibilities of Value Maintenance and Value Increment of State-Owned Assets

To prevent the loss of state-owned assets and realize their value maintenance and value increment, the first is to solve system and responsibility problems fundamentally. The state-owned asset supervision and administration organization, as the investor of state-owned assets, must well perform responsibilities of the investor, perfect corporate governance structure and define operation responsibilities of state-owned assets to provide systematic guarantee for value maintenance and value increment of state-owned assets. The second is to achieve the separation of administrative functions from enterprises management. The state-owned asset supervision and administration organization should fully respect rights and interests enjoyed by the enterprise according to the law and it can't directly interfere with production and operation activities of the enterprise and can't excessively control the enterprise to make the enterprise truly become market player and corporate body who operate independently. The third is to intensify social supervision. State-owned asset condition and state-owned asset supervision and administration works shall be released to the public regularly according to the law and transparence shall be continuously improved to actively create favorable social supervision environment.

5 Conclusions

Following three main conclusions can be reached by studying state-owned asset supervision and administration: the first, an essential way to strengthen state-owned asset supervision and administration is in legislation; the second, the state-owned asset supervision and administration legal systems covering the whole process from purchase, use and disposal of state-owned assets must be established and perfected, which also covers every aspect of state-owned asset management; the third, the efficiency level of state-owned assets is closely related to the supervision and administration works.

References

1. Zhou, B.: Considerations on Strengthening State-owned Asset Supervision and Administration. Reform and Opening up 274, 36–37 (2007)
2. Jiao, Y.: Discussion on State-owned Asset Supervision and Administration. Economist 286, 222–223 (2009)
3. He, H.: Problems existing in State-owned Asset Supervision and Administration System and Countermeasure Analysis. China Chief Financial Officer 294, 118–119 (2010)
4. Huang, B.: Development Trend of State-owned Asset Administration System after Releasing the People's Republic of China Enterprise State asset Law. China Development Observation 282, 50–52 (2009)

On the Enterprise Human Capital Investment

XuYing Jiang, ZhenKai Xie, and JiChun Chen

Dept. of Air Force Finance, Xuzhou Air Force College, Xuzhou, China
jiangziya@163.com

Abstract. In modern enterprises, human capital, as one of production factors, is increasingly playing an important role. Same as the physical capital, human capital investment also has a high investment risk. Therefore, it is concerned much by business managers, who gradually start to establish correct human capital investment philosophy. This paper describes present situation of human capital of Chinese enterprises, analyzes potential risks of enterprise human capital investment, and discusses measures avoiding the risks of enterprise human capital investment on this basis.

Keywords: Human capital, investment risk, enterprise revenue.

1 Introduction

With fast development of knowledge economy, business managers are increasingly aware that people are the most important resource for enterprises. Talents will bring great asset to enterprises and human capital investment is getting more and more important in modern human resource management. More and more enterprises start to establish human capital investment philosophy. However, there is blindness during human capital investment for many enterprises from the lack of correct understanding for human capital and investment, which results in higher or lower human capital investment of enterprises. This doesn't maximize the role of employees but influences development strategy of the whole enterprise. It can clearly be seen that how to correctly plan enterprise human capital investment directly relates to existence and development of enterprises.

2 Present Situation of Enterprise Human Capital

2.1 Total Human Capital Investment Is Inadequate

Compared with many foreign large companies, investment for human capital by Chinese enterprises lags far behind the investment for physical capital. Meanwhile, human capital level of managers and technicians is low and current human capital stock is seriously inadequate for Chinese enterprises, and investment on reserve of backup personnel is relatively small for most enterprises, which will badly affect future existence and development of enterprises.

D. Jin and S. Lin (Eds.): Advances in ECWAC, Vol. 1, AISC 148, pp. 79–83.
springerlink.com © Springer-Verlag Berlin Heidelberg 2012

2.2 Structure of Enterprise Human Capital Is Unreasonable

The first, structure of enterprise human capital is unbalanced. The youth of our country who received higher education only makes up a small proportion and talents are mainly concentrated in government organizations coupled with Chinese traditional employment concept. Therefore, professional technicians and high quality managers for scientific research and product innovation are deficient.

The second, structure of human capital stock is unbalanced. In high-tech areas and large medium enterprises, human capital stock has exceeded actual demands whereas many small medium enterprises almost have no human capital stock due to reasons of geographical position or economic power.

The third, loss of human capital of high level is serious. There is talent outflow phenomenon for most undertaking units of high-tech projects and the outflow personnel are normally greater than the inflow personnel. According to the estimate, over 70% middle and higher level technicians, managers and skilled workers of foreign-invested enterprise and privately-run enterprise are from state-owned enterprise in recent years.

2.3 Waste Phenomenon for Enterprise Human Capital Is Serious

Enterprises blindly emphasize high education level and high professional title when recruiting talent. They take interest in inviting high-level talents by offering a high salary one the one hand but neglect development and utilization of existing talents on the other hand, so that vicious circle of talent deployment is generated. High talent consumption of only paying attention to recruiting high-level talents by offering a high salary but neglecting enthusiasm of existing talents causes the enormous waste of human capital of enterprises.

Enterprise Lacks Effective Competition. As Chinese enterprises lack marketability concerning management style of human capital and reasonable ideology for employee encouragement and development and fail to establish one effective competitive mechanism, so that initiative of employees can't be fully mobilized, thereby restricting their enterprising spirit and creativity.

3 Analysis of Risks of Enterprise Human Capital Investment

3.1 Environmental Risks

The first, risks of policies and regulations. Change of national rules and regulations and adjustment of macroeconomic policies and of industry standards and industry policies will bring certain risks to human capital investment of enterprises.

The second, risks of scientific and technical innovation. Enterprises can't accurately predicate external scientific and technical innovation and change when carrying out human capital investment concerning certain new technology due to non-symmetry of information, which results in un-competitiveness of new technology and accelerated depreciation and devaluation of the human capital originally invested, or even worthless, so that expected investment effectiveness can't be obtained.

The third, risks of market. Market changes unpredictably so that it is difficult to predict demands. Actual effect of investment on current human capital will be greatly reduced when market demand changes.

The fourth, risks of force majeure. Force majeure covers risks of seismic hazard, war and death or work ability loss of invested target due to sudden accident, etc and will bring great risks to human capital investment once it occurs due to its irresistance.

3.2 Management Risks

The first, risks of investment subject. It refers to risks which are caused because investment subject fails to completely and accurately grasp all information of invested target due to non-symmetry of information or human capital investment risks which are caused because initiative and creativity of invested target can't be fully displayed even it is selected properly due to unscientific and unreasonable incentive and restraint mechanism of enterprises.

The second, risks of moral. Efficiency of human capital also will be affected by willingness of invested target after its formation. If the employee is not interested in the work, he will not work hard or even be slack in work, which will make human capital can't be fully functioned. Or the employee is disloyal to the enterprise and sells out commercial secret, which also will bring heavy loss to the enterprise.

The second, risks of outflow of talent. Under the increasingly fierce talent competition, it is easy to create risks of outflow of talent if the original enterprise fails to satisfy material and spiritual demand of invested target due to his wish. This is the main risk faced by human capital investment of enterprises at present.

4 Measures Avoiding Risks of Enterprise Human Capital Investment

4.1 Defining Own Business Strategy of the Enterprise

In order to better play the role of human capital, enterprises have to provide employee with larger space for actively creating greater value and benefit by establishing good environment. Therefore, enterprises must carry out human capital investment according to own business strategy and vigorously combine human capital investment with own business strategy to lower investment risks. Only in this way can better give full play to human capital advantages [1].

4.2 Formulating Occupational Planning of Enterprise Human Resources

Human resource planning is the basis of human capital investment of enterprises. How human resources give play of profit making, the key is in reasonable allocation of talents [2]. Human resource planning can help enterprises select talents who are suitable to demands of own production and operation whereas various talents also can find out working posts that are suitable to own features and through which the role also can be fully played. In this way, development objectives of enterprises can be consistent with personnel development objectives of employees, thus, win-win relationship between the enterprise and the employee can be established, thereby forming benefit or even common destiny to better realize operation and development objectives of enterprises.

4.3 Establishing Risk Assessment Mechanism of Enterprise Human Capital Investment

Scientifically carrying out human capital investment risk assessment and establishing effective risk assessment mechanism are important links of investment decisions. Enterprises should carry out human capital investment assessment according to new technical invention, new production development, knowledge update and uncertain factors, strengthen control of human capital investment plan for uncertain factors and timely grasp probability distribution of occurrence of uncertain factors to avoid or reduce financial loss of investment caused by uncertain factors [3].

4.4 Establishing Good Human Capital Incentive Mechanism

To obtain better human capital investment effect, it is required to establish and improve incentive mechanism of enterprises. The first is to adhere to the principle of combining material incentive with moral incentive. Simulative income distribution system shall be established and favor shall be given to high quality excellent staff who made great contributions on the basis of fairness to make wage level of enterprises competitive externally and make employees feel fair; moreover, moral incentive also should be strengthened and self realization of employees in the work shall be supported to strengthen their sense of belonging [4]. The second is to adhere to the principle of combining positive incentive with negative incentive. Behavior of employees shall be restrained by perfecting rules and regulations and signing labor contract, etc to lower moral risks and reduce losses caused by outflow of talents.

4.5 Valuing Loyalty of Employees to the Enterprise

Enterprises should select the job seeker who recognizes corporate values during recruitment of staff to obtain optimum matching between technology, knowledge, ability, interest and hobby of the candidate and qualification requirements of the post, thereby making employees feel a sense of belonging, motivating their job enthusiasm and making them form natural loyalty to the enterprise.

4.6 Building Harmonious Enterprise Cultural Environment

Enterprise culture is the spirit of enterprises. It has strong cohesive force, guiding force and influence power, which helps to mobilize employee initiative, stimulate employee creativity and reduce human capital loss. Therefore, moral risks of human capital investment can be lowered and initiative of human capital can be fully played by building a healthy and harmonious work environment and enterprise cultural environment with independent innovation and team spirit, and resisting erosion of decadent ideas and backward consciousness to employees through excellent enterprise cultural.

5 Conclusions

Following main three conclusions can be reached by studying enterprise human capital investment: the first, enterprise human capital investment should be vigorously combined with own business objectives. This can effectively guarantee the consistency between the occupational planning of human resources of the enterprise and its business objectives to realize win-win between both parties; the second, good incentive mechanism is the basis for talents giving full scope to initiative and creativity. The creative incentive mechanism can give full play to potential of talents and create more wealth for the enterprise; the third, healthy and harmonious work environment is the prerequisite for cultivating human resource loyalty and building team spirit.

References

1. Zhang, P.: Enterprise Human Capital Investment and Competitiveness Improvement. CO-Operative Economy Science 296, 30 (2009)
2. Lun, R.: Discussion on Human Capital Management and Innovation of Private-owned Enterprises under the New Situation. Business Economic Research 287, 33–34 (2009)
3. Que, C., Zhou, B.: Theoretical Analysis of Risks of Enterprise Human Capital Investment and Countermeasures. Journals of Fujian Agriculture and Forestry University 268, 57–60 (2009)
4. Xiao, X.: On the Human Capital Management of Knowledge-based Enterprises. Enterprise Vitality 297, 65 (2008)

Strengthening Financial Centralized Management, Improving Enterprise Group Management Benefit

JiChun Chen, XuYing Jiang, and ZhenKai Xie

Dept. of Air Force Finance, Xuzhou Air Force College, Xuzhou, China
jiangziya@163.com

Abstract. Financial centralized management is a new financial organization and management mode which is based upon centralizing financial accounting and financial management works of the enterprise group in a unified way under the network environment. It changes financial management environment and management concept of enterprise group, broadens range and content of financial management and poses new challenges to financial management of the enterprise group along wide application of information technology. This paper describes connotation of financial centralized management of enterprise group, analyzes advantages of financial centralized management and discusses problems to be noticed in strengthening financial centralized management of the enterprise group.

Keywords: Finance, Centralized management, Enterprise group, Benefit.

1 Introduction

With unceasing expansion of scale of collectivized operation of the enterprise, the economic environment also changes fundamentally, such as expansion of assets scale, elongating of financing chain and widening of regional distribution, etc. These factors make the emphasis of operation and management of enterprise group are in improving fund utilization rate and lowering enterprise operation risk and also make "centralized management, shared service" become the core of financial management of the enterprise group [1]. The centralized management is the preferred mode for financing of the enterprise group and the operation and management benefits of the enterprise group is to be effectively improved through centralized use and management of financial resources.

2 Connotation of the Enterprise Group Financial Centralized Management

The enterprise group financial centralized management is a management mode of "unified management, centralized accounting and unchanging of Three Powers" that gradually changes from extensive management to intensive management. The financial centralized management establishes perfect financial data system and information sharing mechanism with the benefit of communication technology of

D. Jin and S. Lin (Eds.): Advances in ECWAC, Vol. 1, AISC 148, pp. 85–89.
springerlink.com © Springer-Verlag Berlin Heidelberg 2012

modem network, thereby implementing the centralized monitoring strategically, integrating the financial internal resources, preventing the decision risks and improving the enterprise management efficiency [2]. The unified management is the financial organization and management which is based upon centralizing financial accounting and financial management works of the enterprise group in a unified way under the network environment; the centralized accounting is the confirmation, measurement, record and report by the financial accounting center of the enterprise group taking the currency as the measuring unit for the fund movement of member units and monitoring for reasonableness, legality and effectiveness of the fund movement. In fact, after implementing financial centralized management, the financial accounting center and the member unit are a principal-agent relationship of financial accounting. The financial accounting center only functions as service and monitoring and it can't substitute internal financial control of member units. The implementation of centralized accounting doesn't transfer the body of legal responsibility, that is, the responsible person of member units still is the accounting responsibility body of the unit and the responsible person of centralized accounting institution is the accounting responsibility body of the accounting institution. The link and the responsible person of the behavior must be affirmed when defining the responsibility of illegal activities to ascertain corresponding responsibilities; unchanging of "Three Powers" is unchanging of the body and the rights in economic activity of member units, who still have decision-making power, examination and approval authority and use right for business transaction activities and finance activities. The financing body of member units still is the unit itself and the responsible person of the unit still is the first responsible person of accounting works.

3 Advantages of the Enterprise Group Financial Centralized Management

3.1 Realizing Sharing of Financial Information Resources

Operation and management decisions of member units of the enterprise group relay more on financial information and data. After centralizing financial resources, the financial information of member units is timely centralized, which is facilitate sharing of financial information resources. This can not only make leaders, decision makers and supervisors timely know financial information and economical operation conditions of member units, but also make directors of headquarters timely obtain financial information and economical operation conditions of member units, even employees' representatives of member units also can timely know financial information of the unit at the corresponding level.

3.2 Improving Efficiency of Fund Utilization

Under the inflation environment, shortage of money has already become important factors that influence the enterprise operation and hinder its development with continuous implementation of tight money policy. On the one hand, funds of member units are clear and accurate through centralized management of funds, which creates objective conditions for reasonable regulating and using internally and production of scale economies. On the other hand, the centralized management can reach the purpose of real-time monitoring for funds. The funds are blood that runs through the whole operating activities of enterprise; therefore, efficiency of fund utilization of the enterprise can be improved and operational risks can be lowered by monitoring financial flow in real time.

3.3 Improving Ability of Financial Management

After realizing sharing of financial information resources, accuracy of financial accounting is promoted and ex ante forecast, concurrent control and ex post analysis ability of financial works in enterprise operation activities is improved. Ex ante forecast, that is, during the feasibility investigation of the project, project feasibility, fund recovery rate and recovery cycle, and rate of return, etc are analyzed from financial perspective so that leaders can decide whether to establish the project; concurrent control, that is, during the implementation of the project, reasonableness, scientificalness and truthfulness and design requirement conformity, etc of budget making and operational effectiveness as well as financial condition of the project are reviewed so that leaders and decision makers timely adjust funds and polices and make project operate more normally; ex post analysis, that is, after completing the project, financial evaluation and acceptance of project are timely carried out to evaluate whether fund utilization of the project is reasonable and scientific and whether there are phenomena of appropriation, misuse, waste or corruption, etc, and evaluate whether the project is acceptable and can be accepted.

3.4 Saving Resources, Improving Labor Productivity

Under the centralized management mode, information and data can be stored in special server at any time for centralized management and any part of financial software management system can be shared with the network construction. On the one hand, human and material resources for collection of financial information will be greatly economized and the effect on financial reporting and financial auditing, etc is especially noticeable. On the other hand, the informatization reduces financial staff and avoids unnecessary waste of financial talent resources, and also saves financial office, transport and data keeping costs, etc and effectively lowers office costs.

4 Problems to Be Noticed by the Enterprise Group during the Implementation of Financial Centralized Management

4.1 Establishing Uniform Enterprise Group Financial Policy

Management activity of enterprise group is essentially the process of information collection, management, selection and transmission. If the enterprise group wants to realize financial centralized management, a complete set of clear and operational financial system must be established, including uniform financial accounting system, uniform investment and financing management and profit allocation system, uniform fund management system and uniform budget system.

4.2 Paying Attention to the Relationship between Centralization and Decentralization

The management mode of decentralized operation gradually becomes the development trend of group management with diversifying of enterprise group. It should be noted that centralized management and decentralized operation are philosophy of the management and there is no necessary relation between the centralization and the decentralization of information processing means. Strengthening management, opening up and revitalizing should avoid repeating the route of "tightening control leading to non vitality and loosening control leading to chaos" to prevent problems of inflexible management or operation, etc. Problems to be solved by financial centralized management are chaotic management and low efficiency. Great attention should be paid to separation of management rights and economic rights under the financial centralized management mode. Financial monitoring and information centralized management should be strengthened; however, operation should be opened up and revitalized and enthusiasm of member units should be fully mobilized.

4.3 Member Units Should Still Strengthen Financial Management

After enterprise group establishes financial settlement center and financial accounting center, the financial accounting center and the unit are a principal-agent relationship of financial accounting. The financial accounting center only functions as service and activity main body and rights of each member unit remain unchanged. Except reimbursement works of the unit, the financial staff, under the leadership of the responsible person, also should continue to well do appropriation budget management, fixed asset management and well do analysis and summary of financial receipts and expenditures; provide leader with decision basis according to accounting information and do other financial management works. Moreover, the financial accounting center should pay attention to the contact with managed units and strengthen account checking and inventory checking works to ensure that account agrees with documents, account agrees with other accounts, account agrees with statements and account agrees with physical inventory. Financial management works of the unit will be further strengthened with the implementing departmental budget, cash concentration and disbursement (CCD) and various policies [3]. In this sense, the burden on the shoulders of financial staff is heavier instead of light after implementing financial centralized management. The idea that financial staff has nothing to do and doesn't not have the ground of use force is incorrect.

4.4 Strengthening Security of Financial Information System

Security is of vital importance for enterprise financial management information system. Centralized management necessarily requires further opening up and data sharing of accounting system, thereby increasing difficulties of security control. Security threats are not only threat inside the company but also threat outside the company. Potential security hazards that the system face must necessarily increase with the increase of opportunities to interfere system by the operator and the user, thus increasing operational risks of the enterprise.

5 Conclusions

Following three main conclusions can be reached by studying enterprise group financial centralized management: the first, unifying group financial system is an important guarantee to realize enterprise group financial centralized management. Financial management activities inside the enterprise group must be standardized by system to have laws to abide by and ensure its smooth implementation; the second, strengthening internal auditing and checking are important means for enterprise group to advance financial centralized management. Trueness of accounting information of member units, effective implementation of various internal control systems, policies and management measures of group to be earnestly implemented and executed by member units and implementation effect of financial centralized management shall be ensured through internal auditing and checking; the third, establishing capital management system by the enterprise group is effective means to implement cash control. Member units open the account in the internal bank. Settlement of all transactions in production and operation activities is handled in the internal bank and fund surplus and deficiency among member units are uniformly regulated with compensation and managed by the internal banks to minimize idle cash balance.

References

1. Zhang, M.: Strategies for Strengthening Enterprise Group Financial Centralized Management. Fortune Today 276, 113–114 (2009)
2. Li, J.: Implementing Financial Centralized Management, Improving Group Management and Control Ability. Coal Mine Modernization 294, 12 (2009)
3. Huang, J.: On the Implementation of Enterprise Group Financial Centralized Management. Commercial Accounting 286, 51 (2008)

Analysis of the Employment Effect of the CDM Forestation and Reforestation Project of Heilongjiang Province

PeiYan Yu* and YingLi Huang**

Northeast Forestry University, Harbin, China
{peiyan775533,yinglihuang336}@yahoo.com

Abstract. Launching forestry activities for carbon sequestration actively and enhancing the forest coverage rate ,which, is important for easing the change of the climate and improving the sustainable development of forestry region. Meanwhile, it is also a key way to resolve the current unemployment problem of the forestry region. According to the *China statistical yearbook* (2002-2009) and China's clean development mechanism carbon sequestration project's successful cases "*Facilitating Reforestation for Guangxi Watershed Management in Pearl River Basin Project* ", using the input-output analysis method, the Shelter forest forestation engineering investment estimation index and the instruction of the National Bureau of Forestry, the direct and indirect employment chances generated from the Heilongjiang CDM Forestation and Reforestation were computed. The research reflects the total direct and indirect employment chances generated by the CDM Forestation and Reforestation are considerable, these chances can actually improve the current unemployment problem of the forestry region.

Keywords: CDM forestation and Reforestation project, Employment, Input-output analysis method.

After stopping forest cut, timber production industry is directly affected, subsequent industry development will be affected as well, and this will cause large quantities of employees facing unemployment problem. According to the survey, the 369 thousand existing workers will have 164 thousand of them facing the unemployment problem, 81281 workers will be affected by the cessation of wood cutting. Although forest tending work can create some job opportunities for these workers, but there are still a large number of workers without a job. The three functions :Developing forestation program for carbon sequestration actively , increasing the forest coverage rate and enhance the forestry operation level ,developing forestry in the adaptation and mitigation of climate change and promoting sustainable development ,are one of the most important contents in forestry construction during the new period in china. At the

* PeiYan Yu (1989-), Undergraduate of Northeast Forestry University.
** YingLi Huang, (1971-), People of Harbin City in Heilongjiang, Province, Doctor, Professor, Main research direction: The forestry carbon collects with the carbon finance.

D. Jin and S. Lin (Eds.): Advances in ECWAC, Vol. 1, AISC 148, pp. 91–96.
springerlink.com © Springer-Verlag Berlin Heidelberg 2012

same time, Developing forestation program for carbon sequestration can improve reemployment of the employees in the forestry region and to some degrees, alleviate the unemployment problem of the forestry region.

1 The Basic Theory of CDM Forestation and Reforestation Project

CDM forestation and reforestation is not only one of the fifteen fields of carbon sequestration CDM, but also one of our country focused projects[1]. The implementation of forestation and reforestation project includes construction period and operational guidance period. During the construction period, many activities like soil preparation, forestation, fertilization and weeding have to be performed. Whereas many activities like prevention of forest fire and protecting the forest, management, wood cutting needs many manpower. The employment chances are hence created and the unemployment problem of the forest workers is solved as well.

2 Analysis of Heilongjiang Province CDM Forestation and Reforestation Project

According to the instruction given by the National Bureau of Forestry, the land selected for carbon sequestration forestation activity should fulfill the following conditions: (1) The land is derelict land and in good condition for forestation since at least 1st Jan 2000. According the practical local situation, the time constraint could be extended to before 1st Jan 2005. (2) The land's ownership must be clear with the certificate of land ownership approved by the country-level government. (3) Suitable for growing of trees, expected to exert large carbon sequestration function. (4) Helpful in protecting the diversity of local creatures, preventing land degradation, promotes the development of local economy and society. The area of wild grass ground of Heilongjiang Province is about 960467.5 hectares according to the data of Heilongjiang Ministry of Land and Resources and hence the total area of land that Heilongjiang Province could use for CDM forestation and reforestation project is 960467.5 hectares. The implementation of forestation and reforestation project includes construction period and operational guidance period. We can get the employment rate according to the analysis of employment chances created by developing CDM forestation and reforestation project during construction period and operational guidance period.

3 Analysis of Employment Chances Created by CDM Project

3.1 Analysis of Direct Employment Chances

According to the working time quota standard of forestation set by the "Shelter forest afforestation engineering investment estimation index" issued by the National Bureau of Forestry, which takes the formal mountain area, cave-shaped land, initial plant density to be 2500 trees per hectares, land size to be 40 cm ×40 cm ×30 cm as the reference. The working time quota per hectares of forestation(cleaning of forestry land+ soil prepration + forest tending, daily tending is not included) is about 71~136 working days, 103.5 working days on average. The total area of land that Heilongjiang Province can use for developing CDM forestation and reforestation project is 960467.5 hectares, thus the total working time quota of the CDM forestation and reforestation project of Heilongjiang Province can be computed using equation (1). If take the standard working time quota of one job vacancy to be 300 working days/year, the direct employment chances created by developing CDM forestation and reforestation project will be increased by 331.36 thousands in short term, which is very significant.

Total working time quota(Ld)=unit area working time quota × total area of forestation per year (1)

The forestry labor productivity will increase as the result of progress of technology and the increase in management standard, the total working time quota will decrease. Based on the working time quota per hectares of past years China's forestation activities, assume the average increase in labor productivity to be 20.14%, then the coefficient of correction of working time quota after the increase in labor productivity will be 79.86%.

Newly increased job vacancies(ED)=[newly increased total working time quota of forestation(working day)/300(working day/person)] ×79.86% (2)

The short term and newly increased employment opportunities created by developing CDM forestation and reforestation of Heilongjiang Province thus could be calculated using equation (1) and (2). Refer to table 1 as follow:

Table 1. The short term and newly increased employment opportunities created by developing CDM forestation and reforestation (hectares,10 thousand people)

The area of land in developing CDM forestation and reforestation project	short term employment opportunities	newly increased direct employment opportunities
960467.5	33.136	26.463

3.2 Analysis of Indirect Employment Chances

Indirect employment, which refers to the employment opportunities brought by other industries through supplying this industry with the middle-procession input of production essential elements, during the process of production. Such as the indirect industries which are in relation to forestry e.g., transportation, commerce, catering and service, ect. As forestation activities belongs to the agriculture section in the input-output table. According to the statistics in China Statistical Yearbook(from 2002 to 2009), the indirect employment vacancies can be calculated using the result of 4000 hectares' forestation in Facilitating Reforestation for Guangxi Watershed Management in Pearl River Basin Project (a successful case in CDM forestation and reforestation project) and the research take the capital in construction period(3020thousand)as ΔX)on the basis of the equation (3).(4).(5)and the indirect employment vacancies during the construction period resulted from 960467.5 hectares ' CDM forestation and reforestation in Heilongjiang Province(from 2003to 2010) Eventually, the indirect employment vacancies in CDM forestation and reforestation in Heilongjiang Province can be conjectured in the future.

$$a_{vj} = \frac{v_j}{x_j} \tag{3}$$

In the equation above, a_{vj} is the parameter of labor pay, v_j represents the jth department labor pay, x_j represents total input of the jth department, this equation represents the total output labor pay by the jth department.

$$\Delta v = \hat{A}_{vj} \Delta X \tag{4}$$

In the equation above, Δv is the increase in labor pay of each department, $\hat{A}_{vj}$ is the diagonal matrix of the parameter of labor pay, ΔX is the increase in the final product supply of the department, this equation means when the total variation of the gross product of the jth department is ΔX , the variation of labor pay of the jth department is Δv .

$$\Delta L_i = \frac{\Delta V_i}{W_i} \tag{5}$$

In the equation above, ΔL_i is the quantity of labor force, ΔV_i is the increase in labor pay of each department, W_i is the average salary of the department, the salary of each department is known, the expected increase quantity of labor force of each department is ΔL_i .

Using equation (3), (4) and (5), the indirect employment vacancies created by the 4000 hectares' forestation of Guangxi forestation and reforestation project, the indirect employment vacancies resulted from 2250 thousand hectares' CDM forestation and reforestation in Heilongjiang Province can be conjectured. The results are shown in table 2 as follow:

Table 2. The indirect employment vacancies created by developing CDM forestation and reforestation during the period of construction(per person)

Project	Guangxi CDM forestation and reforestation project (4000 hectares)	the developing of CDM forestation and reforestation in Heilongjiang Province (960467.5 hectares)
Time	Indirect employment vacancies	
2003	1353	324878
2004	1207	289821
2005	900	216105
2006	793	190413
2007	628	150793
2008	589	141429
2009	525	126061
2010	457	109643
2011	398	95796
2012	346	82975

3.3 Analysis of Direct and Indirect Employment Chances

Based on the forest management standard quota set by the Shelter forest afforestation engineering investment estimation index, "take the normal mountain area, concentrated woodland with convenient transportation as the reference, the forest management standard quota per person per year is set to 150 hectares." The total area of land that can be used by Heilongjiang Province CDM forestation and reforestation project is 960467.5 hectares, this will create 6403 direct job opportunities of management of forest for carbon sequestration. The cost of Facilitating Reforestation for Guangxi Watershed Management in Pearl River Basin Project during operational guidance period was 19.68 million dollars(ΔX during operational guidance period). The indirect employment opportunities created by Heilongjiang Province CDM forestation and reforestation project could be calculated using 4000 hectares forest created by Guangxi Province CDM forestation and reforestation project. The indirect employment opportunities created by developing CDM forestation and reforestation of Heilongjiang Province thus could be calculated using equation (3), (4) and (5). Refer to table 3 as follow:

Table 3. The indirect employment opportunities created by developing CDM forestation and reforestation during operational guidance periodin Heilongjiang Province(per person)

Indirect employment vacancies	2006	2007	2008	2009	2010	2011	2012
	162541	139912	120686	107629	93606	81747	70807

From the table the enormous direct and indirect employment opportunities created by the CDM forestation and reforestation project of Heilongjiang Province could be conjectured.

4 Conclusion

Developing CDM forestation and reforestation will create many employment opportunities for Heilongjiang Province, this improves the current unemployment problem of forest workers. Meanwhile, it plays a vital role in the decreasing of the greenhouse gases and make a contribution to making strategies by the authorities when facing the climate change and, it is significant for economic transition in the forestry region in Heilongjiang Province. In conclusion, the CDM forestation and reforestation of Heilongjiang Province during construction period creates 331.36 thousands of direct employment opportunities. According to the statistic of the "China Statistical Yearbook", the indirect employment opportunities were calculated as 109643(2010), 95796(2011), 82975(2012). The indirect employment opportunities during operational guidance period were calculated as 93606(2010), 81747 （2011） ,70807 （2012） . This could help predicting that in the future, the CDM forestation and reforestation of Heilongjiang Province will create more employment opportunities, Heilongjiang Province will be able to accomplish economic transition of state-owned forestry and at the same time, the forest workers problem will be solved. Therefore Heilongjiang Province should focus more on developing the CDM forestation and reforestation project.

Acknowledgments. This research was supported by the Central University Fundamental Research Funding No. DL11GC03 and Science Research funding Heilongjiang Province No. LC201039 .

Reference

1. Information on, http://www.carbontree.com.cn/NewsShow.asp?Bid=5162

Analysis of the Game between Human Beings and Climatic Environment Based on Two-Stage Dynamic Chicken Game Model*

Liu Mengqiao, Fu Xingxing, Zhu Bo, and Huang Yingli**

Northeast Forestry University, Harbin, Heilongjiang Province, China
yinglihuang336@yahoo.com

Abstract. Recently, climate disasters, especially global warming aroused extensively concern about the climatic environment problems of human beings. This paper applied two-stage dynamic chicken game model to analyze the relationship between the human beings and climatic environment, and pointed out that climatic environment chose the survival benefit while human beings preferred to choose material benefit under this frame of game theory. Benefit divergence made the two parties of the game make different strategy choices in two stages. While the strong measures by human beings could lead to the negative benefit return in the end. Thus this paper proposed the developmental strategies of low carbon economy to cope with climatic environment problems, so as to search for the new approach of the harmony between human beings and climatic environment.

Keywords: dynamic chicken game, benefit choice, low carbon economy.

1 Introduction

In 2011, the World Meteorological Organization (WMO) reported that the year 2010 is almost certain to rank the top 3 warmest years since the beginning of instrumental climate records in 1850 [1]. These problems lead to a profound reflection of existing patterns between human beings and climatic environment.

Meanwhile, the dynamic chicken game describes a dilemma which humans and climatic environment facing now. Initially, the relationship of the game could be summarized that humans had to adapt climatic environment ex parte. However, it transformed into an interrelationship along with human evolution which is a typical model of chicken game, and the tactics are changed with time. Therefore we take the model of dynamic chicken game.

* This research was supported by the Central University Fundamental Research Funding No. DL11GC03 and Science Research funding Heilongjiang Province No. LC201039.

** Corresponding author.

D. Jin and S. Lin (Eds.): Advances in ECWAC, Vol. 1, AISC 148, pp. 97–100.
springerlink.com © Springer-Verlag Berlin Heidelberg 2012

2 Interest Analysis of the Game between Human Beings and Climatic Environment

In the game theory, each player would realize their own benefits maximally. To the climate, its benefit is going together with all kinds of bionts for a long time; we call it "survival interests". Nevertheless, to the humans, there are also material interests except survival interests. Humans always neglect the survival interests and prefer material interests.

To enable rational players to pursue collective interests, we need a game rule—nature law to solve the problem that human will neglect the importance of collective interests.

Under the threat, there are two measures to take so that man values the survival interests: Firstly, give human more menace. Secondly, add material interests to the survival interests so that the new interests exceeds the pure material interests.

Weighing up the two methods, the former will lead us to a dilemma. So we should find reasonable material interests and add them to the survival interests so that humans will inevitably pursue the survival interests.

3 Analysis of Strategy Choice between Climatic Environment and Human Beings

The game relationship between climatic environment and human beings interacts with each other and adapts to each other as well. When the climatic environment is normal, the adaptation of man means the positive effect on climate, namely reasonable protection and use. When climatic disasters occur, the adaptation of man means men can do nothing useful. Next is the representation of expanded model and strategy model combined with chicken game to specifically quantify the relationship of both sides. (Table 1 and Figure 1).

Table 1. Strategy model between climatic environment and human beings

The first game		
H\C	T	y
T		10, −15
Y		20, 5
The second game		
H\C	T	y
t		
y	−15, 5	

Note: H: human beings; C: climatic environment; t: tough; y: yield

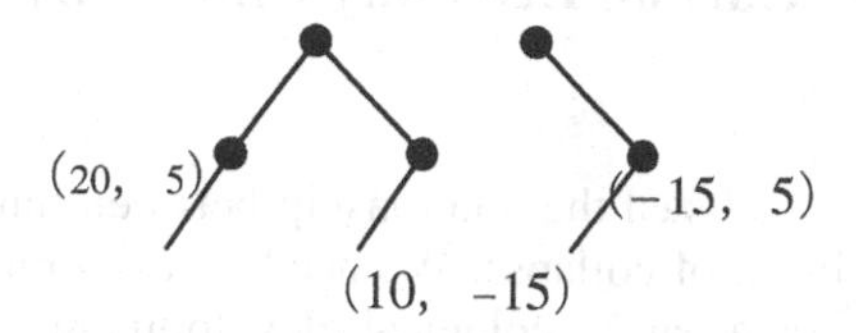

Fig. 1. Expanded model of two-stage chicken game

As Figure 1 shown, when C is the coward, H can choose t or y strategy but C can only choose y strategy in the first game. Then H faces decisions, H takes action first and C acts according to the decision of H. In the discussion below, Ht and Cy means the profit H and C can gain when H choose t strategy and C choose y strategy, Hy and Cy' means the profit H can gain when H choose Y strategy and C choose Y strategy.

According to the strategy model, win-win strategy (y,y) exists and the profit is (20,5).Each player pursues to maximize their own value, H should choose (y,y) strategy between two choices. Then the collective interests is expressed as hy+cy'=20+5=25>0. The interests are dispersed in decades or even centuries for the game are dynamic. But when H choose(t,y) strategy, the interests Ht=10<Hy=20,but the profit from t strategy can be gain in a short time which means it can meet the material need of men greatly. As a result, H will choose t strategy and C will choose y strategy passively.(There are few disasters in the long history of human beings.) Obviously, when H choose t strategy, Cy=-15<cy'=5 and the collective interests Cy+Ht=-15+10=-5<0.It meet the opinion that men will pursue to maximize their own value, so their won't care whether the collective interests is positive or not.

On the other hand, H can only choose y strategy and C can only choose t strategy in the second game. H acts after C. In the discussion below, Hy and Ct means the profit H and C can gain when H choose y strategy and C choose t strategy. The choice can only be(y,t) according to the model and the profits is(-15,5),the collective interests Hy+Ct=-15+5=-10<0.(C is positive because climate can recover after disasters) .

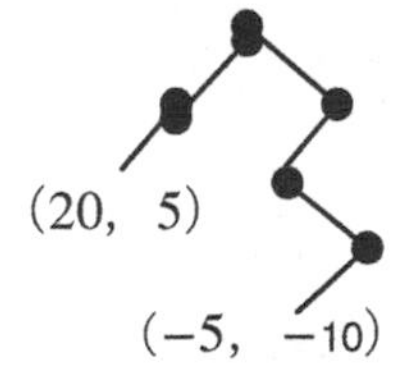

Fig. 2. Whole expanded model of two-stage chicken game

The second game is the result of H's choice of t strategy in the first game. According to Figure 2, the comprehensive profits of each player and the collective interests are the loss. the profits of H in the twice games can be expressed as10+(-15)=-5, the same of C can be expressed as(-15)+5=-10,and the profits of the collective interests can be expressed as(-10)+(-5)=-15[. It is adverse for each player and the collective in the long term. As a result, it is ineffective for H to choose t strategy. It will lead to the loss for both climatic environment and human beings finally if human beings choose to be tough all the time.

4 The Way of Low Carbon Economy Chosen by Game Model Analysis

In view of what we have analyzed the relationship between humans and environment has already been at the brink of collapse. We need a development pattern which can unifies the economic, social and ecological development, namely lower carbon economy development, a way that added material benefits to survival benefit so as to reach the harmonious coexistence of environmental protection finally.

According to the chicken game, humans can not continue to take "tough" measures. Therefore, humans should treat with the exploit and use of new energy more rationally as well as pay more attention to its potential problems existing in the exploit and use, and follow the natural rules to avoid new disasters.

Reference

[1] World Meteorological Organization (WMO) statement on the global climate in 2010 (2010), http://www.wmo.int/pages/mediacentre/press_releases/pr_904_en.html

Von Mises Uniform Distribution Inference for CPI of Shaanxi Province from 2006 to 2010 in China*

Yu Zheng[1], Juan Zhang[2], and Zongke Hou[3]

[1] School of Management, Xi'an Polytechnic University, No.19 Jinhua South Road, 710048 Xi'an, China
zyki2003@163.com

[2] Applied Technology College, Xi'an Polytechnic University, No.19 Jinhua South Road, 710048 Xi'an, China
zhangjco@163.com

[3] Northwest No. 2 Cotton Group Co., LTD., No. 4 Renmin West Road, 712000 Xianyang, China
houzke@sohu.com

Abstract. The uniformity of Consumer Price Index (CPI) of Shaanxi Province could be tested by using the Hodges-Ajine testing method. The result, that population is submitting to uniform distribution, was obtained. Correspondingly, the uniformity of CPI indicates that the general price level was tolerance although it increased in the past five years.

Keywords: Uniform Distribution, Hypothesis, Hogdges-Ajine Test, CPI, Shaanxi Province.

1 Introduction

The directional data is referred to the random test result of angle or direction, which was defined within the range of 0° to 360°[1]. The fact that 0° and 360° are identical angles, and that for example 180° is not a sensible mean of 2° and 358°, provides one illustration that special statistical methods are required for the analysis of some types of data. Other examples of data that may be regarded as directional include statistics involving days of the week, months of the year, compass directions, dihedral angles in molecules, orientations, rotations and so on.

2 Directional Satistics Model

2.1 Von Mises Distributions

The equivalent in circular statistics of the Gaussian or normal distribution in conventional statistics is the von Mises distributions. This distribution on the circle is probably the best known distribution from the field of directional statistics. The

* Supported by Education Department of Shaanxi Provincial Government of China (09JK118).

D. Jin and S. Lin (Eds.): Advances in ECWAC, Vol. 1, AISC 148, pp. 101–107.
springerlink.com © Springer-Verlag Berlin Heidelberg 2012

distribution is widely used to model angular data, where an angle is represented by a two-dimensional unit vector.

The von Mises distribution $M(\mu_0,k)$ has probability density function[2,3]

$$f(\theta;\mu_0,k) = \frac{1}{2\pi I_0(k)} e^{k\cos(\theta-\mu_0)} . \tag{1}$$

where I_0 denotes the modified Bessel function of the first kind and order 0, which can be defined by

$$I_0(k) = \frac{1}{2\pi}\int_0^{2\pi} e^{k\cos\theta} d\theta \cdot \tag{2}$$

The parameter μ_0 is the mean direction, and the parameter k is known as the concentration parameter. The mean resultant length ρ is $A(k)$, where A is the function defined by

$$A(k) = I_1(k)/I_0(k) \cdot \tag{3}$$

Note that $M(\mu_0+\pi,k)$ and $M(\mu_0,-k)$ are the same distribution to eliminate this indeterminancy of the parameter μ_0, k, it is usual to take $k \geq 0$.

If k=0, the distribution is uniform, and for small k, it is close to uniform. If k>0, the distribution becomes very concentrated about the angle μ_0 with k being a measure of the concentration.

2.2 Primary Statistic

Suppose that we are given unit vectors $X_1,\cdots,X_n$ with corresponding angles θ_i, $i=1,\ldots n$. The mean direction $\overline{\theta}$ of $\theta_1,\ldots,\theta_n$ is the direction of the resultant $X_1+\cdots+X_n$ of $X_1,\cdots,X_n$. It is also the direction of the centre of mass $\overline{X}$ of $X_1,\cdots,X_n$. Since the Cartesian coordinates of X_i are $(\cos\theta_i,\sin\theta_i)$ for $i=1,\ldots n$, the Cartesian coordinates of the centre of mass are $(\overline{C},\overline{S})$, where

$$\overline{C} = \frac{1}{n}\sum_{i=1}^{n}\cos\theta_i , \quad \overline{S} = \frac{1}{n}\sum_{i=1}^{n}\sin\theta_i \cdot \tag{4}$$

Therefore θ is the solution of the equations $C = \overline{R}\cos\overline{\theta}$, $S = \overline{R}\sin\overline{\theta}$ (provided that R>0), when the mean resultant length R is given by

$$\overline{R} = (\overline{C}^2 + \overline{S}^2)^{1/2} . \tag{5}$$

The resultant length R is the length of the vector resultant $X_1+\cdots+X_n$. Thus

$$R = n\overline{R} . \tag{6}$$

For grouped data,

$$\bar{C}=\frac{1}{n}\sum_{i=1}^{m}\cos\theta_i',\quad \bar{S}=\frac{1}{n}\sum_{i=1}^{m}\sin\theta_i',\quad R=[(\sum f_i\cos\theta_i')^2+(\sum f_i\sin\theta_i')^2]^{1/2}. \tag{7}$$

where n is data amount, m is grouped amount, f_i is the frequency within No. i group, and θ_i' is the median value within No. i group.

3 Inference for CPI of Shaanxi Province Submitting to Von Mises Distributions

3.1 Statistical Data Sourcing and Processing

Statistical data used in this paper is obtained from CPI monthly bulletin of Shaanxi Provincial Bureau of Statistics. The data ranges from each month of 2006-2010 (Table 1). In order to eliminate influence caused by accidental factors on calculating result, monthly average of CPI ranging from 2006 to 2010 are used in this paper. Table 2 indicates the adjusted data and statistic calculating.

Table 1. Monthly Data of CPI From 2006 to 2010

Year	Jan.	Feb.	Mar.	Apr.	May	Jun.	Jul.	Aug.	Sep.	Oct.	Nov.	Dec.
2006	102.9	100.8	101.3	101.3	101.3	101.0	100.6	101.2	102.0	101.1	101.6	102.7
2007	101.4	102.8	103.3	103.3	103.9	105.3	106.7	107.3	106.2	107.0	107.2	107.3
2008	108.4	109.8	109.0	109.2	108.2	106.9	106.1	105.4	105.0	104.3	103.1	101.4
2009	101.2	99.5	100.2	99.9	100.5	100.3	100.0	99.8	100.2	100.1	101.7	102.9
2010	102.1	103.5	102.7	103.2	103.0	103.1	103.8	104.7	104.8	105.3	106.2	105.3

Table 2. Average Monthly Data of CPI and Statistic Calculating

Month	θ_i	θ_i'	f_i	Month	θ_i	θ_i'	f_i
Jan.	(0°,30°]	15°	103.20	Aug.	(210°,240°]	225°	103.68
Feb.	(30°,60°]	45°	103.28	Sep.	(240°,270°]	255°	103.64
Mar.	(60°,90°]	75°	103.30	Oct.	(270°,300°]	285°	103.56
Apr.	(90°,120°]	105°	103.38	Nov.	(300°,330°]	315°	103.96
May	(120°,150°]	135°	103.38	Dec.	(330°,360°]	345°	103.20
Jun.	(150°,180°]	165°	103.32	Σ			1242.06
Jul.	(180°,210°]	195°	103.44				

3.2 Hypothesis of CPI Submitting to Von Mises Distributions

If frequency submits to von Mises distributions, CPI of Shaanxi province has seasonal fluctuation could be considered as the uniformity test of the directional data. That is, sample from population $M(\mu_0,k)$, denoting

$$H_0: k=0 \quad \text{against} \quad H_1: k>0$$

In practice, μ_0 is usually unknown, it could be proved that the statistic of grouped data $2n\overline{R}^2$ approximately submitted to $\chi^2_\alpha(2)$ distribution[4-6]. Note

$$2n\overline{R}^2 = \frac{2}{n}R^2 = \frac{2}{n}[(\sum f_i \cos\theta_i)^2 + (\sum f_i \sin\theta_i)^2] \cdot \tag{8}$$

Thus, under confidence level $(1-\alpha)$, testing critical region is $2n\overline{R}^2 > \chi^2_\alpha(2)$. While the null hypothesis H_0 is rejected, we can have interval estimate of the mean direction to ascertain confidence interval of CPI peak.

Under confidence level $\alpha = 0.05$, Using (11), $2n\overline{R}^2 = 0.00352 < \chi^2_\alpha(2) = 5.991$, hence the hypothesis H_1 is clearly rejected, and the null hypothesis H_0 is accepted. That is, CPI of Shaanxi province has an unobvious seasonal fluctuation, which submits to uniform distributions.

4 Tests of Uniformity

4.1 The Hodges-Ajine Test

Let $N(\theta)$ denote the number of observations in the semicircle centred on θ, i.e. in the arc $(\theta - \pi/2, \theta + \pi/2)$. The Hodges-Ajne test rejects uniformity for large values of $\max_\theta N(\theta)$,or, equivalently, for small values of

$$m = \min_\theta N(\theta) \, . \tag{9}$$

This test, was introduced by Ajne (1968), using *n-m*, the maximum number of observations in a semicircle, as the test statistic. Bhattacharyya & Johnson (1969) pointed out the connection with the bivariate sign test of Hodges.

A combinatorial argument shows that

$$\Pr(m \le t) = 2^{-n+1}(n-2t)\binom{n}{t}, \quad t < \frac{n}{3} \cdot \tag{10}$$

a formula obtained by Hodges. Tables for *m* are given by Hodges and Klotz.

For large *n*, it can be shown that

$$\Pr(\frac{n-2m}{n^{1/2}} \le t) \cong \frac{4t}{\sqrt{2\pi}} \exp\left\{-\frac{t^2}{2}\right\} \cdot \tag{11}$$

The approximate 5% and 1% values of $(n-2m)n^{0.5}$ are 3.023 and 3.562, respectively.

4.2 Ajine's An Test

Since the expected value of $N(\theta)$ under uniformity is *n*/2, it is intnitively reasonable to reject uniformity for large values of

$$A_n = \frac{1}{2n\pi}\int_0^{2\pi}\left\{N(\theta)-\frac{n}{2}\right\}^2 d\theta \cdot \tag{12}$$

The following computational formula for A_n is

$$A_n = \frac{n}{4} - \frac{1}{n\pi}\sum_{j=2}^{n}\sum_{i=1}^{j-1}\min\{\theta_{(j)}-\theta_{(i)}, 2\pi-[\theta_{(j)}-\theta_{(i)}]\} = n\left(\frac{1}{4}-\frac{\overline{D}_0}{2\pi}\right) \cdot \tag{13}$$

where $\overline{D}_0$ is the circular mean difference.

The large-sample asymptotic null distribution of A_n under uniformity can be obtained by

$$\lim_{n\to\infty}\Pr(A_n > \alpha) = \frac{4}{\pi}\sum_{k=1}^{\infty}\frac{(-1)^{k-1}}{2k-1}\exp\left\{-\frac{(2k-1)^2\pi^2}{2}\alpha\right\} \cdot \tag{14}$$

Some upper quantiles of the large-sample asymptotic null distribution of A_n are given in Table 3. Upper quantiles of A_n for various values of n were given by Stephens. His table shows that for $n \geq 16$, the upper 5% quantile of A_n is within 0.01 of the value given in Table 3.

Table 3. Upper Quantiles of An

α	0.10	0.05	0.025	0.01
A_n	0.516	0.656	0.797	0.982

A generalization of A_n to

$$A_n(t) = \frac{1}{2n\pi}\int_0^{2\pi}\{N(t,\theta)-nt\}^2 d\theta \cdot \tag{15}$$

where $N(t,\theta)$ denotes the number of observations in the arc $(\theta - t\pi, \theta + t\pi)$ and $0<t<1$, was considered by Rothman and by Rao. Note that $A_n(1/2)=A_n$.

5 The Hodges-Ajine Test of Shaanxi CPI Submitting to Uniform Distribution

H_0: Population submits to uniform distribution, against H_1: Population does not submit to uniform distribution.

Hypothesis of Hogdges-Ajine test works as follows: (i) Examine the dots on a unit circle which correspond to angles $\theta_1, \theta_2, \ldots \theta_n$. (ii) Draw a line *l* which divides the above dots into two parts and line *l* passes the center of the circle. (iii) Denote the number of the data in the smaller one as *M*. (iv) Let line *l* circumrotate around the center of the circle, the value of *M* will change correspondingly. (v) Denote the minimum value of *M* as *m*. (vi) When *m* is less than a given number, refuse H_0.

According to Daniels H. E.[7], if H_0 exists, the discrete distribution of m under given conditions can be derived. Based on Hodges, the critical value of Hogdges-Ajine test can been calculated[8].

When Hogdges-Ajine Test is used to analyze, the following points must be taken into account: dividing the circumference equally into even number r ($r=2k$) parts, and making the data in each part concentrate on the middle of this part, denoting the numbers of each part as $f_1,\ldots,f_j$, therefore $\sum f_i = n$ and

$$A_j = \sum_{i=1}^{k} f_{j+i} \ (j = 0,1,2,\cdots,r-1), \quad m = \min_j A_j \ (j = 0,1,2,\cdots,r-1). \tag{16}$$

Denoting $B = \max_j |A_j - A_{j-k}| \ (j = r-1, r-2, \ldots k)$.Using equations (16), $A_j + A_{j+k} = n$, we have

$$B = \max_j (n - 2A_j, 2A_j - n) = n - 2m \quad (n \geq 2m). \tag{17}$$

When B is equal to or bigger than a given number, hypothesis of uniformity is not in existence. According to David H. A. and Newell D. J.[9], the statistic variable m^* is

$$m^* = \frac{B}{\sqrt{n}} = \sqrt{n} - \frac{2m}{\sqrt{n}}. \tag{18}$$

The critical value m_0^* of corresponding test refusing range $m^* > m_0^*$, as shown in Table 4. According to Table 2, r=12, k=6, n=1242.06, A_j is calculated based on equation (16), the result are indicated in Table 5 (j=0,1,2,…,11).

Table 4. $P(m^* \geq m_0^*) = \alpha$

α/k	2	3	4	5	6
0.01	2.81	2.93	3.00	3.05	3.09
0.05	2.23	2.37	2.45	2.50	2.53

Table 5. The Values of A_j of Shaanxi Province CPI

j	A_j	j	A_j
0	619.86	6	621.48
1	620.10	7	621.24
2	620.50	8	620.84
3	620.84	9	620.50
4	621.02	10	620.32
5	621.60	11	619.74

The value of m can be calculated from Table 5, $m = \min_j A_j = 619.74$. Using quations (17) and (18), We can further obtain B=1.86, m^*=0.053.Comparing with Table 4, $m_0^* = 2.53 > m^* = 0.053$.Therefore, H_0 can be accepted that uniformity hypothesis is in existence.

6 Conclusions

CPI of Shaanxi Province, is the main indication reflecting the degree of inflation or deflation in regional economy, which general price level change can be mirrored by the change of CPI. Although CPI of Shaanxi Province has been increased for successive months since June, 2007, even its maximum was as high as 9.8% in February 2008, the uniformity of CPI indicates that since the 11th Five-Year Plan of Shaanxi Province has been implemented, the overall price level has gently remained increase, and regional economy has developed healthy.

References

1. Mardia, K.V.: Statistics of Directional Data. Academic Press, London (1972)
2. Mardia, K.V., Dryden, I.L.: Shape distributions for landmark data. Adv. Appl. Prob., 742–755 (1989)
3. Mardia, K.V., Jupp, P.E.: Directional statistics. John Wiley & Sons, New York (2000)
4. Zheng, Y., Zhang, J., Yang, H.-W.: Application of Von Mises Distribution in Shaanxi Province Insurance Premium. Statistics & Information Forum 26, 28–31 (2011)
5. Zheng, Y., Zhang, J., Hou, Z.: Analysis of China Import Cotton Price Seasonal Fluctuation. China Cotton 36, 38–39 (2009)
6. Zheng, Y., Zhang, J., Wang, W.: Von Mises Unimodal Distribution Inference for RTAC of Shaanxi Province. In: The 3rd International Conference on Information Science and Engineering (ICISE 2011), Yangzhou, China, September 29, October 1, pp. 2244–2246 (2011)
7. Daniels, H.E.: A Distribution-free Test for Regression Parameter. Ann. Math. Statist. 25, 499–513 (1954)
8. Hodges Jr., J.L.: A Bivariate Sign Test. Ann. Math. Statist. 26, 523–527 (1955)
9. David, H.A., Newell, D.J.: The Identification of Annual Peak Periods for a Disease. Biometrics 21, 645–650 (1965)

Research on Flexible Connection of Battery Pack

ChanMing Chen[1], ChengXi Luo[1], and ZhenPo Wang[2]

[1] School of Software, Beijing Institute of Technology, Beijing, China, 100081
[2] National Engineering Laboratory of Electric Vehicles, Beijing Institute of Technology, Beijing, China, 100081
{hanming,lcxl,wangzhenpo}@bit.edu.cn

Abstract. To try to eliminate inconsistency among cells in battery pack which decreased the cycle life of battery pack heavily, a new equalizing approach which could connect the cells flexibly was discussed. A flexible connection system for battery pack was invented, which composed of voltage monitoring circuit, controlling chip and switching devices. An experiment of discharge was carried out with the system. The result showed that the system equalized voltage of cells with high efficiency and high precision, greatly extended the discharging time of battery pack. It was also analyzed how to meet power requirement of electric vehicle with the system.

Keywords: Battery pack, flexible connection, equalization, electric vehicles.

Introduction

Cycle life of battery pack was far less than that of a cell, which led to rising costs and degradation of performance of battery pack. It seriously hampered the development of electric vehicles [1] [2] [3]. According to the experiment of literature [3], cycle life of battery pack which was series-wound by 10 cells would be about 200 times with only security protection. In contrast, the cycle life of the cell would be more than 500 times. If increasing the number of cells, the cycle life of battery pack would be even shorter.

Inconsistency among cells, as well as gradually expanding of the inconsistency caused by driving condition, was the major factor that caused cycle life of battery pack far less than a cell [1]. Because of insuring security of series-wound battery pack, its capacity was base on "short-board principle". That is, the minimum capacity of cells in the pack determined the overall capacity. After working for long time, the inconsistency would be great. So the minimum capacity of cells would be much less than the average capacity. Accordingly, the overall capacity would be much less than the average, so the cycle life of battery pack would be much shorter than the average[3] [4].

To reduce the inconsistency, equalization became a solution to try to make cells in pack consistent as possible.

1 Overview on Equalization of Battery Pack

Equalization technology of battery pack can be divided into two types, which are chemical equalization and physical equalization. Chemical equalization focused only

D. Jin and S. Lin (Eds.): Advances in ECWAC, Vol. 1, AISC 148, pp. 109–114.
springerlink.com © Springer-Verlag Berlin Heidelberg 2012

on the battery pack itself, with no circuit additional. But the practicality and security of this approach was far from the requirements. Physical equalization mainly relies on external equalization circuit. Performance of the circuits directly affects the effects of equalization. Physical equalization can be divided into two types, energy dissipation model and non-energy dissipation model [5].

Energy dissipation equalization system consisted of resistors to discharge some current. Although it was simple, but it consumed some energy and heated the system. It was not suitable for electric vehicle.

The research on non-energy dissipation equalization focused on switched-capacitor and DC-DC convertor nowadays. Switched-capacitor technology used combination of switch and capacitor to transfer energy in the adjacent cells. The loss of energy of this technology was little, but it had some defects. For example, its reliability could not be guaranteed without sensors, its efficiency of equalization was low. It was not suitable for high current and fast equalizing [5] [6].

According to its structure, DC-DC convertor equalization system can be divided into the centralized and distributed. The centralized system included a transformer with multiple output which transferred energy to the cells of lower voltage. It was efficient to equalize cells. But it was difficult to match the secondary winding accurately. The difference of voltage of cells caused by leakage inductance of transformer was difficult to compensate too. So it was hard to be equipped for an electric vehicle on road. In the distributed system, each cell was connected in parallel with an equalizer circuit. It also had the defects as the centralized system [5] [6] [7].

2 Flexible Connection System of Battery Pack

The ultimate purpose of the equalization of battery pack is to try to keep capacity of cells consistent during charging and discharging, or to control the difference in a limited range. If the rigid joint between cells in a series-wound battery pack could be broken through, some cells with lower voltage could be disconnected timely to pause during discharging (or some cells with higher voltage could be disconnected timely to pause during charging), which could compensate the difference of capacity of cells too. The battery pack could be equalized then. Helped by flexible connection, some extreme cells could be disconnected from the battery pack, and stop working temporarily. When their status is close to other cells, they could be connected to the battery pack again timely. The flexible connecting is carried out automatically and intelligently by control chip all the time during working.

The working voltage of a cell would vary from beginning to end of discharging. So, the voltage of a battery pack with many cells in series would vary larger during working. Fox example, working voltage of a li-ion cell varies from 3.8v to 3.0v. Then working voltage of a battery pack with 100 cells in series would vary from 380v to 300v. The minimum voltage of the battery pack is only 80% of the maximum voltage. Some research showed that there was only one or few cell in a battery pack which was "short board" [8]. If 2 cells is disconnected, only 2% voltage of the total would be loss. And it absolutely does nothing with the driving. During driving, extreme cells could be disconnected when non-peak power is enough to drive. So, flexible connection could match electric vehicle.

A battery pack with 4 cells in series was experimented here with a flexible connection system. The experimentation was focused only on discharging.

2.1 Components of Flexible Connection System

The flexible connection system consisted of three parts, namely, voltage monitoring module, controlling module and the switching module. Structure of the system was shown in Figure 1.

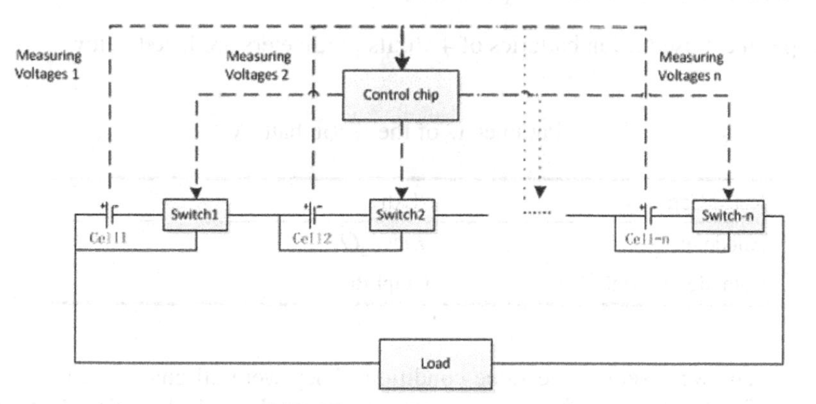

Fig. 1. System diagram of flexible combination

Voltage monitoring module collected the voltage of each cell real-timely. The sampling interval was 10um. And this function had been achieved by most of the battery management system now. Controlling module made decision on the status of all switches. Based on the voltage of each cell, it outputted controlling signal to switches according to the controlling strategy which had been programmed into control chip in advance. Most of control chip available could be adopted. The experimental system used a chip of FreeScale MC9S12XS128 MAL. The switching module consisted of field effect transistors and their driving circuits. The experimental system used MOSFET switches which model was IRF2807. Its driving circuit was very simple with low cost.

2.2 Controlling Strategy and Program

The goal of this experiment was to test the equalizing effect of flexible connection. Voltage of cells in the battery pack was the object to be equalized. The main controlling principle was to make voltage of cells equal as soon as possible. So the controlling logic was relatively simple. The specification was set as below: the difference of voltage of cells should be within 0.02V, discharging should be stopped immediately once voltage of any cell below 2.8V. The process of controlling program was as follows:

The system compared each cell's voltage every second. If the voltage of any cell is falling to the specification (such as 0.02V) below the highest voltage of all cells by, then the cell would be disconnected from the battery pack. If there were several cells which voltages were all lower than the highest by the specification, the two cells which voltage are the lowest would be disconnected from the battery pack. If voltage of any cell reaches 2.8V, the discharging of the battery pack should be stopped.

3 Discharging Experiment of Flexible Connection

3.1 Scheme and Data of the Experiment

The experiment used li-ion batteries of 4Ah. Its parameters are listed below.

Parameters of the li-ion battery

Rated Capacity	4[Ah]
Anode material	$LiMn_2O_4$
Cathode material	Graphite

The 4 cells were kept at the same condition. They were all charged with constant current of 5A, but the cut-off voltages were set separately as 4.1V, 4.0V, 3.9V, 3.8V. The charging were stopped immediately once reaching the cut-off voltages. After kept still for 5 minutes, the cells connected in series were discharged with constant current of 8.3A. The discharging was stopped once the voltage of any cell dropped to 2.8V. The voltage-time curves of all cells were shown in figure 2.

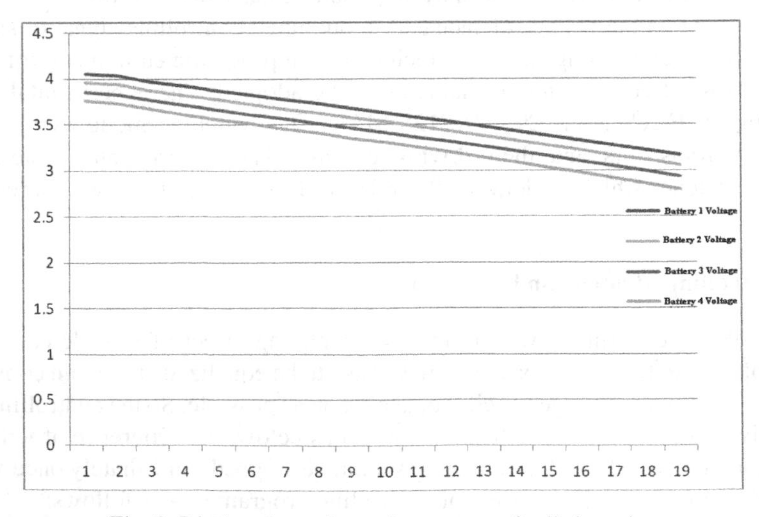

Fig. 2. Discharging voltage-time curves of cells in series

Then the 4 cells were kept at the same condition too. They were all charged with constant current of 5A, but the cut-off voltages were set separately as 4.1V, 4.0V, 3.9V, 3.8V. The charging were stopped immediately once reaching the cut-off voltages. After kept still for 5 minutes, the cells connected flexibly by the system were discharged with constant current of 8.3A. The discharging was stopped once the voltage of any cell dropped to 2.8V. The voltage-time curves of all cells were shown in figure 3.

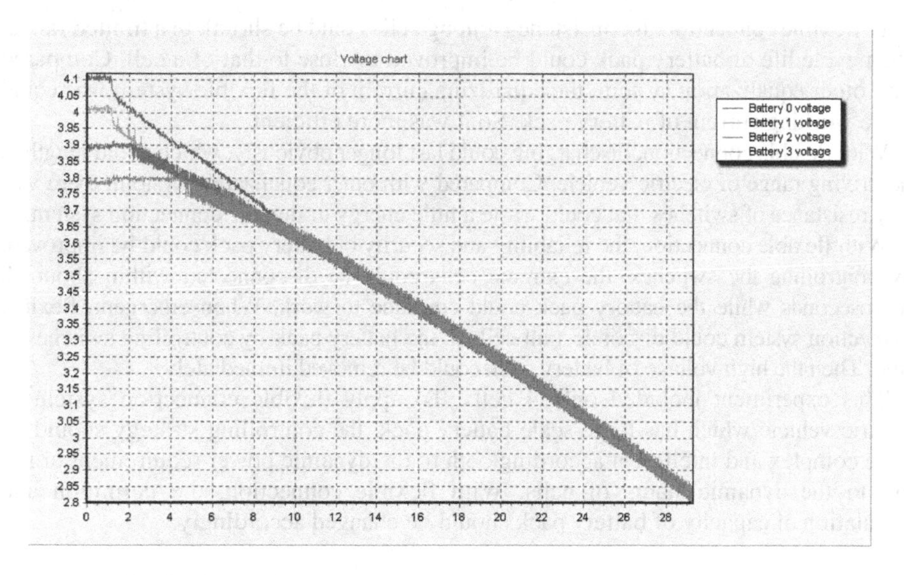

Fig. 3. Discharging voltage-time curves of all cells connected flexibly

3.2 Analysis on Result of Experiment

1. Comparison of Inconsistency in Voltage. Comparing the discharging curves of figure 2 and 3, it was obvious that flexible connection system had significant equalizing effect. In figure 2, the difference of voltage of the 4 cells became larger from 0.298V at the beginning to 0.370V at the end of discharging. It continued to enlarge all the time during discharging. In figure 4, the difference of voltage of the 4 cells was 0.325V at the beginning, but it shrank to 0.034V at the end of discharging. It continued to shrink all the time during discharging. Especially, it shrank to 0.040V 10 minutes after beginning.

The controlling strategy of the system was carried out by software. It could be reprogrammed when the rule, goal and environment change. So it could be intelligent to meet the specification of speed and accuracy of equalizing. The result of experiment showed that flexible connection system could quickly achieve a higher accuracy of equalization.

2. Comparison of Duration of Discharging. Based on the same charging and discharging mode, the discharging of figure 2 lasted 19 minutes, but that of figure 3 lasted 28 minutes. The discharging of flexible connection has been prolonged by 47%. The result showed that "short board principle" of battery pack in series has been broken through. With flexible connection, each cell in the battery pack could be fully discharged as possible, which could lengthen the driving range of electric vehicle.

4 Conclusion and Discussion

With flexible connection, inconsistency among cells could be shrunk to a limited range. Then cycle life of battery pack could be improved to close to that of a cell. Compared with other equalization system, the equalizing current of the flexible system was equal to the working current of battery pack. So it was more efficient.

With flexible connection, discharging could last longer obviously, which could lengthen the driving range of electric vehicle. Compared with other equalization system, there was only resistance of switches that could waste a little energy in flexible connection system.

With flexible connection, the reliability and security of battery pack could be improved. By controlling the switches, the extreme cells could be disconnected within about 10 microseconds while the battery pack could continue to work. When emergent, flexible connection system could disconnect all cells in the battery pack by controlling switches in time. Then the high voltage of battery pack could be removed immediately.

This experiment included only 4 cells. To apply flexible connection system to electric vehicle which has large-scale battery pack, the controlling strategy should be more complex and intelligent according both to the dynamic power requirement of EV and to the dynamic status of cells. With flexible connection, the definition and calculation of capacity of battery pack should be changed accordingly.

References

1. Wang, Z.-P., Sun, F.-C., Lin, C.: An Analysis on the Influence of Inconsistencies Upon the Service Life of Power Battery Packs (2006)
2. Ke, W.-L., Miao, P.-C.: Overcharge/overdischarge and the failure of lead-acid battery pack for electric bicycles (2005)
3. Jiang, X.-H.: Study on Battery Management System of Lithium-ion Batteries (Shanghai Institute of Microsystem and Information Technology), pp. 14–14 and 69–70 (2008)
4. Wu, Y., Jiang, X.-H., Xie, J.-Y.: The reasons of rapid decline in cycle life of Li-ion battery (2009)
5. Lei, J., Jiang, X.-H., Xie, J.-Y.: The status quo of development of equalization circuit of Li-ion batteries, pp. 62–63 (February 2007)
6. Chen, J.-J.: Research of Equalization for Series-connected Lithium-ion Batteries. Zhejiang University
7. Yin, Z., Zhang, P.-B., Yang, Y.-G., Hu, Y.-M., Cheng, Y.: Charging Technique of Vehicle's Li-ion Battery. Internal Combustion Engine & Powerplant No. 3 (2010)
8. Ke, W.-L., Miao, P.-C.: Overcharge/overdischarge and the failure of lead-acid battery pack for electric bicycles (2005)

Labor Mobility and Regional Economic Growth: Evidence from China

JianBao Chen[1,2] and Yang Chen[1]

[1] Department of Statistics, School of Economics, Xiamen University, Xiamen 361005, P.R. of China
jbjy2006@126.com
[2] Fujian Key Laboratory of Statistical Sciences, Xiamen University Xiamen 361005, P.R. of China
chenyangxmu@126.com

Abstract. Labor mobility freely plays an important role in resources allocation effectively, productivity improvement and economic growth. This paper uses non-parametric panel model to study the impact of labor mobility on regional economic growth in China from 1997 to 2009. The results are summarized as follows: (a) considered from whole country, labor input (output) has positive (negative) influence to economic development, the contribution of labor input is higher than that of labor output; (b) the contribution of labor input to economy of eastern and western regions are positive but negative influence to middle region, the contribution of labor output to economy of eastern and western regions are negative but positive influence to middle region.

Keywords: Labor Mobility, Regional Economic Growth, Non-parametric Panel Model.

1 Introduction

According to the famous economist Lewis's theory, there exist two distinct sectors of economy in a developing country, one is the industrial sector of city capitalists, and the other is the traditional rural agricultural sector of low productivity. When the profits of capitalists are used to investment, the production of industrial sector will be expanded, and the surplus labor forces attracted from agricultural sector will be increased, the development trend continues until all rural surplus labor forces to be transferred to the industrial sector. Starting from 80s, especially into 90 years later of last century, a large number of rural labor forces from central and western regions in China have input to eastern region, mainly in the coastal areas. Statistical data shows that there are nearly 130 million flowing agricultural population in China. It is of interesting to study how big influence of such labor mobility to these regions and whole country.

As regard to the influence of labor mobility to economic growth, there are many published papers. Some related results were given by Cai and Wang(1999), Fan, Wang and Shen(2004), Ao(2005), Yao and Li(2006), Temp and Wobmann (2006), Hsieh and Klenow (2007), Vollrath (2009) and etc.. They adopted different econometrical methods to do empirical analysis and obtained some valuable conclusions. As we know that

D. Jin and S. Lin (Eds.): Advances in ECWAC, Vol. 1, AISC 148, pp. 115–119.
springerlink.com © Springer-Verlag Berlin Heidelberg 2012

nobody has used this method to do empirical analysis for this problem.Comparing to the traditional parametric panel model, the non-parametric panel model doesn't need to assume the model form speciffically; the model building is obtained directly from data. Therefore, this paper tries to use non-parametric panel model to study the relations between labor mobility and regional economic growth in China, and hopes to find some inside phenomena.

2 Methodology

In order to eliminate heteroscedasticity, all variables mentioned in this section are transformed as natural logarithms. The form of non-parametric panel model is as follows:

$$\ln Y = \alpha + g(\ln K, \ln(HL), \ln(lin), \ln(lout)) + \varepsilon \tag{1}$$

Here Y indicates GDP; let K and L be physical capital stock and labor force respectively, H is the human capital level which is replaced by "the average education years". Therefore, HL represent human capital stock; Lin and $Lout$ represent labor input and labor output respectively; ε are independent identically normal distribution with mean 0 and variance σ^2; g is an unknown smoothing function.

Let $y_{it} = \ln Y_{it}, x_{it} = (\ln K_{it}, \ln HL_{it}, \ln lin_{it}, \ln lout_{it})^T$, by using Taylor expansion at a fixed point $x = (x_{1,} x_2, \cdots, x_4) = (\ln K, \ln HL, \ln lin, \ln lout)$, we have

$$y_{it} = \alpha_i + g(x) + (x_{it} - x)'\phi(x) + \varepsilon_{it}, i = 1, \cdots, n, t = 1, \cdots, T \tag{2}$$

Here ε_{it} includes the remainder term after partial linearization; $\phi(x) = (\phi_1(x), \phi_2(x), \phi_3(x), \phi_4(x))'$ is a column vector, $\phi_i(x) = \partial f(x) / x_i (i = 1, 2, 3, 4)$ is the output elasticity of variable x_i. What we are interested in is the estimate of $\phi(x)$. In order to make the model (2) more concise, we can transform it as follows:

$$y_{it} - \overline{y}_i = (x_{it} - \overline{x})\phi(x) + \varepsilon_{it} - \overline{\varepsilon} \tag{3}$$

Here $\overline{y_i} = \Sigma_{t=1}^T y_{it} / T$. The estimate of $\phi(x)$ can be obtained by local linear kernel estimation method which was proposed by Ullah and Roy (1998). The estimate expression of $\phi(x)$ is

$$\hat{\phi}(x) = (\sum_{i=1}^{n} \sum_{t=1}^{T} (x_{it} - \overline{x}_{ik})'(x_{it} - \overline{x}_{ik}) K((x_{it} - x) / h))^{-1} \sum_{i=1}^{n} \sum_{t=1}^{T} (x_{it} - \overline{x}_{ik})'(x_{it} - \overline{x}_{ik}) K((x_{it} - x) / h) \tag{4}$$

where $K(\bullet)$ is a kernel function and h is the window width.

We can calculate the non-parametric estimates at the sample mean $\bar{x}$ to obtain the average estimate elasticity of each factor, where $\bar{x} = (\bar{x}_1, \bar{x}_2, \bar{x}_3, \bar{x}_4)$; $\bar{x}_r = \frac{1}{nT}\sum_{i=1}^{n}\sum_{t=1}^{T} x_{r,it}$, $r = 1,2,3,4$ and $x_{r,it}$ is the value of variable x_r at the point (i,t).

2.1 Empirical Analysis

Data Source.The date used in this paper are collected from "China Statistical Year book (1997~2009)", "Nationwide Census 2000", "Nationwide Population Spot Check 2005" and "Registered Permanent Residence Transfer (1997~2009)".

In the following empirical analysis, we will use the related variable observations of 29 provinces in mainland of China (Sichuan and Chongqing are treated as one province and except for Tibet) from 1997 to 2009. Among these 29 provinces, 11 provinces belong to eastern region, 9 provinces belong to middle region and the other 9 provinces belong to western region.

2.2 Empirical Results

We can obtain the impact of labor mobility to national economic growth from the empirical analysis as follows: According to the paper of Ullah and Roy (1998), we take the kernel function as Gauss distribution and the optimal window width is $h = aN^{-1/11}$, we build three groups corresponding to non-parametric panel models at different window widths $(a = 4, 4.2, 4.5)$. The nonparametric estimation results of models are given in Table 1.

Table 1. The average output elasticity estimation of four explanatory variables

Coefficient estimates	Different choices of window width		
	a=4	a=4.2	a=4.5
$\hat{\phi}_1$	0.5343	0.5342	0.5341
$\hat{\phi}_2$	0.7894	0.7934	0.7982
$\hat{\phi}_3$	0.1314	0.131	0.1306
$\hat{\phi}_4$	-0.0086	-0.0091	-0.0097

By observing the estimates of regression coefficients in Eq. (3), we find the following results: (a) the estimates of coefficients for physical capital stock, human capital stock and labor input in all models are positive, this fact implies that these three factors have positive impacts on national economic growth. While the estimate of coefficient for labor output in all models are negative and this reveals that labor output has negative impacts on national economic growth; (b) compared with average output elasticity estimates for physical capital stock and human capital stock, the average output elasticity estimates for labor input and output are smaller.

In the same way, we can obtain the impacts of labor mobility on economic development of eastern, central and western regions respectively. The results are summarized in Table 2.

Table 2. The average output elasticity estimation of four explanatory variables in eastern, central and western regions $(a = 4.2)$

Coefficient estimates	Three regions in China		
	Eastern region	Central region	Western region
$\hat{\varphi}_1$	0.4245	0.5632	0.7986
$\hat{\varphi}_2$	0.8786	0.779	0.146
$\hat{\varphi}_3$	0.1085	-0.0181	0.017
$\hat{\varphi}_4$	-0.1523	0.0431	-0.1602

According to Table 2, we find the following facts: (a) in all three regions, compared with average output elasticity estimates for physical capital stock and human capital stock, the average output elasticity estimates for labor input and output in are smaller; (b) the contribution of labor input to economy of eastern and western regions are positive but negative influence to middle region, the contribution of labor output to economy of eastern and western regions are negative but positive influence to middle region.

3 Conclusion

Based on the panel data of 29 provinces, municipalities and autonomous regions in China from 1997 to 2009, this paper uses non-parametric panel models to study the influences of labor input and output to economic growth considered from whole country,western region, central region and estern region respectively.

Our empirical results can be summarized as follows: (a) considered from whole country, labor input (output) has positive (negative) influence to economic development, the contribution of labor input is higher than that of labor output. This indicates that Chinese government should take some corresponding policies to encourage labor forces transfer freely; (b) the contribution of labor input to economy of eastern and western regions are positive but negative influence to middle region, the contribution of labor output to economy of eastern and western regions are negative but positive influence to middle region. The government in different regions may make corresponding policies according to these finds.

References

1. Ao, R.J.: Manufacting comcentration, labor mobility and marginalization of central region. Nankai Economic Studies 1, 61–66 (2005) (in Chinese)
2. Cai, F., Wang, D.W.: The sustainability of China's economic growth and labor contributions. Economic Reseach 10, 62–68 (1999) (in Chinese)
3. Fan, J., Wang, L.J., Shen, L.J.: Industrial concentration and the trans-region flow of rural labor force. Management World 4, 22–29 (2004) (in Chinese)
4. Hsieh, C.T., Klenow, P.J.: Misallocation and manufacturing TFP in China and India. NBER Working Paper. w13290 (2007)

5. Temple, J., Wobman, L.: Duanlism and cross-country growth regressions. Econominc Growth 11, 187–288 (2006)
6. Ullah, A., Roy, N.: Nonparametric and Semiparametric Econometric of Panel Data. In: Ullah, A. (ed.) Handbook of Applied Economic Statistics, pp. 579–604. CRC Press Inc., Boca Raton (1998)
7. Vollrath, D.: How important are dual economy effects for aggregate productivity? Journal of Development Economics 88, 325–334 (2009)
8. Yao, L.R., Li, L.: Labor migration, indusrial agglomeration and regional disparity. Finacial Studies 8, 135–143 (2006) (in Chinese)

A Link Quality Assessment Model for WSNs Based on BDCT-SVM

Linlan Liu[1,2], Chao Zang[1,2], Jian Shu[1,3], Yangfan Ge[1,2], and Yuhao Zhou[1,2]

[1] Internet of Things Technology Institute, Nanchang Hangkong University, Nanchang, China
[2] School of Information Engineering, Nanchang Hangkong University, Nanchang, China
[3] School of Software, Nanchang Hangkong University, Nanchang, China
linda_cn68@yahoo.com, chaozang.cn@gmail.com,
shujian@jxjt.gov.cn, geyangfan@163.com, youthterm@hotmail.com

Abstract. A link quality assessment model named BDCT-SVM based on support vector machine and decision tree is proposed for wireless sensor networks in this paper. The communication area is divided into effective region, transitional region and clear region by experiments in the first place. Then a 4 level link quality decision tree is made to have five grades of link quality. The radial basis function is chosen as the kernel function in BDCT-SVM model and the k-fold cross validation method is used to optimize parameters according to the CCI, RSSI and PRR value by statistics. Experimental results show that the BDCT-SVM model proposed in this paper is reasonable and more accurate.

Keywords: Wireless sensor networks, Link quality assessment, Support vector machine, Radial basis function.

1 Introduction

Wireless sensor networks (WSNs) are composed of large number of cheap miniature sensor nodes which are deployed in the monitoring area [1]. The communication medium between nodes is by way of electromagnetic wave, which is often unreliable and random due to the harsh environment, resulting loss of packets. It's very important to evaluate the current link quality quickly, accurately and efficiently, so that the routing protocols can take relevant strategies in time to keep the entire network working steadily and efficiently. The researchers have proposed a variety of link quality assessment mechanisms for WSNs, but none of them has been widely accepted because the communication links are time variable [2,3], directional [4,5] and asymmetric [5].

Support vector machine (SVM) is a kind of intelligent machine learning theory using small sample extraction methodology based on data mining and artificial intelligence. This paper applies SVM classification methodology to the assessment of link quality in WSNs. The link quality level is divided according to the theory of SVM in classification and the analysis of communication link characteristics in WSNs. By doing this, the link assessment module based on binary decision classification tree (BDCT) and SVM has been bulit. The results show that the BDCT-SVM method is an accurate and efficient link quality assessment methodology.

D. Jin and S. Lin (Eds.): Advances in ECWAC, Vol. 1, AISC 148, pp. 121–127.
springerlink.com © Springer-Verlag Berlin Heidelberg 2012

2 SVM Methodology

Suppose the training data set is $\{x_i, y_i\}_{i=1}^N$, where $x_i \in R^n$, $y_i = +1$ when x_i is in front of hyperplane, $y_i = -1$ when x_i is on the opposite of hyperplane. Assume that the hyperplane equation for classification is

$$\omega \cdot x + b = 0. \tag{1}$$

Where ω is a normal vector on the hyperplane, b is a constant offset. The goal of this training is to obtain the normal vector ω_0 on the optimal hyperplane and the optimal constant offset b_0 when the isolation edge between positive case and negative case is maximized. So the spacing between positive case and negative case can be deduced as follows

$$dis = \frac{\omega}{\|\omega\|} \cdot (x_1 - x_2) = \frac{2}{\|\omega\|}. \tag{2}$$

For any of the (x_i, y_i) ,there's $\omega \cdot x_i + b \leq -1, y_i = -1$ or $\omega \cdot x_i + b \geq 1, y_i = 1$, so the problem of seeking the optimal hyperplane can finally be attributed to solve a quadratic programming problem

$$\min \frac{\|\omega\|^2}{2}, \quad \text{s. t. } y_i \cdot (\omega \cdot x_i + b) \geq 1, \ \forall i = 1,2,\ldots,N. \tag{3}$$

Additionally, the relaxation factor ξ can be introduced to obtain a balance between empirical risk and generalization performance when the positive case and negative case can not be completely separated by the optimal hyperplane [6]. The classification plane $\omega \cdot x + b = 0$ at this time will follow the condition below

$$y_i \cdot [(\omega \cdot x_i) + b] \geq 1 - \xi_i, \ i = 1,2,\ldots,n. \tag{4}$$

When $0 < \xi_i < 1$, sample point x_i can be classified correctly; when $\xi_i \geq 1$, sample point x_i will be misclassified. So the penalty term $C\sum_{i=1}^n \xi_i$ is introduced to the process of getting minimized objective $\frac{\|\omega\|^2}{2}$ and the hyperplane optimization problem will become

$$\min(\frac{1}{2}\|\omega\|^2 + C\sum_{i=1}^n \xi_i), \text{ s. t. } y_i \cdot [(\omega \cdot x_i) + b] \geq 1 - \xi_i \ , \ \xi_i \geq 0. \tag{5}$$

Where C is the penalty factor. The optimal result can be obtained by using Lagrange multiplier method. Then the following function is established

$$J(\omega,b,a)=\frac{1}{2}\omega^T\omega-\sum_{i=1}^{N}a_i[y_i\,(\omega\cdot x_i+b)-1]. \tag{6}$$

Where α_i is the Lagrange multiplier. Then we can get the wolfe duality [6] from the original optimization problem by getting the partial derivatives of ω and b:

$$\max Q(a)=\sum_{i=1}^{N}a_i-\frac{1}{2}\sum_{i=1}^{N}\sum_{j=1}^{N}a_ia_jy_iy_jx_i^Tx_j\,. \tag{7}$$

This is a quadratic function optimization problem and it must have a unique solution. If a_i^* is the optimal Lagrange multiplier, then the optimal result from the original problem will be

$$\omega_0=\sum_{i=1}^{N}a_i^*y_ix_i\,,\;\; b_0=1-\omega_0x^{(s)}\,. \tag{8}$$

Where $x^{(s)}$ is the support vector and the sum can only be carried out from support vectors. Finally, the optimal classification function will become

$$f(x)=\text{sgn}\{\sum_{i=1}^{N}a_i^*y_iK(x_i,x)+b^*\}. \tag{9}$$

Where the *kernel function* $K(\cdot)$ needs to meet *Mercer* condition. The *radial basis function* (RBF) has been chosen as the kernel function in this paper:

$$K(x_i,x)=\exp\{-\gamma\,\|\,x_i-x_i\,\|^2\},\;\gamma>0. \tag{10}$$

We can know that the building of SVM classifier is the optimization of parameters (Eq. 8) essentially.

3 The Building of BDCT-SVM Link Quality Assessment Model

3.1 The Analysis of Link Characteristics

For this paper, the spatial characteristics of communication links in WSNs were studied first. The experiments were conducted in the open outdoors and the hardware platform is *Telosb* [7] series nodes produced by Crossbow company. About 1000 data packets were sent continuously from the *TX* node to the *RX* node each time and the *packet reception rate* (PRR) value was calculated at the same time. The distance between nodes was gradually increased at a rate of 0.5 m and the experiment was repeated many times, the experimental scenario is shown in Fig. 1.

Fig. 1. Experimental scenario

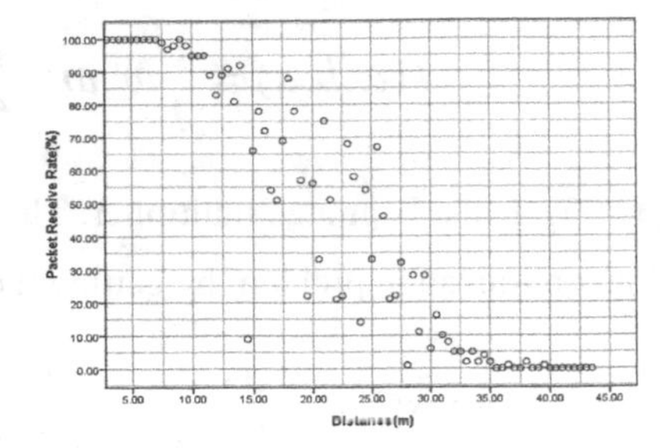

Fig. 2. Scatter plot of each PRR value

Then we got the scatter diagram on the relationship between PRR and distance through the above experiment (shown in Fig. 2). We can know that: the PRR will be more than 90% when the communication distance is less than 12 m in this particular environment; the PRR will be less than 10% when the communication distance is more than 33 m; there will be no law when the communication distance is between 12 m and 33 m. According to the views proposed in [8], the communication area within 12 m can be classified as the *effective region*, the communication area without 33 m as the *clear region* and the communication area between 12 m and 33 m as the *transitional region*.

3.2 The Building of BDCT-SVM Model

A formulaic expression would be given in a link quality assessment model in the past. For example, the link quality assessment model (FPRR) based on *link quality indication* (LQI) and *receive signal strength indication* (RSSI) has been given in [9].

$$PRR = \begin{cases} 95(\%) & (RSSI < 65) \\ -0.0001577694\,347132523\,LQI^3 + \\ 4.7084252735\,99874\,LQI \\ -214.9115458068\,067(\%) & (65 <= RSSI <= 95) \\ 5(\%) & (RSSI > 95) \end{cases} \quad (11)$$

LQI is defined as symbol correlation, more precisely, it should be called *chip correlation indicator* (CCI) [10]. CCI is considered to be some kind of metric used to detect chip error rate. The PRR value is calculated by LQI and RSSI in the FPRR link quality assessment model, which will be considered as the current link quality. However, link status will be considered as good when PRR is greater than a certain threshold in a practical application, then the transmission can be established. On the contrary, the link status will be judged as bad and this information should be quickly relayed to the upper layer protocol, so that the appropriate strategy can be taken in time. In order to improve the efficiency of the SVM classifier, we should design the link quality assessment model reasonably. According to the analysis of link characteristics above, three communication areas should be separated in the first place, then more details of the link quality classification can be given according to a specific application requirement. We design our link quality assessment model named BDCT-SVM as follows.

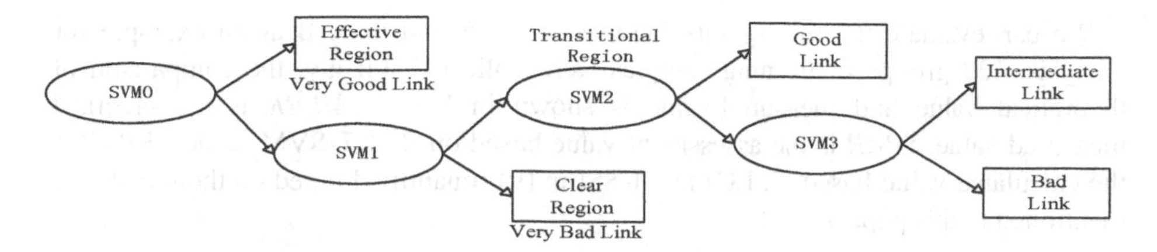

Fig. 3. BDCT-SVM link quality assessment model

In Fig. 3, we define the PRR value between 90% and 100% as *very good link*, the PRR value between 70% and 90% as *good link*, the PRR value between 40% and 70% as *intermediate link*, the PRR value between 10% and 40% as *bad link*, the PRR value between 0% and 10% as *very bad link*. So the *very good link* level corresponds to the *effective region*, the *good link, intermediate link* and *bad link* level correspond to the *transitional region*, and the *very bad link* level corresponds to the *clear region* in this paper. We also choose *CCI* and *RSSI* as two kinds of properties of BDCT-SVM model in our design.

4 Experiment and Analysis

The experiment was built based on the solutions mentioned above in the open outdoors and *Telosb* [7] series nodes produced by Crossbow company were used. The experimental scenario is shown in Fig. 4.

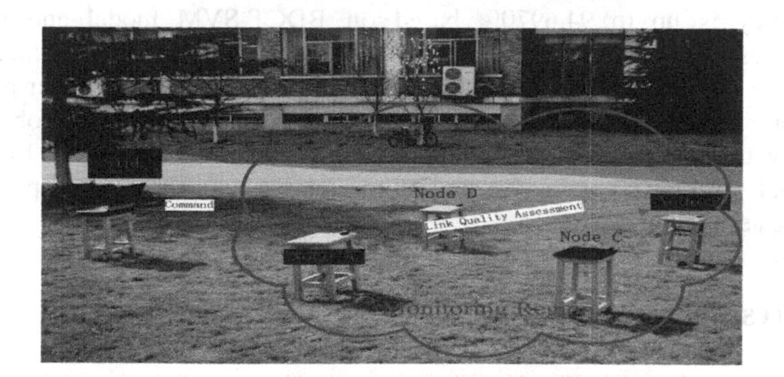

Fig. 4. Experimental scenario

After data acquisition command from Sink node been broadcast to the whole network, the probe requests between nodes will begin.When a certain amount of data packets are collected successfully, the parameter ω_0 and b_0 (Eq. 8) for *SVM0~SVM3* will be obtained by machine training in Sink node. Then the building of BDCT-SVM link quality assessment model (Eq. 9) will be done after the SVM parameters were sent to the entir network.For the penalty factor C and parameter γ in SVM, the *k-fold cross validation method* [6] will be used.

We can evaluate the link quality between node A and node B as an example for analysis. 132 groups of training samples were collected. Finally, the comparison of theoretical value and measured value is shown in Fig. 5. *MPRR* is the practical measured value, *SPRR* is the assessment value based on BDCT-SVM model, *FPRR* is the calculated value based on LQI and RSSI in [9] (quantified based on the thresholds mentioned in this paper).

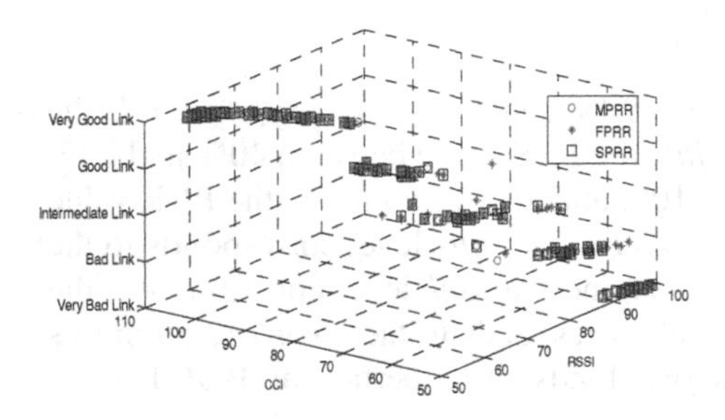

Fig. 5. Comparison of assessment value and measured value

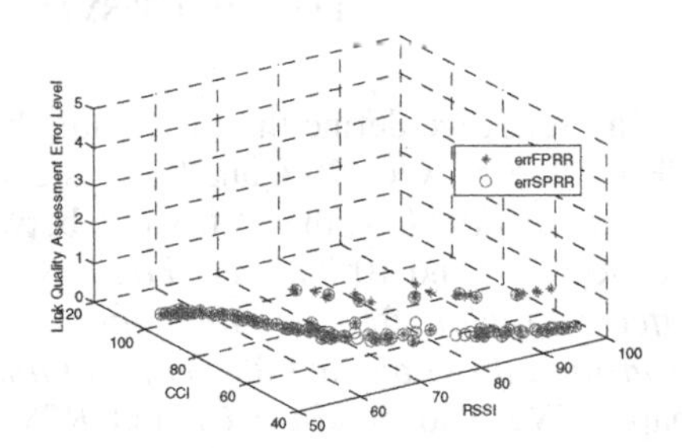

Fig. 6. Error comparison between two models

In Fig. 5, it can be seen that the assessment value based on BDCT-SVM model is very close to the measured value, but the assessment error gotten by FPRR model in [9] is relatively greater. In Fig. 6, we can see that the assessment error in FPRR and BDCT-SVM model is up to one error level at the same time, but the assessment accuracy is up to 94.6970% based on BDCT-SVM model and it's only 81.0606% based on FPRR model. So the BDCT-SVM link quality assessment model is reasonable and can provide better link selection service for upper-layer protocols. Furthermore, the BDCT-SVM model can get more accuracy with less probe packets compared with routine link assessment mechanism based on the statistical PRR value, which avoids excessive energy cost by sending a large number of probe packets and extends the network lifetime.

References

1. Sun, L., Li, J., Chen, Y., Zhu, H.: Wireless Sensor Networks. Tsinghua University Press, Beijing (2005)
2. Woo, A., Tong, T., Culler, D.: Taming the underlying challenges of reliable multihop routing in sensor networks. In: 1st International Conference on Embedded Networked Sensor Systems, pp. 14–27. ACM Press, New York (2003)
3. de Couto, D., Aguayo, D., Chambers, B.A., Morris, R.: Performance of multihop wireless networks: Shortest path is not enough. In: 1st Workshop on Hot Topics in Networks, pp. 83–88. Computer Communication Review, New York (2003)
4. Zhou, G., He, T., Krishnamurthy, S., Stankovic, J.A.: Impact of radio irregularity on wireless sensor networks. In: 2nd International Conference on Mobile Systems, Applications and Services, Boston, MA, pp. 125–138 (2004)

5. Reijers, N., Halkes, G., Langendoen, K.: Link layer measurements in sensor networks. In: 2004 IEEE International Conference on Mobile Ad-hoc and Sensor Systems, pp. 224–234. IEEE Press, New York (2004)
6. Deng, Y., Tian, Y.: Support Vector Machine. Science Press, Beijing (2009)
7. TelosB Information, http://www.xbow.com
8. Sun, P., Zhao, H., Luo, D., Zhang, X., Zhu, J.: Study on measurement of communication quality in wireless sensor networks. Journal on Communications 28, 14–22 (2007)
9. Liu, L., Fan, Y., Shu, J., Yu, K.: A Link Quality Prediction Mechanism for WSNs Based on Time Series Model. In: 2010 Symposia and Workshops on Ubiquitous, Autonomic and Trusted Computing, pp. 175–179. IEEE Press, New York (2010)
10. Liu, L., Li, J., Shu, J., Wu, Z., Chen, Y.: CCI-Based Link Quality Estimation Mechanism for Wireless Sensor Networks under Perceive Packet Loss. Journal of Software 5, 387–395 (2010)

The Influence of Randomness on Network Formation

Yanjun Fang

Business School of Shandong Normal University, Jinan, Shandong, China, 250014
fangyanjun@126.com

Abstract. It presents a dynamic model of network formation where nodes with that form links in two ways: some are connected randomness, and others are connected based on current structure of the network. In the formation model the deleting links are also considered. The degree distribution and degree exponent are obtained by mean-field approach theory. The parameters which represent the process of randomness, and the degree distribution are set up. The different network formation is discussed based on the change of parameters.

Keywords: network formation, degree distribution, cumulative degree distribution.

1 Introduction

In 20th century, it is found that the complex network is a paradigm to describe the complex systems. Complex network is composed by nodes and links between nodes. Nodes represent the agents, links present the relationship between agents[1]. Lots of complex systems can be described by complex network such as electric network, social relation networks, transit network and so on. There are two aspects in the research on networks. One aspect is concerned on the structure and characters of networks, another aspect is concerned on the dynamic behavior in networks. Network formation is developed quickly after ER model which is shown by Erdǒ and Rényi[2], and the most famous is WS model [3]and BA model[4]. It is shown two kinds of linking: randomness and preferential attachment[5,6]. Lots of experts change the probability of preferential attachment to shown more network structure [7,8,9,10,11]. In the paper, it is consider the randomness and preferential attachment, it is also consider the deleting in network formation, and to analyze the relationship between the parameters and the network structure.

2 Parameters Setup and Network Structure Model

There are two kinds of linking: random and attachment based on current structure of the network. It is contain deleting of node in the network formation based on Jackson's network structure[12]. The network formation will comply with rules as following.

a) adding new vertex: adding a new vertex along with variation of time, let N_t represents the total number of nodes in time t;

D. Jin and S. Lin (Eds.): Advances in ECWAC, Vol. 1, AISC 148, pp. 129–134.
springerlink.com © Springer-Verlag Berlin Heidelberg 2012

b) randomness: the new node links m_r old nodes stochastically, the probability which one old node is linked is p_r;

c) structural attachment: then new node links m_n nodes which is the neighbor of the linking old nodes, the probability is p_n;

d) deleting: links between old nodes which is linked new node is deleted stochastically, the probability of deleting is $p_d (0 < p_d << 1)$.

Using mean-field theory, in time t+1, the probability of the node i with degree $k_i(t)$ adding a link is as following:

$$\frac{\partial k_i}{\partial t} = \frac{p_r m_r}{t} + p_n \left(\frac{m_r k_i}{t} \right) \times \left(\frac{m_n (1 - p_d)}{m_n (p_r m_r + p_n m_n)} \right) - \frac{k_i p_d}{t} \tag{1}$$

Let $m = p_r m_r + p_n m_m$ which represents the total links after adding new node, then function(1) can be showed as:

$$\frac{\partial k_i}{\partial t} = \frac{p_r m_r}{t} + \frac{p_n m_n k_i (1 - p_d)}{mt} - \frac{k_i p_d}{t} \tag{2}$$

The network degree is $d_0 = m$ if we supposed that each node gives m links when it adds the network, we also supposed that the ratio of the probability between linking on randomness and on structure is $r = p_r m_r / p_n m_n$, then function (2) can be shown as following:

$$k_i(t) = \left[m + \frac{rm}{1-(2+r)p_d} \right] \left(\frac{t}{t_0} \right)^{\frac{1-(2+r)p_d}{1+r}} - \frac{rm}{1-(2+r)p_d} \sim t^{\beta} \tag{3}$$

Then the degree distribution can be shown as following：

$$P\{k_i(t) < k\} = 1 - P\left(t_i > t \left(\frac{m + \dfrac{rm}{1-(2+r)p_d}}{k + \dfrac{rm}{1-(2+r)p_d}} \right)^{\frac{1+r}{1-(2+r)p_d}} \right) \tag{4}$$

and

$$P\left(t_i > t\left(\frac{m + \frac{rm}{1-(2+r)p_d}}{k + \frac{rm}{1-(2+r)p_d}}\right)^{\frac{1+r}{1-(2+r)p_d}}\right)$$

$$= 1 - P\left(t_i \le t\left(\frac{m + \frac{rm}{1-(2+r)p_d}}{k + \frac{rm}{1-(2+r)p_d}}\right)^{\frac{1+r}{1-(2+r)p_d}}\right) \quad (5)$$

$$\sim 1 - \frac{t(1-p_d)}{m + t(1-p_d)}\left[\frac{m + \frac{rm}{1-(2+r)p_d}}{k + \frac{rm}{1-(2+r)p_d}}\right]^{\frac{1+r}{1-(2+r)p_d}} \quad (6)$$

$$P(k,t) = \frac{\partial P\{k_i(t) < k\}}{\partial t} \sim \left(\frac{m + \frac{rm}{1-(2+r)p_d}}{k + \frac{rm}{1-(2+r)p_d}}\right)^{-\left(1+\frac{1+r}{1-(2+r)p_d}\right)} \quad (7)$$

Based on the mean-field theory, the function of accumulated degree distribution is as following:

$$F_t(k) = 1 - \left(\frac{m + \frac{rm}{1-(2+r)p_d}}{k + \frac{rm}{1-(2+r)p_d}}\right)^{\frac{1+r}{1-(2+r)p_d}} \quad (8)$$

It is supposed $1-(2+r)p_d > 0$ that to ensure the growth of the network, it means:

$$p_d < \frac{1}{2+r} \quad (9)$$

We turn function (4) as following in order to discuss the relationship among r, p_d and $F(k)$ conveniently.

$$\log(1-F(d)) = \frac{1+r}{1-(2+r)p_d} \times \left[\log\left(m+\frac{rm}{1-(2+r)p_d}\right) - \log\left(k+\frac{rm}{1-(2+r)p_d}\right)\right] \quad (10)$$

3 Results

Then we can discuss the relationship among r, p_d and $F(k)$ through function (10) as following.

a) when $r \to 0$, $\log(d)$ shows linearity which is to agree with degree distribution of scale-free network, and degree distribution index will be weaken following the p_d increasing.

b) when $r \to \infty$, network is turned to ER random network which the links between nodes are uniform randomness.

We can see the results visually in figure 1.

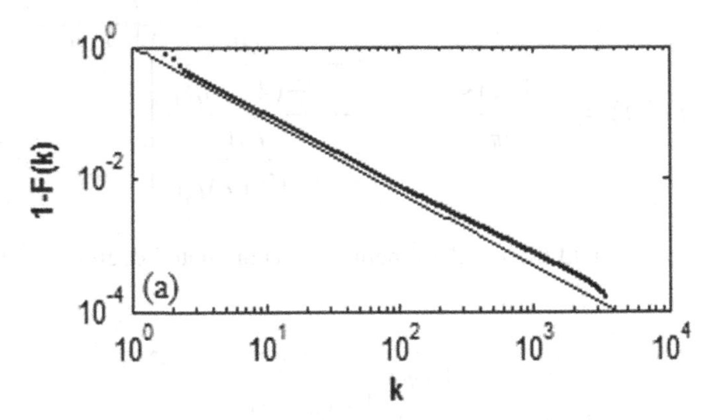

Fig. 1. Cumulative degree distribution: (a): $m_r = m_n = 10$, $p_r = 0$, $p_n = 1$, $p_d = 0.05$; straight line represents theoretical results, thick represents simulated result. (b): $m_r = m_n = 10$, $p_r = p_n = 1$, $p_d = 0.05$; straight line represents theoretical results, thick represents simulated result. (c): when p_d is fixed, the change of cumulative degree distribution following the change of r. (d): when $r = 0$, the change of cumulative degree distribution following the change of p_d.

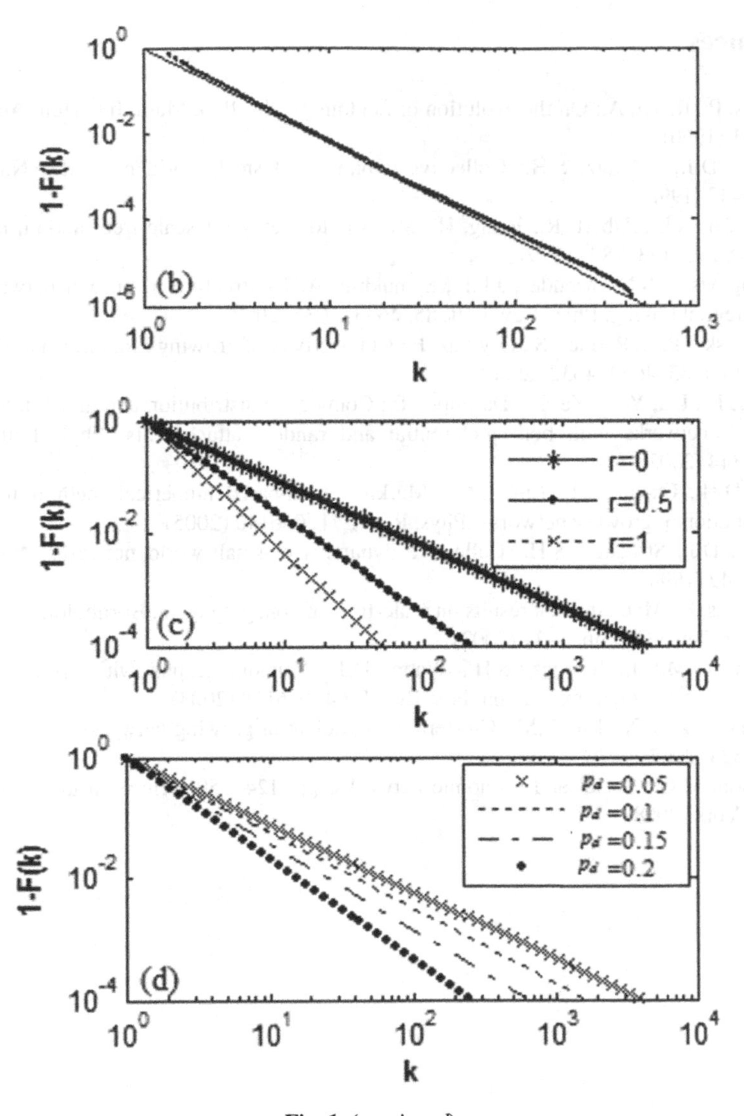

Fig. 1. *(continued)*

4 Conclusions

There are two kinds of links which are randomness and preferential attachment in network formation, there are also deleting in network formation. We use to setup parameters and mean-field theory to learn the relationship between parameters and network structure, and it can help to analyze the network structure and the characters of network.

References

1. Erdős, P., Rényi, A.: On the evolution of random graphs. Pub. Math. Ins. Hun. Aca. Sci. 5, 17–61 (1960)
2. Watts, D.J., Strogatz, S.H.: Collective dynamics of small-world networks. Nature 393, 440–442 (1998)
3. Barabási, A.L., Albert, R., Jeong, H.: Mean-field theory for scale-free random networks. Phy. A 272, 173–187 (1999)
4. Dorogovtsev, S.N., Mendes, J.F.F., Samukhin, A.N.: Structure of growth networks with preferential linking. Phys. Rev. Lett. 85, 4633–4636 (2000)
5. Krapivsky, P.L., Redner, S., Leyvraz, F.: Connectivity of growing random networks. Phys. Rev. Lett. 85, 4629–4632 (2000)
6. Liu, Z.H., Lai, Y.C., Ye, N., Dasgupta, P.: Connective distribution and attack tolerance of general networks with both preferential and random attachments. Phys. Lett. A 303, 337–344 (2002)
7. Shi, D.H., Chen, Q.H., Liu, L.M.: Markov chain-based numerical method for degree distribution of growing networks. Phys. Rev. E71, 036140 (2005)
8. Watts, D.J., Strogaze, S.H.: Collective dynamics of small-world networks. Nature 393, 440–442 (1998)
9. Bollobas, B.: Mathematical results on scale-free random graphs. In: Bornholdt, S., Schuster, H.G.: Wiley-VCH, pp. 1–34 (2002)
10. Newman, M.E.J., Strogaz, S.H., Watta, D.J.: Random graphs with arbitrary degree distribution and their application. Phy. Rev. E 64, 026118 (2003)
11. Shi, D.H., Zhu, X., Liu, L.M.: Clustering coefficient of growing networks. Physica A 381, 515–524 (2007)
12. Jackson, M.O.: Social and economic networks, pp. 124–150. Princeton university press, New York (2008)

An Empirical Research on the Correlation between Market Sentiment and Returns of Stocks

Jijiao Jiang and Tongtong Sun

School of Management, Northwestern Polytechnical University,
Xi'an, China
jjj_leon@nwpu.edu.cn, suntongtong2006@126.com

Abstract. From the point of view of behavior finance, this paper submits the empirical evidence of the correlation between market sentiment and returns of stocks. Considering the realistic situations of China's stock market, we adopt the historical data of 27 close-end funds from 2006 to 2010.The empirical results show that market sentiment can generate significant influence on the volatility of fund income and the latter also play an important role in the form of the former, which suggest that China Stock Market has not reached weak-form market.

Keywords: market sentiment, return volatility, closed-end fund.

1 Introduction

With China's rapid economic development, research on financial instability has transferred from "Mutual Relations between Financial System and Real economy" to "analysis on the instability of stock market" [1]. Various factors had effects on stock market and contribute to the instability of stock market, the investors play a decisive role in the process of instability transfer, specifically changes in group investors' psychology and Investment behavior based on these changes. These changes shall be deemed to be liable to the instability and Financial Crisis [2]. As is known, Chinese Securities Market evolves from the transformation of economy which determines its natural indigenization Characteristics and this has profound impact on the investors. As a result, there is cognitive bias in China investors which is with Chinese characteristics [3]. Therefore, the problem also exists in market sentiment and return volatility. Compared with the developed stock market, the conclusion which deduces from the study of behavioral finance in emerging markets has both theoretical and practical importance.

Based on this viewpoint, this paper is to explore the relationship between market sentiments and return volatility with Chinese stock market background.

Following the introduction, the remainder of this paper performs by three parts. Section 2 contains a brief literature review on the relationship between investors' sentiment and stock returns volatility at home and abroad. In section 3, data, methodology and empirical results are offered. Section 4 provides the conclusions of this paper.

D. Jin and S. Lin (Eds.): Advances in ECWAC, Vol. 1, AISC 148, pp. 135–139.
springerlink.com © Springer-Verlag Berlin Heidelberg 2012

2 Literature Review

Fisher and Statman [4] adopt American association of individual investors index as sentiment index to find that the investors' sentiment has negative correlation with future income. Brown and Cliff [5] adopt investors wisdom index to obtain that the sentiment has significant relationship with the return of the same period stock. In China, some domestic scholars employ investors' emotional indirect index ,such as discount of closed-end fund, entrepreneurs' index, derivatives trading index, market's confidence index, the first day revenue of the IPO, liquidity of assets, and to discuss the correlation between market sentiment and returns of stocks. Yumei Jiang and Mingzhao Wang [6] employ discount of closed-end fund, consumer confidence index, the growth rate of investor's accounts and the rate of turnover to be investor's emotional indirect index to demonstrate the influence on the return of stock from the investor's sentiment. Lixu Chi and XintianZhuang [7] use capital flow of each stock which is obtained by censusing the data of heavy warehouse stock as investors' sentiment index and find that the sentiment of investors' has influence on the evaluation to the stock price. FeiXue [8] deems that consumer confidence index is fitter than discount of closed-end fund to be investors' sentiment index. We also choose turnover rate as one of sentiment index considering the speculative characteristics of China. Liyan Han and Yanran Wu [9] deem the new accounts of exchange per month should be identified as one of sentiment index in that the new accounts of exchange per month and the most important is that it can intuitively reflect the sentiment of investor.

So, we choose turnover rate, consumer confidence index and the new accounts of exchange per month to construct investors' emotional composite index. Then, we utilize Granger causality test to analyze the relation between market sentiment and returns of stocks.

3 Date and Empirical Results

3.1 Sample Data

By December 31, 2010, there are 882 funds formally operating and 82 are close-end funds. Fund issuance is accelerating expansion in recent two years. As a result, the longer the period of sample, the smaller the sample size. Generally speaking, the data can reflect a cycle from three to five years and the data is also used by Granger model for empirical test. According to total number and the length of sample time, 27 closed-end funds which have been operating five years are adopted and the sample period is from January, 2006 to December, 2010. We choose the net fund value per week which comes from CCER (Chinese Stock Market price and return database). We process the return of funds and index which have been expanded 100 times for fear of error. The rate of return is defined as r_t:

$$r_t = ln(\frac{N_t}{N_{t-1}}) \times 100 \tag{1}$$

r_t denotes the rate of fund at t . N_t Means the net value of fund at t and N_{t-1} is at t-1.

3.2 Sentiment Index of Market

Turnover rate comes from Shanghai Stock Exchange Statistical Yearbook and the new accounts of exchange every month comes from China Stock Registration and Settlement Statistical Yearbook.

First, we find out the correlations among turnover rate, consumer information index and the new accounts of exchange every month. The results are showed as follows:

Table 1. The correlation of sentiment index

variable 1	TURN	CCI	NA
TURN	1	0.3*	0.546**
CCI	0.3*	1	0.253
NA	0.546**	0.253	1

According to the results, turnover rate significantly associated with the new accounts of exchange every month, which suggest that they are commonly influenced by the sentiment of market. There is no other correlation among those variables, which means that they can be used to construct sentiment index. The next, we adopt principal component analysis to judge the weight of every variable and the sentiment composite index is constructed as follows.

$$\mathrm{SENT}_t^N = 0.840\mathrm{TURN}_t + 0.612CCI_t + 0.818NA_t \tag{2}$$

The research on the sentiment of the China Exchange Market should consider "Chinese characteristics" --investors' cognitive biases, considering the singularity of China Stock Market which is in the process of economic transformation, especially the influence on the sentiment of market from China culture. Against that background, this paper tries to adopt Granger causality test to analyze the sample. We reveal the relation between the sentiment of market and the volatility of fund return by using Eview5.1

3.3 Empirical Results

According to theoretical model, the emotion changes of noise traders can affect the return by two routes when they obtain a signal. One route is the emotion leading to miscalculate the intrinsic value of the securities and the other is the leading to revise the volatility of return by noise trader risk. There exist volatility cluster characteristics in the distribution of High-frequency time series and the volatility can be revised by GARCH (Generalized Autoregressive Conditional Heteroskedasticity) model. But in this paper, it refuses unit root and presents significance at 1% significant level .Because the rate of the return presents smooth and steady by ADF test (the value of the statistic equals -5.213). Therefore, this paper reflects the mutual influence between market sentiment and the return of fund by Granger causality test.

The following table presents the result of Granger causality test and the lagging number can be made by AIC and SCI information criterion.

Table 2. The results of Granger test

Lags	Note	F-statistic	P	Result
One month	QX does not Granger Cause BD	0.05649	0.81302	No
	BD does not Granger Cause QX	2.3457	0.13136	No
Three months	QX does not Granger Cause BD	2.22881	0.09659***	Yes
	BD does not Granger Cause QX	1.27386	0.29370	No
Six months	QX does not Granger Cause BD	2.23496	0.05936***	Yes
	BD does not Granger Cause QX	0.83023	0.55367	No
Twelve months	QX does not Granger Cause BD	4.05547	0.00219*	Yes
	BD does not Granger Cause QX	0.81821	0.63123	No
Eighteen months	QX does not Granger Cause BD	2.30673	0.21736	No
	BD does not Granger Cause QX	72.3581	0.00042**	Yes

Note ; QX : market sentiment , BD : fund profit ; $^{*}p<0.05$; $^{**}p<0.01$; ***p<0.1

The empirical results suggest that there exists significant one-way leading role and Granger causality test exists. In China Stock Market, the sentiment can significantly influence the volatility of return. Hence, when investors are optimistic, the earnings will increase significantly and when investors are pessimistic, revenues reduce significantly, which proofs that if market is optimistic, all investors will make a profit and if market is pessimistic, investors will lose money. And the sentiment of market has significantly positive reinforcement to the volatility of the return. There exists certain lagging period. As a result, the drastic change of the market sentiment can lead more acutely volatility of the return. Also, the result suggests China Stock Market has not been weak form efficient market.

4 Conclusion

Against the background of China's stock market, this paper empirically researches the relationship between the sentiment of market and the fluctuation of the fund's return. On the one hand, the empirical result has an important role in improving the normalization of Chinese security market and guides the decision-making behavior of investors. On the other hand, fund managers can adjust behavioral finance strategies to the shock effect rule of China.

This paper mainly does the research in the view of the return series' distribution, which doesn't refer to the mutual influence between stock prices. And it is very important to further research the relationship between market sentiment and income fluctuation in different industries.

Acknowledgments. This study was supported by the Humanities and Social Science Foundation of Ministry of Education of China under grant 08JC630064.

References

1. Yeh, Y.H., Lee, T.S.: The interaction and volatility asymmetry of unexpected returns in the greater China stock markets. Global Finance Journal 1, 129–149 (2000)
2. Torngren, G., Montgomery, H.: Worse than chance? Performance and confidence among professionals and laypeople in the stock market. Journal of Behavioral Finance 3, 148–153 (2000)
3. Pei, P., Zhang, Y.: The empirical test on cognitive bias of China stock investors. Management World 12, 12–21 (2004)
4. Fisher Kenneth, L., Statman, M.: Investor sentiment and stock returns. Financial Analysts Journal 2, 16–23 (2000)
5. Brown, G., Cliff, M.: Investor sentiment and Asset Valuation. Journal of Business 2, 405–439 (2005)
6. Jiang, Y., Wang, M.: Investor sentiment and the Cross-Section of stock returns. Economy Management 10, 134–140 (2009)
7. Chi, L., Zhuang, X.: Investor Sentiment in Chinese Stock Market. Journal of Management Science 3, 79–87 (2010)
8. Xue, F.: The empirical test on the choice of the investors sentiment in China. World Economic Outlook 14, 14–17 (2005)
9. Han, L., Wu, Y.: The mystery of investors' sentiment and Initial public offerings–Suppression or premium price. Management World 3, 51–61 (2007)

Chinese OFDI Risk Analysis and Control

YuDuo Lu, DongYang Qiu, and Qiao Song

School of Economics, Dalian University of Technology,
No.2 Linggong Road, Ganjingzi District,
Dalian City, Liaoning Province, P.R. C., 116024
luyuduo@163.com, qdyang2006@sina.com, Songqiao2010@163.com

Abstract. Lack of awareness on the risks is an important reason lead to the enterprises loss in OFDI. This paper analyses and classifies the mainly risks of Chinese enterprises ODI. Then based on the risk classification puts forward the OFDI risk prevention measures and strategies, emphasizes study, combination and application of various policies to deal with those risk.

Keywords: ODI, risk classification, risk control.

1 Introduction

China's foreign direct investment has a late start, but s great increase. However, due to the complexity of the international business environment and lack experience of OFDI, investment risk is one of the most important reasons that caused many Chinese OFDI enterprises suffered so much loss. The data (2009) of the Ministry of Commerce showed that 65% of outward investment of Chinese enterprises was at a loss. Thus, accurately identify and control the risks of investment to reduce or eliminate the risk as much as possible is very important.

2 The Risk Classification

There are various methods to classify different risks. This article will classify the mainly foreign direct investment risks of Chinese enterprises according to their sources into two levels that are the national level and corporate level.

2.1 The National Level Risks

That refers the differences or changes of the political, economic, social, legal, cultural or natural environment of the host country result in the possibility of economic benefits or loss to investors. This kind of risk is the hardest to predict and prevent in OFDI. The national level risks include political risk, cultural risk and macroeconomic risk.

(1) The political risks: refer that in foreign direct investment, since the abnormal changes in power, policy and legal aspects of the host government, lead to the deterioration of the investment environment, and then result in economic

D. Jin and S. Lin (Eds.): Advances in ECWAC, Vol. 1, AISC 148, pp. 141–146.
springerlink.com © Springer-Verlag Berlin Heidelberg 2012

losses to the investment activities of foreign investors. In OFDI activities, the political risks of host country include the risk of nationalization; the risk of war; policy change risks; transfer of risk; sanctions risks; and so on.

(2) The cultural risks: refer that the cultural differences between the home country and the host country, such as language, religion, time, ethnicity, values, customs, cultural traditions, lead to the risks. American management expert D. A. Felix once said: "Generally, the failure of cross-cultural marketing, almost all ignored the cultural differences just because basic or subtle understanding and experience of the results incurred."

(3) The macroeconomic risks: in the process of outward investment the enterprises will face many macroeconomic factors which are disadvantage of investing and developing and then bring investors the macroeconomic risks. The macroeconomic risks usually are in the forms of the host of economic recession, inflation, fluctuations of exchange rate and interest rate.

2.2 The Enterprise Level Risks

They are also known as the non-systematic risks. Changes in market conditions and the factors of production, management, decision-making cause the possibility of economic losses of companies. The enterprise level risks include operational risks, financial risks, human resource risks, technology risks and etc.

(1) The operational risks: operational risks are the risks that the company needed resources not be met, or the imperfect transport system may cause production and management of the transnational business can't be normal operated.

(2) The financial risks: In the investment process of the enterprise, the contents of the financial activities involved a very complicated process, include the financing, the use of funds, the capital recovery, income distribution and other aspects of the organic links and the links of the various activities are likely to bring about risks. In carrying out the process of outward investment, due to many uncertain factors, the financial risks faced by enterprises are more complicated.

(3) The human resource risks: the human resource risks are the uncertainties that come from changes in the recruitment, vocational training, mobility, quality of staff and working attitude, labor allocation and labor productivity of the foreign-invested enterprises. In the outward direct investment, human resource risks are mainly in organizational collaboration risks, employee communication risks, health risks, employee turnover risks, and several other aspects.

(4) The technical risks: Mille (1992) holds that technological innovation influences product of an industry itself or the process, but at the same time, it also has brought new uncertainty to the industry, because it may disrupt the pattern of competition and cooperation that the enterprises have already established.

(5) The governance risks are additional risks that the absence of norms of corporate governance or the loss of governance mechanism. In China, as modern governance mechanism missing or imperfect, Chinese state-owned

enterprises "internal control" is serious; the governance of private enterprises or family enterprises exist many irregular aspects. The difference of Chinese business practices, modes of operation, and so on, result in that the risk management the primary risk of Chinese OFDI enterprises.

3 OFDI Risk Prevention Measures of Chinese Enterprises

The goal of risk prevention is to eliminate the uncertainty or try to minimum the fluctuations of expected value. During the initial period of OFDI, investors are face with various risks, not only those kinds analyzed above, while they also need detailed classification and deeper assessment.

3.1 Prevention of Political Risk

Political risk is very common in OFDI. As a non - market risk, political risk directly affects the strategic objectives and implementations of enterprises investing aboard.

(1) Strengthen the risk evaluation and prediction

Some international agencies periodically take some investigation in different countries and regions on political risk, and make a rating survey report according to the level of political risk, then announced to the public. These data have an important reference value to firms to determine the outward investment region, the portfolios and the investment scale. In addition, firms must conduct its own evaluation of the size of political risk in objective country. Including assessment: the stability of the host country's government, prestige and ability to continue in power of the current leaders, coups in history and the duration, relationship with its neighbors, the existing of unresolved territorial disputes or other international issues and so on. After finished the primary investment in the host country, transnational enterprise need to establish a risk forecast system of the country in order to detect the precursor of the imminent risk and take appropriate measures before the outbreak of the risk to avoid unnecessary loss of life and property as much as possible.

(2) Protection investment agreement

Before investing in foreign country, negotiating with the local government and signed series investment agreements is an effective way to prevent political risk. We can take advantage of the outward investment protection law and policy of home country namely these bilateral and multilateral investment protection agreements which specified responsibilities and rights between home and host countries. Agreements to take preventive measures and reduce loses of political risk that get agree with the host should adequately include about these terms(especially in some areas with potential conflicts): the basis of transfer pricing, the rights to exports to third countries, the obligations to establish public projects, spare parts from import and local sources, formulation of staff employment system, arbitration clause of labor dispute, terms about discarding the investment and etc.

(3)Risk diversification strategy
Risk diversification strategy is a very useful measure to control various kinds of risk, certainly it is suitable for political risk. The economic globalization and specialization provide the foundation of decentralizing management and investment. One side, we can arrange capital and materials for OFDI to invest in different countries, different industries, and different products to reduce the risk. When a transnational enterprise arranges its distribution of production, more often it doesn't follow a principle of a complete (no matter large or small) production organization but follow the principle that highly specialized division of labor to its subsidiaries in different areas. These subsidiaries cooperate to complete the entire production process. Even if a subsidiary engaged in part of the process was closed for the nationalization of the host country, it is also worthless to the host country, because they can't get all the production processes. On another side, interest conflict is one of the most possibilities to arouse political risk. If the outward enterprises strengthen the integration of main interests with the host countries which increased the costs of nationalization of the host government policy, it will significantly reduce political risk. The OFDI enterprises should consider joint ventures, diversification of the investment main body and transfer part of risk to the host country interests. For example, appropriately increase the proportion of local procurement of raw materials, employ the local workers, and cultivate the potential common interests to link the benefits of the local interest groups of the host country to the outward enterprises. Furthermore, different investment forms can avoid or reduce political risk in some degree. Risks of different form of investment projects are quite different. They can transform the form of investment such as convert direct investment into indirect investment, when the direct investment with a threat of political risk. Sometimes these transforms are very effective.

Of course there are still many any other methods to prevent political risk, For instance control the essential technologies and processes or product distribution channels, legal remedy and so on.

3.2 Prevention of Exchange Rate Risk

Another risk that all of the transnational enterprises must confront like political risk, is exchange rate risk. With the implementation of the exchange rate reform of China, Chinese enterprises must pay more attention to exchange rate risk.

Foreign exchange risk management is flexible and diverse. From technical measures point, we have a broad options: choose and use flexible currency of settlement, try to use local currency-denominated to settlement when export goods, build a Multi-currency Management Center if there are huge cross-border capital flow every year, net settlement, Leads or Lags, matching management etc. Enterprises should continuously collect information of foreign exchange risk according to the spot and forward foreign exchange risk of different currencies reflected in the foreign capital flows and foreign exchange risk positions table reports. Based on a variety of risk analysis, assess the impaction on projects of outward investment. It is also necessary to gather and search information related to tax income of foreign country, exchange control, and establish an effective exchange market intelligence system. Most operational strategies are also useful to deal with exchange risk. The principle of diversification is well-proven approach: scatter production and sales markets, production facilities and sources of raw materials around the world. Financial instruments as hedging or forward transactions in outward financial markets are available to reduce or avoid foreign exchange risks.

3.3 Investment Decision and Operation Risk Controls

OFDI is a continuous dynamic process. This process comes with the change of critical risk to deal with. Enterprises have to response to additional risk different from domestic environment. In the early of the investment, once the investment decided get implemented, any change of the original plan will be a large cost. Before make the crucial investment decision, enterprises need to make a comprehensive analysis of host country's economic development situation though investigation on the ground, expert consultation to determine the appropriate location of investment and the industry of investment.

When the initial investment is successful, the following operation risks as production risk, marketing risk, human resource risk, capital risk, financial management risk will be more complex than domestic due to complexity and objectivity of cross-cultural management. Cross-culture management is the key to maintaining the stability of foreign investment in the long term. We should reinforce the concept of cross-cultural management and strengthen the capacity of cross-cultural management. Prevent the characteristic of host country language, values, thinking forms and other cultural factors arousing obstacles in the management. Emphasis on learning and understanding the knowledge relate to the host country's language, culture, economic and legal. At the same time, adopt different types of measures to different cultures, search for the integration point for different cultures, built a common principle for employees from different cultural backgrounds, and thereby maximized potential value of enterprises. Take the localization strategy into account in the premise that there is no or less affect on the core interests. Participate social service is also a good measure to integrate into the society and appeal the host pay more attention to the investment project.

4 Conclusion

OFDI risk analysis and control must base on the balanced costs of risk management and the size of risk. Some measures such as diversification can prevent various risks which are relatively prevalent in dealing with outward complex investment environment. It is more important, enterprises can make a combination of these measures appropriately to compensate for outward investment risk in the absence of risk analysis. The implement of first OFDI plan was a learning process, especially for Chinese enterprises with less experience on OFDI. During the study of how to overcome outward investment risk, Chinese transnational enterprises consolidate and promote the status in the global industrial chain and their international influence has been strengthened.

References

1. Qi, Z.: China's foreign direct investment risk identification and prevention. International Economic Cooperation 4, 53–56 (2010) (in Chinese)
2. Jinjarak, Y.: Foreign direct investment and macroeconomic risk. Journal of Comparative Economics 35(3), 509–551 (2007)
3. UNCTAD: World Investment Report (2010)
4. Information of Chinese OFDI over the years on, http://www.mofcom.gov.cn

Implementation of a Chaos-Based Encryption Software for Electroencephalogram Signals

Chin-Feng Lin* and Chun-Yuan Chang

Department of Electical Engineering,
National Taiwan Ocean University
Taiwan, R.O.C.
lcf1024@mail.ntou.edu.tw

Abstract. In the paper, a chaos-based electroencephalogram (EEG) encryption software was developed using Microsoft Visual Studio and C# programming. Two level chaos-based encryption mechanisms are implemented for clinical EEG signals. A chaos logic map, initial value, and a bifurcation parameter of the chaos logic map are used to generate level I chaos-based encryption bit streams. Several clinical EEG signals are tested, and the encryption effect is superior.

Keywords: Chaos, encryption, software, implementation, Microsoft Visual Studio, C#.

1 Introduction

In recent times, chaos-based encryption theory and mechanisms have been extensively studied, and they have been applied to the encryption of multimedia signals such as audio, images, and video; medical signals such as electrocardiogram, and electroencephalogram signals; and ordinary data. Chaos-based encryption has the following features: it is highly unpredictable, it is sensitive to the starting point and the type of chaotic logistic map used, and different starting points and chaotic logistic maps generate different chaotic sequences for encryption. The concepts of chaos-based encryptions have already been discussed in detail in previous studies [1]. Murali *et al* [2], Yen *et al* [3], Li *et al* [4], and Man *et al* [5], have proposed various chaos-based encryption mechanisms for data, images, video, and audio, respectively. In our previous studies [6-9], we have discussed various chaos-based encryption mechanisms for electrocardiogram, and electroencephalogram (EEG) signals, and for mobile telemedicine. Here, we report the software implementation of a chaos-based encryption mechanism using Microsoft Visual Studio and C# programming. We also discussed the encryption accuracy, speed, and robustness of the proposed implementation.

* Corresponding author.

D. Jin and S. Lin (Eds.): Advances in ECWAC, Vol. 1, AISC 148, pp. 147–150.
springerlink.com © Springer-Verlag Berlin Heidelberg 2012

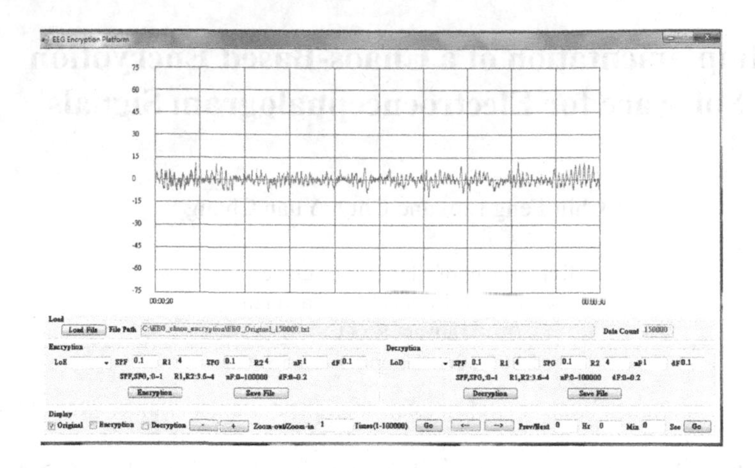

Fig. 1. The implementation of the chaos-based level I encryption software for EEG signals

2 Chaos-Based Encryption Mechanism

The section describes the implementation of the level I chaos based EEG encryption software.

The chaos-based EEG level I encryption is given as follows.

Step 1: Select a chaotic logistic map type CMT, the starting point SP_F, the length L_F for an encrypted clinical EEG bit streams.

Step 2: Generate a chaotic sequence with length L_F.

$$x_{n+1} = CMT(x_n)\,;\, x_0 = SP_F\,;\, n=\{1,2,\ldots,L_F\} \tag{1}$$

Step 3: The chaos-based encryption bit streams is defined as

$$CBEBS = \{y_n\} \quad n = 1,2,\ldots,L_F \quad y_n = \begin{cases} 1 & x_n \geq 0.5 \\ 0 & x_n < 0.5 \end{cases} \tag{2}$$

Step 4:Deliver EEG medical signal bit streams *EEGS* with length L_F.

$$EEGS = \{eeg_1,\ldots,eeg_{L_F}\}$$

Step 5:Generate encrypted clinical EEG signal bit streams *GEEG*.

$$\begin{aligned} GEEG &= EEGS \oplus CBEBS \\ &= \{eeg_1 \oplus y_1,\ldots,eeg_{L_F} \oplus y_{L_F}\} \\ &= \{GEEG_1, GEEG_2,\ldots,GEEG_{L_F}\} \end{aligned}$$

'$\oplus$:' exclusive or gate operation. (3)

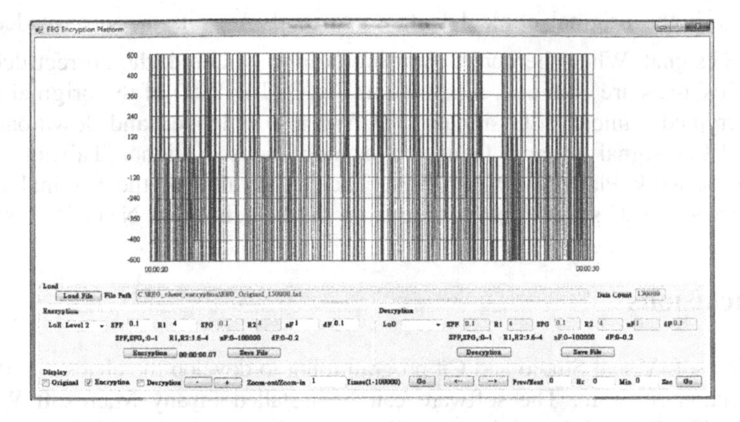

Fig. 2. The encrypted level I EEG signals

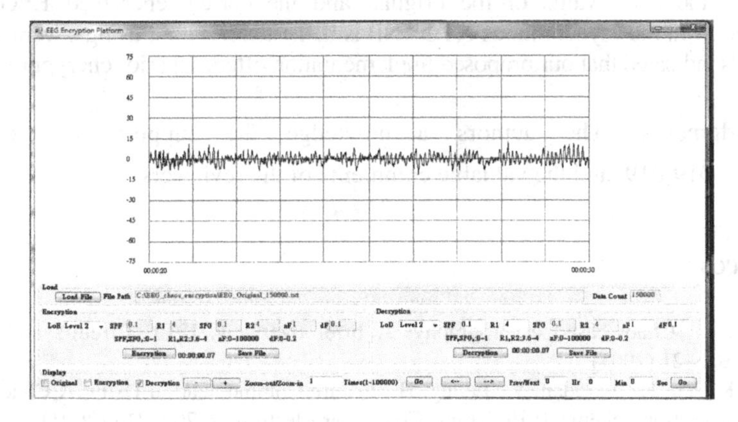

Fig. 3. The correct decrypted level I EEG signals, when the correct starting point is entered

3 Software Implementation and Discussions

We use Microsoft Visual Studio and C# programming to develop the chaos-based level I EEG encryption software, as shown in Fig. 1. This software can be installed in any Microsoft Windows environment. Fig. 2 shows encrypted level I EEG signals; clearly, the encrypted signal are unrecognizable. The distribution of the correct decrypted and the original signal is discussed in the term of the percent root-mean-square difference (PRD). The *PRD* value is defined as

$$PRD = 100 \times \sqrt{\frac{\sum_{i=1}^{L}(X_{ori}(i) - X_{dec}(i))^2}{\sum_{i=1}^{L} X_{ori}^2(i)}} \tag{4}$$

where X_{ori} is the original clinical EEG signal and X_{dec} is the correct decrypted clinical EEG signal. When the correct starting point, is entered, the correct decrypted level I EEG signals are obtained, as shown in Fig. 3. The PRD of the original and the correct decrypted clinical EEG signals is 0. We also uploaded and downloaded the encrypted EEG signal to and from the asynchronous National Taiwan Ocean University Network Platform (NTOUNP). The PRD value of the original and the correct encrypted EEG signal downloaded from the asynchronous NTOUNP is 0.

4 Conclusions

We use Microsoft Visual Studio and C# programming to develop the chaos-based level I EEG encryption software. The software can be installed in any Microsoft Windows environment. The PRD of the original and the correct decrypted clinical EEG signals is 0. In addition, the PRD value of the original and the correct encrypted EEG signal downloaded from the asynchronous NTOUNP is 0. Test carried out using several clinical EEG signals indicated that our proposed implementation offers superior encryption.

Acknowledgments. The authors acknowledge the support of the NSC 100-2221-E-019-019, and the valuable comments of the reviewers.

References

1. Kocarev, L.: Chaos-based cryptography: A brief overview. IEEE Circuits and System Magazine, 6–21 (2001)
2. Murali, K., Yu, H., Varadan, V., Leung, H.: Secure Communication Using A Chaos Based Signal Encryption Scheme. IEEE Trans. Consumer Electronics, 709–714 (2001)
3. Yen, J.C., Guo, J.I.: Efficient hierarchical chaotic image encryption algorithm and its VLSI Realisation. IEE Proc.-Vis. Image Signal Process, 167–175 (2000)
4. Li, Y., Liang, L., Su, Z., Jiang, J.: A new video Encryption Algorithm for H.264. IEEE ICICS, pp. 1121–1124 (2005)
5. Man, K.P., Wong, K.W., Man, K.F.: Security Enhancement on VoIP using Chaotic Cryptography. In: IEEE Conference, pp. 3703–3708 (2006)
6. Lin, C.F., Chung, C.H., Lin, J.H.: A Chaos-based Visual Encryption Mechanism for Clinical EEG Signals. Medical & Biological Engineering & Computing, 757–762 (2009)
7. Lin, C.F., Chung, C.H.: A Chaos-based ECG/EEG Visual Encryption Methods and Devices, Taiwan, R.O.C. patent, I 338845
8. Lin, C.F., Chang, W.T., Li, C.Y.: A Chaos-based Visual Encryption Mechanism in JPEG2000 Medical Images. J. of Medical and Biological Engineering, 144–149 (2007)
9. Lin, C.F., Chung, C.H., Chen, Z.L., Song, C.J., Wang, Z.X.: A Chaos-based Unequal Encryption Mechanism in Wireless Telemedicine with Error Decryption. WSEAS Transactions on Systems, 49–55 (2008)

Analysis of County Economic Development Disparity in Guangdong Province Based on BP Artificial Neural Network

Qinglan Qian and Yingbiao Chen

School of Geographical Sciences, Guangzhou University, 510006,
Guangzhou, China
qianlynn@21cn.com

Abstract. 67 counties (cities) in Guangdong Province have been selected as the research area, and 17 indicators including GDP and gross industry output regarding economic aggregate, economic level, industrial structure, development speed and foreign investment utilization. Evaluation indicator system is established based on these 17 indicators, and a comprehensive evaluation of the disparity of county economic development in Guangdong is conducted by using BP Artificial Neural Network. The result shows that: There is a significant disparity in the county economic development in Guangdong, as is presents a decrease pattern from the Pearl River Delta to the outer area of east, west and north of Guangdong. The evaluation result of BP Neural Network can effectively avoid the possible error brought by subjective weight, therefore enhance the accuracy of the result and better estimate the disparity of the county economic development.

Keywords: County Economy, BP Artificial Neural Network, Spatial Disparity, Guangdong Province.

1 Introduction

The county economy, also as known as the county-level economy, is the regional of China's administrative, As a symbol of an independent financial, t the basic structure of the county town, township and village levels of the linkages and the ratio. It's through two-way feedback that the human flow, capital flow, material flow, information flow to show the economic system of overall function [1]. At present there are many research results to analysis the county economic development disparity using of traditional quantitative analysis methods [2]-[7]. The traditional social statistical and spatial analysis methods were used for measure the county economic development disparity, According to the latest statistics of the 67 counties (cities) in Guangdong, using artificial neural networks, the author analysis quantitatively he county economic development disparity in Guangdong in 2009. It's amid to analysis preliminary for disparity and reasons of the county economic development, and to reference for the coordinated development of regional economies.

D. Jin and S. Lin (Eds.): Advances in ECWAC, Vol. 1, AISC 148, pp. 151–156.
springerlink.com © Springer-Verlag Berlin Heidelberg 2012

2 Research Methods and Data Processing

2.1 BP Artificial Neural Network

Artificial neural network is an abstraction and simulation for some features of a natural human brain or neural network. BP neural network is a typical artificial neural network model and an MLP structure, which is composed of several layers of neurons. In this article, with the MATLAB neural network toolbox, which provides weights initialization, learning and training, simulation and other functions for establishing arbitrary input and output neurons of the BP network, BP neural network has some preparation process to achieve. We will put the grading standards of the level for a sample input, the evaluation level as the network output. BP network can modify weight by learning, find the complexity inherent correlation between evaluation index and assessment level, and conduct a comprehensive evaluation of the level of economic development with the network model.

2.2 Data Sources and Model Construction

In this article, as the 67 counties (cities) in Guangdong for the study object,. The author makes a measure of the county economic development in Guangdong from five aspects including the total economy, economic development, economic structure, development speed and foreign investment (Table 1).To build a neural network, There are 17 second indicators of county(city) for the input neurons, the level of economic development for the output neurons. The second indicators data is directly or calculatedly obtained by "The Guangdong Statistical Yearbook 2010" and "Guangdong Yearbook 2010". It is deal with the statistical by Maximum standardization to set the statistical range from zero to one. It is considered that the distribution patterns of each indicator data are mostly skewed distribution, should not be used equally spaced linear interpolation method. The author apply natural break for data classification, and build artificial neural network for training data. We will be divided into three levels of economic development, four is on behalf of highest levels of economic development, three is on behalf of higher, Two is on behalf of Medium, one is on behalf of low level.

Table 1. The Indicators System of the Measured County Economic Disparity

Target layer	First Class Indicator	Second Class Indicator
The county economic development disparity	Total Economics	GDP（X1） Total Investment in Fixed Assets （X2） Gross Industrial Output Value （X3） Saving Deposits by Urban and Rural Residents at the Year-end （X4） Total Retail Sales of Consumer Goods （X5） Revenue （X6）

Table 1. *(continued)*

Economic Development	Per Capita GDP (X7)
	Wages of Fully Employed Staff Workers (X8)
	Secondary Industry to GDP (X9)
Economic Structure	Tertiary Industry to GDP (X10)
	Proportion of Non-agricultural Population (X11)
	Growth rate of GDP (X12)
	Growth rate of Primary Industry (X13)
Development Speed	Growth rate of Secondary Industry (X14)
	Growth rate of Tertiary Industry (X15)
Foreign Investment	Total Exports (X16)
	Foreign Loans (X17)

To set up models, based on BP neural network, the county economic development disparity which has three steps including BP neural network model construction, BP neural network training and BP neural network, algorithmic process is as follows. (Fig 1)

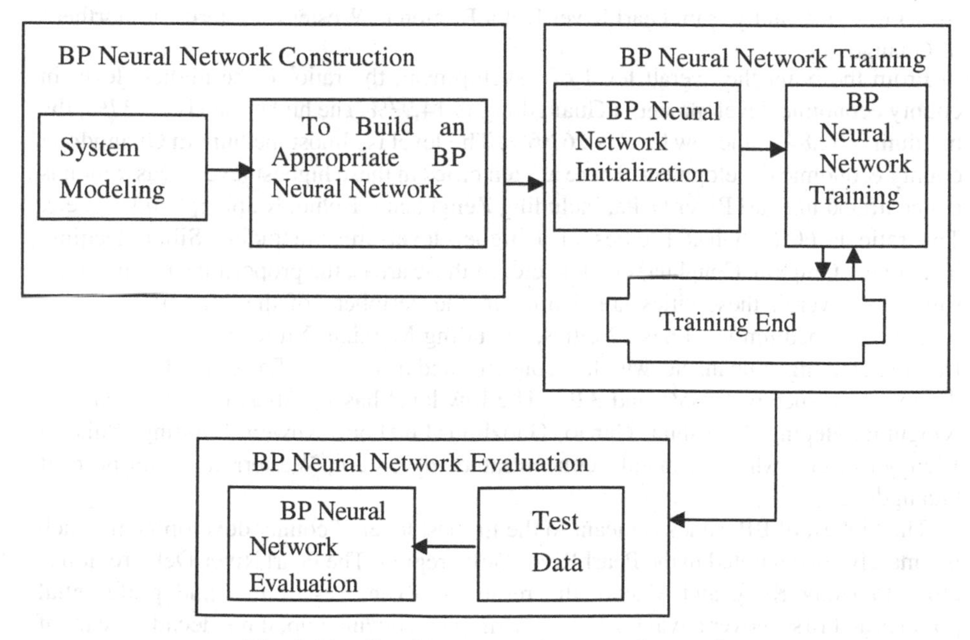

Fig. 1. The Algorithm Flow of BP neural network

15 samples are selected for training, we will assign first indicators of 15 samples to data by expert scoring method and let the raw normalized data enter the network, when the training error is as the small as the possible, the normalized data in the network to obtain the evaluation Of BP value. The network design parameters are that the initial value range from zero to one, the basic learning rate is 0.09, the largest training batch is 5000, the maximum error is 0.000001.

3 Results

According to the evaluation index system, we will evaluate the county economic development in Guangdong Province in 2009.When the network training reaches the standard, based the raw normalized data in the network on he "Guangdong Province Statistical Yearbook 2010" and "Guangdong Yearbook 2010" .We make use of BP neural network model to simulate the level of economic development of 67 counties (cities) in Guangdong, to obtain BP evaluation value.

The level of county economic development will be spatial visualization by ArcGIS spatial analysis module (Fig2). According to BP values, It's divided into four levels, including highest, higher, medium and low level. Greater than 2.00 of BP value is on behalf of the highest, the range of 1.63 to 2.00 represents the higher level, the value from 1.25 to 1.63 is on behalf of medium, less than 1.25 represents low level. If the BP value is higher, the county economic development level is higher. As follow as figure 2, In 2009,the level of county economic development in Guangdong Province is still showing significantly from Pearl River Delta Region to Western, Eastern and northern of Guangdong.

From the point the overall level of development, the ratio of the highest level in county economic development in Guangdong is 14.93%.The higher one is 25.37%, the medium is 32.84%, the low level is 26.86%. The level is almost medium in Guangdong county economic development. There are ten cities in these highest level areas which is concentrated in Pearl River Delta, including Zengcheng, Boluo, Kaiping, Huidong. etc. The ratio is 60%. It has 17cities in a higher level area, including Sihui, Deqing, Yangchun, Qingxin, Conghua, Fogang. etc. In these areas, the proportions is almost the same, but overall these cities are mainly in the periphery of the Pearl River Delta region, The medium level has 22 cities, including Meixian, Xinfeng, Zijin, Lianping, Lufeng, Haifeng, Yunan, etc. which is concentrated in Western, Eastern of Guangdong, the ratio is respectively 45% and 32%. The low level has 19 cities, including Puning, Xingning, Heping, Fengshun, Nanao, Gaozhou, Leizhou,, Xuwen, Luoding, Shixing Wengyuan, etc, which is mainly concentrated in Eastern, Western and Northern of Guangdong.

The highest of BP values is meant to the highest level of county development which is mainly concentrated in the Pearl River Delta region. The Pearl River Delta region is close to Hong Kong and Macao, the regional advantages of that and preferential policies and first-mover advantage of reforming and opening up, three decades years of reforming and opening up, the developed rapidly economic of big city, like as

Guangzhou, Shenzhen, Dongguan, etc, the strong industrial base, the more optimization industrial structure, the more invested and gradually improved traffic and other large infrastructure, that is an important growth pole in the regional economic development ,.The important reason is that the Pearl River Delta lies in the forefront of province by the huge radiated driven .

The higher of BP values is meant to the higher level of county development which is mainly concentrated in the periphery of the Pearl River Delta region. With the process of regional economic integration acceleration, especially the regional transportation improvement, the tertiary industries like as tourism has been developed vigorously as the back garden of the Pearl River Delta, relying on the convenience of traffic, that is improved the county development.

At this stage the medium and low level in county economic development areas are mainly concentrated in eastern, western and northern of Guangdong. In the process of the county development, due to geographic conditions, natural conditions, transport infrastructure and original weak economic foundation is restricted, the total economy is small, industrial development is relatively backward, foreign investors is weakly attractive. the industrial structure is also being in energy, raw materials and other low value-added industries, there is still more large disparity in the county's economic development level which is compared with the Pearl River Delta.

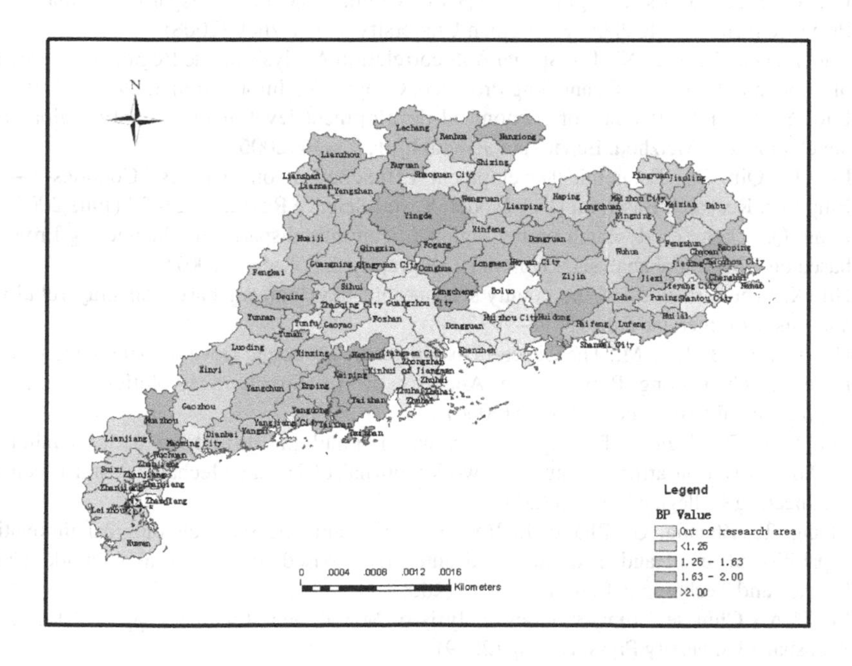

Fig. 2. The region's differentiation of the BP value of County economic development in Guangdong Province in 2009

4 Conclusion

(1) There are still large differences in the county's economic development of Guangdong. It is show that the level of country economic development is significantly from the Pearl River Delta region to the periphery, the eastern, western and northern of Guangdong to decrease.

(2) The evaluation results of BP neural network can be better simulated different in the country economic development. Compared with traditional research methods, the BP neural network can be more effective to avoid errors by subjectively and to improve accuracy of measurement. BP artificial neural network approach is a valid evaluation method of regional economic disparities.

Acknowledgements. The authors are grateful for the support provided by National Natural Science Foundation of China (No. 41071078). The authors are grateful for the support provided by National Natural Science Foundation of Guangdong (No.9151008901000187).

References

1. Liu, C.: Study on the changing and types of County Economic Disparity in Guangdong Province from 1997 to 2006. Zhongshan University, Guangzhou (2008)
2. Gao, F., Guo, Z., Wei, X.: The Spatial Autocorrelation Analysis on the Regional Divergence of Economic Growth in Guangdong Province. Geographic Information 4, 29–34 (2010)
3. Luo, Y.: Spatial difference of economical development level among county regions and development in Meizhou. Economic Geography 1, 32–36 (2006)
4. Liu, L., Qin, W.: Economic Development Efficiency among Chinese Counties——An Empirical Research Based on DEA Model. Social Science Research, 23–27 (June 2009)
5. Zeng, Q., Chen, Z.: Research on the regional economic disparity in Guangdong Province based on GIS spatial analysis. Economic Geography 4, 558–561 (2007)
6. Hu, X.: The strategies of the county's economic development based on gray relational analysis. Commercial Times 3, 98–99 (2008)
7. Hu, X., Luo, B., Luo, M.: The General Disparities in Economic Development at the County Level in Guangdong Province–An Analysis on Its Structural Evolution. Journal of Agrotechnical Economics 6, 76–80 (2006)
8. Sun, X., Li, Z., Chen, F.: The application research of multiple attributesynthetical evaluation methods based on artificial neural network. Journal of Henan Mechanical and Electrical Engineering College 63, 61–63 (2003)
9. Zhang, T., Zhuang, G., Zhang, J.: Research on evaluating the technological innovation capability of small and medium -sized enterprises based on BP neural network. China Science and Technology Forum 2, 53–57 (2009)
10. MATLAB Chinese forum. 30 cases analysis of Matlab neural network, pp. 1–10. Beijing Aerospace University Press, Beijing (2009)

Interference Suppression of Complex Pairing Access in CDMA System Based on Network Coding

Xiaoguang Zhang, Yonghua Li, Jiaru Lin, and Li Guo

Key Lab of Universal Wireless Communications, Ministry of Education
Beijing University of Posts and Telecommunications
Beijing, China
{xiao_guang831129,liyonghua,jrlin,guoli}@bupt.edu.cn

Abstract. The complex pairing access of multiple users based on network coding is discussed in this paper. In the system, through Network Coding and registration matrices, the same spreading code is allocated to each pair of users. New decision regions are considered. Then we proposed new interference reducing approach for the scenario, comfortable for relay and mobile station. Simulation results demonstrate the performance of the proposed scheme. The capacity and BER are improved with interference suppression.

Keywords: network coding, two way relay, interference suppression.

1 Introduction

Recently, application of network coding in wireless two way relay channel (TWRC) is absorbing for its improvement in throughput [1-3]. The realization of network coding also emerge endlessly. And they can be classified to three catalogs: amplify-and-forward (AF) [3 4], decode-and-forward (DF) [5 6] and denoise-and-forward (DNF) [7 8]. However, the complex pairing access based on network coding is rarely discussed, especially its interfere reducing.

Fig. 1shows a complex pairing model. The transmitters on both sides exchange information and the communication relation is shown by dash lines. All the exchangings are finished in two timeslots. The same spreading code is allocated to a pair of transmitters which exchange information. For clarity, we take some notes to register the complex pairing relation. Let $t_{ij}=1$ denote the left transmitter 'i' exchanges information with the right one 'j'. $t_{ij}=0$ denotes they have no exchanging. Their shared spreading code is expressed as c_{ij} c_{ij}. Aided by TWR with network coding, the exchanging process is finished in two time slots:the first time slot: the spreaded signals are multi-accessed to the relay. The second time slot: the processed signals are broadcasted to the transmitters.

Using network coding, the received signals spreaded by the same spreading code are superposed together. They are decoded, made decision and remapped to symbols for broadcasting.

D. Jin and S. Lin (Eds.): Advances in ECWAC, Vol. 1, AISC 148, pp. 157–163.
springerlink.com © Springer-Verlag Berlin Heidelberg 2012

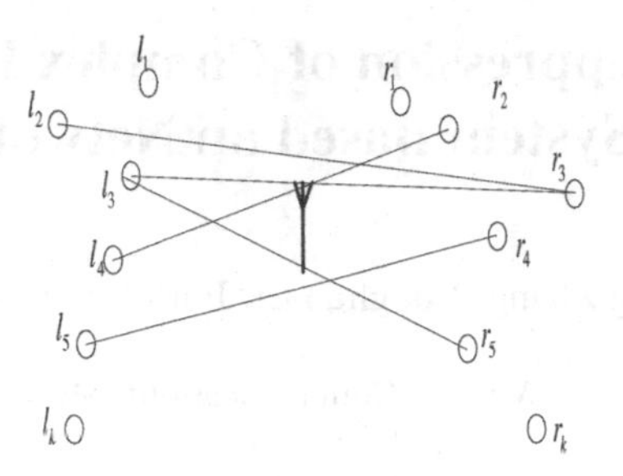

Fig. 1. Complex Pairing Illustration Aided by TWR

2 System Model

For the system model, we consider the slow flat fading channel. In complex pairing scenario, we assume there are m transmitters on the left and n ones on the right. The left transmitter l_i exchanges information with several right transmitters. Instead of spreading signal $l_i (i=1\sim m)$ with corresponding spreading codes $c_{ij}(t_{ij} \neq 0)$ one by one, we use $\sum_{j=1}^{n} c_{ij}(t_{ij} \neq 0)$ to spread l_i . Likewise, for the right user $r_q (q = 1 \sim n)$, spread r_q with $\sum_{p=1}^{m} c_{pq}(t_{pq} \neq 0)$.

Users on the both sides transmit the spreaded bit-frame to TWR in time-slot1. For simplicity of discussion, we consider BPSK modulation. So the received signals on TWR can be expressed as:

$$\begin{aligned} g_{up} &= \sum_{i=1}^{m} \sum_{j=1}^{n} \sqrt{\alpha_i p_{l_i}} l_i c_{ij} + \sum_{i=1}^{m} \sum_{j=1}^{n} \sqrt{\alpha_j p_{r_j}} r_i c_{ij} + n_0 \\ &= \sum_{i=1}^{m} \sum_{j=1}^{n} (\sqrt{\alpha_i p_{l_i}} l_i c_{ij} + \sqrt{\alpha_j p_{r_j}} r_i c_{ij}) + n_0 \end{aligned} \tag{1}$$

where n_0 denotes the AWGN (additive white Gaussian noise), α_α denotes the Channel fading coefficient, p denotes the chip-power for the chips of spreaded signal. Let 'm' denote the despreaded signal by TWR, for a pair of transmitters $l_{i'} - r_{j'} (i'=1\sim m, j'=1\sim n)$, we have:

$$\inf = \frac{\sum_{i=1}^{m} \sum_{j=1}^{n} (\sqrt{\alpha_i p_{l_i}} l_i c_{ij} + \sqrt{\alpha_j p_{r_j}} r_i c_{ij}) c_{i'j'}^{T} c_{ij}}{L_c} \tag{2}$$

where the 'inf' denotes the interference of other pairs, Lc is the length of the spreading code.

Based on network coding, TWR makes a decision whether $l_{i'}$ and $r_{j'} (\in \{+1,-1\})$ have the same polar. Let $d_{i'j'}$ denote $d_{i'j'}$ the decision result. The threshold has been derived as [5 6]:

$$I^{th} = \max\left(\left|\sqrt{\alpha_{i'} p_{l_{i'}}} l_{i'}\right|, \left|\sqrt{\alpha_{j'} p_{r_{j'}}} r_{j'}\right|\right) \quad (3)$$

and the rule can be expressed as:

$$\begin{cases} \text{same sign: } d_{i'j'} = 1 & -I^{th} < m_{i'j'} < I^{th} \\ \text{opposite sign: } d_{i'j'} = -1 & \text{others} \end{cases} \quad (4)$$

Then the decision result $d_{i'j'}$ is spreaded by TWR for boradcasting. Instead of spreading $d_{i'j'}$ with $c_{i'j'}$ one by one, relay check the nonzero elements $d_{i'j'} = 1$, superimpose the corresponding elements $c_{i'j'}$ and then spread '1' by $\sum_{d_{i'j'}=1} c_{i'j'}$, spread '-1' by $\sum_{d_{i'j'}=-1} c_{i'j'}$ in the same way. So we have:

$$g_{broad} = 1 \times \sqrt{p} \sum_{d_{i'j'}=1} c_{i'j'} - 1 \times \sqrt{p} \sum_{d_{i'j'}=-1} c_{i'j'} \quad (5)$$

where p is the power of spreading code chip. In the downlink, the process of dispreading by users can be expressed as the following equation:

$$\begin{cases} u_{kt} = \dfrac{g_{down} c_{kt}^T}{L_c} = \pm\sqrt{\alpha_k p} + \inf_{kt} + n_0 \\ \inf_{kt} = \dfrac{c_{kt}^T (1 \times \sqrt{\alpha_k p} \sum_{d_{ij}=1} c_{ij} - 1 \times \sqrt{\alpha_k p} \sum_{d_{ij}=-1} c_{ij})}{Lc} & (c_{ij} \neq c_{i'j'}) \end{cases} \quad (6)$$

where c_{kt}^T is the vector transposes of spreading code.In other words, despreading with different $c_{kt}(t = 1 \sim n)$ can get different bit-information of l_k XORed. where α_k is channel fading coefficient in the downlink to the left kth user. $\inf_{kt}$ denotes interference from other pairing-relation and n_0 denotes noise of wireless channel. Without considering interference suppression, the despreaded bit-frame u_{kt} in Eq.is Gaussian random variables with mean $\pm\sqrt{\alpha_k p}$, and we could derive the reasonable decision rule for l_k l_k:

$$\hat{b}_{kt}(t = 1 \sim n) = \begin{cases} 1 & u_{kt} > 0 \\ -1 & u_{kt} < 0 \end{cases} \quad (7)$$

where '1' means l_k has the same polar with $r_t = l_k$,'-1' means the opposite decision result $r_t = -l_k$.

3 Interference Suppression at the Relay and Users

In the multiple users pairing scenario aided by TWR, all the despreaded bit-frame have been used to make a decision directly without interference suppression whether on the relay or at the user sides. So when the complexity and number of pairing-relation increases, interference and BER increase rapidly (that is MAI of the CDMA). Interference also has an intuitive display in the abrove formulations. When we consider one pairing-relation $l_k - r_t$, all other users introduce interference for quasi orthogonality of spreading codes. Quantitative descriptions of multi users' interference on TWR and at user sides are given in (2) and (6) respectively. For m-sequence, the quasi orthogonality is $\frac{c_{kt}^T c_{ij}}{Lc} = \frac{-1}{Lc}$. Based on the equations we can reduce interfere. For $l_{i'} - r_{j'}$ of $\sqrt{\alpha_i p_{l_i}} l_i + \sqrt{\alpha_i p_{l_i}} l_i$ interference 'inf' in (2) can be got after despreading. To estimate the interference, we use $c_{i'j'}$ to do periodic cross-correlation operation with $c_{ij} : \frac{c_{i'j'} c_{ij}}{Lc} (c_{i'j'} \neq c_{ij})$ $\mathbf{c_{ij}}$, multiply it with the corresponding elements in XORed-matrix M, superimpose the result together, and invert the polar of the result to get :

$$-\frac{\sum_{i=1}^{m}\sum_{j=1}^{n}(\sqrt{\alpha_i p_{l_i}}\, l_i + \sqrt{\alpha_j p_{r_j}}\, r_i) c_{i'j'}^T c_{ij}}{L_c} \quad (c_{ij} \neq c_{i'j'}) \tag{8}$$

Then we could derive the quasi-pure $\sqrt{\alpha_{i'} p_{l_{i'}}} l_{i'} + \sqrt{\alpha_{j'} p_{r_{j'}}} l_{j'}$ by inserting Eq. into Eq.(1).Similar to the relay, so does the interference at the user sides. To this end, opening the elements of the coding-matrix C to users is necessary, but it is not permitted for the traditional CDMA for the information safety. After taking Network Coding in the system, though the elements of coding-matrix C are opened to the current study user, it is impossible to know the bit-information of other users not pairing with it for no reference. In other words, the BER for knowing bit-frame of other users is 50%.

The mechanism of interference suppression on the relay is suitable for users: for the left kth user l_k $\mathbf{l_k}$, we suppress interference against per elements in the kth row of pairing-matrix A which represents the pairing users to it, separately. First, after despreading g_{down} with c_{ij}, we set XORed-matrix M^{l_k} for l_k, and then for the kth row of M^{l_k}, using periodic cross-correlation $\frac{c_{kt}^T c_{ij}}{Lc} (c_{ij} \neq c_{kt}, t = 1 \sim n)$ to multiply the corresponding despreaded XORed bit-frame $m_{ij} = \pm\sqrt{\alpha_k p}$ to estimate the interference as:

$$\frac{c_{kt}^T (1 \times \sqrt{\alpha_k p} \sum_{d_{ij}=1} c_{ij} - 1 \times \sqrt{\alpha_k p} \sum_{d_{ij}=-1} c_{ij})}{Lc} \quad (c_{ij} \neq c_{i'j'}) \tag{9}$$

4 Iterative Algorithm of Interference Suppression

In order to suppress interference, we use spreading code to despread g_{up} and g_{down} to estimate the interference, but unfortunately the elements are also not pure for the MAI(multiple-access interference). So we call the previous result after interference suppression quasi-pure.

To achieve better interference suppression, we need to make the despreaded superposing signals purer. We think this problem as interference suppression against elements of these matrex. On the bi-directional relay, we suppress the interference for other elements to get the purer XORed $\sqrt{\alpha_i p_{l_i}} l_i + \sqrt{\alpha_j p_{r_j}} l_j (c_{ij} \neq c_{kt})$ bit-frame $\sqrt{\alpha_{i'} p_{l_{i'}}} l_{i'} + \sqrt{\alpha_{j'} p_{r_{j'}}} l_{j'}$. The operation can be expressed as :

$$\frac{g_{up}c_{ij}^T}{Lc} - \frac{\sum_{c_{vw} \neq c_{ij}} (\sqrt{p_v \alpha_{l_v}} l_v + \sqrt{p_w \alpha_{r_w}} r_w) c_{i'j'}^T c_{vw} + \sigma_{n_0}^2 c_{i'j'}^T c_{ij}}{Lc} = \sqrt{p_i \alpha_{l_i}} l_i + \sqrt{p_j \alpha_{r_j}} r_j + n_0 \tag{10}$$

And then replace the elements m_{ij} by Eq.(10). So we could get purer $\sqrt{\alpha_i p_{l_i}} l_i + \sqrt{\alpha_j p_{r_j}} l_j (c_{ij} \neq c_{kt})$ to replace the original ones, and the purer one can be inserted into Eq.(1) again. After several loops, insert the latest $\sqrt{\alpha_i p_{l_i}} l_i + \sqrt{\alpha_j p_{r_j}} l_j (c_{ij} \neq c_{kt})$ into Eq. (10)to suppress the interference to $l_{i'} - r_{j'}$.

5 Simulation Result

In this part, we present the simulation result of the designed scheme and the performance of interference suppression. All the simulations take m-sequence with Lc=31 as spreading code.

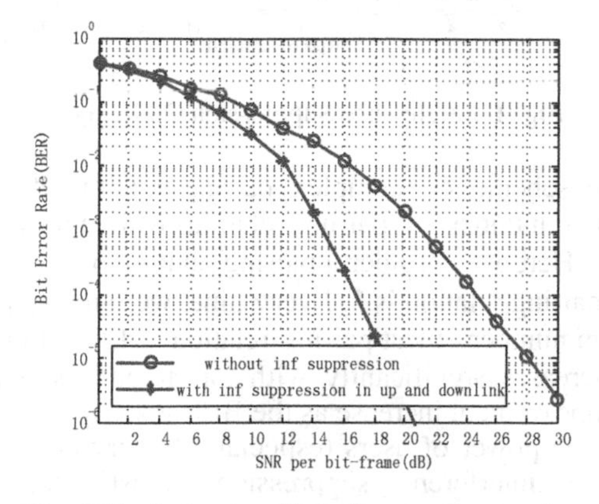

Fig. 2. Performance of Interference Suppression

In order to check the performance of interference suppression, we set a pairing relation of $l_1-r_{1\text{-}6}, l_2-r_{7\text{-}12}, l_3-r_{13\text{-}18}$. the channel is Gaussian.In Fig 2, we can observe the interference suppression providing a remarkable BER improvement after one loop.

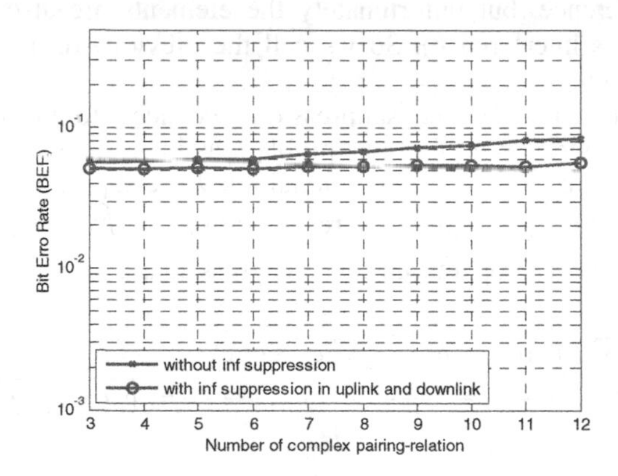

Fig. 3. Supported Number of System

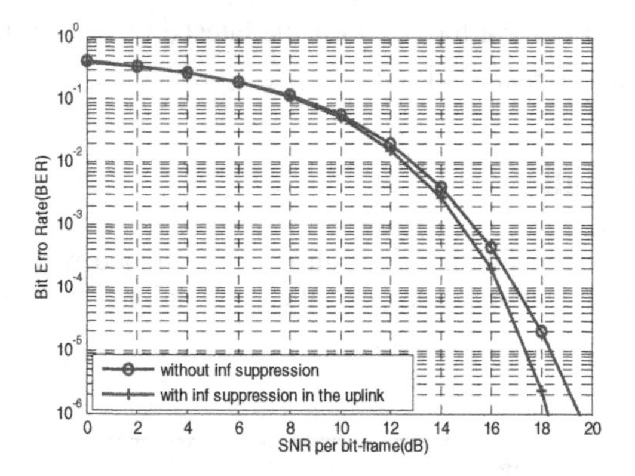

Fig. 4. Interference Suppression On Relay

Under the pairing scheme for complex pairing-relation of multiple users, we compare the supported number of pairing-relation with and without interference suppression in system. Here we consider a pairing relation of $l_{1\text{-}5}-r_{1\text{-}5}$ SNR=10dB, and the initial number of pairing relation is 3. Every time, we randomly set a zero element to 1, and the maximum number of test pairing-relation is 12. In Fig 3, we can see the supported number increases significantly with interference suppression, under the same BER, and the improvement increase as the BER goes up.

Then considering the power of users (especially battery-powered mobile phone), we test improvement of interference suppression against uplink on the relay. The tested pairing-relation is taken as Fig. 1, SNR is 0~30dB. As the previous theory analysis, in Fig 4 with interference suppression, the BER improve obviously, after one loop. But the performance of interference suppression in both up and down links is improved as we expect theoretically.

6 Conclusion

This paper discusses the problem of multiuser pairing relation with Network Coding.We propose the model for the scenario for processing simplicity. In the uplink, at the user sides, we design a new way to spread bit-frame for all the pairing-relations only one time, and it is also suitable for the spreading on the relay for broadcasting. Then, we present the decision rule and spreading method with Network Coding, we discuss the new decoding way correspondingly. Then We introduce a new interference reducing approach and derive the expression of interference suppression on TWR and at the user sides. Finally, we design the iterative algorithm foritand attribute it to an iterative problem under the requirement of QOS in the uplink and downlink. It is meaningful for the non-battery-powered users. From the simulation of eighteen pairing-relations, we can sure the proposed scheme for this multi-access problem is reasonable, and the interference suppression way is effective.

Acknowledgment. This work was supported by National Basic Research Program of China (2009CB320401), National Natural Science Foundation of China (60972075, 61072055), Key Scientific and Technological Project of China 2010ZX03003-003-01, and Fundamental Research Funds for the Central Universities (2009RC0116).

References

1. Ahlswede, R., Cai, N., Li, S.-Y.R., Yeung, R.W.: Networkinfomation flow. IEEE Trans. Inform. Theory 46(4), 1204–1216 (2000)
2. Li, S.-Y.R., Yeung, R.W., Cai, N.: Linear Network Coding. IEEE Trans. Inform. Theory 49(2) (February 2003)
3. Zhang, S., Liew, S., Lam, P.: Hot topic: Physical-layer Network Coding. In: Proceedings of the 12th Annual International Conference on Mobile Computing and Networking, pp. 358–365. ACM, New York (2006)
4. Popovski, P., Yomo, H.: Wireless Network Coding by amplify-and-forward for bi-directional traffic flows. IEEE Commun. Lett. 11(1), 16–18 (2007)
5. Chen, M.: Resource management for wireless ad hoc networks. Ph.D.dissertation. The Pennsylvania State University, University Park, PA (2009)
6. Chen, M., Yener, A.: Multiuser Two-Way Relaying:Detection and Interference Management Strategies. IEEE Trans. on Wireless Communications 8(8) (August 2009)
7. Koike-Akino, T., Popovski, P., Tarokh, V.: Denoising maps and constellations for wireless network coding in two–way relaying systems. In: IEEE GLOBECOM, New Orleans (November–December 2008)
8. Lu, K., Fu, S., Qian, Y., Chen, H.-H.: SER Performance Analysis for Physical Layer Network Coding over AWGN Channels. In: IEEE GLOBECOM, New Orleans (November–December 2009)
9. Kim, C.-J., Kim, Y.-S., Jeong, G.-Y., Mun, J.-K., Lee, H.-J.: SER analysis of QAM with space diversity in Rayleigh fading channels. ETRI J. 17, 25–35 (1996)
10. Proakis, J.: Digital Communications. Mc Graw-Hill, New York (1989)

Motor Fault Diagnosis Based on Decision Tree-Bayesian Network Model

Yi-shan Gong and Yang Li

School of Information Science and Engineering, Shenyang University of Technology, Shenyang 110870, China

Abstract. Motor is widely used in various industries, at the same time, it leads to motor fault diagnosis along with the rapid development of technology. Traditional motor fault diagnosis methods have not quickly and accurately diagnose the motor faults. Therefore, by analyzing the characteristics of the decision tree and the advantages of Bayesian network in dealing with uncertain information, which advances to use the decision tree combining with Bayesian network to diagnose motor fault, so that the diagnosis can be more accurately and quickly. This paper also describes the model structure and the basic ideas of decision tree and Bayesian network, combines the advantages of the two, and solves the uncertainty of diagnosis information effectively. It has practical significance.

Keywords: decision tree, Bayesian networks, fault diagnosis, uncertainty.

With the rapid development of modern science and technology, machinery and equipment is increasingly towards large-scale, complex, high-speed, heavy, continuous, comprehensive, high-class and other highly automated direction, at the same time, the development of economic construction and a higher level of electrification, electrical equipment has been widely used in various fields of industrial production. As the costs of the modern machinery equipment in normal state and the loss of downtime, the proportion of the cost increases, equipment failure or loss caused by accidents is increasing, the importance of equipment maintenance business is increasingly becoming a prominent issue[1],[2]. In recent years, with the development of the Bayesian network technology, it provides a new way to solve the uncertain problem, therefore, this paper presents a way of a decision tree and Bayesian network combining for motor fault diagnosis.

1 Decision Tree Model

1.1 The Definition of Decision Trees

Decision tree is an inverted tree structure, which consists of internal nodes、leaf nodes and edges components. One of the top nodes is called root. Decision tree classification method uses top-down recursive way, it comparisons attribute value of the internal nodes of the tree and judges the down branches from the node based on

D. Jin and S. Lin (Eds.): Advances in ECWAC, Vol. 1, AISC 148, pp. 165–170.
springerlink.com © Springer-Verlag Berlin Heidelberg 2012

different property values, and reaches a conclusion from the leaf node of the tree[3],[4].Therefore, it is a conjunctive rule from the root to a leaf node, whole grain decision tree corresponds to extract expression rules.

1.2 The Construction of Decision Tree

Constructing a decision tree requires a training set and some examples, each example with some attributes and a category tags to describe. Constructing a decision tree classifier is usually divided into two steps: the generation and pruning trees[5], [6]. Tree generation is a top-down recursive method. Such as multi-tree, the structure of thinking is that if the training examples in a collection case are similar, then as a leaf node, the node content that is the category tag. Otherwise, choosing a property based on some strategy, according to the value of each attribute, the example set is divided into several subsets, so that all the examples of each subset in the property on the same property value. Then it recursively processes each subset in turn.

2 Bayesian Network Model

2.1 The Definition of Bayesian Network

A Bayesian network is a directed acyclic graph (DAG), it makes up with representing the variable nodes and these nodes connected to the side. Bayesian network model consists of network nodes by a random variable (can be discrete or continuous) set, network nodes with a causal relationship to the edges of the collection and use of a conditional probability distribution between nodes[7],[8]. Node represents a random variable which is the characteristics described of process、event、states and other entities; side represents the probability dependencies between variables. The assumptions of cause and the data of results are used node, the causal relationship between variables by the directed edges between nodes, it is use of digital encoding description of the effect from one variable to another variable.

2.2 The Advantage of Bayesian Network

(1)Bayesian network methods have a solid theoretical foundation[9],[10].Bayesian network is based on probabilistic reasoning, reasoning results have higher persuasive. The expression of conditional independence of Bayesian network can effectively express the relationship between equipment failures.

(2)Bayesian network has sophisticated probabilistic reasoning algorithms and development software.

(3)Bayesian network is more suitable for the expression of fault diagnosis problems. Bayesian network can improve the network structure and parameters by practice and learning, improve fault diagnosis capability.

(4) Bayesian network has a strong ability to learn.

3 Decision Tree-Bayesian Network Model

3.1 The Definition of Decision Tree-Bayesian Network Model

By analyzing their own characteristics of decision tree and Bayesian networks, that using only the decision tree model or Bayesian network model for motor fault diagnosis, are not making the diagnosis quickly and accurately, therefore, combining the advantages of the two, it proposes a decision tree-Bayesian network model for motor fault diagnosis. The model is a top-down tree, the tree root is the failure of representation, a leaf node is the electrical fault. Meanwhile, it finds the child node by Bayesian network, in the finding process, it can be able to quickly determine the child node, thereby removing unnecessary nodes, significant savings in system operation time, improves system efficiency.

3.2 Decision Tree-Bayesian Network Model Construction Algorithm

The decision tree model into a decision tree-Bayesian network model described as follows:

(1) All the basic events of decision tree correspond to the nodes of decision tree-Bayesian network model, due to the decision tree-Bayesian network model is with top-down recursive way, so the basic events of decision tree correspond to the root following nodes, each decision node、status node and result of decision tree correspond to decision node、status node and result of decision tree-Bayesian network model, the same event can merged into a single node. Meanwhile, according to the known conditions and logical relationship between each node of decision tree gives a priori probability.
(2) The priori probability of each basic event in the decision tree directly assigned to the corresponding node of decision tree-Bayesian network model as a priori probability.
(3) According to the logic of the decision tree to structure decision tree-Bayesian network model. First, using the arc to connect the decision node、status node and result node, second, it determine the status and decision-making type of decision tree-Bayesian network model based on the physical meaning of each node of decision tree model.
(4) Relationship between nodes of decision tree expresses a conditional probability table of corresponding node of decision tree-Bayesian network model.

3.3 The Structure and Transformation Relationship of Decision Tree-Bayesian Network Model

Decision tree from root to leaf node corresponds to a path of a conjunctive rule, and whole grain decision trees correspond to a set of extract expression rules. Therefore, all the events of decision tree are connected by association, and according to the characteristics of Bayesian network, that changes the decision tree model to decision tree-Bayesian network model, the specific conversion shown in Figure 1.

P1(F=1|A=1,B=1)=1;P2(F=1|A=1,B=0)=0;P3(F=1|A=0,B=0)=0;P4(F=1|A=0,B=1)=0;P5(F=1|C=1)=1 P6(F=1|C=0)=0.

The relationship of P5 and P1、P 2、P3、P4 is or, for example:

P(F=1|A=1,B=1||C=0)=1, P1(F=1|A=0,B=0||C=1)=1

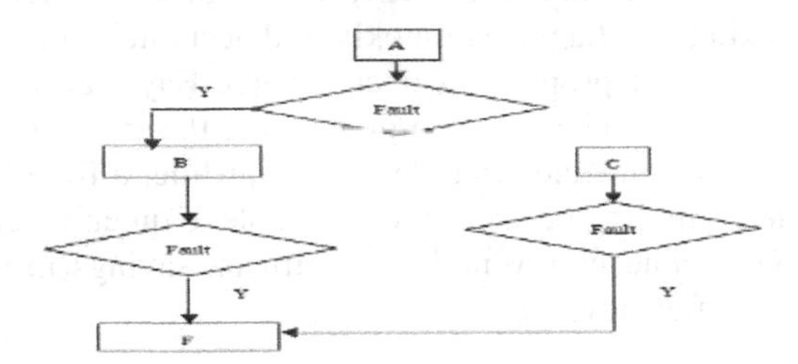

Fig. 1. Transforming relationship of decision tree structure

3.4 Example Simulation of Decision Tree-Bayesian Network Model

According to the transformation method from decision tree model to decision tree-Bayesian network model, the motor fault were constructed decision tree model and decision tree-Bayesian network model, it used model transformation algorithms to reason, and ultimately got the motor failure. By analyzing the characteristics of the motor winding insulation failure, it got a relational table of common fault feature and fault of motor. The relation table was shown in Table.1.

Table 1. Relational table of common fault feature and fault of motor

Order number	fault	The characteristic of fault
F1	Stator winding insulation breakdown	Generator insulation resistance[C1] 、DC voltage winding[C2] 、winding DC leakage current[C3]、AC voltage series resonant output voltage[C4]
F2	Stator phase short circuit	Generator insulation resistance[C1] 、DC voltage winding[C2] 、winding DC leakage current[C3]
F3	Stator winding hollow conductor plug	Stator winding temperature[C6]
F4	Stator and rotor leakage	Generator insulation resistance[C1] 、DC voltage winding[C2] 、winding DC leakage current[C3]
F5	Stator end bad welding	DC resistance[C5]
F6	Rotor winding short circuit	Rotor current[C7]

According to the relationship table, it can analyze the causal relationship between the motor winding insulation failure, combine with the characteristics of decision tree, finally arrive at decision tree model of motor winding insulation failure, the decision tree model is shown in Figure 2.

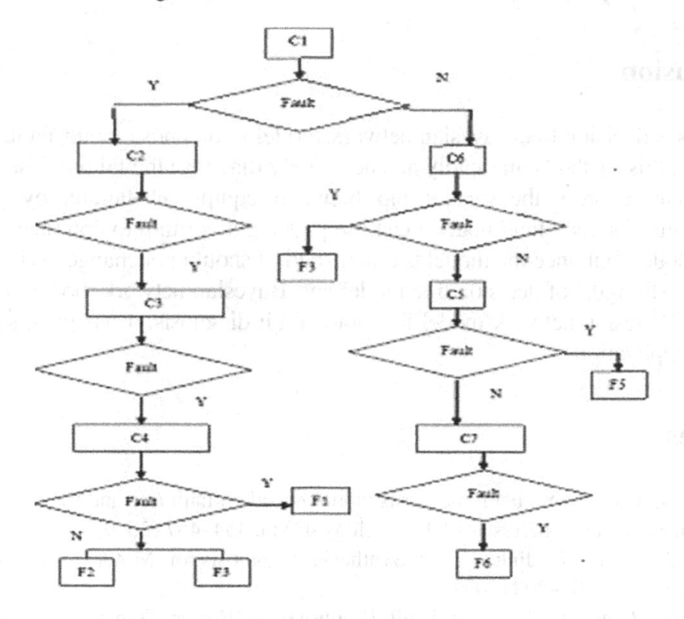

Fig. 2. Decision tree model of motor fault diagnosis

By analyzing the decision tree model of motor winding insulation failure, combined with transformation algorithm of decision tree-Bayesian network model, it can draw the decision tree-Bayesian network model of motor winding insulation failure to diagnose the motor winding insulation fault. the decision tree-Bayesian network model of motor winding insulation failure is shown in Figure 3.

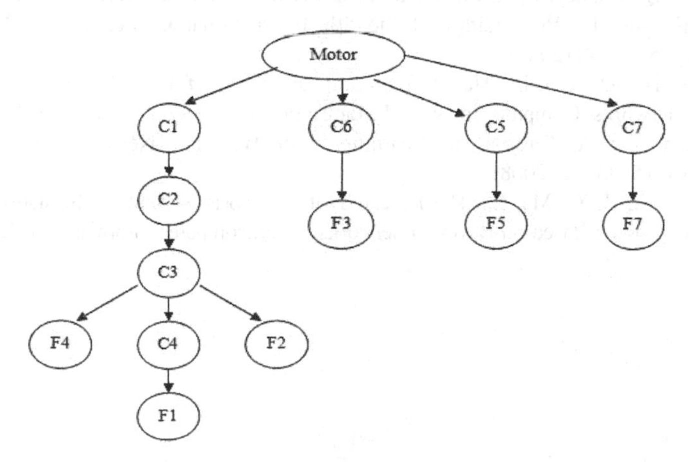

Fig. 3. Decision tree- Bayesian network model of motor fault diagnosis

By constructing a decision tree-Bayesian network model for fault reasoning, and analyzing the probability of each node, then it uses the Bayesian network inference algorithm to reason according to the obtained probability value, as a result, it determines the fault generated by motor.

4 Conclusion

This paper uses decision tree-Bayesian network model to diagnose motor fault, compared to other ways, this method can rapidly and accurately diagnose the failure. Using decision tree model can express the relationship between equipment failure, by giving the probability value for each fault node, it can use pruning algorithm to optimize to filter out some of the nodes, but once the model is established, it should not change, so by analyzing the respective strengths of decision tree model and Bayesian network model, it presents a decision tree-Bayesian network model for motor fault diagnosis, it has great significance for practical applications.

References

1. Gong, Y.-S., Gao, Y.-Y.: Fault diagnosis model based on fault tree and Bayesian networks. Journal of Shenyang University of Technology 4(31), 454–457 (2009)
2. Guo, F.-Y., Jiang, J., Jing, S.-X.: Synthesis Diagnosis of Motor Faults. Coal Mine Machinery 3(30), 206–208 (2009)
3. Gao, W.-S., Qian, Z., Yan, Z.: Fault Diagnosis of Power Transformer Using Neural Network of Decision Tree. Journal of Xi'an Jiaotong University 33(6), 11–16 (1999)
4. Chen, W.-W.: Decision Support System and Development. Journal of Shenyang University of Technology (2000)
5. Li, Y., Guo, J.-N.: Research on Pruning Algorithm of Decision Tree. Henan Science 27(3), 320–323 (2009)
6. Wang, S.-S., Sun, J.-Y., Li, L.-L.: Fault Decision Tree model in the Application of Expert System. Computer Applications 25(s), 293–294 (2005)
7. Lan, P., Ji, Q., Looney, C.G.: Information Fusion with Bayesian Networks for Monitoring Human Fatigue. In: Proceedings of the 5th International Conference on Information Fusion, pp. 535–542 (2002)
8. Wang, W.-H., Zhou, J.-L., He, Z.-Y.: Fault diagnosis of Complex System Based on Bayesian networks. Computer Integrated Manufacturing Systems 2(10), 230–234 (2004)
9. Heckerman, D.: A Tutorial on Learning with Bayesian Networks. Computational intelligence 15, 33–82 (2008)
10. Zhu, G., Zhou, Z.-X., Ma, L.: The Research of Diagnosis System on Insulation Status of Generator Winding Based on Bayesian networks. Microcomputer Information 1(22), 166–168 (2006)

Design and Realization of an Intelligence Mobile Terminal on Emergency Response System for Sudden Affairs Based on Android

Zhiwen Nan[1], Qingping Meng[2], Kehao Wang[3], and Houqin Su[1]

[1] School of Computer Science & Technology, Donghua University, Shanghai 200051 China,nan_zhi_wen_666@163.com, China,suhq@dhu.edu.cn
[2] City Comprehensive Management and Emergency Joint Action Center of Minxing District, Shanghai 201108, China
[3] Wonders Information Co., Ltd. R&D Center, Shanghai 201112, China

Abstract. Based on the analysis of the present status of current domestic and international city emergency system and smart phone, the development and application of smart phone on Android 2.2 Mobile platform have been discussed mainly in this paper, a technical solution combining with an intelligent mobile terminal emergency response system software architecture has been studied and presented, and through integration innovation, the intelligent mobile terminal which can integrate application functions such as GIS information location, voice recording, taking a picture, camera shooting, HTTP transmission and etc has been realized originally. Through GPRS or WCDMA technology, files and data can be transmitted to the emergency response command center so that an emergency response can be effectively realized in time to reduce disaster losses, and has been successfully applied in practice.

Keywords: Mobile platform, Smart Mobile Phone, GIS System, HTTP protocol, Android 2.2, GPRS.

1 Introduction

As the city expanding, the situation suffered from disasters has become more and more serious, the establishment of efficient integrated Urban Emergency System is a inevitable choice letting a city into the modern, digital management. London as early as in 1992 has set up a computer aided emergency command system [1], the Federal Emergency Management Agency (FEMA) has also established a nationwide state emergency rescue system [2], since China paid attention to the establishment of emergency mechanism in 2003, people gradually began to realize the sudden disaster can often bring us the huge loss, thus causing us to realize the importance and urgency of establishing emergency mechanism. No matter what the emergency, before and after the event happening, the information can be sent to the command center timely and effectively, which has an important role in decision-making, therefore, the emergency system based on intelligent phone terminals can play its special role.

Android is one of the most popular smart phone operating system, Android smart phone now has become personal information management, entertainment and enterprise

D. Jin and S. Lin (Eds.): Advances in ECWAC, Vol. 1, AISC 148, pp. 171–177.
springerlink.com © Springer-Verlag Berlin Heidelberg 2012

application expanded platform. At present the city emergency system is mainly designed in PC applications in domestic, while rarely research in the smart phone application, based on this, the system's terminal module was described and designed in the paper based on Android 2.2 platform mainly from the point of the emergency mobile terminal system.

2 The System Overall Software Architecture

Seen from the emergency system's intelligent mobile terminal application platform reference architecture [3] shown in Fig. 1, the whole system is divided into three parts: intelligent mobile terminals, transmission network and Web server. Mobile terminal module complete data acquisition and information feedback, such as from the server to obtain the latest instructions, tasks, events, and mobile terminals can send command, task, event feedback information to the emergency Web server; transmission network by GPRS, CDMA, 3G wireless network is mainly responsible for data transmission, such as the mobile terminal instructions, tasks and events feedback information including live audio, camera, video accessories, can be transfer to the emergency Web server timely and successfully; Emergency Web Server completes data reception and processing, and can return the latest relevant information such as the latest instructions, tasks, events, etc.. to the smart phone terminal, to realize the two-way information transmission. GIS server provide location database, through the smart phone sending GPS coordinates to the server at a regular time or manually , keep abreast of the mobile terminal's location and updates.

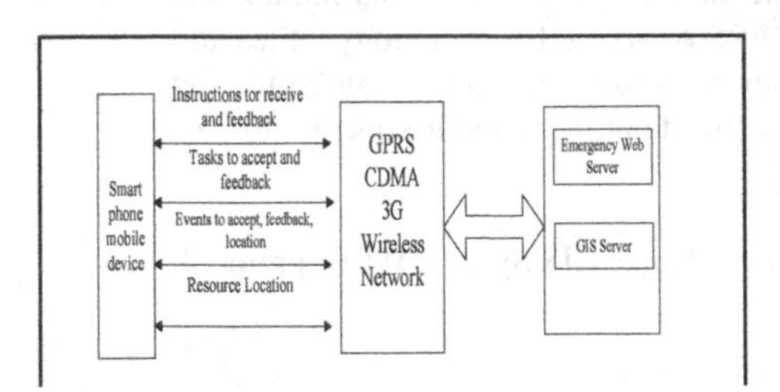

Fig. 1. Reference Architecture Design

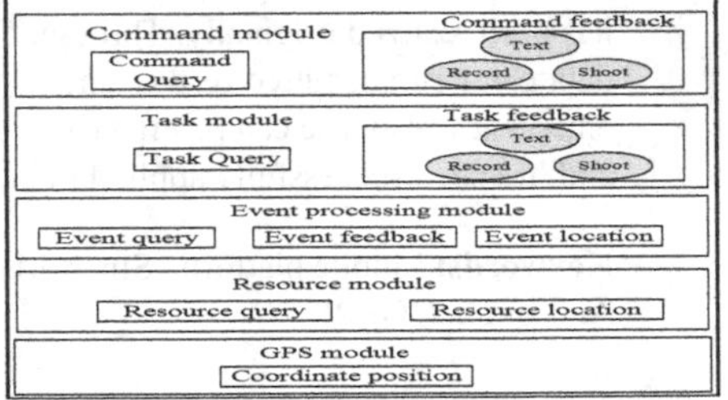

Fig. 2. System function reference Model View

3 System Function Reference Model

According to requirements and technical index, from the perspective of software system design, the system function reference model [3] based on Android 2.2 platform shown in Fig. 2 is divided into five modules.

Based on the reference model views, mobile phone terminal can realize the following application objectives:

a) Command processing: get the latest directive, check the instructions details, and feedback the instructions in accordance with the disaster site, including live recordings, photographs, video attachments and text messages sent back to the Emergency Command Center server, so that the emergency commanders keep track of emergency scene dynamic.

b) Task processing: get the latest task, view task details, has similar features as commands.

c) Event handling: get the latest emergency current events, event location, that is, by passing a parameter (such as event ID) to the GIS system, open GIS system's relevant page, enter into GIS system's event handling process.

d) Resources to deal with: query latest resources in the emergency system, resource location, that is, enter into GIS system's resources display screen, including the resources' dynamic display, position display, and display the information near the resource.

e) GPS: access to GPS information, the geographic longitude and latitude data, display them on the phone, the present location can be known by displaying GPS coordinate information.

4 Key Technology Implementation

4.1 XML Analytical Model Based on XmlPullParser

Mobile terminal and server communicates in XML data formats, mobile phone terminal get XML data returned by the server by accessing server interfaces ,then through the XML analytical technology system, and thus to generate new data. on the Android platform, three kinds of XML Pull analytical techniques: SAX(Simple API for XML), DOM(Document Object Model), can be adopted to parse XML.

SAX takes a event driven streaming analytical technology, analysis from the beginning of the file to the end of the document, not suspend or backward; DOM need to load all the contents of the XML file into memory, which consume large memory; while Pull parser works by setting a marker to stop parsing in advance when meeting the conditions[4], thereby greatly reducing the analysis time and saving system's memory, which is particularly important and optimized for the mobile devices that their Android system run slowly and has lower memory, so Pull parser can provide some performance advantages and ease of use. Based on this, the system takes the XmlPullParser in XML Pull technology to parse XML. The server and client communication events XML format defined as shown in Figure 3, the specific parsing algorithm flow is shown in Figure 4.


```
<?xml version="1.0" encoding="UTF-8" ?>
- <events>
  - <event id="20">
      <title>Emergence event</title>
      <eventDesc>Event description</eventDesc>
    </event>
  - <event id="21">
      <title>Emergence event</title>
      <eventDesc>Event description</eventDesc>
    </event>
  </events>
```


Fig. 3. Defined events XML format examples

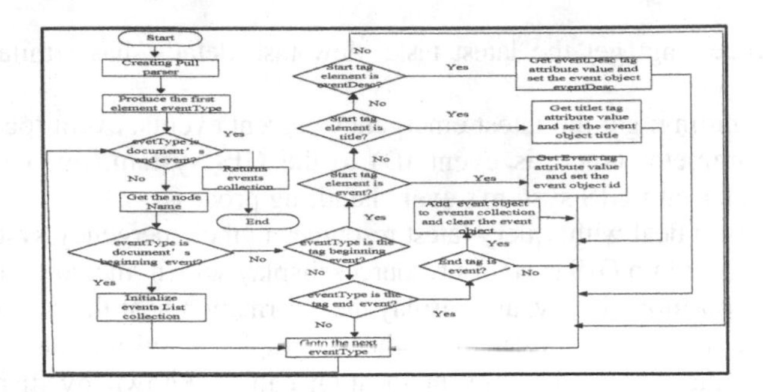

Fig. 4. XmlPullParser resolution algorithm chart

4.2 GPS Functionality to Achieve

For smart phone, generally with GPS module, through opening GPS,the GPS information including latitude, longitude and altitude, direction, speed and related data can be acquired, the measured data is displayed at different time through the program design, so that we can clearly know oneself coordinates information. The program need to call android.location.LocationManager Class in Android SDK2.2 to get GPS services, LocationManager object provide variety of ways to get the latitude and longitude coordinates and other data. the ways to get the positioning data are as follows:

For longitude and latitude, getLatitude () returns latitude data, getLongitude() returns longitude data; hasAltitude() determine whether there is elevation data, and getAltitude() return elevation data by metres.

When GPS information database connection open and update, get real time GPS information through the open operation, display these information on the screen according to the need and send them timely and effectively to the emergency command center.

4.3 Recording, Shooting to Achieve

When the GPS is running, keep GPS running in the background, the updating data is shown according to specified time intervals, open recording, photography program when dealing with the emergency scene, can save audio, image, video information, the saved results information require to associated with the GPS information, that is, file name should be included GPS longitude and latitude information, so that later uploaded to the server, the emergency command center can clearly know emergency persons' the site location.

The android.provider.MediaStore Class's corresponding multimedia control behavior can be triggered to finish recording, shooting, then call the relevant file storage function to complete the function that save data to SDCARD in Android phone ,the key implementation code of the specific capture (recording is similar) are as follows:

```
//create directory on SD card
File dir = FileUtils. CreateSDDir (path);
//create documents saved to SD card in the specified directory
```

```
File File = FileUtils. CreateSDFile (path + fileName);
Uri uri = Uri. FromFile (file);
//go into the camera view, and save pictures
Intent imageCaptureIntent=new Intent(MediaStore.
ACTION_IMAGE_CAPTURE)
ImageCaptureIntent.PutExtra (MediaStore.EXTRA_OUTPUT, uri);
StartActivityForResult(imageCaptureIntent,ClientSettings.RESULT_CAPTURE_I
MAGE);
```

4.4 Http Connection and Transfer Mode

HTTP transmission is widely applied in data transmission because of high efficiency, in Android, file upload can make use of HttpPost and HttpGet class to encapsulate post and get request, then send post or get request by execute method and return back data responsed by the server. Whether HttpPost and HttpGet way, the following three steps must be passed through to visit HTTP resources [5]:

(1) Create HttpGet or HttpPost object, pass URL object into HttpGet or HttpPost object by the construction method.

(2) Send GET or POST request by the DefaultHttpClient Class's execute method, and returns HttpResponse object.

(3)Returns the response information through the HttpResponse interface's getEntity method, and with a corresponding processing, If take HTTP POST method to submit request, the HttpPost class's setEntity method need to be taken to set the request parameters.

We need to send the taken pictures, audio or video information to the server through the GPRS, CDMA, 3G wireless network in the emergency environment , which will be a deal with special module to complete the data transmission, thereby generate the HTTP server connection and sending module in this situation, in this module adopting HTTP protocol, is technically a bright spot, at the same time, Taking GPRS or CDMA, etc, multithreading technology such as Handler and Thread can be adopted in time to handle asynchronous message, which can achieve ideal transmission effect, its specific file transfer mode as shown in Figure. 5.

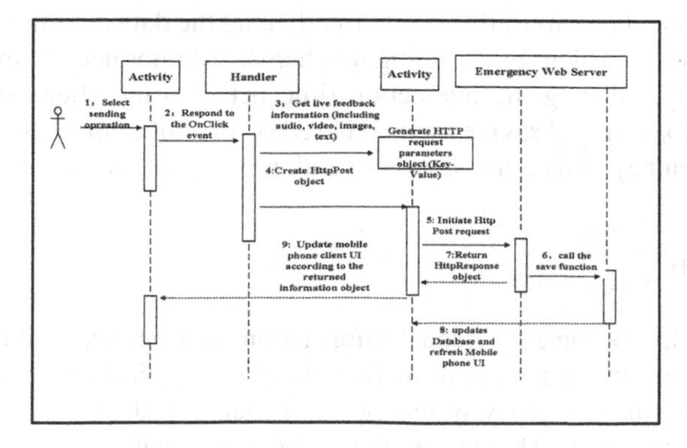

Fig. 5. The file transfer mode based on HTTP

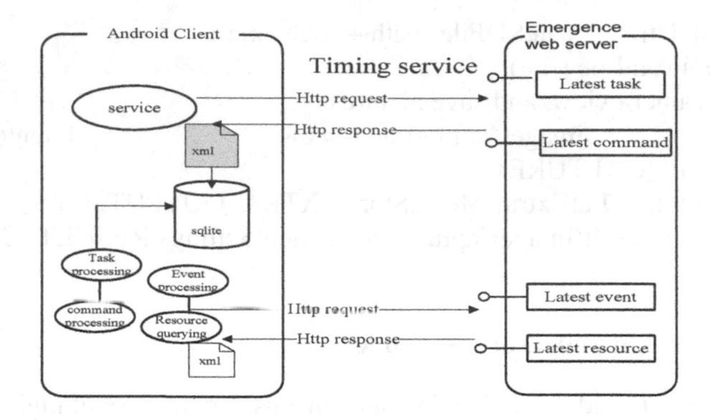

Fig. 6. Information storage and access model

4.5 The Method of Access Server-Side Data by Sqlite

Based on the actual needs of the emergency services, two different client-side to access server data modes have been presented in the paper: to read and store server-side data based on Sqlite mechanism, and directly to request data from the server. Taking into account emergency events and resource have dynamics and real-time traits in the practical application, so Android client directly access the server's appropriate interface to get the XML data source and stored as an XML file on the client, thereby when the client's other business logic needs access to information, only read from the XML file directly, without re-request to the server, as a result of improving data transmission speed.

Considering the instructions to scene commander sent by the command center are relative stability in a certain period , in order to improve data transmission efficiency between the server and the client as much as possible, the client take indirect access to Sqlite database (a lightweight database that memory is small and the processing speed is quick), the realization principle is shown in Figure.6, the specific step are as follows: when start Android client, a service in the background every once in a while (not fixed, independent setting) timing request to the server, then get the XML data sources, parse them and save them into the Sqlite database. When the scene disposal persons request latest instructions and missions, the client directly read the data in Sqlite without access to the web server, equivalent to a caching mechanism, do not need to every request to the server, greatly reducing the interaction time between the client and server. The server-side data loading and saving method based on the Sqlite mechanism, can greatly improve the efficiency of data transmission, and greatly improve the user experience.

5 Conclusion

Android devices has become a personal information management platform, a portable entertainment platform, and a platform for enterprise applications in the future, the mobile terminal urban emergency system prototype integrated with GIS, GPS, GPRS or 3G is a new concept, new and high technology emergency platform technology solution, solve the problem of traditional emergency moving difficultly, improve the speed of dispatching persons and supplies, win time for rescue work. The system provides various kinds of visual real-time information, and assists rescue persons in business operations, thus can take the scientific rescue. With smart phone, mobile communications, network technology developed, in the near future, Android mobile devices have very broad application prospects in the urban emergency system as well as some related departments, such as civil defense, public security, municipal, health and epidemic prevention and so on.

References

1. Hougham, M.: London Ambulance Service Computer aided Dispatch System. International Journal of Project Management 14(2), 103–110 (1996)
2. Wang, L.: Kehua. U.S. national emergency system. The Global Technology Economy (9), 49–51 (2003)
3. Len, B., Che, L.(trans.): Software architecture practice. Qinghua University Press (February 2004)
4. Different xml parser ways, http://blog.csdn.net/ichliebephone/archive/2010/10/31/5977959.aspx
5. Li, N.: Everyone play happy network. Electronic Industry Press (July 2010)

Scale Development of Customer Psychological Contract

Yan Ma[1,3], Junye Deng[2], Jinjin Hao[1], and Yong Wu[4]

[1] School of Business Administration Northeastern University, Shenyang, 110004, China
[2] Qingdao Binhai University, Qingdao 266555, China
[3] School of Economics & Trade Qingdao Technological University, Qingdao 266520, China
[4] Linyi Water Conservancy Prospecting & Designing Institute, Linyi, 276000, China
mayanmnk@163.com

Abstract. This article reviews the relative literatures about the customer psychological contract to develop psychological contract scale. Customer psychological contract has become an important and difficult problem of service enterprises, and it can reflect customers' value and the deepest level of customers' needs. Based on the existing related researches, we expatiate the scale of customer psychological contract. And this scale is developed in view of the retailing as the research object. This article uses the widely accepted tool for the development of measurement methods, and develops a reliable and effective scale of customer psychological contract.

Keywords: Customer, Psychological Contract, Scale.

1 Introduction

Psychological contract is a series of reciprocal obligations assumed by two sides. The study of psychological contract in the field of organizational behavior has become mature, and most empirical studies show that high levels of psychological contract can improve the loyalty of the employees to their organization, and extend time of staff retention. And studies have found that this kind of psychological mutual expectation exists not only in between employees and organization, there is also a similar mutual expectation between customers and merchants (Roehling, 1997). In the field of service research, although some literatures discuss the concept, dimensions of customers psychological contract, and the relation with customer satisfaction, loyalty, (Luo Haicheng, Yang Lin, Yu Shibing, etc.), understanding of their formation mechanism still needs further. This paper mainly discusses the customers' belief perception about the content of mutual responsibility and obligations. Then exploratory factor analyses were used to explore the dimensions of the responsibilities of enterprises and customers respectively. Finally, confirmatory factor analysis and validity test were made for the customer psychological contract scale.

2 Customer Psychological Contract

In order to develop a more effective measurable scale of customer psychological contract, we studied the measurement of some similar staff's psychological contract.

D. Jin and S. Lin (Eds.): Advances in ECWAC, Vol. 1, AISC 148, pp. 179–183.
springerlink.com © Springer-Verlag Berlin Heidelberg 2012

At present, the measurement on the psychological contract is mainly carried out from three aspects. Understanding various measurement methods of the psychological contract can help us how to measure and in which way we should start to measure. The three measuring angles are: the first angle is oriented from the content of psychological contract, which measures the reciprocal responsibility of individual and the other party; the second angle is to compare the attitudes towards dimensions from the characteristics of psychological contract, such as the subtle, and not documented attitude; the third angle measures the fulfillment, change and contrary of the psychological contract from the overall evaluation.

So far, there hasn't been a mature and well-established customer psychological contract scale. Therefore, this paper will have an exploratory study on the content and evaluation of psychological contract and develop a customer psychological contract scale which is more suitable for the relationship between enterprises and customers. Firstly, customers' question items about the responsibilities of enterprises and customers were collected, which were classified by the method of induction. Then the collected subjects were assessed by experts, getting a real psychological contract scale about the relationship between customers and enterprises.

3 Scale Development

The critical incident technique was used in this study, according to the results of focus interviews and an original scale of customer psychological contract was made after the discussions with the scholars of specialty in this field.

According to the classified results of interviews and critical incident, researchers dealt with the 656 collected cases wholly and arranged the descriptions of the repeated and same meaning; finally they summed up 18 enterprise responsibilities, 15 customer responsibilities, and 33 measurable projects in total. We simplified the initial designed measurable projects in order to assess the reliability and validity of customer psychological contract scale. First these formed items were carried on customer interviews to test the applicability of the measurable projects, that is, to evaluate the suitability of the scale content. Study found that some projects do not accord with the customers' shopping experiences. Questionnaires were conducted among 30 students, asking them to judge whether these items accord with the concept of customer psychological contract and those gotten the low score were deleted to ensure that the content of our scale have a higher content validity. Research group arranged these 33 items gotten through the depth interviews and coding analysis of open questionnaire results and some relevant literatures, trying to make the language simple and easy to understand. They also discussed the unclear and ambiguous items, verified the coded materials further and found out and modified the inappropriate items.

4 Analysis

The whole scale was carried out the reliability analysis through the statistical software, SPSS16.0, and overall reliability index Cronbach's α of the scale was 0.866, which shows basically there are no sentences problems in questionnaires. 33 measuring projects of customer psychological contract scale were done the exploratory factor analysis by the software, SPSS16.0 to test the basic structure of 33 items in responsibilities of enterprise and customer. The numerical value of KMO is 0.861 and 0.805 respectively, passing Bartlett's spherical inspection at the same time, which shows that the data has a good condition of factor analysis. The data was done the exploratory factor analysis by the method of maximum orthogonal rotation through the principal component analysis, and the results show that the enterprise responsibility in the concept of customer psychological contract includes four factors whose eigenvalue is higher than 1, and the four factors all explain the variance of 62.74%. Among the items, EO1, EO2, EO6 and EO10 were deleted, fourteen items remained. The customer responsibility includes 4 factors whose eigenvalue is higher than 1 and the 4 factors all explain the variance of 62.74%, and fifteen items remained. The concrete results are shown in table 1.

Table 1. Result of Exploratory Factor Analysis

Items	Factor Load	Cronbach's α
Enterprise Responsibility		
Prices of commodities are authentic and reliable.	0.781	0.722
A market releases real information of promotion.	0.703	
A market doesn't publish false propagandas.	0.634	
Service staff serves with smile.	0.776	0.796
When customers select commodities, service staff can guide them with patient.	0.786	
When customers pay for the commodities, cashier staff serves with courtesy.	0.741	
Market's environment is clean and comfortable.	0.614	
Air condition flows well and temperature condition is pleasant in a market.	0.614	
A market treats every customer fairly.	0.712	0.741
A market organizes customers' checkout line.	0.710	
Service staff doesn't judge people solely by their appearance.	0.814	
A market can handle complaints with patience.	0.720	0.777
A market provides timely and effective service of maintenance.	0.676	
A market should meet the customers' requirements of return and exchange of goods as possible as they can.	0.742	
Customer Responsibility		
I'm polite to the service staff.	0.618	0.839
I take the labors of service staff seriously.	0.811	

Table 1. (*continued*)

I believe the service staff and I are equal.	0.828	
I haven't litter in markets	0.655	
I uphold consciously the service order.	0.519	
I use the facilities in markets reasonably.	0.771	
Concession and discounts make me patronize here again.	0.727	0.686
Serving quickly and shopping conveniently make me patronize here again.	0.615	
Excellent quality, reasonable price and good service are one of the reasons that I patronize here.	0.804	
I can suggest the	0.857	0.727
When finding problems, I'll put forward my proposals to the market.	0.873	
I can cooperate with the market to do investigative activities.	0.561	
I often tell my friends about the market.	0.765	0.772
I recommend the market to my friends.	0.791	
I'd like to talk about the market with my friend.	0.826	

5 Discussion

The development of this scale discusses mainly the responsibilities of service enterprises and customers, using the approaches of exploratory factor analysis, confirmatory factor analysis and multiple data analysis. In the process of confirmatory factor analysis, the reliability coefficients of the factors are all above 0.7 and the overall reliability of the scale is higher than 0.80. The above results analyses show that the scale has good internal consistent reliability and the reliability of the scale thus be tested. The result of data analysis indicated that customer psychological contract scale has good reliability and validity.

Customer responsibility for enterprise mainly reflects the aspect of customers' civil responsibility. Under the current consumption situation, high level of consumption is regarded as a major thrust of society. Customers pay more attention to consumer culture and focus on the level of consumer spending impact on the natural environment and on their personal development. More and more customers' demand has been beyond their own interests, considering the widespread influences of social consumption, which are the customer citizenship behaviors many scholars are studying now.

In this article, retail industries were chosen for research, through the discussion of the responsibilities of enterprises and customers, the four dimensions of responsibilities of both enterprises and customers were put forward further. To help enterprises better understand customer's psychological needs and provide practical guidance. In the future study, the general applicability of our psychological contract scale can be tested more extensively and thorough empirically.

References

1. Oliver, R.L.: Cognitive, Affective, and Attribute Bases of the Satisfaction Response. Journal of Consumer Research 20(3), 418–430 (1993)
2. Rousseau, D.M.: Psychological and Implied Contracts in Organizations. Employees Rights and Responsibilities Journal 2(2), 121–139 (1989)
3. Rousseau, D.M.: New Hire Perception of Their Own and Their Employers Obligations: A Study of Psychological Contracts. Journal of Organizational Behavior 11(5), 389–400 (1990)
4. Rousseau, D.M.: Perceived Legitimacy & Unilateral Contract Change: It Takes a Good Reason to Change a Psychological Contract. In: Symposium at the SIOP Meetings, SanDiago (1996)
5. Roehling, M.V.: The Origins and Early Development of the Psychological Contract Construct. Journal of Management History 3(2), 204–217 (1997)
6. Herriot, P., Manning, W.E.G., Kidd, J.M.: The Content of the Psychological Contract. British Journal of Management 8(2), 151–162 (1997)
7. Robinson, S.L., Kraatz, M.S., Rousseau, D.M.: Changing Obligations and the Psychological Contract: A Longitudinal Study. Academy of Management Journal 37, 137–152 (1994)
8. Churchill, G.A.: A Paradigm for Developing Better Measures of Marketing Construct. Journal of Marketing Research 16(1), 64–73 (1979)
9. Bowen, D.E.: Interdiseiplinary study of serviee: Some Progress, some Prospects. Journal of Business Research 20(l), 71–79 (1990)

The Categorization and Consequences of Customer Misbehaviors

Yan Ma[1,3], Junye Deng[2], and Guangwei Fan[3]

[1] School of Economics & Trade Qingdao Technological University, Qingdao 266520, China
[2] Qingdao Binhai University, Qingdao 266555, China
[3] School of Business Administration Northeastern University, Shenyang, 110004, China
mayanmnk@163.com

Abstract. Customer misbehavior has become a more and more important topic both in reality and theory, with few researches especially in China. This paper classifies general customer misbehavior at service encounter in restaurants by critical incident technique. The results show that customer misbehavior's object can be the enterprises, customer-contact employees and customers (other customers and even customers themselves). In addition, this paper discusses the consequences of customer misbehaviors from its direct and indirect effects, and enterprises should pay more attention to the latter one.

Keywords: customer misbehavior, service encounter, critical incident technique.

1 Introduction

Competition in hospitality services has become more serious. In order to improve core competition and benefit ability, enterprises turn to the ideology of "customer-centered" and "consumer is god". However, customers don't behave just as enterprises expect. In fact, customer misbehavior is prevalent in the consumption of services, such as drunken trouble-makers, shouting, disrespecting to the service employees, and so on. Customer misbehavior descends service delivery efficiency and social civilization, results in direct or indirect economic losses of service enterprises, and brings pressure or bad feelings to customer-contact employees and other consumers in the same service encounter. Therefore, studying and analyzing customer misbehavior is an important and difficult problem of service enterprises, which also attracts the attention of researchers.

In exploring and describing failed service encounters, theorists have employed a variety of terms and phrases including "deviant consumer behavior", "aberrant consumer behavior", "problem customers", "inappropriate behavior", "consumer misbehavior", and "jaycustomers" [1]. In 1994, Christopher Lovelock coined the term jaycustomers to refer to dysfunctional customers who deliberately or unintentionally disrupt service in a manner that negatively affects the organization or other customers [2]. Subsequently, Bitner, Booms, and Mohr found empirical support for Lovelock's

D. Jin and S. Lin (Eds.): Advances in ECWAC, Vol. 1, AISC 148, pp. 185–189.
springerlink.com © Springer-Verlag Berlin Heidelberg 2012

jaycustomers [3]. Because of the close customer contact and economic importance [4], quantities of studies are mostly based on food and beverage industries.

Although there is a quantity of outstanding researches in customer misbehavior, there are few researches in China which has huge consumer market. As Chinese customers behave differently from foreign customers, it is still important to study Chinese customer misbehavior. Therefore, this paper uses depth interviews and critical incident technique to classify general customer misbehavior, as a basis for the further consequences discussion.

2 Literature Review

2.1 Forms of Customer Misbehavior

Possibly the best-known categorization is from the anecdotal work of Lovelock [5], who identifies six service-based jaycustomers. Contrasting typologies are also offered by Fullerton and Punj [6], Harris and Reynolds [7], and most recently. However, although these classifications offer notable insights into the diverse varieties of customer misbehavior, such studies lack empirical support. Furthermore, the focal concentration of empirical research regarding the listing or categorization of jaycustomer behaviors has been on individual forms. While the need for a development of a typology of general or all-embracing jaycustomer behaviors has been recognized and forwarded by a small number of researchers [8], such efforts have been anecdotal or conceptual in nature [9], or have emerged as part of wider research [3]. This has been to the detriment of empirical research that concentrates on the formulization of a typology of jaycustomer behaviors that incorporates insights from both service personnel and customers.

2.2 Consequences of Customer Misbehaviors

Consequences of customer misbehavior range from the mundane financial costs [10], to spoilt consumption experiences for other customers [5], to extreme cases of service personnel homicide [11]. Speciafically, the effects of customer misbehaviors are wide-ranging affecting employees, firms, and fellow customers [1]. Yet, despite the grave implications of customer deviance, research in this area is in its infancy and tends to be exploratory [12].

3 Research Design and Methodology

Critical incident technique is appropriate during exploratory research where concepts or phenomena are not entirely clear or understood [13], and the data collected by critical incident technique meets the criteria established by Ericsson and Simon in providing meaningful and reliable data with respect to cognitive processes [14]. In addition, due to the open-ended nature of questions, critical incident technique arguably generates unequivocal dialogue and thus, rich and real data, as respondents are given the opportunity to provide a thorough and in-depth depiction of their own experiences (Stauss and Weinlich, 1997).

Therefore, we conducted depth interviews among 10 undergraduate and 8 graduate students to collect critical incidents of customer misbehaviors in restaurants. According to the interview's feedback, interview outlines were changed to clarify the description and then designed the open questionnaire. Questionnaires were distributed to 248 interviewees and 194 questionnaires returned. After excluding employees' misconduct and other invalid questionnaires, 143 valid questionnaires were received.

3.1 Categorization of Customer Misbehavior

Most customer misbehavior has not the only object, so this paper classifies customer misbehavior by the key object (see Table 1).

Table 1. forms and frequency of customer misbehaviors

Object		Customer misbehavior	Frequency		Percentage
enterprises	items	get out of bills	7	24	29.2%
		take away cutlery	4		16.7%
		take away other items	2		8.3%
	aids	graffiti on the wall	5		20.8%
		damage plastic tablecloth	2		8.3%
		liter	1		4.2%
	corporate image	blackmail	3		12.5%
customer-c ontact employees		not with the staff	15	44	34.1%
		not respect service staff	12		27.3%
		drunkenness	9		20.5%
		verbal abuse	6		13.6%
		molesting young waitress	1		2.3%
		assault service staff	1		2.3%
customers	other customers	jump the queue	24	75	32.0%
		crowd	10		13.3%
		shout	12		16.0%
		quarrel	9		12.0%
		smoke	7		9.3%
		steal	3		4.0%
	customers themselves	waste	6		8.0%
		hard to eat	4		5.3%

Table 1 shows that the number of customer misbehavior based on customers is largest, the second one is on customer-contact employees and the last one is on enterprises. Furthermore, customer misbehavior based on customers concludes two parts: on other customers and on customers themselves, and the frequency of the former one is bigger than any other, a total of 65 incidents.

3.2 Consequences

Customer misbehavior has direct and indirect effects on enterprises. The direct effect means that, customer misbehavior, such as theft and damage to items and aids, causes economic loss and financial cost. The indirect effect means that, on the one hand, customer misbehavior can put pressure on employees and force them to quit, resulting in more cost of recruitment and training; on the other hand, customer misbehavior smashes other customers' image of enterprises and may reduce frequency of interaction and spend less money in the enterprises, even make passive word of mouth. Enterprises maybe can endure direct effect of customer misbehavior, however, indirect effect causes much more observed and latent cost and more difficult to control.

In the catering business, customer misbehavior, such as jumping the queue, shouting, smoking and quarrelling, not only takes a bad impression on others, but also affects other customers' dining mood. If customer encountered misbehavior of other customer during consumption, regardless of the perpetrators of misbehavior or the witness, he would be unsatisfactory. Therefore, he will vent his dissatisfaction through contempt, intimidation or assault of the service employees on the psychological and even physical abuse; while enterprises want to keep all of the customers and obey the service philosophy of "customer is God". If the service employees can't stand the two aspects of pressure, they may have to leave for other jobs. At the same time, enterprises have to face the problem to recruit and train new employees. Furthermore, customer misbehavior will lower the morale of the team, resulting in decreased efficiency of service delivery.

4 Conclusion

This paper interviewed customers to collect critical incidents of customer misbehavior in restaurants. The results of content analysis show that customer misbehavior consists of three parts: as the object of the enterprises, customer-contact employees and customers. Customer misbehavior based on customers concludes the object of other customers and customers themselves, and the frequency of the former one is bigger than any other one.

Customer misbehavior brings direct effects and indirect effects to enterprises, from directly economic cost to indirectly employees and customers' turnover. In that case, enterprises should pay more attention to the latter one.

Despite this paper classified customer misbehavior in restaurants and analyzed its consequence qualitatively, the classification and effects in other industries may be different. It's necessary to discuss customer misbehavior in more industries and find the similarity and difference, which can support deeper implications to enterprises.

References

1. Harris Lloyd, C., Reynolds, K.L.: The Consequences of Dysfunctional Customer Behavior. Journal of Service Research 6(2), 144–161 (2003)
2. Lovelock Christopher, H.: Product Plus: How Product and Service Equals Competitive Advantage. McGraw-Hill, New York (1994)
3. Bitner, Booms, B.H., Mohr, L.: Critical Service Encounters: The Employee's Viewpoint. Journal of Marketing 58, 95–106 (1994)
4. Hartline Michael, D., Ferrell, O.C.: The Management of Customer-Contact Service Employees: An Empirical Investigation. Journal of Marketing 60, 52–70 (1996)
5. Lovelock Christopher, H.: Services Marketing: People, Technology, Strategy, 4th edn. (2001)
6. Fullerton Ronald, A., Punj, G.: Repercussions of Promoting an Ideology of Consumption: Consumer Misbehavior. Journal of Business Research 57(11), 1239–1244 (2004)
7. Harris, L.C., Reynolds, K.L.: Jaycustomer Behavior: An Exploration of Types and Motives in the Hospitality Industry. Journal of Services Marketing 18(5), 339–357 (2004)
8. Moschis, G.P., Cox, D.: Deviant Consumer Behavior. Advances in Consumer Research 16, 732–737 (1989)
9. Fullerton, R.A., Punj, G.: Choosing to Misbehave: A Structural Model of Aberrant Consumer Behavior. Advances in Consumer Research 20, 570–574 (1993)
10. Krasnovsky, T., Lane, R.C.: Shoplifting: a review of the literature. Aggression and Violent Behavior 3(3), 219–235 (1998)
11. Huefner, J.C., Hunt, K.H.: Consumer retaliation as a response to dissatisfaction. Journal of Consumer Satisfaction, Dissatisfaction and Complaining Behavior 13, 61–82 (2000)
12. Reynolds Kate, L., Harris Lloyd, C.: Dysfunctional Customer Behavior Severity: An Empirical Examination. Journal of Retailing (2009)
13. Walker, S.J., Truly, E.: The critical incidents technique: philosophical foundations and methodological implications. Paper presented at the AMA Winter Educator's Conference, San Antonio, TX (1992)

Evolution Supporting Class-Cluster Data Model for PLM

Huawei Zhong, Guangrong Yan, and Yi Lei

School of Mechanical Engineering &Automation, Beihang University, Beijing, China
timothyzhong@gmail.com

Abstract. To meet the requirement of Product Lifecycle Management (PLM) system phased implementation and frequently modification, a PLM Class-Cluster data model based on Object-Relation data model was proposed to support system evolution. A set of meta models were defined as semantic criterion and a four layers Class-Cluster model consist of meta models, class models, object models and cluster models was build . Class models were used to describe business data and relations of enterprise. Cluster models were used to reorganize contents and structures of data objects, support different requirements of data schema at different stages of product lifecycle. Model-driven method was employed to refract modification of data models quickly to system. The data models were proved to meet the system evolution requirements through a practical application.

Keywords: Product Lifecycle Management, PLM, evolution, data model, cluster.

1 Introduction

Product Lifecycle Management (PLM) has been employed by more and more enterprises to ensure the information of product can be managed and used effectively from its conceptualization to its destruction or recovery, then to improve core competitiveness of the enterprises. PLM systems are gaining acceptance for managing all information of products throughout the full lifecycle [1]. The functions and contents of PLM are various for different enterprises, so it is difficult to build a uniform set of software to suit all enterprises [2]. However PLM system is usually phased implementation starting from the design department to the whole company, and it is frequently modified during the implementing and running time. So it is important that the PLM system can evolve effectively to adapt for the modification.

The evolution of PLM system can be divided into changing of business process level and data information level. Workflow model can well meet the frequently changing and reuse requirements of business processes level [3]. At data information level PLM system can be abstracted as a set of interrelated information and how to safely create, modify and access the information. Traditional data models such as Entity-Relationship (ER) models and Object-Oriented (OO) models are static, so they cannot adapt for different environments effectively.

D. Jin and S. Lin (Eds.): Advances in ECWAC, Vol. 1, AISC 148, pp. 191–196.
springerlink.com © Springer-Verlag Berlin Heidelberg 2012

A data model named Class-Cluster Data Model (CCDM) is proposed in this paper based on the Conceptual Clustering Mechanism (CCM) and object-relation model to support the evolution of PLM system. First meta-model is defined as the unified standard to establish CCDM based on OO meta-model, then a hierarchical structure CCDM is built to break the evolution of PLM as atomic operations on four levels of CCDM. PLM system based on CCDM is proved have adaptability that reduces the workload of system changes significantly, shortens the implementation cycle, and reduces maintenance costs.

2 Evolution of Data Models

Data model is abstracted from all information in PLM. Evolution of the data model has four dimensions: data scope, data structure, data granularity and product lifecycle.

Evolution of data scope means new data type added when the implement of PLM system along with more functions added. The iteration upgrades of the process cause frequent changing of pre-defined data model or adding, updating or deleting data attributes. For example when project management installs to the system, models of project, user and department are added to the data model.

Evolution of data structures are changes of data association including horizontal and vertical changes. The horizontal refer to the associations change to one to one, one to many or many to many. The vertical are semantics changes to the relationship of sibling, father-son and so on. For example in product structure management the original product-part hierarchy added module as child of product, then the father of part changes from product to module.

Evolution of data granularity refers to the gradual improvement when content of PLM becomes more detail and some unstructured data needs to change into structured data to manage more accurately. The data model is not predefined before PLM implement, and is gradually refined when the system implemented in-depth. For example information of the product's features and structure is saved as a Word document in a vault, and late in the project it need to be converted to structured data stored in the database to facilitate information analysis and statics.

Evolution of the product lifecycle means the individual product is configured based on the general structure of the product family and its content and structure are different at different stages like require, design, engineer , manufacture, assemble, sale, maintenance in the lifecycle[2]. Participants can browse the various stages within the different views of the data content and structure.

3 Cluster Class Data Models

General domestic and international PLM software used object-oriented data model [4]. The objects with the same set of attributes and behaviors can be abstracted as a class. The instances of one class are homogeneous and their structure cannot be changed after being initialized. However in PLM system a data object during the life cycle has corresponding changes in the properties and functions, which the traditional object-oriented data model cannot achieve. CMM allow homogeneous or heterogeneous objects dynamic gathered into cluster. The same part of the contents of the data objects is defined as clusters. A cluster is a dynamic combination of semantic-related objects whose properties, methods, roles and actor are dynamically added or removed [5]. A cluster is a set composed of a collection of subset objects, which can be described as: $C = \bigcup P_i (1 \le i \le n)$. Subsets of a cluster can be divided into different clusters with roles. A cluster has properties and methods just like a class does. The objects can express their characteristics by included by different clusters with different roles, which means that cluster adapt flexibility to the dynamic structure under uncertain environments. Clusters are introduced to the PLM system to support the implementation and maintenance process to dynamic evolution.

3.1 Meta-model of CCDM

A sematic language has been built to form CCDM which is defined by a meta-model data to provide the universal semantic description and constraint expression. We extend the meta-model based on the object-oriented model and the core structure is shown in Fig. 1.

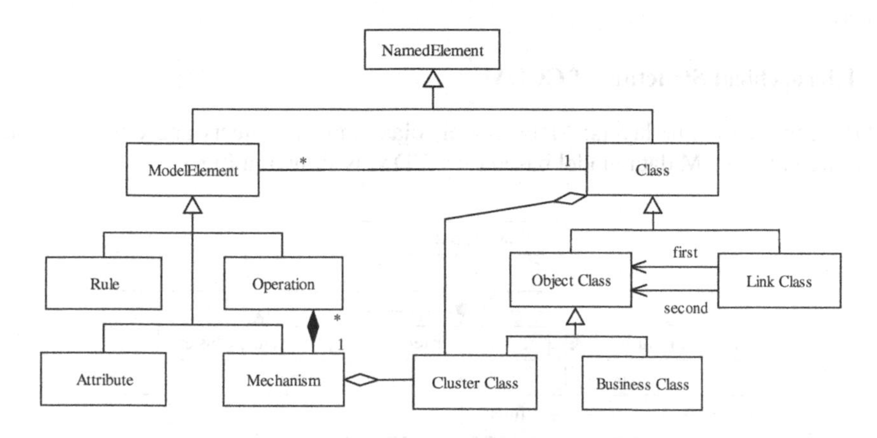

Fig. 1. Core structure meta-model of CCDM

Business Class (BC) is a set of objects of PLM that share the common attributes and structure. BC can be expressed as: $b := < bid, name, A(b), R(b) >$, where bid is an identity, $\forall b1, b2 \in b,\ b1 \ne b2 \Leftrightarrow bid1 \ne bid2$. $A(b)$ is a set of attributes of the BC which consists of General Attributes (GA) and Reference Attributes (RA). GA is $\forall ga \in GA, ga := < name, dtype, atype, value >$ and RA is $\forall ra \in RA$, $ra := < name, bid', dtype, RAL >$. GA can express simple data types such as number, string, or date. RA is used to represent complex data structure type or aggregation of data objects. $R(b)$ is a set of rules including object and class rules which constraint the structure and values of BC and their instances, which is described as $\forall r \in R(b),\ r := < name, type, content, A'(b) >$.

Link Class (LC) is used to link two objects in PLM. Link is $l := < lid, oid1, oid2, A(l), R(l) >$ where: lid is identity of the LC. $oid1, oid2$ are identities of the two objects which can be written as $l.oc1$ and $l.oc2$. $A(l)$ is a set of attributes of LC, which described as $\forall a \in A(l), a := <name, dtype, value>$. $R(l)$ is a set of rules between the links to constraint either the link can be created or not. Links can be divided into three groups according to their origin and nature: General Links (GLC), Composite Links (CLC) and Reference Links (RLC). $l.oc1$ and $l.oc2$ of GLC are equal such as the link of parts and drawings. While in a CLC $l.oc2$ is a part of $l.oc1$ like that a products contains several parts. RLC is used to realize the RA of BC to reused the data or aggregate the complex objects.

Cluster Class (CC) is a set of clustered objects of business data. CC is subclass of *Object Class*, so CC shares the common natures of class: (i) every object of cluster has a unique identity; (ii) the child cluster inherent the attributes and behaviors of parents. Meanwhile cluster has some special natures: (i) clusters depend on the classes. Objects of cluster is only integrated from the existing class objects and cannot be instanced by the cluster.(ii) all objects of cluster can be included or excluded from the cluster. Cluster definition as $c := <cid, name, A(c), P(c), M(c)>$,where *cid* is the identity. *A(c)* is an interface to access the cluster members. *P(c)* is the subsets of the cluster.

3.2 Hierarchical Structure of CCDM

CCDM consist of four layers: Meta model, class model, objects and cluster model. The structure of PLM data model based on CCDM is shown in Fig.2.

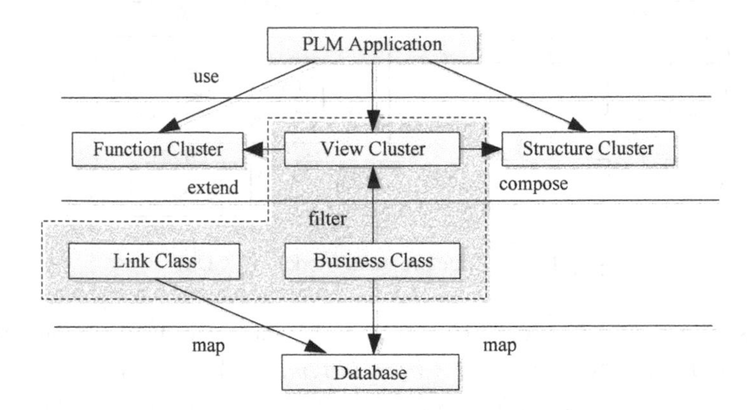

Fig. 2. Structure of PLM data model based on CCDM

Meta models describe the inner relations of elements in CCDM, and define the interface to access the model. Class model is built according to the industry of companies include BC and LC. Objects are the data created and maintained by the PLM, and the content is store in database.

The class model is mapping to database physical model. The types of BC attributes include internal, basic and dynamic. Internal attributes are used to maintain the system functions. Basic attributes are template for the area of industry based on the knowledge. Dynamic attributes are used to meet the iterative evolution needs of individual companies during the running time. The internal and basic attributes are relatively stable, and dynamic will be modified frequently. To reduce the impact of data structure increased by the modification, the business class is mapped to two tables in the database which can be described as:

$$\begin{aligned}&bc \rightarrow table(sta\_bc)\{bid, objid,\\&\{a.name \mid \forall a \in A(bc) \wedge (a.dtype = internal \vee a.dtype = basic)\}\}\\&\vee table(dyn\_bc)\{classid, objid, attriid, attritype, attrivalue\}\\&\vee attrid \mapsto \{a.id \mid \forall a \in A(bc) \wedge a.dtype = dynamic\}\end{aligned}$$

sta_bc is a static table with fields mapping the internal and basic attributes. dyn_bc is a dynamic table and with records correspond the dynamic attributes. This ensures that the business class when the evolutions occurring the database structure remains relatively stable.

Link class is mapped to a single table including the identity of the objects and type, and some other attributes.

$$lc \rightarrow table(lc)\{lid, olid, oid1, objid1, oid2, objid2, \{a.name \mid \forall a \in A(lc)\}\}$$

GLC and CLC share the same schema by the semantic constraints, so they can be interconverted actually without any affect to the structure. Combination of the constraints associated with higher than normal association, generally associated with portfolio into the evolution associated with the need to ensure there will not loop.

View Cluster (VC) can transform or filter objects of subsets to cluster objects with the same structure. A subset can be set to several VC to expose the necessary attributes to the outer through different mapping functions. VC can simplify the content of objects in various stages, and protected the unnecessary information not to be accessed.

Structure Cluster (SC) can aggregate objects of subset to a hierarchical structure. SC use BC or VC as the root and gather all objects linked with the root to form a composite cluster object.

Function Cluster (FC) defines a set of mechanisms to make objects of subset have a group of domain behaviors. The objects added to the FC have the privilege to execute these behaviors, and they do not have the behaviors once excluded from the FC.

4 Summary

The PLM CCDM consisting of meta-models, class models, object models and cluster models was proposed to support system evolution. Class models were used to describe business data and relations of enterprise. Cluster models were used to reorganize contents and structures of data objects, support different requirements of data schema at different stages of product lifecycle. CCDM can be easily defined to organize various forms of data and will not damage the original data structure frequently. CCDM can reduce the workload of system changes significantly and speed up PLM implementation.

References

1. Fan, Y., Huang, S.: Overview of Product Lifecycle Management. Computer Integrated Manufacturing Systems 10(1), 1–14 (2004)
2. Li, X., Qi, G., Liu, H., et al.: Incremental-convergent method for development and implementation of product lifecycle management system. Computer Integrated Manufacturing Systems 13(12), 2427–2432 (2007)
3. Qiu, Z.M., Wong, Y.S.: Dynamic workflow change in PDM systems. Computers in Industry 58, 453–463 (2007)
4. Sudarsan, R., Fenves, S.J., Sriram, R.D., et al.: A product information modeling framework for product lifecycle management. Computer-Aided Design 37, 399–1411 (2005)
5. Huang, L., Chen, H., Zheng, Q., et al.: A New Dynamic Data Model for Object-Oriented Database Systems. Journal of Software 12(5), 735–741 (2001)

Quadtree-Based Gridfile: A New Grid for Spatial Data

ZhongJie Zhang[1,2], DePeng Zhao[1], and DeQiang Wang[1]

[1] Dalian Maritime University, 1 Linghai Road.Dalian, 116026
[2] Shandong Polytechnic University, University Technology Park, Western NewCity, Jinan, Shandong Province, 250353
Termyzhang@gmail.com

Abstract. This paper presents a new spatial index structure – quadtree-based gridfile. The new gridfile combined with the advantages of quadtree and gridfile. According to the algorithm of full quadtree, we can calculate the result quickly. New algorithm to overcome the original algorithm, let gridfile can effectively index line objects and surface objects. Our paper gives the cache quadtree structure and the key algorithm in detail. This paper presents the principles and methods of calculation on the new hierarchical grid.

Keywords: Spatial index, Gridfile, Quadtree, Full quadtree.

1 Introduction

Spatial index is a data structure, which lined up the space object, according to their location, shape and other spatial relations. Spatial index use a wide range in spatial databases, GIS systems, etc. More representative of the spatial index is the R-tree, quadtree, grid files and space sorting. Practical application of integrated systems may use several indexing methods.

In this paper, we study the index theory of quadtree and gridfile, and proposed a quadtree-based gridfile. It overcomes some shortcomings of the gridfile.

2 The Gridfile and Quadtree

The basic idea of gridfile is to divide the k-dimensional space into some small blocks, every block is associated with one buck. Space object was placed in the buck which associated with its block. Figure 1 is an $M \times N$ grids, it also has the same $M \times N$ bucks. The grids can also be subdivided until the meet the required precision. The advantage of the grid index is, the algorithm is simple and easy to understand, Its configuration flexibility and faster. The algorithm itself also supports file storage structure, and does not require another storage method. The disadvantage of the grid index is only suitable for point data, line and area data will lead to reduced efficiency.

The quadtree regarded the space as its root node. The node was divided into four parts by a point in it. Every divided part is th subnode. The subnodes were divided recursively, until the node meets the requirements. Quadtree structure is simple; you can quickly query the data. The middle node can cover its subtree nodes, and it can store line or area object.

D. Jin and S. Lin (Eds.): Advances in ECWAC, Vol. 1, AISC 148, pp. 197–202.
springerlink.com © Springer-Verlag Berlin Heidelberg 2012

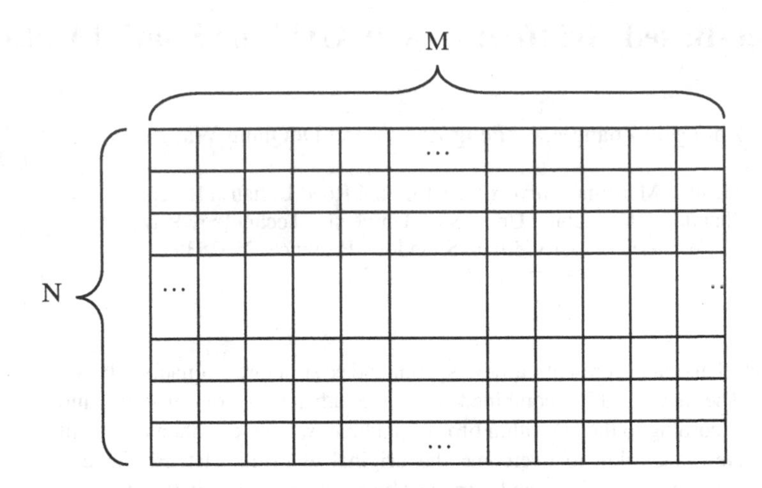

Fig. 1. Traditional Gridfile

In this paper, the quadtree and the grid combined each other, form a hierarchical quadtree-based grid index.

3 The Hierarchical Quadtree-Based Gridfile

3.1 Full Quadtree

The root of quadtree contains the entire space; the space is divided into four parts, which are the four sub-nodes. Similar to the binary tree, Quadtree is also a sequential storage structure. For a quadtree with d layers, if all its leaf nodes in the d layer and every node on other layers have four children. The quadtree was called full quadtree. The sequential storage structure of quadtree is, all of its nodes are put into a vector. Root on the front, followed by placing the second layer of elements, storing each element in this turn, until all elements are placed in the vector. Position in the vector is index code of the node. Figure 2 is a full quadtree, it has 3 layers, and number on the node is its index code.

An full-quadtree, its i-layer has 4^{i-1} nodes. If the quad-tree has d layers, it owns a total of $\sum_{i=0}^{d-1} 4^i$ nodes. For any node, if its index is p, the following relations [1]:

The index code of parent node is: $$\left\lfloor \frac{p-1}{4} \right\rfloor \quad (1)$$

The index codes of sub-nodes: $$4p+i, \quad i=1,2,3,4 \quad (2)$$

The layer of node: $$\lfloor \log_4(3p+1) \rfloor \quad (3)$$

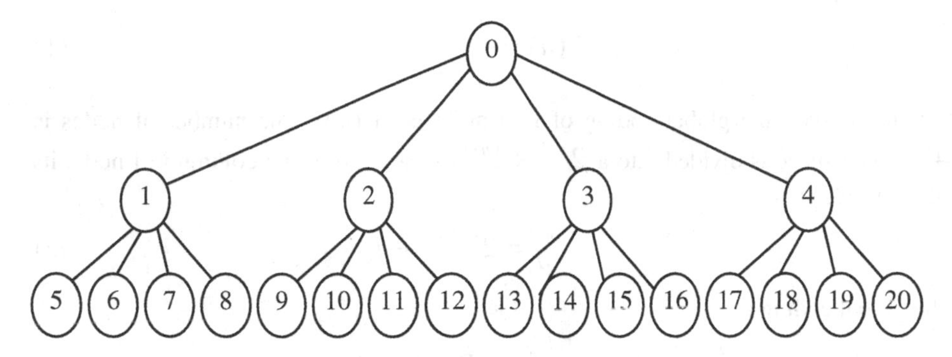

Fig. 2. A 3-tier full quadtree

3.2 The Structure of New Gridfile

In the paper, our gridfile is a hierarchical gridfile. In each layer, space is divided in the same way as quadtree. The first layer is one; the second is four, and so on. The code of every block is the same with the node in quadtree. Figure 3 is a three-layer hierarchical gridfile.

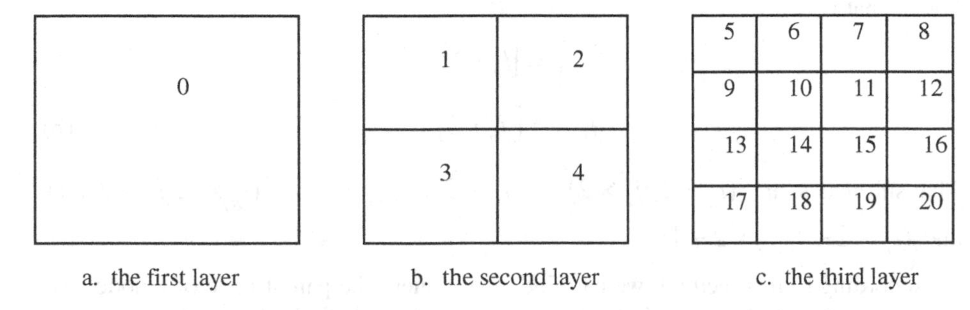

Fig. 3. Three-layers Gridfile

Each Layer of quadtree is usually coded by z-order. For quick calculations, we use the natural line order, not z-order. Therefore, we cannot use equation (1) and (2) compute node. Only codes in a layer have changed, equation (3) is still valid. The total number of nodes in each layer does not change.

Each node has two codings, which is layer coding and global coding. For layer coding, the first node is zero. The coding of Figure 3(c) is from 0 to 15. The layer coding plus base coding of layer, the result is its global coding. For the n-layer, all nodes before this layer form a full quadtree with n-1 layers. The quadtree has $\sum_{i=0}^{n-2} 4^i$ nodes, it is $\frac{4^{n-2}-1}{3}$. The base code of the n-layer is

$$B(n) = \frac{4^{n-2} - 1}{3} \tag{4}$$

For the n-layer, the global coding of first node is of B(n). The number of nodes is 4^{n-1}. The space is divided into a $2^{n-1} \times 2^{n-1}$ square. For row i column j of node, its layer coding is

$$q_{i,j}^{n} = 2^{n-1} i_n + j_n \tag{5}$$

The global coding is

$$p_{i,j}^{n} = B(n) + q_{i,j}^{n}$$

According to equation (4) and (5), it is

$$p_{i,j}^{n} = \frac{4^{n-2} - 1}{3} + (2^{n-1} i_n + j_n) \tag{6}$$

Layer coding is very simple. If we know the layer coding of a node, it is easy to find the layer coding of its parent node or sub-node. Based quadtree segmentation, we can easily draw that the node located on (i,j) has its parent node on ([i/2],[j/2]) of the front layer. That is

$$i_{n-1} = [i_n / 2] \tag{7}$$

$$j_{n-1} = [j_n / 2] \tag{8}$$

The sub-nodes are $(i_n \times 2, j_n \times 2)$, $(i_n \times 2 + 1, j_n \times 2)$, $(i_n \times 2, j_n \times 2 + 1)$ and $(i_n \times 2 + 1, j_n \times 2 + 1)$.

According to this method, we can find in the query the parent node of a node, you can also find its child nodes. In the application system, the link of a quadtree need not be stored. The same way with the ordinary grid, only a vector is stored. Based on the above formula, we can follow the quadtree approach to access the vector.

4 Algorithm

4.1 Insertion

The basic operation of new gridfile is to find out a block which contains the orthogonal region. In the new gridfile, we divided the average grid for rapid calculation of the corresponding block.

First, find out all blocks in the deepest layer occupied by the orthogonal region. This can be calculated through the lower left and upper right corner of occupied blocks. First, find out all blocks in the deepest layer occupied by the orthogonal region.

Insertion algorithm is to find just the grid that contains the specified object. We first find the object area of outsourcing orthogonal region, and calculate the block corresponding to the orthogonal region. You can put the object into the buck of the block.

4.2 Query

Query is to find all objects contained by the orthogonal region. These objects may be in many different layers, the situation is more complicated.

First, we find out all blocks in the deepest layer occupied by the orthogonal region. These blocks and their ancestor is the result of query.

5 Experimental Results

We test the new gridfile in a map, which has 48750 point objects and 34320 regional objects. Testing performed in Java environments; we implement a general gridfile and a new gridfile.

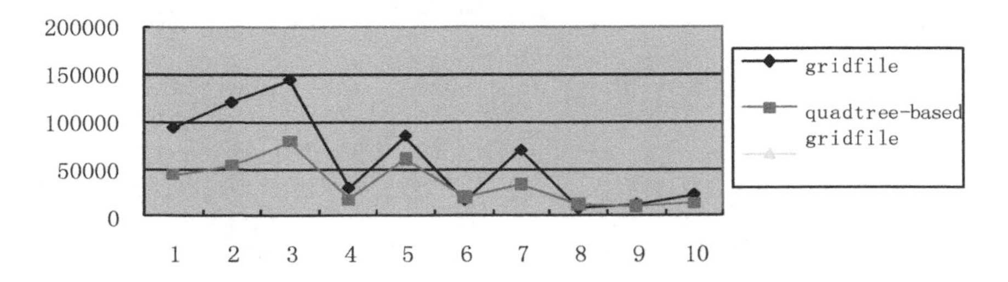

Fig. 4. The Performance of the Gridfile and Quadtree-based Gridfile

We query about 100 times, each point is the sum of 10 operations. As can be seen from Figure 4, when the query range is large, Quadtree-based grid is superior to normal gridfile. The main reason is, in the new gridfile, each object is indexed only once this avoids a lot of duplicate objects appeared in search results.

References

1. Finkel, R.A., Bentley, J.L.: Quad trees: A Data Structure for Retrieval on Composite Keys (1974)
2. Nievergelt, J., Hinterberger, H.: The Grid File: An Adaptable, Symmetric Multikey File Structure (1984)
3. Wang, X., Wang, H.: Data structure for the fast display of spatial objects in ECDIS (1999)
4. Zhou, Y., He, J.-N., Tu, P.: Technique of auto-selection multi-layers grid spatial index (2005)
5. Xia, Y., Zhu, X., Li, D.: Indexing Technology in Spatial Information Multi-Grid (2006)
6. Mahran, S., Mahar, K.: Using Grid for Accelerating Density-Based Clustering (2008)

7. Zhang, L., Tang, L.-W.: Study of Organizing Seam less Great Capacity Spatial Data Based Quadtrere (2011)
8. Zhao, Y., Sun, W., Wang, J.: Research on Map Showing Algorithm Based on Embedded GIS (2011)
9. Sun, X.-G.: The Research of Creating Method of Spatial Index in the Navigable Database (2008)

Research and Implementation of Application Server Compatible with IPv4 and IPv6

Qi Xiong and Yucai He

School of Computer Science and Technology
Hunan University of Arts and Science, Changde, PRC
abcxq@21cn.com

Abstract. Nowadays, most networks are constructed based on IPv4.But with the IP addresses are becoming less and less, networks based on IPv6 are becoming more and more. So, some application programs based on IPv4 are needed to be improved to suit IPv4/IPv6 environment. The difference between IPv4 and IPv6 socket is first compared in the paper. Then a FTP program based on IPv4/IPv6 is designed by dual stack software. Syetem test shows that FTP server can support IPv4/IPv6, and the basic functions such as login, authentication of passwords, list of sending files, download files, upload file and delete files were implemented.

Keywords: IPv6, FTP server, Socket, Dual stack.

1 Introduction

The biggest issue of IPv4, the internet technology which is used wildly nowadays, is that the resources of network IP addresses are in short supply day by day. With the number of internet users increase rapidly, the IP address of IPv4 will soon be used up. IPv6 is an next generation IP protocol which adopts 128-bit address that differs to IPv4, which adopts 32-bit address, and that means infinite address space. In the future IPv4 must be displaced by IPv6. But it is impossible to upgrade the entire networks from IPv4 to IPv6 in a short time. There will be a period when the two version of the protocol co-exist. It is very important to interconnect systems of supporting different versions of IP.

2 Comparison of IPv4 with IPv6

From the literature[1],we can see there are many conspicuous differences between socket address structure of IPv4 and IPv6. But two general function named getaddrinfo() and getnameinfo () was introduced into IPv6.These functions are protocol-independent.Function getaddrinfo( ) translates the name of a service location or a service name to a set of socket addresses. It is described as follows:

getaddrinfo (in const char *nodename, in const char *servname,in const struct addrinfo *hints, out struct addrinfo **res);

D. Jin and S. Lin (Eds.): Advances in ECWAC, Vol. 1, AISC 148, pp. 203–207.
springerlink.com © Springer-Verlag Berlin Heidelberg 2012

The first parameter of getaddrinfo( ) is host name or IP address which is denoted by IPv4 or IPv6 .The second one is services name which is indicated by port NO or existent services name ,such as ftp,http, etc .The third parameter is a pointer to addrinfo structure which is filled some return information. The structure addrinfo is described as follows:

```
struct addrinfo {
        int   ai_flags;          /* address information */
        int   ai_family;         /* AF_xxx */
        int   ai_socktype;       /* type of  socket */
        int   ai_protocol;       /* 0 or IPPROTO_xxx for TCP or UDP */
        socklen_t  ai_addrlen;   /* length of address */
        char  *ai_canonname;     /* canonical name for nodename */
        struct sockaddr  *ai_addr; /* binary address  for socket  address  structure */
        struct addrinfo  *ai_next; /* next structure in linked list */
   };
```

Another socket function getnameinfo() is used to translate address to name. Initialized socket structure will sent to it and name or services corresponding to address or port will be returned.

3 Making Application Compatible with IPv4 and IPv6

3.1 Special Address in IPv6

From those principles showed above ,according to the compatible support of socket API, we try to modify IPv6 FTP server ,which was designed in literature [2],to make it compatible with IPv4.We need two special IPv6 address to do this.

IPv4-mapped IPv6 Address. If an node, which have only IPv4 address, uses the IPv4-mapped IPv6 address to access IPv6 server. The IPv4 address is encoded into the low-order 32 bits of the IPv6 address, and the high-order 96 bits hold the fixed prefix 0:0:0:0:0:FFFF. These addresses can be generated automatically by the getaddrinfo() function.

IPv6 Wildcard Address. The source IP address of UDP packets and TCP connections will be assigned by bind() function .In this paper ,we use wildcard address to make the kernel to select the source address for the server. With IPv4, we use the symbolic constant INADDR_ANY in the bind() function().With IPv6, we use a global structure variable named "in6addr_any" which is assigned by system and initialized as a constant. The extern declaration for this variable is defined in <netinet/in.h>:extern const struct in6_addr in6addr_any.

3.2 Dual Stack

In order to support both IPv4 and IPv6 on Linux server, we adopted dual stack, which can handle both IPv6 and IPv4 client .A TCP listening socket for IPv6 is created, put into dual stack mode, and bound to port 21. This dual-stack socket can accept connections from IPv6 and IPv4 TCP clients which are both connecting to port 21. Different process is adopted for different connection by IPv6 socket .For IPv6 client, datagram are exchanged between client and server directly without any processing. For IPv4 client, IPv6 address, which is corresponding to its IPv4 address, is returned to IPv6 sockets by the system call accept (). The work process is shown as Fig.1.

4 Application

According to the detail illustration of FTP Server Function Module in literature [2], we illustrate some modifications about some mainly function modules.

Socket creating function ftp_creat_srv(). A interface ,which can listen any FTP connection request from clients, is created for FTP server by this function. Function process is as follows:

i IPv6 address structure is first defined, in which IPv6 address and TCP port of the server is filled.

ii Iinitializing structure of socket :

```
memset(&hints,0,sizeof(hints)) ;
hints.ai_family=PF_UNSPEC;              /*IPv4 or IPv6*/
hints.ai_socktype=SOCK_STREAM;            /*byte stream*/
hints.ai_protocol=IPPROTO_TCP;           /* TCP protocol */
hints.ai_flags=AI_NUMERICHOST;           /* numeral IP address */
rc=getaddrinfo(NULL,"21",&hints, &res) ; /* resolution analysis local address*/
```

iii Setting the parameter of listening port by socket () function.It is defined as follows:

```
#include <sys/types.h>
#include <sys/socket.h>
int socket (int family, int type, int protocol)
argument
family: AF_INET        //Used in IPv4 network
        AF_INET6       // Used in IPv6 network
        PF_UNSPEC      //Used in both IPv4 and IPv6 network
type:   SOCK_STREAM
        SOCK_DGRAM
        SOCK_RAW
Protocol: 0
```

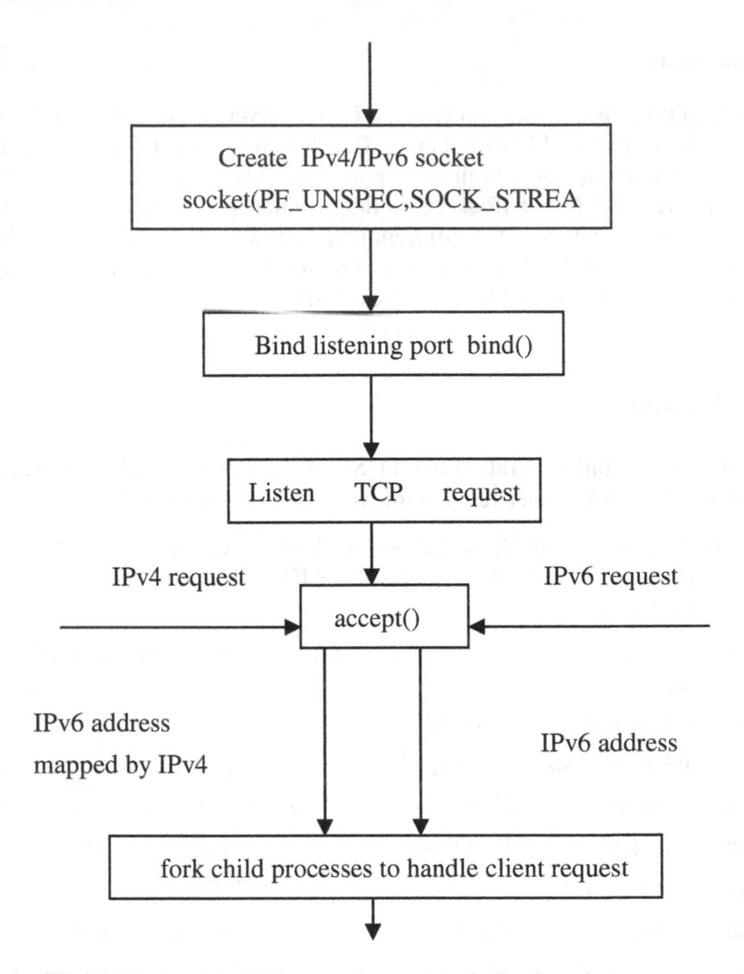

Fig. 1. Work process of connection request in Dual stack server

If an error occurs -1 will be returned; otherwise the return value is a descriptor referencing the socket. In order to make FTP server compatible with IPv4 and IPv6 ,we need the setting as follows:

```
s_send= socket(PF_UNSPEC, SOCK_STREAM, 0);
```

iv bind() is used to bind a name to a socket

```
rc= bind(s_send, res->ai_addr, res->ai_addrlen);
```

v listen() is used to listen for connections on a socket

```
listen(s_send,5);
```

Circulating listen function ftp_loop (). Some child processes can also be generated by this function to handle some connection request and FTP commands sent from clients. The design process is as follows :

i Each client request is sent to accept () function by FTP server while it is listening the port and connection is established.

new_fd=accept(s_send,(struct sockaddr *)&sin, &sin_len); /*accept connection*/

ii After connection is established, IPv6 address or mapping IPv6 address of client and TCP port is returned and showed by getpeer-name() function ;

iii Fork() function is called to generate child processes by the server and handle those connections.FTP server continue listening to deal with other requests from clients;

iv Ftp_ctrl_conn_handler() function handle FTP commands sent from clients and lead the process into the FTP command parser.

5 System Test

We implemented that kind of FTP server in Linux system which was installed dual stack. We can login in FTP server from client with both IPv6 and IPv4 address.From the result of test, we can see the FTP server is fully compatible with IPv4 and IPv6 according to a method illustrated in this thesis. Some basic functions, such as user login in, password verify, file list transfer, files transfer, etc.

6 Conclusions

The paper analyzed the difference between IPv4 and Ipv6 and implemented a FTP server which compatible with IPv4 and IPv6 according to the reference [2].In fact, during the transition period from IPv4 to IPv6, IPv4 node and IPv6 node must coexist in the same network for a long time. In the future,when we design server programs,we should let the programs compatible with IPv4/IPv6 according to this mothod methioned in the paper.

Acknowledgments. This work was supported by Scientific Research Fund of Educational Department of Hunan Province(NO.08C609) and Scientific Research Fund of Hunan University of Arts and Science (NO.JJZD0903).

References

1. Gilligan, R.E., Thomson, S., Bound, J., et al.: Basic socket interface extensions for IPv6. RFC 2553 (1999)
2. Wang, Y.-L.: Research and Implementation of the Application Server Based on IPv6. Computer Engineering & Science 32(12), 12–14, 18 (2010)
3. Hinden, R., Deering, S.: IP Version 6 Addressing Architecture, RFC 2373 (July 1998)
4. Xiong, Q., Zhang, H., et al.: Design of streaming media cluster server based on MPI. Applied Mechanics and Materials 63-64, 643–646 (2011)

Student Achievement Databases Assist Teaching Improvement

Xinhui Wu, Junping Yang, and Changhai Qin

Department of Electronic Information and Electric Engineering,
Anyang Institute of Technology, Anyang, 455000, China
cockwuxh@163.com

Abstract. In most colleges and universities, students' performances of basic courses are assessed usually by scores. Therefore, student achievement database, especially the student scores included, can help teachers and the departments assess their students objectively and give the students oriented educations. This paper used a large achievement database for the early identification of students with low performances. Scores from 1392 Electrical and Electronic Engineering students registered in courses were studied during a period of 4 years. Students with two or more courses failed were included into low-performance group and others were included into high-performance group. ROC curves were built to identify a cut-off average score in the first semesters that would be able to predict low performances in future semesters. In the next 4 years, we put more focus on the students with low performance in their first semesters and promoted relevant pedagogical strategies to affect their future achievements.

Keywords: ROC curves, linear regression, course trend, performance assessment, student achievement database.

1 Introduction

In colleges and universities, students' performances of basic courses are assessed usually by scores. Although the association between scores in specialized courses and professional achievements is still controversial, a strong relationship has been found between the scores of students in basic science courses taken in the first year and the subsequent performance of these students during and beyond colleges [1]. Therefore, student achievement database, especially the student scores included, can help teachers and the departments assess their student in an objective perspective in early stages and give the student oriented education in accordance with his or her individual performance [2]. Properly conducted overall student assessment is a great tool to highlight student professions [3].

ROC (Receiver Operating Characteristic) analysis originated from statistical decision theory in the 1950s, which was first to access the radar signal observation ability. To sum up, ROC curve is to access the classified patterns of two variables and the area under the ROC curve can evaluate the prediction accuracy. Recently, many educators adopt ROC curves to predict the students' achievements and analyze the risk level of the students with low performance [4].

D. Jin and S. Lin (Eds.): Advances in ECWAC, Vol. 1, AISC 148, pp. 209–214.
springerlink.com © Springer-Verlag Berlin Heidelberg 2012

In this paper, scores from Electrical and Electronic Engineering students registered in courses are studied during a period of 4 years to identify the trend for each course. ROC curves were built to identify a cut-off average score in the first semesters that would be able to predict low performances in future semesters. And the study result is used as a guidance to help the next grade students in the next 4 years. We put more focus on the students with low performance and promoted relevant pedagogical strategies for these individuals to affect their future achievements and helped some of them achieve professional success when graduated.

2 Theoretical Analysis and Experimental Verification

In this paper, a study is taken retrospectively from 2003 to 2006 of all students registered in the required courses according to the curriculum of Electrical and Electronic Engineering Department, Anyang Institute of Technology which is a very typical science and engineering college in China. Within this college of technology, the assessment of student performance is using scores from 0 to 100 , stored in the student achievement databases. The data consists of the student numbers and scores in courses taken between the first semester and eighth semester which is the final semester of their professional training period. The study identification number is an access to all the scores obtained from a single individual in each course and the relevant student's personal information. Electrical and Electronic Engineering courses in China last for eight semesters. The first six are to learn basic science subjects. The last two semesters focus on supervised practice, known as graduation design and practice. During the first, second and third semesters students are enrolled in four specialized course per semester: in the forth, fifth and sixth semesters students study five courses per semester. Every professional course lasts only for one semester. Totally, 1392 undergraduates registered in the Electrical and Electronic Engineering Department, from 2003-2006. We used two different approaches to assess two major objectives of the study in detail as follows.

Firstly, Scores obtained in the first semester was used to test the prediction of the students' academic performance in the second and sixth semesters. According to Table 1, total number of students registered in the first six semesters from 2003 to 2006 is 1392. However, data from 8 classes were excluded, because their scores in the 5th , 6th, 7th or 8th were not accessible. Only 10 classes were included (the number of students n= 759).

Table 1. Classes registered in the study period and the number of courses in each semester

Classes first registered in	Semesters (2003-2006)
2003 (5 classes, 389 students)	1st, 2nd, 3rd, 4th, 5th, and 6th
2004 (5 classes, 370 students)	1st, 2nd, 3rd, 4th, 5th, and 6th
2005 (4 classes, 325 students)	1st, 2nd, 3rd, and 4th
2006 (4 classes, 308 students)	1st, and 2nd

We tested correlations between the mean scores of these students in the first semester and their scores in the second or sixth semester to demonstrate if the performance in the beginning was consistent with the performance in the remainder of the course. The correlations were calculated using the Pearson coefficient. Linear regression was used for the estimation of the causal relation (relative risk) between the performances in the first and second or sixth semesters.

Secondly, we calculated the cut-off values for the scores obtained in the first semester that could predict a low performance in the second or sixth semesters using Receiver-Operator Characteristics (ROC) curves [5]. The cut-off values represent the score mean in the first semester that presents the highest sensitivity, specificity and likelihood ratio for predicting a low performance in the second or sixth semesters. Data from all the 18 classes were included (the number of students n= 1392).

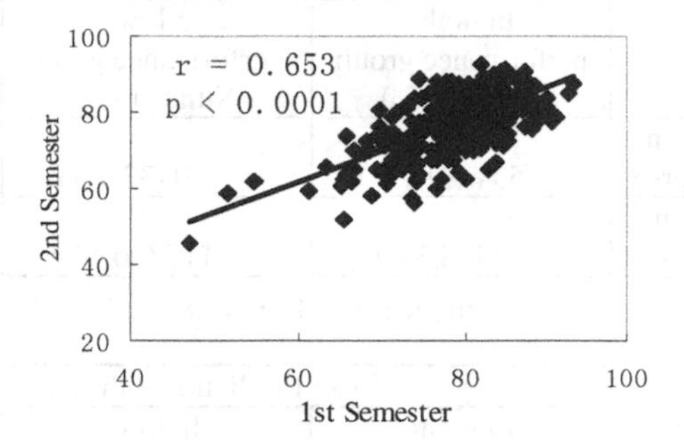

a. Correlation between scores of the students from the first semester

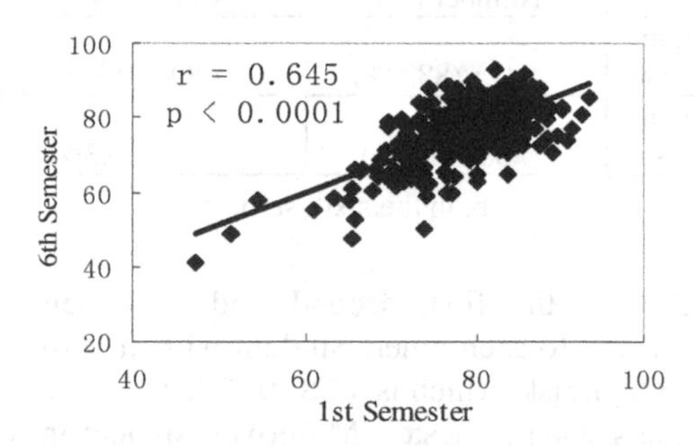

b. Correlation between scores of the students from the sixth semester

Fig. 1. Correlation between scores of the students from the first and second or sixth semesters

A summary of the sampling approaches used for the different analyses are shown in Fig. 1. All data were analyzed using the software Matlab6.5 and Microsoft Excel 2010. Differences were considered significant when $p<0.001$. According to Fig. 1, the global score obtained from each student in the first semester of medical school was correlated with the scores obtained in the second or sixth semesters using the Pearson coefficient ($n = 759$). Linear regression was used to illustrate the general trend of the correlations. In Fig 1a, r equals to 0.653, 95% CI is ranging from 0.609 to 0.678, and the p is less than 0.0001. From the data above, there is a conclusion that according to the large number of scores from student achievement database, the mean scores obtained by students in their first semester and those in their second semester have a close correlation statistically. Meanwhile, it is interesting that the correlation between the mean scores in the first semester and those in the sixth semester is positive according to Fig 1b, in which r equals to 0.645 (95% CI: 0.594-0.669, p<0.0001). Usually, an individual student's performance should be accessed in comparison with other students in the same semester. Thus, in each semester, students were classified into two groups: low-performance group (the students who failed in 0-1 course) and high-performance group (the students who failed in more than 1 course).

Table 2. Prediction of student score performances using scores from the first semester

		Model Prediction: 2nd semester		
		In high-performance group Number (%)	In low-performance group Number (%)	Relative risk
Gold Standard Truth	Failing in 0-1 course	534 (91.7%)	60 (33.9%)	4.08 (3.59-4.43)
	Failing in >1 course	48 (8.3%)	117 (66.1%)	

a. in the second semester

		Model Prediction: 6th semester		
		In high-performance group Number (%)	In low-performance group Number (%)	Relative risk
Gold Standard Truth	Failing in 0-1 course	475 (88.8%)	89 (39.7%)	3.54 (2.89-3.94)
	Failing in >1 course	60 (11.2%)	135 (60.3%)	

b. in the sixth semester

Table 2 shows that, in the first, second and sixth semesters, the students' performance correlated well to each other. Student with at least two courses failed in the first semester had a high risk, which is 4.08 (95% CI: 3.59-4.43), of having at least one course failed in the second semester. Moreover, student in the low-performance group in the first semester had a higher risk of being in the lower quartile in the sixth semester with a relative risk of3.54 (95% CI: 2.89-3.94).

Furthermore, we did analysis using ROC curves, shown in Fig 2a. The area under the curve equals to 0.759 approximately, the sensitivity is about 91% (95% CI: 85%-93%), the specificity is 66% (95% CI: 61%-97%) and the likelihood ratio equals to 2.71 (p<0.0001), which revealed that an average score in the first semester which can be a good cut-off value to classify the failed and passed courses and predict a low performance in the second semester is 75. Likewise, Shown in Fig 2b, the area under the curve equals to 0.703 approximately, the sensitivity is about 90%(95% CI: 82%-93%), the specificity is 59%(95% CI: 55%-61%) and the likelihood ratio equals to 2.2 (p<0.0001), which also revealed that an average score in the first semester of 75 can be a good cut-off value to classify the fail and passed courses and predict a low performance in the sixth semester.

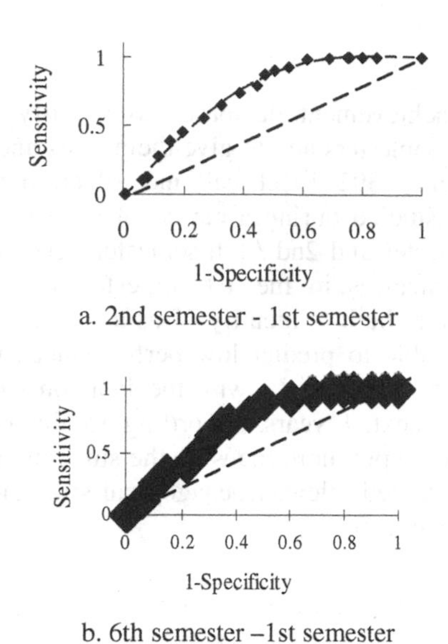

a. 2nd semester - 1st semester

b. 6th semester –1st semester

Fig. 2. Establishing Cut-off score by building ROC curves

3 Application and Measures of Teaching Improvement

The discussion above shows that it's highly necessary for the teachers and colleges to give a good favor to the students with poor performances in the first semester. From 2007 to 2010, we apply the theory above to help the student with difficulty in the Electrical and Electronic Engineering Department. The students who got a poor performance in the first semester and had a trend to fail in later semesters were particularly focused on. We teachers in the department helped the students who need help from the following aspects: make sure they had frequent attendance to the professional courses [6]; encourage and assist them to improve their study skills; help them make up missed lessons that could influent their later studies; and take tutorial method. We monitored 20 students who had poor performance trends each year during the 4 years observation period. With the teaching methods taken, of the first 20 students, 14 students were up to the high-performance group in the second semester and 17 students were up to the high-performance group in the sixth semester. Of the second 20 students, the result is 13 and 19 respectively. Of the third 20 students, 15 students were up to the high-performance group in the second semester. This study show that efficient teaching methods and pedagogical arrangement is able to decrease the possibility of keeping low performance in later semesters for the students failed in the first semester.

4 Conclusions

This paper used large achievement databases to identify the students with low performance in their first semesters and to give them assistance to help them enhance performance. Scores from 1392 Electrical and Electronic Engineering students registered in courses are studied during a period of 4 years. First, linear regression analysis between 1st semester and 2nd / 6th semester verified the close relationship between students' performances in the 1st semester and in the later semesters. Moreover, ROC curves were built to identify 75 as a cut-off average score in the first semesters that would be able to predict low performances in future semesters, by calculating the AUCs and relative risks with the Pearson Correlation Coefficient p less than 0.0001. In the next 4 years, according to the theoretical analysis and experimental verification, we put more focus on the students with low performance in the first semester and promoted relevant pedagogical strategies for these individuals to affect their future achievements.

References

1. Hojat, M., Paskin, D.L., Callahan, C.A., Nasca, T.J., Louis, D.Z., et al.: Components of postgraduate competence: analyses of thirty years of longitudinal data. Med. Educ. 41, 982–989 (2007)
2. Abdel-Salam, T., Kaufftnann, P., Williamson, K.: A case study: do high school GPA/SAT scores predict the performance of freshmen engineering students? Frontiers in Education, S2E–7 (2005)
3. Roach, A.T., Elliott, S.N.: The Influence of Access to General Education Curriculum on Alternate Assessment Performance of Students with Significant Cognitive Disabilities. Educational Evaluation and Policy Analysis 28, 181–194 (2006)
4. Wang, X., Shu, P., Cao, L.: Performance Evaluation with Optimization Strategy for Support Vector Machine Based on ROC Curve. Computer Science 37, 240–242 (2010)
5. Wei, X., Zhou, Y.: A New Performance Categories Evaluation Method Based on ROC Curve. Computer Technology and Development 20, 240–242 (2010)
6. Millis, R.M., Dyson, S., Cannon, D.: Association of classroom participation and examination performance in a first-year medical school course. Adv. Physiol. Educ. 33, 139–143 (2009)

Study on Development and Application of Online FAQ in Experimental Computer Courses

Shuzhen Li

Department of Management Engineering, Jiangxi Blue Sky University, Nanchang China
{330098,gaokaoren}@126.com

Abstract. For the on-line question-answering system based on ASP.net and c # technology WEB applications, and to the Internet for media to realize remote education of a teaching platform. System mainly discusses the design development, design and development of the basic process of the main use asp.net technology and c #, SQL technology, according to object-oriented development principle, based on B/W/S three layers of the structure of the system development model system. For the server to IIS, SQL server 2005 for database platform, providing a communication between teachers and students, mutual discussion of network space, make the students to gain knowledge more proactive. This system including instant answer management and control, document resources sharing management, problems and solutions of the management, doubt message backend database information management and maintenance of these basic modules.

Keywords: Online FAQ, C#, Computer Courses.

1 Introduction

Computer experiment online FAQ system USES sql2005[1-5] + asp.net for design development platform, mainly for the computer room provides a use of the existing equipment course teaching, the teachers and students interact FAQ discussion of system. Through the corresponding functional modules realize the real-time problems or message, experiment contents file such as the question and answer the resources sharing of upload and download, and related knowledge to expand, the teachers and students in front of the computer to complete the entire process teaching, and make students get rid of the previous shy psychology, active and involved in learning activities to network, effectively aroused the enthusiasm of student's study. Students and students, teachers and students between discussions between the already active thinking, also pulled close distance of each other, promote the teaching activities. This system based on network teaching support platform teaching mode human machine interaction, the teachers and students to interact with the advantage, manifests the humanist, the features of autonomous learning, the breakthrough time space boundary, and extending the classroom space, expand thought method, to cultivate good thinking, high-quality comprehensive personnel, to the requirements of the development of education in the information age [6-9].

D. Jin and S. Lin (Eds.): Advances in ECWAC, Vol. 1, AISC 148, pp. 215–220.
springerlink.com © Springer-Verlag Berlin Heidelberg 2012

2 Demand Analysis

First of all, according to the characteristics of computer laboratory, online FAQ function module can use existing LAN topology structure, in the network environment in the information input and transmission shows, and stored in the database server host. Through the control of conversational style in group chat and one-to-one communication both private conversation switches between conversational style.

Secondly, in the process of teaching, there must be related with experiment course experiment guidance and experiment contents and related documents. For free installation FTP server realize the trouble, file sharing in the question-answering system should also include the document resources of the upload and download function module, realize the mutual information related to the relay.

Again, online FAQ modules, main is to achieve similar QQ group chat of instant communication function, the inside of the present DuoRen speech, cross speech, more content and more messy discussed the phenomenon, of a doubt unfavorable to deep discussion. Therefore, in the system, we should be set at ordinary times the BBS used a similar message function modules, and contains published doubt, doubt, doubt information management reply etc. Function.

In addition, in the process of teaching, to deepen the students' understanding of the content of the experiment, to broaden the application of knowledge, it is necessary to set up the knowledge expansion module, including related knowledge portal website links, application examples show, technical article, etc. For the more convenient processing class affairs, notice the students in the class the related information, and system should also include the class that the recent announcement function, the class arrangement. In order to ensure the safety of the database, and teaching go smoothly, the system also need to manage the identity of the students, the simplest thing is to visit the system requirements of the students, the identity of the login operation, and put the identity verification in the system the use of other function in the system of data stored for later need to confirm a student status call operation.

Finally, this system provides background data management, should let the personnel management authority has the database operations, including the management of the students' information, upload files management, doubt, links, notice the management functions such as specific functions, management for presenting the data source of data for a visit, add, edit, and delete operation. It is convenient to make unified management system to control and display.

3 Feasibility Analysis

On demand online FAQ system makes the feasibility: students in computer classroom convenient to study and discussion, may also manage the affairs of the class, notice. Make the teaching interaction between teachers and students, energy save time. Great have to improve students learning initiative, and promote the development of it on teaching.

Technical feasibility: at present, IT has developed a mature and network technology used widely, hardware, CPU dual-core chips become popular configuration, development operation required by the system. The system requirements, the memory requirements can completely meet. To realize this system is technically feasible.

Economic feasibility: this system development the tools needed to main is a we commonly used individual host, plus a set of common software VS2005 + SQL2005 to existing computer classroom of local area network as operation transmission environment can. To realize the system cost less, and is feasible.

4 Overall Design Based on System Architecture

Overall planning: Online FAQ system is a typical asp.net database development application, the system shows the front desk, background module data management module two main components.

Front desk display module: This module provides real-time speech input and display; file uploads and downloads of resources, questions and answer, published related website access, view class notice, and identity verification login function modules.

Background management: The module is mainly to the front desk database management, mainly including student management, file management, doubt management, link management, notice management.

5 Chart of Whole System Function Structure

User successfully log on, if verification is administrator, it may be chosen identity into the front desk show, which is just as figure 1 or background management, which is just as figure 2, if is ordinary users, the orientation to the system the front desk home page.

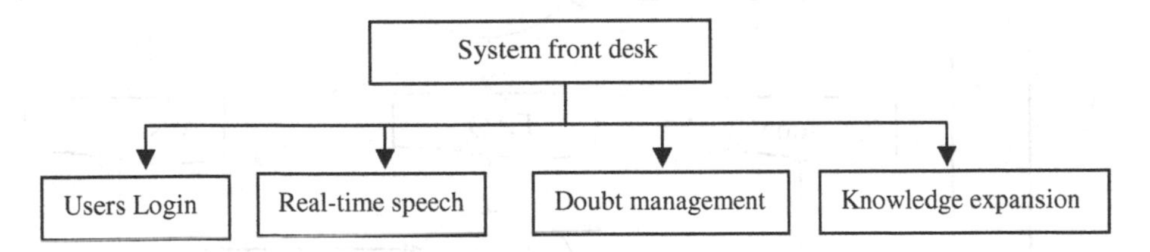

Fig. 1. Front Desk Function Structure

On the home page, the default is immediate page, also can speak through pages link to file sharing or view the doubt page.

In real-time speech page, which can choose to speak the color of the text, switch group chat and private conversation mode. In the file sharing page, shows the existing file list, right click the file name can choose to download, and click upload files button, into the file upload page.

In check, can express doubt questions page or click on the corresponding question, into the detailed answer page. Each page, to return to the home page contains links to button

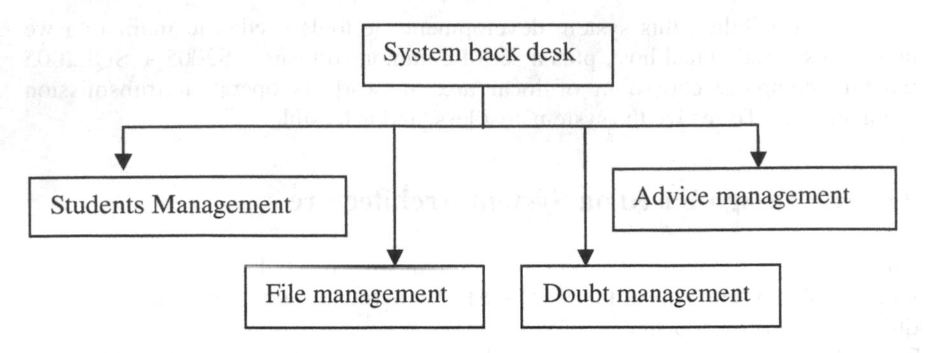

Fig. 2. Back Desk Function Structure

6 Business Data Flow

According to the investigation to the user, this system has three kinds user use, respectively are students, management and teachers. The three users with different variety of identity permissions, administrators responsible for management system accounts and system function; other and all Teachers can't manage the user but can update maintenance FAQ, modify password, and etc. This system flow is mainly as shown in figure 3.

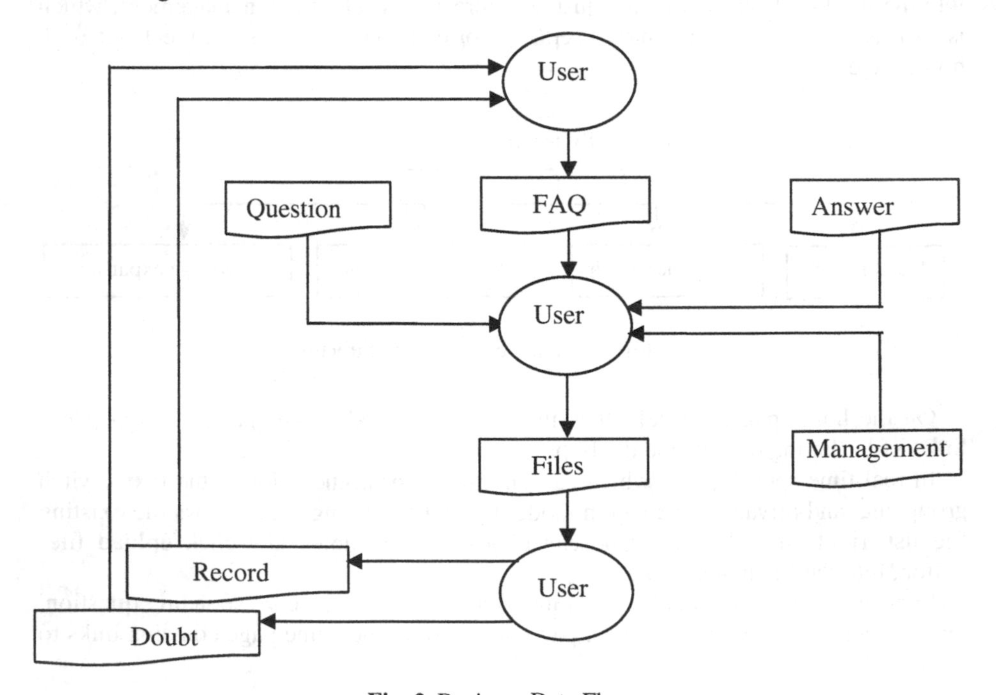

Fig. 3. Business Data Flow

7 Result

7.1 Module of Real-Time Speech

Home page is for the module.aspx face amount, the header for custom control WebUserControl.aspx, which is inside and including the function module links and displaying the system time and other information. For real-time speech interface between, mainly display information and send information, footer for backstage login link, when the admin log in for, the link control by gray, not with a available, click on that link into the background login page.

7.2 Module of File Management

This function module is including documents show download page SaveFile.aspx and file upload page Chuanshu.aspx with two pages, and saving resource file folder. Through this module, easy to download the data such as the experimental contents file.

7.3 Module of Doubt Management

This module mainly page that lists of questions include showthread.aspx, a question and the corresponding page showmessages.aspx, which reply list published, new questions edit thread page.aspx. This module mainly realizes the doubt published and reply. Delete function included in the background management pages.

7.4 Module of User Management

This module pages for admin.aspx, main is to use SqlDataSource control to realize application and the database links, grid view data control display, edit, delete data information, Form View control students realize data into function.

Acknowledgments. Mrs. Yan helps me for the languages. There are other colleagues, hereon, to thank. The authors are grateful for the anonymous reviewers who made constructive comments.

References

1. Shao, L.S., Liu, Ma, H.J.: Asp.net (c #) practice course, pp. 252–298. Tsinghua university press, Beijing (2007), Good Increased
2. Ma, Y.H., Su, G.Y., Yuan, Y.: Asp.net 2.0 network programming from foundation to practice, pp. 350–400. electronic industry press, Beijing (2007)
3. Wang, B.J.: Asp.net website construction experts. Beijing Tsinghua university out club., pp. 97–126 (2005)
4. Hao, G., Yuan, Y.G., Ji, Y.: Asp.net server control development technology and examples, pp. 196–215. People'S posts and telecommunications publishing house, Beijing (2005)
5. Chen, G.J.: Proficient in asp.net 2.0 typical module design and implementation, p. 83. People'S posts and telecommunications publishing house, Beijing (2007)

6. Feng, S.: Assumption of Building Students Safety and Protection Mechanism in Colleges and Universities. Journal of Hunan International Economics University 02, 65–69 (2010)
7. Qin, F., Hou, S.: Design and Implementation of ERP System Based on SOA. Microcomputer Information 01, 153–155 (2011)
8. Cui, L.: General Speech Signal Processing Platform Based on C# Language. Shandong university (2009)
9. Cui, K., Peng, X.: A Model of Web Service Combination. Computer & Digital Engineering 01, 168–172 (2011)

The Research about Transmitting Mechanisms and Channel Models of Wireless Sensor Semantic Network

Gu Xiao Jun

School of Mathematics and Computer Science, Ningxia University,
Yinchuan, 750021, China
gxjns@163.com

Abstract. A Sensor Semantic Network (SSN) is a sensor network including semantics of sensory data and context information, and relationships between the semantics by using Semantic Web technologies. Even though much research has been activated on SSN, there is little activity abort transmitting mechanisms and channel models of wireless sensor semantic network. In this paper, a basic Channel of Wireless sensor semantic network is studied and the effects of distance, the wireless communication is shown. The paper presents a generic wireless sensor semantic network way and identifies the role and function of sensor semantic network. The paper provides the guidance on the wireless communication management methods used to solve wireless communication issues. The transmitting mechanisms and channel models of wireless sensor semantic network is studied in this paper.

Keywords: Sensor Semantic Network, wireless communication, Mechanisms, EBGN.

1 Introduction

In every Wireless SSN a transmitter and a receiver node is needed for communication. Radio waves start from a certain point, typically the transmitter antenna, and with a certain level of power or energy. However, since the energy is finite, at any distance from the transmitter the original energy is now spread over a much larger space. The signal is said to be attenuated. If the signal is too weak, the receiver node cannot correctly receive the transmitted information. If this happens, the receiver lies outside the coverage area of the transmitter. The fundamental issue behind the coverage problem is the phenomenon of fading. Fading can be considered to be the time variation of the channel strengths due to multi-path fading, as well as path loss via distance attenuation. Because receiver nodes can be mobile, the possibility of this happening increases greatly and methods to prevent out-of-coverage issues must be implemented.

A significant disadvantage of Wireless SSN is interference. Tse et al. [1] illustrate that a transmitter–receiver pair can often be thought of as an isolated point-to-point link, and that wireless users communicate over the air and there is significant

D. Jin and S. Lin (Eds.): Advances in ECWAC, Vol. 1, AISC 148, pp. 221–226.
springerlink.com © Springer-Verlag Berlin Heidelberg 2012

interference between them. The interference can be between transmitters communicating with a common receiver (e.g., uplink of a cellular system), between signals from a single transmitter to multiple receivers (e.g., downlink of a cellular system), or between different transmitter-receiver pairs (e.g., interference between users in different cells). If the effects of interference between the isolated point-to point link is not mitigated the signal could be completely lost.

As highlighted with fading and interference, Wireless SSN has some disadvantages compared to a wired network. However, the advantages of Wireless SSN surpass the disadvantages. The key advantage is mobility, i.e., the ability of the transmitter–receiver pair to move around and still avail of wired network functionality and semantics while in the coverage of the Wireless SSN. Planning and network management techniques for fading and interference can overcome problems like coverage. Cell coverage varies as it is dependent on the wireless access technology in use. This cluster represents a semantic area in which mobile users have the freedom to move/roam and gain network access simultaneously.

Movement in this semantic area has brought new challenges in wireless communication at all levels of the network. If relating a Wireless SSN to the Open System Interconnected (OSI) stack, mobility primarily affects the physical and sensor access layers. Handling mobility at this level requires the network to handle mobile devices at a radio level. However, current Wireless SSN allow applications, including real-time and non-real-time, to run over Internet Protocol (IP) which brings issues in maintaining TCP/IP connections while the user moves from base station to base station. The Transmission Control Protocol (TCP) relies on the source and destination IP address to create a checksum; if either IP address change (which can happen if changing base station, network, etc.) then the checksum fails. If the checksum fails then the TCP connection is broken. Resolving this issue in a Wireless SSN can be achieved by providing mobility management at the IP layer and above.

Underlying all these management techniques are the mobility characteristics of the mobile device in the Wireless SSN. It is the movement of the mobile device that causes the management technique to be executed. Therefore, it is imperative that the movement is correctly understood. In a real-world scenario movement occurs at a multitude of levels: global movement, intercontinental, intercity, and on a smaller scale-campus, street or office. Furthermore, mobility can occur in groups or individually. Naturally, there is a speed associated with each of the movement levels; for example, in an office the speed of a mobile device is limited to walking speeds-whereas with a vehicle on a road the speed of the mobile device is now limited by road speed conditions. Associated with mobile device speed is mobile device density (measured in devices/m2).

Mobility management techniques must therefore be able to accommodate all these wireless technologies and be able to handle mobility speeds in all geographical areas. A single solution is not feasible so a multi-task solution is currently used. Before describing mobility management solutions, the Channel of Wireless SSN is explained to understand how conventional terminal mobility management works. A reference model of a generic future wireless SSN is presented that supports a multi-task mobility management technique.

2 Channel of Wireless SSN

The Channel of Wireless SSN is the physical medium that supports the transmitting of all signals in a Wireless SSN. In contrast with wired communications systems, the transmitted signals often do not reach their destination directly due to the presence of obstructions in the line-of-sight path. Obstacles in a communications link lead to the occurrence at the receiver of different replicas of the transmitted signal. This phenomenon, known as multi-path transmitting, weakness, delays and distorts the transmitted signal in a time-varying fashion. In addition to multi-path transmitting, the motion between the transmitter and the receiver also has a negative effect on the received signal strength. Thus, the characterization and understanding of the mobile radio channel is essential to combat these distortion effects and to acquire a desirable system performance, which in turn can satisfy the required quality of semantic (QoS).

An accurate estimation of multi-path channel behavior allows for improving the bit error rate using equalization techniques at the receiver, optimization of the maximum capacity, and effectiveness of the shared medium access mechanisms. Furthermore, the design of a mobile radio system that satisfies the stipulated QoS metrics under realistic Channel of Wireless SSN conditions constitutes a major challenge, especially when semantic requirements demand very high data rates coupled with high-speed mobility.

Large scale fading Wireless SSN represents the average signal power attenuation or the path loss caused by the motion of the mobile device over large areas (usually of the order of the wireless SSN cell size). This phenomenon occurs when the receiver is shadowed by the presence of prominent terrain contours (e.g., buildings, trees, hills, etc.).

Different approaches for modeling the path loss may be considered depending on the transmitting environment. The simplest model corresponds to the direct line-of-sight transmission between transmitter and receiver, referred to as free space loss [2]. The principle of energy conservation establishes that the integral of the power density over any closed surface surrounding the transmit antenna must be equal to the transmit power. As the integration area is a circle of radius d centered at the transmitter position, the received power collected by the receiver antenna is obtained according to the transmission equation [2] as:

$$P_{Rx}(d) = P_{Tx}G_{Tx}\frac{1}{4\pi d^3}A_{Rx} + 1 = \frac{P_{Tx}G_{Tx}G_{Rx}}{L_{FS}+1}, \tag{1}$$

where PTx is the transmitter power, LFS is the free-space path loss and GTx is the transmit antenna gain in the direction of the receive antenna. The parameter ARx is the effective area of the receiver, which follows the relationship ARx = λ2/4πGRx, with GRx being the gain of the receive antenna and λ = c/fc being the wavelength with c the speed of light and fc the carrier frequency.

More realistic models take into account the effects caused by reflections, scattering and diffraction. A simple model that considers the reflective effect of the earth's surface is named plane earth loss [3]. The received power is computed for the situation where only line-of-sight ray and ground-reflected wave exists, obtaining

$$P_{Rx}(d) = \frac{P_{Tx}G_{Tx}G_{Rx}}{L_{PE}+1} - p \approx P_{Tx}G_{Tx}G_{Rx}(\frac{h_{Tx}h_{Rx}}{d^2})^3 \tag{2}$$

where h_{Tx} and h_{Rx} are the height of the transmitter (mobile device) antenna and the receiver (base station) antenna, respectively. This path loss model is valid for distances larger than $d \geq 4h_{Tx}h_{Rx}/\lambda$. In this model, the received power decays with a factor d^{-4}, resulting in a very inaccurate measure for different types of transmitting sensor (rural, office, etc.). The frequency dependence of the path loss is also not taken into account in this model.

Small-scale fading describes the rapid fluctuation in signal amplitude and phase that occurs in the received signal over a small period of time or a short traveled distance (on the order of a wavelength). These types of fluctuations are manifested in two different manners, time-spreading of the signal due to multi-path transmitting and time-variant behavior of the channel caused by the Doppler effectiveness.

Multi-path fading is caused by the presence of multiple replicas of the transmitted signal that occur at the receiver due to reflections against obstacles. Multiple copies of the transmitted signal sum together in either a constructive or destructive manner depending on the phase of each partial wave. These additions create fading dips in the received signal power and distort the frequency response characteristics of the transmitted signal. These distortions are linear and need to be combated at the receiver by applying equalization and diversity techniques.

The multi-path Channel of Wireless SSN can be initially classified according to their coherence bandwidth. The coherence bandwidth represents the range of frequencies over which the channel response is flat, meaning that all the frequency components are equally attenuated. Flat fading occurs when the signal bandwidth is less than the channel coherence bandwidth. In this situation, the presence of deep fades due to multi-path can result in deep fades in the signal over all its bandwidth. On the other hand, when the signal bandwidth is greater than the channel coherence bandwidth, the multi-path fading is considered to be frequency selective. That is, multi-path fades are unlikely to affect the signal over its entire bandwidth without experiencing fades in received energy.

3 Experiment Background: Wireless Communication Networks

As wireless technologies evolve, the demand for better and faster communication systems also increases. Since its introduction in the 1990s, the growth of the business of commercial cellular systems has been rapid. The number of wireless communication users as well as the spectrum of the available semantics has increased at an unexpected rate. The main reason for this growth was the newly introduced notion of terminal, semantic and user mobility.

Existing terrestrial wireless SSN are based on the cellular concept as outlined by MacDonald [13]. The network structure is composed of a fixed core network with wireless last hops between radio station, and mobile devices. Different cell size environments such as macro-ell, micro-cell is grouped together as they are dependent on the wireless access technologies. The wireless SSN cell architecture is comprised of these environments tiled and layered to create a tiered cell structure.

The core network system or simply the core network (CN) is responsible for maintaining information (such as identity, location, authentication, and billing) about the users in the network. This information is stored in registers or a database.

The core network provides access to other backbone networks-Internet, circuit switched, etc. When a call arrives for the mobile device, the CN interrogates its registers and processes the command request.

The access network system comprises base stations (BS) and base station controllers (BSC). The BS is the last fixed wired link in the wireless SSN, that is, the BS provides a radio link to the mobile device. The geographical area the BS is in provides radio coverage of an area called a cell. The size of the cell is dependent on the power of the BS and the height of the antenna. A BSC has control over several BSs and thus clusters a group of cells together; this cluster of cells forms an area known as the location area. Combining all the location areas together provides the network's semantic area.

4 Conclusion

In the course of the accelerated contention of wireless technologies with wired technologies, broadband wireless semantic has become a reality, and wireless Internet is attainable. However, QoS and cost remain as deficiencies of wireless systems. Despite the freedom of mobility, in the data arena wireless technologies have enjoyed limited popularity to speak of. The true advantage of mobility in the context of broadband semantics is exemplified by the capability to deliver location specific semantics to the user.

Without any doubt, next-generation Wireless SSN will be considerably more complex than today's second/third-generation wireless systems. Global roaming and the ability to access the Internet and data resources everywhere are among the key driving factors that will have profound impact on how these networks will be managed. In this changing environment, operators will introduce new semantics and more powerful and efficient ways of doing business by integrating new technologies with existing ones. This is already beginning to happen with mergers of WLAN networks and cellular networks. Not only will the network elements and communication devices evolve but so too will the management systems and way of managing. Mobility management has widely been recognized as one of the most important and challenging problems for next-generation networks. New management strategies will be needed to enable seamless access to Wireless SSN and mobile semantics.

References

1. Alves, A., et al.: Web services business process execution language version 2.0. (2006), http://www.oasisopen.org/committees/documents.php
2. Arkin, A., Askary, S., Fordin, S., Jekeli, W., et al.: Web Service Choreography Interface (WSCI) 1.0. W3C (2002), http://www.w3.org/TR/wsci/
3. Chan, P.: Best Route Finding specification (2006), http://www.cse.cuhk.edu.hk/pwchan/BRF.doc

4. Chan, P., Lyu, M.: Developing aerospace applications with a reliable web services paradigm. In: Proc. of IEEE 22nd International Conference on Advanced Information Networking and Applications, Okinawa, Japan, March 25-28 (2008)
5. Chan, P.P.W., Lyu, M.R., Malek, M.: Making Services Fault Tolerant. In: Penkler, D., Reitenspiess, M., Tam, F. (eds.) ISAS 2006. LNCS, vol. 4328, pp. 43–61. Springer, Heidelberg (2006)
6. Chan, P., Lyu, M., Malek, M.: Reliable web services: Methodology, experiment and modeling. In: Proc. of IEEE International Conference on Web Services, Salt Lake City, Utah, USA, July 9-13 (2007)
7. Chan, P.P.W.: Building Reliable Web Services: Methodology, Composition, Modeling and Experiment. PhD thesis, The Chinese University of Hong Kong, Hong Kong (December 2007)
8. Dong, W.-L., Yu, H., Zhang, Y.-B.: Testing bpel-based web service composition using high-level petri nets. In: Proc. of the 10th IEEE International Enterprise Distributed Object Computing Conference (EDOC 2006), October 2006, pp. 441–444 (2006)
9. Jones, S.: Toward an acceptable definition of service (service-oriented architecture); IE[7] Thagard, R. P.: The best explanation: Criteria for theory choice. The Journal of Philosophy (1978)
10. Edwards, J., Handzic, M., Carlsson, S., Nissen, M.: Information Management Research and Practice: Visions and Directions. Information Management Research & Practice 1(1), 49–60 (2003); IEEE Transactions on Software 22(3), 87–93 (2005)
11. Bai, X., Dong, W., Tsai, W.-T., Chen, Y.: XMLbased automatic test case generation for network information processing testing. In: Proc. IEEE International Workshop on Service-Oriented System Engineering 2005, pp. 215–220 (2005)
12. Bertolino, A., Frantzen, L., Polini, A., Tretmans, J.: Audition of Web Services for Testing Conformance to Open Specified Protocols. In: Reussner, R., Stafford, J.A., Ren, X.-M. (eds.) Architecting Systems with Trustworthy Components. LNCS, vol. 3938, pp. 1–25. Springer, Heidelberg (2006)
13. Blum, A.L., Furst, M.L.: Fast planning through planning graph analysis. Artificial Intelligence 90(1-2), 279–298 (1997)
14. Coalition, T.O.S.: Owl-s: Semantic markup for network information processing (2003), http://www.daml.org/services/
15. Heckel, R., Mariani, L.: Automatic Conformance Testing of Web Services. In: Cerioli, M. (ed.) FASE 2005. LNCS, vol. 3442, pp. 34–48. Springer, Heidelberg (2005)

Self-adapting Unified Software Test for Network Information Processing

Liu Fu Xiang

School of Mathematics and Computer Science, Ningxia University,
Yinchuan, 750021, China
liufx@nxu.edu.cn

Abstract. We present an self-adapting approach to generate unified the SLS(Symbolic Labeled System) for network information processing. The semantics of the network information processing are defined using the Inputs, Outputs, Effect (IOE) paradigm. For cach network information processing, our approach produces testing goals high are refinements of the network information processing using set of fault methods. A new planner component accepts these testing goals, along with an initial state of the network information processing definitions to generate a sequence of network information processing invocations as a test case. We study the techniques used in our software test approach. These results indicate that the approach described here leads to a large number of remains in effort with comparable results for requirements coverage and fault detection effectiveness.

Keywords: Network information processing, Software test, Method-Based, processing fault Methods.

1 Introduction

Service oriented architectures (SOA), and network information processing in particular, offer the promise of easier system integration by providing standard protocols for data exchange. Network information processing are the foundations that enable the dynamic discovery and binding of network information processing. Network information processing are specified using standards such as OWL-S and XML [6] and coreLOS of service descriptions in terms of Inputs, Outputs, Outputs, and Effects (IOE).

Testing of such network information processing poses challenges because the persistent state of the world (in terms of domain instances) of the network information processing needs to be accounted for in order to derive the test suite. Functional unified testing is an activity which ensures that the implementation conforms to some functional specification. It is distinct from standards unified testing which is concerned with the issues of compliance with standards such as WS-I[1].

A functional unified test case for a network information processing is a concatenation of three parts: 1) A Set Up sequence of invocations of network information processing operations from a given initial world state S0 to bring the network information processing into a suitable world state S, 2) a test invocation of a

D. Jin and S. Lin (Eds.): Advances in ECWAC, Vol. 1, AISC 148, pp. 227–232.
springerlink.com © Springer-Verlag Berlin Heidelberg 2012

single network information processing operation in state S to produce a new world state S_, and 3) a Verification Sequence which verifies that the world state S_ is reflected in the implementation of the network information processing. An expected value for each output parameter of the network information processing operation is associated with each step of the test case. During the test execution phase, network information processing operation is invoked with values chosen for each input parameter, and the actual values returned by the implementation of the network information processing are compared with the expected output to produce a verdict of success or failure.

To address the issue of end user testing, our technique elicits another artifact, called test template, which represents the manner in which the end use deploys and navigates the network information processing in its presentation layer. These test templates are used to transform the test cases generated through the planning process into an executable set of test cases which can be run using available GUI test execution automation tools such as Rational Function Tester [8].

In particular, the specific contributions of this paper are:

1. A novel technique to generate a set of effective test cases for network information processing defined as IOE.
2. An approach to leverage the network information processing during end user testing
3. Results of an extensive case study in an industrial environment to illustrate the applicability of our technique.

These results indicate substantial effort savings using our approach over the traditional practice without loss of fault detection effectiveness.

2 Software Test for Semantic Network Information Processing

Our software test approach uses the process for generating test cases from the IOE description of the network information processing. To facilitate the testing process, an initial state of the world is also specified. First, our software test approach derives test objectives for each TG pair for each operation (called Testing Goals). The derivation of testing goals leverages a set of fault methods each of which represents a best practice in the testing literature. These fault methods include boundary values for world state attributes and the input parameters; cardinality constraints for each collection valued world domain method attribute and the notion of fault sensitization. Fault sensitization is a technique based on the intuition that each testing goal should target exactly one interesting aspect of the operation's behavior. Otherwise, potential faults may be masked. In network information processing, the need for fault sensitization arises because of the overlapping preconditions between two pairs of a given operation.

These testing goals, along with the network information processing IOE description, and an initial state of the world are input to planner component for generation of the set up sequences. The planner and other components of in our testing approach also rely on a suite of constraint solvers for various domains including linear arithmetic, String, Boolean, Enumerated, Set, and List. The planner extends the well-known Graph-plan planning algorithm [5] to address the creation and deletion of instances in the world. Other enhancements to the Graph-plan algorithm include ability to handle numeric, string, and collection valued parameters on the operations (as is typically required in a network information processing world).

Simulation of the test sequence generated so far results in a modified world state. In the next step, our software test approach generates a verification sequence to validate that this world state is also reflected in the internal system state of the implementation. To do so, our approach derives a set of mutant world states based on the expected world state using various fault methods such as 1) No Instance Creation - which negates the effects of instance creation, 2) No Attribute Update - which negates the effect of modifying an attribute of an instance, and 3) Normal Effects in Exception - which augments the effects associated with an exceptional behavior with each of the individual effects in the successful behavior of the same operation. This last mutant enables us to generate test cases which ensure that no unwarranted behavior is exhibited by the implementation. We exploit the planner to generate a sequence which distinguishes the mutant world state from the expected one.

The resulting logical test case is in a execution harness independent format and cannot be executed directly. In order to produce executable test case, we allow the user to specify a template which contains the necessary code fragments for network information processing operation. This template contains placeholders for input and output parameters so that the corresponding actual values can be substituted from the logical test case.

3 Generating Testing Goals

The IOE method is analyzed to generate testing goals for each TG pair in each operation. Our approach applies several refinements to the precondition of a TG pair in order to accomplish this task. These refinements are based on fault methods which represent common programming errors and best test practices. Here we provide an example of applying the cardinality refinement to the pre conditions. The principle behind the cardinality refinement derives from the quantifiers (both existential and universal) present in the preconditions of operation. A universal quantifier implies iteration over a collection valued domain method entity D such as instances of a class or members of an association) to check for a certain property. An interesting testing goal in this situation is to check the behavior when the collection D is empty to begin with. Other testing goal which is commonly practiced requires the collection to have cardinality of one. Lastly, to make testing process manageable another testing goal treats all the other cardinalities (≥2) as same. We encode this practice in our testing goal generation and derive from each universal quantified expression over collection D, three independent testing goals by conjoining the original flow condition with one of the following: COUNT(D)=0, COUNT(D)=1, COUNT(D)≥2. Similarly, for an existential quantified guard condition, we derive two independent testing goals by conjoining one of COUNT(D)=1, COUNT(D)≥2. Note that the case where the collection is empty is not relevant in this situation since at least one instance of the collection is expected to satisfy the property being checked.

In certain cases, the precondition may not have any quantification, yet the network information processing operation effect may have an update which creates an instance of a class. For example, the precondition for Create Order is a quantifier free expression in Order. However, Create Order includes an update effect which leads to creation of a new instance of Order. From a testing point of view, one would like to test such creation behaviors under different conditions of preexisting instances of Order. In such circumstances, our analysis derives three independent testing goals obtained by conjoining the original guard with one of the following: COUNT(D) = 0, COUNT(D) = 1, COUNT(D)≥2.

4 Case Studying

In this Section, we report the findings from a case study during end user testing stage of an industrial application. The subject of the case study is an SOA application for online management of re-usable assets. For proprietary reasons, we refer to the application only as Product-A. Product-A is a consumer of five network information processing each serving an purpose.

We divided the study in two iterations. The first iteration consisted of method the Asset Management network information processing and in the second iteration all the remaining four services were viewed. The requirements coverage, coverage efficiency and fault detection efficiency of the test suite thus generated was measured against our benchmark. The first iteration was used to calibrate the measurement process and also to identify if our software test approach itself had any missing requirements. Results of the first iteration indicated that our approach lacked the capability of handling NULL value for the optional variables.

During iteration two, we created test templates for each service operation using Rational Function Tester (RFT) which was the choice of test execution automation by the testing team which was doing manual test design. The test cases created using this approach were imported into RFT. The case study was designed for an "in-vivo" application and therefore like most other situations one could not have access to the entire set of development artifacts. Furthermore, data regarding the developer testing of individual services was also not available.

Our limitation was that we had access to only a version of the application and its XML description. Thus we had to allow various approximations in order to guess the actual measure. For instance, the requirements for the application were measured by counting the operations and by identifying various failure modes of them. The assumption was that the application did not have any missing requirements (since it is "very stable" it may be safe to assume so). Thus if we are able to test for all the failure causing scenarios, we are able to test for all behavioral requirements. Also, one would ideally measure effectiveness by running the two set of test cases on an identical version of the application, but such an experiment could not conducted since older buggy versions of Product-A were not available.

During the first iteration, our approach did not support NULL valued optional variables. Consequently, the Self-adapting test suite had smaller requirements coverage and it found one less defect than the benchmark. However, the improvements in effort spent using the two approaches are encouraging for both the iterations. The results from the second iteration indicate the significant improvements in effort without any loss of requirements or fault detection coverage. Out of the seven faults detected during iteration 2, four were logical in nature such as: implementation not handling incorrect parameters, or not checking for creation of instances with duplicate keys, or not ensuring that the deleted entities were indeed deleted from the system state. The last defect underscore the importance of state verification sequences - without one this defect may have gone unnoticed. The remaining three faults were due to GUI navigation - for example cancel button leading to infinite loop. Even though the IOE method of the network information processing does not contain any information about GUI navigation, the test template portion exercised the typical GUI navigation scenarios (such as pressing Cancel or Back buttons). Furthermore, we were able to simulate the developer testing scenario by creating a test template to produce executable test cases which consisted of SOAP messages which could be directly applied to the network information processing implementation through an execution engine. Thus, we could reuse the same IOE information for both the developer and end user testing and obtain even more significant effort reductions. Further, these results also imply that if our software test approach was used during the development testing using the IOE paradigm, several logical faults that were encountered by the end user testing team would have been caught earlier in the life cycle.

5 Summary and Future Work

We have presented an self-adapting approach to generate functional unified the SLS for network information processing which are defined using the Inputs, Outputs, Effect(IOE) paradigm. For each network information processing, our approach produces testing goals using a set of fault methods. A novel planner component accepts these testing goals to generate a sequence of network information processing invocations as a test case. Another salient feature of our approach is generation of verification sequences. Lastly, our technique allows generation of executable test cases which can be applied to the various interfaces through which the network information processing may be accessed. We have described our technique through an example network information processing. We also presented results which compare two approaches: an existing traditional approach without the formal IOE information and the IOE-based approach reported in this paper. These results indicate that the approach described here leads to substantial savings in effort with comparable results for requirements coverage and fault detection effectiveness during both the development and end user testing.

We would like to extend our work in several directions. For example, our current work assumes only atomic network information processing. We would like to extend our approach to composite network information processing (also defined using IOE paradigm). We are interested in comparing our approach to those developed for testing descriptions (which represent compositions of atomic network information processing). We would also like to extend our approach to a more on-the-fly approach where the test cases are not produced a priori but are generated in an adaptive manner.

References

1. Alves, A., et al.: Web services business process execution language version 2.0. committees/documents.php (2006), http://www.oasisopen.org/
2. Arkin, A., Askary, S., Fordin, S., Jekeli, W., et al.: Web Service Choreography Interface (WSCI) 1.0. W3C (2002), http://www.w3.org/TR/wsci/
3. Chan, P.: Best Route Finding specification, (2006), http://www.cse.cuhk.edu.hk/pwchan/BRF.doc
4. Chan, P., Lyu, M.: Developing aerospace applications with a reliable web services paradigm. In: Proc. of IEEE 22nd International Conference on Advanced Information Networking and Applications, Okinawa, Japan, March 25-28 (2008)
5. Chan, P.P.W., Lyu, M.R., Malek, M.: Making Services Fault Tolerant. In: Penkler, D., Reitenspiess, M., Tam, F. (eds.) ISAS 2006. LNCS, vol. 4328, pp. 43–61. Springer, Heidelberg (2006)
6. Chan, P., Lyu, M., Malek, M.: Reliable web services: Methodology, experiment and modeling. In: Proc. of IEEE International Conference on Web Services, Salt Lake City, Utah, USA, July 9-13 (2007)
7. Chan, P.P.W.: Building Reliable Web Services: Methodology, Composition, Modeling and Experiment. PhD thesis, The Chinese University of Hong Kong, Hong Kong (December 2007)
8. Dong, W.-L., Yu, H., Zhang, Y.-B.: Testing bpel-based web service composition using high-level petri nets. In: Proc. of the 10th IEEE International Enterprise Distributed Object Computing Conference (EDOC 2006), October 2006, pp. 441–444 (2006)
9. Jones, S.: Toward an acceptable definition of service (service-oriented architecture). IEEE Transactions on Software 22(3), 87–93 (2005)
10. http://lsdis.cs.uga.edu/projects/meteors/XMLs/examples/purchaseOrder.wsd
11. Bai, X., Dong, W., Tsai, W.-T., Chen, Y.: XMLbased automatic test case generation for network information processing testing. In: Proc. IEEE International Workshop on Service-Oriented System Engineering 2005, pp. 215–220 (2005)
12. Bertolino, A., Frantzen, L., Polini, A., Tretmans, J.: Audition of Web Services for Testing Conformance to Open Specified Protocols. In: Reussner, R., Stafford, J.A., Ren, X.-M. (eds.) Architecting Systems with Trustworthy Components. LNCS, vol. 3938, pp. 1–25. Springer, Heidelberg (2006)
13. Blum, A.L., Furst, M.L.: Fast planning through planning graph analysis. Artificial Intelligence 90(1-2), 279–298 (1997)
14. Coalition", T.O.S.: Owl-s: Semantic markup for network information processing (2003), http://www.daml.org/services/
15. Heckel, R., Mariani, L.: Automatic Conformance Testing of Web Services. In: Cerioli, M. (ed.) FASE 2005. LNCS, vol. 3442, pp. 34–48. Springer, Heidelberg (2005)

Exponential Stability of Numerical Solutions to Stochastic Competitive Population Equations with Markovian Switching

Hai-ming He[1,*] and Qi-min Zhang[2]

[1] Department of Scientific Research, Ningxia University,Yinchuan 7500021
[2] School of Mathematics and Computer Science, Ningxia University, Yinchuan 750021
hehaim@nxu.edu.cn

Abstract. The aim of this paper is to introduce a class of stochastic competitive system with Markovian switching. By using Bark holder-Davis-Gundy's lemma, Ito's formula, some special inequalities and some criteria are obtained for the exponential stability of stochastic competitive system with Markovian switching.

Keywords: Exponential stability, Stochastic competitive system, Strong solution, Numerical solutions, Markovian switching.

1 Introduction

Recently, stochastic modeling with Markovian switching has received a great deal of attention [1-6]. Ronghua Li [6] studied Convergence of numerical solutions to stochastic age-dependent population equations with Markovian switching. Above mentioned single-species research, In fact, we should consider the influences of multi-population.

Consider the following stochastic age-dependent population equations with Markovian switching

$$d_t u = -\frac{\partial u}{\partial a}dt - \mu_1(a,u)udt - f(a,t)udt - u\int_0^A c_1(x,a)v(x,t)\,dxdt$$
$$+f_1(u,r(t)) + g_1(u,r(t))dw_s, \qquad in\ Q,$$

$$d_t v = -\frac{\partial v}{\partial a}dt - \mu_2(a,v)vdt - g(a,t)vdt - v\int_0^A c_2(x,a)u(x,t)\,dxdt$$
$$+f_2(v,r(t)) + g_2(v,r(t))dw_s, \qquad in\ Q,$$

---

[*] Corresponding author.

D. Jin and S. Lin (Eds.): Advances in ECWAC, Vol. 1, AISC 148, pp. 233–239.
springerlink.com © Springer-Verlag Berlin Heidelberg 2012

$$u(a,0)=u_0(a), v(a,0)=v_0(a), \quad \text{for } a\in[0,A]$$
$$u(0,t)=\int_0^A \beta_1(a)u(a,t)da, \quad \text{for } t\in[0,T]$$
$$v(0,t)=\int_0^A \beta_2(a)v(a,t)da, \quad \text{for } t\in[0,T] \tag{1}$$

where $Q=(0,A)\times(0,T)$. $u(a,t),v(a,t)$ denotes two species population density of age a at time t, $\mu_1(a,u)$ and $\mu_2(a,v)$ represent age and density specific mortality which is the death rate of each species in an infinitesimally small age interval similarly, $\beta_1(a)$ and $\beta_2(a)$ represent age specific fertility. The coupling terms represent the interaction between the species with kernels $c_1(x,a)$, $c_2(x,a)$, $f_1(u,r(t))+g_1(u,r(t))dw$ and $f_2(v,r(t))+g_2(v,r(t))dw$ denote stochastically perturbed, $\dot{w}$ is white noise.

In this paper, we shall extend the idea from the paper [6] to stochastic competitive system with age-structure and from the papers [7] to the exponential stability of stochastic competitive system with Markovian switching. We prove that the numerical solution shall be exponentially stable in mean square.

2 Model Description and Preliminaries

Throughout this paper, let (Ω,F,P) be a complete probability space with a filtration $\{F_t\}_{t\geq 0}$ satisfying the usual conditions (i.e., it is increasing and right continuous while F_0 contains all p-null sets).

Let $r(t)$, $t\geq 0$, be a right-continuous Markov chain on the probability space taking values in a finite state $S=\{1,2,\ldots,N\}$ with the generator $\Gamma=(r_{ij})_{N\times N}$ given by

$$p\{r(t+\Delta)=j\,|\,r(t)=i\}=\begin{cases} r_{ij}\Delta+o(\Delta) & if i\neq j,\\ 1+r_{ij}\Delta+o(\Delta) & if i=j.\end{cases} \tag{2}$$

Where $\Delta>0$ .Here $r_{ij}\geq 0$ is the transition rate from i to j if $i\neq j$ while $r_{ii}=-\sum_{i\neq j} r_{ij}$.

We assume that the Markov chain $r(\cdot)$. is independent of the Brownian motion W_t. It is well known that almost every sample path of $r(t)$. is a right-continuous step function with a finite number of simple jumps in any finite subinterval of R_+.

Let $V = H^1([0,A]) = \{\varphi \mid \varphi \in L^2([0,A]), \frac{\partial \varphi}{\partial a} \in L^2([0,A])$,where $\frac{\partial \varphi}{\partial a}$ are generalized partial derivatives$\}$. V is the Sobolev space . $H = L^2([0,A])$ such that $V \to H \equiv H' \to V'$. V' is the dual space of V .We denote by $\|\cdot\|$, $|\cdot|$ and $\|\cdot\|_*$ the norms in V , H and V' respectively; by $\langle \cdot,\cdot \rangle$ the duality product between V , V' , and by $(\cdot,\cdot)$ the scalar product in H , and m a constant such that $|x| \le m\|x\|$,. $\forall x \in V$, $\|B\|_2$ denotes the Hilbert-Schmidt norm, i.e. $\|B\|_2^2 = tr(BWB^T)$.

For system (1) the discrete approximate solution on $t \in \{0, \Delta, 2\Delta, ..., N\Delta = T\}$ is defined by the iterative scheme

$$
\begin{aligned}
U_t^{n+1} &= U_t^n - \frac{\partial U_t^{n+1}}{\partial a}\Delta - \mu_1(a, U_t^n)U_t^n\Delta - f(a,t_k)U_t^n\Delta - AU_t^n c_1(x,a)v(x,t_k)\Delta \\
&+ f_1(U_t^n, r_k^\Delta) + g_1(U_t^n, r_k^\Delta)\Delta w_k, \\
V_t^{n+1} &= V_t^n - \frac{\partial V_t^{n+1}}{\partial a}\Delta - \mu_1(a, V_t^n)V_t^n\Delta - f(a,t_k)V_t^n\Delta - AV_t^n c_1(x,a)u(x,t_k)\Delta \\
&+ f_2(V_t^n, r_k^\Delta) + g_2(V_t^n, r_k^\Delta)\Delta w_k,
\end{aligned} \tag{3}
$$

with initial value $U_t^0 = u(0,a), U^n(t,0) = \int_0^A \beta(t,a)U_t^n da,\ r_k^\Delta = r(k\Delta)$, $n \ge 1$. Here, U_t^n is the approximation to $u(t_n, a)$, for $t_k = k\Delta$, the time increment is $\Delta = \frac{T}{N} \ll 1$, Brownian motion increment is $\Delta w_k = w(t_{k+1}) - w(t_k)$.

For convenience, we first define two-step functions

$$
Z_t = Z(t,a) = \sum_{n=0}^{N-1} U_t^n 1_{[n\Delta,(n+1)\Delta]}, \quad \bar{r}(t) = \sum_{n=0}^{N-1} r_n^\Delta 1_{[n\Delta,(n+1)\Delta]}(t),
$$

$$
W_t = W(t,a) = \sum_{n=0}^{N-1} V_t^n 1_{[n\Delta,(n+1)\Delta]},
$$

Where 1_G is the indicator function for the set G. Then we define

$$U_t = u_0 - \int_0^t \frac{\partial U_s}{\partial a} ds - \int_0^t \mu_1(a, Z_s) Z_s ds - \int_0^t f(a,t) Z_s ds$$
$$-Z_s \int_0^t \int_0^A c_1(x,a) v(x,t) dx ds + \int_0^t f_1(Z_s, \bar{r}) ds + \int_0^t g_1(Z_s, \bar{r}) dw_s$$

$$V_t = v_0 - \int_0^t \frac{\partial V_s}{\partial a} ds - \int_0^t \mu_2(a, W_s) W_s ds - \int_0^t g(a,t) W_s ds$$
$$-W_s \int_0^t \int_0^A c_2(x,a) u(x,t) dx ds + \int_0^t f_2(W_s, \bar{r}) ds + \int_0^t g_2(W_s, \bar{r}) dw_s, \quad (4)$$

$$U_0 = u(0,a), U(t,0) = \int_0^A \beta(t,a) U_t da, U_t = U(t,a), \bar{r}(0) = i_0,$$

$$V_0 = v(0,a), V(t,0) = \int_0^A \beta(t,a) V_t da, V_t = V(t,a), \bar{r}(0) = i_0, \quad (5)$$

It is straightforward to check that $Z(t_k,a) = U_t^k = U(t_k,a)$, $W(t_k,a) = V_t^k = V(t_k,a)$.

As the standing hypotheses we always assume that the following conditions are satisfied:

(i) $\varepsilon \le \int_0^A c_1(s,a) u(x,t) dx \le M$,

$\varepsilon \le \int_0^A c_2(s,a) v(x,t) dx \le M$, $\varepsilon \le f(a,t) \le k$

(ii) there exists positive constants K such that $x, y \in C, i \in S$,

$|f(i,y) - f(i,x)| \vee \|g(i,y) - g(i,x)\|_2 \le K \|y - x\|_c$,

(iii) $\mu(t,a), \beta(t,a)$ are continuous such that

$0 \le \mu_0 \le \mu(t,a) \le \bar{\alpha} < \infty,\ 0 \le \beta(t,a) \le \bar{\beta} < \infty$..

3 Several Lemmas

In this section, we shall give several lemmas l for the following main results.

Lemma 3.1. Under the assumptions above, if (u_t, v_t) is a global strong solution to Eq.(2), then there exist constants $\tau > 0, C_1 > 0$ such that

$$E\left(|u_t|^2+|v_t|^2\right)\le C_1E\left(|u_0|^2+|v_0|^2+k(\delta)\right)e^{-\tau t} \qquad \forall t\ge 0. \tag{6}$$

The proof of this lemma is seen to Theorem 4.1 in [7].

Lemma 3.2. Under the assumptions above, then

$$E\sup_{t\in[0,T]}\left(|U_t|^2+|V_t|^2\right)\le C_2, \tag{7}$$

where C_2 depends on p_0 and T, but not upon Δ.

The proof of this lemma is similar to that of lemma 3.1 in [8].

Lemma 3.3. Under the assumptions above , then

$$E\sup_{t\in[0,T]}\left(|U_t-Z_t|^2+|V_t-W_t|^2\right)\le C_3\Delta\left(\sup_{t\in[0,T]}E\left(|U_t|^2+|V_t|^2\right)\right). \tag{8}$$

Proof. For $\forall t\in[0,T]$, there exists an integer k such that $\forall t\in[k\Delta,(k+1)\Delta)$, applying condition (i), we have

$$|U_t-Z_t|^2\le 6\left|\int_{k\Delta}^t\frac{\partial U_s}{\partial a}ds\right|^2+6\left|\int_{k\Delta}^t\mu_1(s,Z_s)Z_sds\right|^2+6\left|\int_{k\Delta}^t f(a,t)Z_sds\right|^2$$

$$+6\left|Z_s\int_{k\Delta}^t\int_0^A c_1(x,a)v(x,t)dxds\right|^2+6\left|\int_{k\Delta}^t f_1(Z_s,\bar r)ds\right|^2+6\left|\int_{k\Delta}^t g_1(Z_s,\bar r)dw_s\right|^2,$$

Using the Doob inequality yield

$$\begin{aligned}&E|U_t-Z_t|^2\\&\le 6\Delta m_1^2\sup_{t\in[0,T]}E|U_t|^2T+6\left(\bar\alpha^2+k^2+K^2+M\right)\Delta\sup_{t\in[0,T]}E|U_t|^2T+6K^2\Delta\sup_{t\in[0,T]}E|U_t|^2T\\&=c_3\Delta\sup_{t\in[0,T]}E|U_t|^2\end{aligned}$$

Similarly we have $E|V_t-W_t|^2\le c_3\Delta\sup_{t\in[0,T]}E|V_t|^2$,we obtain (8). □

Lemma 3.4. For any $t\in[0,T]$

$$E\int_0^t\left|f(\bar r(s),Q_s)-f(r(s),Q_s)\right|^2ds\le c_4\Delta+o(\Delta),$$

$$E\int_0^{t_k}\left\|g(\bar r(s),Q_s)-g(r(s),Q_s)\right\|_2^2ds\le c_5\Delta+o(\Delta), \tag{9}$$

The proof is similar to that in [9].

Lemma 3.5. Under the assumptions above, then

$$\sup_{t\in[0,T]} E\left(\left|u_t - U_t\right|^2 + \left|v_t - V_t\right|^2\right) \le c_{5,T}\Delta \sup_{t\in[0,T]} E\left(\left|U_t\right|^2 + \left|V_t\right|^2\right). \tag{10}$$

Proof. Using Ito's formula, and (ii),

$$E \sup_{t\in[0,T]} \left|u_t - U_t\right|^2$$

$$\le \left(A^2\overline{\beta}^2 + 3\overline{\alpha} + 2K^2 + 3k + 3M + 1 + 2\mu_2\right) \int_0^t E \sup_{r\in[0,T]} \left|u_r - U_r\right|^2 dr$$

$$+\left[c_3\left(\overline{\alpha} + k + M + K\right) \sup_{s\in[0,T]} E\left|U_s\right|^2 + c_4 + c_5 + \mu_2 c_5\right]\Delta + o(\Delta) + \frac{1}{2} E \sup_{s\in[0,T]} \left|u_s - U_s\right|^2$$

$$\le D_1\Delta \sup_{s\in[0,T]} E\left|U_s\right|^2 + D_2 \int_0^t E \sup_{r\in[0,T]} \left|u_r - U_r\right|^2 dr,$$

where μ_1, μ_2 are two positive constants,

$$D_1 = 2TC_3\left(\overline{\alpha} + k + M + K\right),$$

$$D_2 = 2\left(A^2\overline{\beta}^2 + 3\overline{\alpha} + 2K^2 + 3k + 3M + 1 + 2\mu_2\right).$$

Similarly we have the same approximation for

$$E \sup_{t\in[0,T]} \left|v_t - V_t\right|^2 \le D_1\Delta \sup_{s\in[0,T]} E\left|V_s\right|^2 + D_2 \int_0^t E \sup_{r\in[0,T]} \left|v_r - V_r\right|^2 ds.$$

The result (10) then follows from the continuous Gronwall inequality with $C_{5,T} = D_1 \exp\left(D_2 T\right)$.

Theorem 2.2. Under hypotheses (i)-(iv) and $v = \alpha/m^2 - \lambda + 2\varepsilon - A\overline{\beta}^2 + 2\mu_0 > 0$ hold, then the Euler method applied to Eq.(1) is exponentially stable in mean square.

The proof is similar to that in [6].

References

1. Arnold, L.: Stochastic Differential Equations, Theory and Applications (1972)
2. Kloeden, P.E., Platen, E.: Numerical Solution of Stochastic Differential Equations. Springer (1992)
3. Mao, X.: Stochastic Differential Equations and Applications, Horwood (1997)

4. Mao, X.: Stability of stochastic differential equations with Markovian swithing. Stochastic Processes and their Applications 79, 45–67 (1999)
5. Mao, X., Matasov, A., Plunovskly, A.B.: Stochastic differential delay equations with Markovian switching. Bernoulli 6, 73–90 (2000)
6. Li, R., Leung, P.-K., Pang, W.-K.: Convergence of numerical solutions to stochastic age-dependent population equations with Markovian switching. Journal of Computational and Applied Mathematics 233, 1046–1055 (2009)
7. Pang, W.K., Li, R., Liu, M.: Exponential stability of numerical solutions to stochastic age-dependent population equations. Applied Mathematics and Computation 183, 152–159 (2006)
8. Li, R., Meng, H., Chang, Q.: Convergence of numerical solutions to stochastic age-dependent population equations. J. Comput. Appl. Math. 193, 109–120 (2006)
9. Zhou, S., Wu, F.: Convergence of numerical solution to stochastic delay differential equation with Markovian switching. Journal of Computational and Applied Mathematics 229, 85–96 (2009)

A Rough-Set Adaptive Interactive Multi-Modulus Algorithm Based on Neural Network

Jing Cao, WenShen Zhao, Ying Gao, and WeiZheng Ren

School of Electronic Engineering,
Beijing University of Posts and Telecommunications, Beijing 100876, China
renwz@yahoo.cn, zhaows@bupt.edu.cn,
gybupt@sina.com, renwz@bupt.edu.cn

Abstract. In order to overcome effects of complex environment and movement on node tracking, and improve tracking precision further, a rough-set neural network-based adaptive tracking algorithm named rough-set adaptive interactive multi-modulus (RSAIMM) was proposed in which tracking filter based on neural network was employed. When target moved between maneuvering and non-maneuvering, rough-set neural network output adaptive feature value automatically online, to adapt to the movement with adequately accurate system variance and maintain high precision tracking of target state. Simulation results showed that compared with traditional interactive multi-modulus (IMM) tracking algorithm, tracking precision of RSAIMM algorithm was improved by 23.15%.

Keywords: WSN, rough-set, neural network, tracking.

1 Introduction

Although neural network integration has become one of effective engineering neural network design methods, there are still many factors that directly affect precision and generalization of neural network in practice, which have limited its further application into multiple fields. Rough-set (RS) theory has great potential in solving issues such as uncertainty and imprecision. Combined with neural network method and applied in tracking model adaptation, RS theory will become an effective way to breakthrough the limitation of traditional methods. In so doing, RS and neural network integration were organically combined, and an adaptive method of intelligent tracking model based on RS and neural network integration was proposed in this paper to solve the issue of tracking model adaptation based on WSN node [1], [2].

2 Adaptive Model Based on RS Neural Network

RS adaptive neural network model was mainly composed of data acquisition, data preprocessing, RS data processing, and integration decision of multiple sub-networks in adaptive neural network (ANN) [3]. RS neural network decision model was shown in Fig.1.

D. Jin and S. Lin (Eds.): Advances in ECWAC, Vol. 1, AISC 148, pp. 241–245.
springerlink.com © Springer-Verlag Berlin Heidelberg 2012

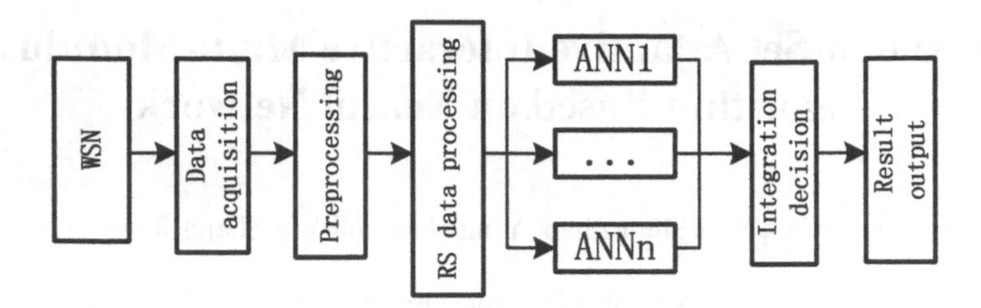

Fig. 1. RS neural network decision model

2.1 Data Acquisition and Preprocessing

In Data acquisition and preprocessing, sensor networks were mainly utilized to acquire and track related movement property information of nodes, then related data signal processing method was utilized to remove noise, process tendency item, extract movement feature and carry out other preprocessing work, thus obtaining samples of original movement model.

2.2 RS Data Processing

There are usually redundant information in original movement model samples obtained by data acquisition and data preprocessing. But neural network fails to judge what knowledge is redundant or useful, and what knowledge plays a greater or smaller role. Thanks to advantages of RS theory in large-amount data processing and redundant information elimination, etc., feature set of movement model which makes greater contribution to movement models was extracted from data samples. So that the amount of neural network training data was reduced and the issue of too large-amount data for neural network algorithm was solved.

2.3 Integration Decision

Firstly, for each neural network, original topological structure of neural network was determined according to the minimum tracking model feature obtained by RS data processing. Then training sample set related to the minimum tracking model feature was used to train each sub-network and mapping relationship between movement feature and tracking model was established. Next, training sample set was used to test integration neural network. Finally, results of model adaptation were obtained through *voting* of each sub-network [4].

3 Node Movement Model

Tracking system of Maneuvering target has been described by target movement model and observation model and nearly all tracking methods are based on model. Commonly used model is described by Bayesian framework of dynamic state space as Equations (1-2) [5].

$$X_k = f_k(X_{k-1}, w_{k-1}) \ . \tag{1}$$

$$z_k = h_k(X_k, n_k) \ . \tag{2}$$

Where, X_k denotes system state, z_k is output observation value, w_{k-1} denotes process noise, n_k denotes observation noise, mapping functions $f(\bullet)$ and $h(\bullet)$ denotes the corresponding process and observation model.

The key of tracking issue is movement model, based on which, numerous model-based filtering methods have been applied. Usually, constant velocity (CV) model (non-maneuvering) has been used to describe weak maneuvering and non-maneuvering point target state. It has been one of oldest simple models usually used for analyzing property of tracking algorithm. However, when target suddenly stops or turns, continuous use of CV model will deteriorate tracking performance, even fails to trap target sometimes. Therefore, constant acceleration (CA) movement model [6] was also joined in the adaptive distributed target tracking research in this chapter.

4 Adaptive Filtering Based on Current Statistical Model

Considering the case of one-dimensional system, discrete state equation (3) [5] was

$$X(k+1) = \Phi(k+1,k)X(k) + U(k)\overline{a}(k) + w(k) \ . \tag{3}$$

Observation equation was

$$Y(k) = H(k)X(k) + V(k) \ . \tag{4}$$

Where, $X(k) = [x(k)\dot{x}(k)\ddot{x}(k)]^T$, $\overline{a}(k)$ was average of random maneuvering acceleration, $W(k)$ was Gaussian white noise with zero-mean variance Q(k), V(k) was Gaussian observation noise with zero-mean variance R(k) and T was sampling period [7].

$$\Phi(k+1,k) = \begin{bmatrix} 1 & T & \frac{1}{\alpha^2}(-1+\alpha T + e^{-\alpha T}) \\ 0 & 1 & \frac{1}{\alpha}(1-e^{-\alpha T}) \\ 0 & 0 & e^{-\alpha T} \end{bmatrix} . \tag{5}$$

$$Q(k) = E[W(k)W^T(k)] = 2\alpha\sigma_a^2 \begin{bmatrix} q_{11} & q_{12} & q_{13} \\ q_{12} & q_{22} & q_{23} \\ q_{13} & q_{23} & q_{33} \end{bmatrix} . \tag{6}$$

When only target position observation data with noise was used in observation equation (4), $H(k) = [1 \quad 0 \quad 0]$, Kalman filter equations were shown in Equations (7-11) :

$$\hat{X}(k/k) = \hat{X}(k/k-1) + K(k)\left[Y(k) - H(k)\hat{X}(k/k-1)\right] . \tag{7}$$

$$\hat{X}(k/k-1) = \Phi(k,k-1)\hat{X}(k-1/k-1) + U(k)\overline{a}(k) \ . \tag{8}$$

$$K(k) = P(k/k-1)H^T(k)\left[H(k)P(k/k-1)H^T(k) + R(k)\right]^{-1} \ . \tag{9}$$

$$P(k/k-1) = \Phi(k,k-1)P(k-1/k-1)\Phi^T(k,k-1) + Q(k-1) \ . \tag{10}$$

$$P(k/k) = \left[I - K(k)H(k)\right]P(k/k-1) \ . \tag{11}$$

5 Simulation Experiments

Target scenes were designed for maneuverable walking personnel targets with uniform velocity, circular movement, slow acceleration or fast acceleration, and tracking simulation was carried out correspondingly. The starting point of targets was between (50m, 100m), WSN localization period $T = 1s$, with x axis and y axis observed independently. Standard deviation of observation noise was designed equally as Var=5m. Variance of state noise was $Q = 10I$. Variance of observation noise was $R = 20I$. Tracking standard deviation was used as performance evaluation index. The number of Monte Carlo simulation was designed as 100.

RS scalar feature set with five reference accelerations was used in RS decision model adaptation. $\alpha = 0.3$ and $\beta = 0.03$ were used for adaptive two-level decision of neural network. Tracking trace and tracking standard deviation of standard interactive multi-modulus algorithm (IMM) were shown in Figs.2-3, respectively. The maximum of tracking standard deviation of IMM was 13.2m. Tracking trace and tracking standard deviation of RSAIMM were shown in Figs.4-5, respectively. The maximum of tracking standard deviation of RSAIMM was 7.3m.

Simulation experiments were employed on Lenevo Kai-day M715E/E7500 computers with dual-core, 2.93GHz clock speed and 3MB cache. Monte Carlo simulation time for 100 periods with IMM algorithm was 20.625s, and computation time with RSAIMM algorithm was 16.156s, which manifested that the computation amount with RSAIMM algorithm was less than standard IMM algorithm owing to faster convergence speed with RSAIMM algorithm.

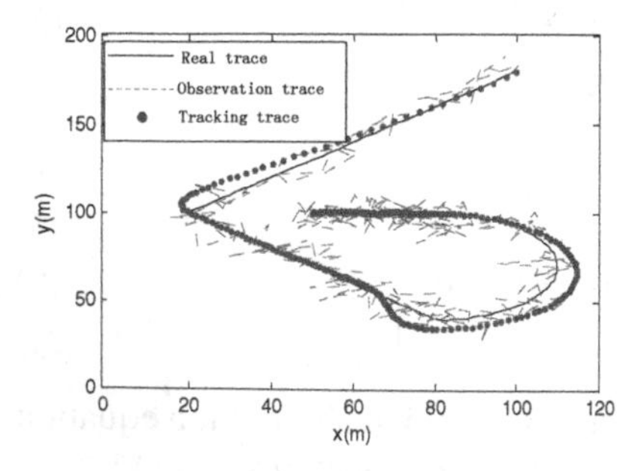

Fig. 2. Tracking trace of IMM

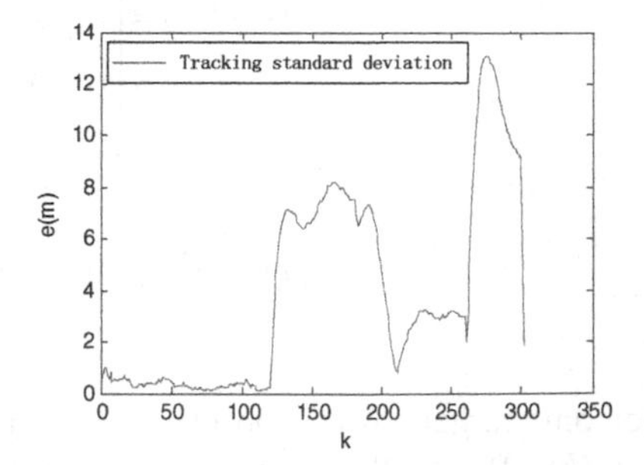

Fig. 3. Tracking standard deviation of IMM

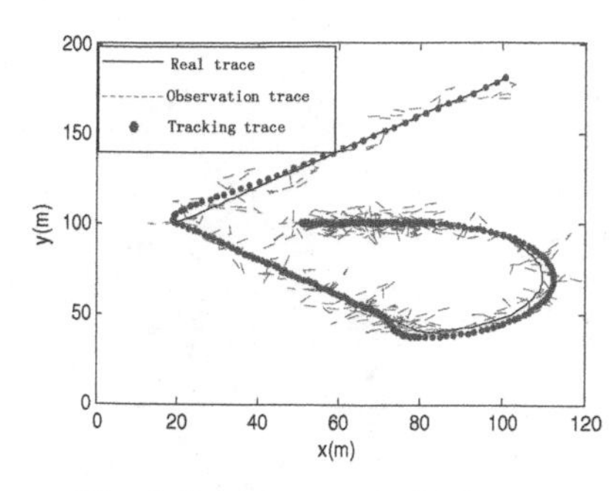

Fig. 4. Tracking trace of RSAIMM

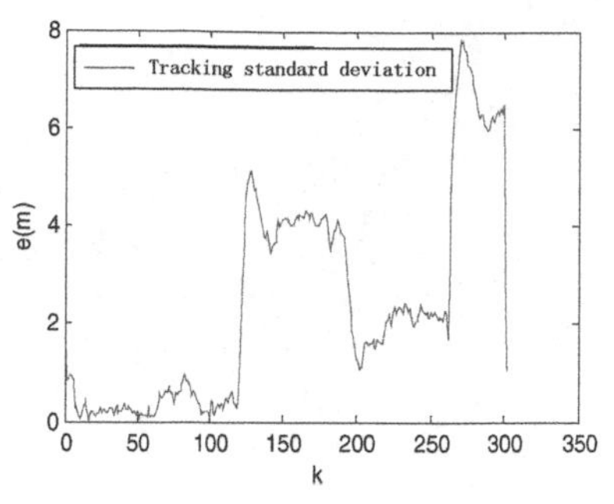

Fig. 5. Tracking standard deviation of RSAIMM

6 Conclusion

In order to effectively monitor the distribution of personnel, vehicles, goods or equipments, etc., eliminate the effect of complex environment on localization and tracking of fast- maneuvering target, and improve tracking precision further, an RSAIMM-based adaptive tracking algorithm with RS neural network double filter parallel structure was proposed. When target moved between maneuvering and non-maneuvering, RS neural network output adaptive feature value online automatically, to adapt to movement of target with adequately accurate system variance and maintain high precision tracking of target state.

References

1. Dai, X., Liu, W., Zhu, Z.: Maneuver target tracking based on fuzzy adaptive IMM algorithm. Ship Engineering 29(3), 1–4 (2007)
2. Dai, X., Liu, W.: Maneuver target tracking based on fuzzy IMM algorithm. Journal of Projectiles, Rockets, Missiles and Guidance 27(1), 34–37 (2007)
3. Cao, L., Cao, C.: The research of diesel engine fault diagnosis with ANN based on rough sets theory. Transactions of CSJCE 20(4), 357–361 (2002)
4. Ling, W., Jia, M., Xu, F., et al.: Optimizing, strategy on rough set neural network fault diagnosis system. Proceedings of the CSEE 23(5), 98–102 (2003)
5. Jing, Z., Xu, H., Zhong, X., et al.: Neural network-based information fusion and parallel adaptive tracking of maneuvering targets. Acta Aeronautica et Astronautica Sinica 16(6), 715–719 (1995)
6. Xue, F., Liu, Z., Qu, Y.: Decentralized algorithm for target passive tracking in wireless sensor networks. Journal of System Simulation 19(15), 3499–3503 (2007)
7. Zhu, Z.: Adaptive IMM tracking algorithm based on fuzzy inference. Journal of Projectiles, Rockets, Missiles and Guidance 28(1), 29–32 (2008)

6 Conclusion

References

Discussion on One Sought-after Skill in Web Development: CMS Themes Design

Jianhong Sun[1], Qun Cai[2], and Yingjiang Li[1]

[1] Engineering College of Honghe University
[2] Science College of Honghe University,
Yunnan Mengzi, 661100, China
sparkhonghe@gmail.com

Abstract. No matter how bad the economy worsens, those who with a sought-after skill always are imperative talents gaped after by market. The content manager system (CMS) themes design is becoming one accepted desirable skill in web development. As a guideline for help those who work in computer related fields preparing knowledge and a expertise skill for keeping a position for living or for help the graduates who have just graduated from colleges or universities to get a good pay job. We discuss the technology of one of the most sought-after skills: CMS themes design in this paper. From the required foundation knowledge to the popular CMS themes develop technology all are described. This work is one results of our research in reflection in education of which objective aim to solve the employment difficulties of students.

Keywords: Web development, content manager system, theme design, CMS, WCMS.

1 Introduction

It is generally believed that everyone hope to master one sought-after expertise skill for a living. Furthermore, it not difficult imagine that even in a time of economic crisis, no matter how bad the economy worsens, some kinds of talents are still imperative for the market. The key point is what kinds of skills are sought-after in the recently years and it will be in the future? It is difficult to answer this question across-the-board. In this paper, we only discuss one recognized sought-after skill, Content Management System (CMS) Themes Design which limited in web development of computer scope.

A CMS offers a way to manage large amounts of web-based information that escapes the burden of coding all of the information into each page in HTML by hand [1]. Actually, for different application, the CMS can be divided several kinds of system. Reference [2] list three familiar of them:

- *Enterprise content management systems (ECMS).* ECMS is the strategies, methods and tools used to capture, manage, store, preserve, and deliver content and documents related to organizational processes. ECM tools and strategies allow the management of an organization's unstructured information, wherever that information exists [3].

D. Jin and S. Lin (Eds.): Advances in ECWAC, Vol. 1, AISC 148, pp. 247–252.
springerlink.com © Springer-Verlag Berlin Heidelberg 2012

- *Web content management systems (WCMS).* WCMS is a software system that provides website authoring, collaboration, and administration tools designed to allow users with little knowledge of web programming languages or markup languages to create and manage website content with relative ease [4].
- *Component content management system (CCMS).* In CCMS, the content is stored and managed at the subdocument (or component) level for greater content reuse.

At most of time, it usually refers to the WCMS when CMS is mentioned without specialize because WCMS is more widely used than the others. In this paper, we focus on WCMS discussion and do not distinguish WCMS and CMS.

2 Background

The computer technology is one of the fastest development disciplines. For catch the heels of computer technology development and the market requirement. The computer knowledge and technique updating is more important than other disciplines. Glen Stansberry, a web developer and blogger presented 10 most sought-after skills in web development with 155 comments on website, "Nettuts+" on Oct 20th, 2008. Even now, they still have important reference value. The 10 skills are [5]:

- Framework knowledge
- Widget development
- *Custom CMS themes*
- CMS Customizations and plugin development
- PSD to XHTML services
- Javascript Plugin creation
- Facebook/MySpace applications
- iPhone applications
- Ecommerce integration
- Flash and Actionscript Knowledge

From the list, we can see Custom CMS Themes is one of them. That because the CMS has been become the fist choice for small-sized and medium-sized web development in recently years. As the CMS widely application, offer Custom CMS Themes service is becoming a pop business.

3 The Template-Based Structure of WCMS

At first, let us introduce two concepts, 'template' and 'theme' which hard to distinguish. For example, in XOOPS, a famous open source CMS, uses the concept of 'theme' and 'template' to further refine presentation. Themes provide the basic structure and overall layout of the whole site and it dictates the common elements found in all pages of the site. Templates control layout and design of page elements. Themes provide gross control of the whole site, while templates provide fine manipulations of page details. For more details about themes and templates see [6], the theme development guide of XOOPS.

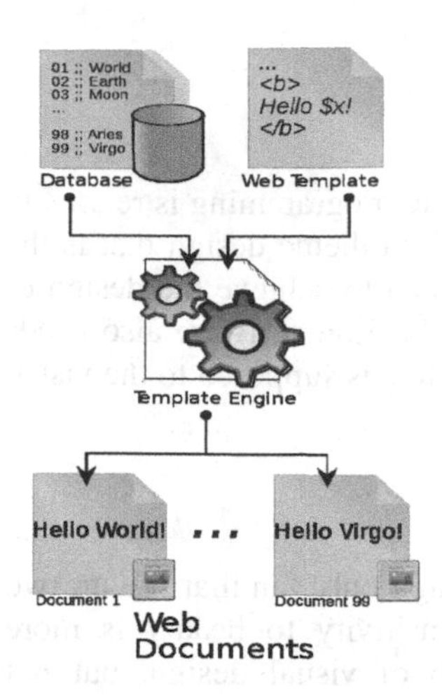

Fig. 1. Content (from a database), and "presentation specifications" (in a web template), are combined (through the template engine) to mass-produce web documents [7]

As a software system, WCMS used to control a dynamic collection of Web material, such as HTML documents, images, flash and other forms of media. A WCMS has many typically features for facilitating document control, auditing, editing, and others management. The template-based structure is one of them. Template-based structure is designed for productivity and control separating the code which provides functionality from the code which makes the content visible to the end user. By separating logic, alternate views can bc introduced without interfering with the content management process. In this structure, a web template engine is designed to process web templates and content information to produce output web documents. The structure is shown as Fig.1. Based on this structure, the programmer can focus on the database design, logical control and functions achievement and the work that how to present the content can left to the theme designer. The theme designers only need focus on building templates without learning much of the internal logic of the system.

At most of cases, it needs not write any code when we employ a WCMS (There are many open source WCMS offering on the Internet.) with wealth of modules to develop a web site, even the theme also has many free themes for choices from the Internet. However, the webs site usually works as a window of a company facing to the world. Most of users do not like their web site has a same theme with others just like the movie stars do not like wear the same cloth with other actors on the same stage. Furthermore, the theme designer of CMS not only needs to familiar some kinds of program languages, such as HTML, XML and CSS, but also he/she must understand the requirement of the theme of the CMS platform. In addition, from the economic cost perspective, the users do not need hire a theme designer. As a result, the situation conduce to that a new kind of career is generating as the development of CMS which is offer customizing CMS themes service.

4 The Required Technologies of CMS Themes Design

To learn the CMS themes design, it needs to understand some basic programming technologies and architectures of specified CMS platform.

4.1 Required Basic Programming Technologies

To be a theme designer usually is expected to be familiar with the following programming languages:

- HTML (and/or) XHTML
- JavaScript
- Cascading style sheets (CSS)

Consider from the programming requirements, the template programming is relatively easy to learn. Furthermore, there is a fairly important job in theme design that is the visual design. The job of theme designer usually is defined to achieve the design of the visual designer via programming. The following application software also needs to learn if the theme designer is going to juggling the job that is supposed to the visual designer. In fact, many freelance is under this situation.

- Photoshop
- Flash

It is noteworthy that the job of visual design is not enough only familiar the up two application software, the ability of originality and sensitivity to beauty is more important. In another word, everyone can do the job of visual design, but not everyone is well.

4.2 Familiar the Architecture of the CMS Theme

Different CMS has different requirement for theme design, the theme designer must understand the architecture of specified CMS theme. As an instance, in this paper, we discuss this issue based on a specified CMS, XOOPS. The Fig.2 shows the directory and of the default theme of XOOPS (\XOOPS\themes\default\) and the HTML files responsible blocks.

XOOPS uses the Smarty Template Engine, so there are some Smarty tags embedded in the HTML code. Smarty tags are identified by the delimiters <{ and }>. As a theme designer of XOOPS, familiar some frequently-used Smarty tags are required.

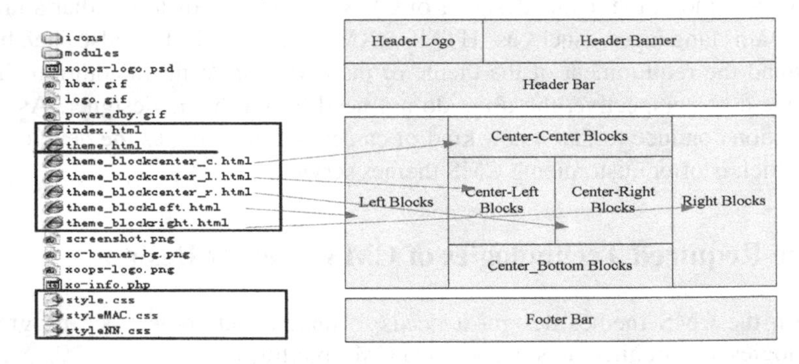

Fig. 2. Directory of the default theme of XOOPS and HTML files responsible control blocks

The XOOPS default theme contains 6 main HTML files to control the representation of responsible blocks, they are: *theme.html, theme_blockleft.html, theme_blockcenter_c.html, theme_blockcenter_l.html, theme_blockcenter_r.html* and *theme_blockright.html* (*index.html* only has a code, '`<script>history.go(-1);</script>`' for back up). Among of them, the *theme.html* is a required file and the others can separate into different files (like the other five files, and these files are optional for request.) or embedded into *theme.html* (like the control file of Header blocks and Footer bar block is embedded into *theme.html*). Let us look at an example; following code is a section of *theme.html*. The fist bold piece code refers to an extra file, *theme_blockleft.html* via a Smarty variable which controls the left block layout. The second bold piece code is the direct embedded code which controls the Footer Bar block layout.

```
......
<!-- Start left blocks loop -->
<{if $xoops_showlblock}>
   <td id="leftcolumn">
   <{foreach item=block from=$xoBlocks.canvas_left}>
    {includeq file="$theme_name/theme_blockleft.html"}>
   <{/foreach}>
   </td>
<{/if}>
<!-- End left blocks loop -->......
......
<!-- Start footer -->
<table cellspacing="0">
<tr id="footerbar">
    <td><{$xoops_footer}></td>
</tr>
</table>
<!-- End footer -->......
```

The following code is the complete code of *theme_blockleft.html*. These codes are optional which can be embedded into *theme.html*.

```
<{if $block.title}>
    <div class="blockTitle"><{$block.title}></div>
<{/if}>
   <div class="blockContent"><{$block.content}></div>
```

From the instance, we can see that the theme designer should familiar the theme architecture of specified CMS first.

5 Introduce Some Popular CMS with Large Market Demand

There are many wonderful CMS available when we are going to build a web site. Some of them are open source web application system. When we are going to be a potential theme designer of CMS, it is good choice to familiar the theme development method of following CMS, which are the most popular currently.

Drupal, Drupal is a free software package that allows users to easily organize, manage and publish the users' content, with an endless variety of customization. As an open publishing platform, it is maintained and developed by many vibrant communities of 630,000+ users and developers. Use as-is or snap in any of thousands of free designs and plug-ins for rapid site assembly. Drupal's content management features make it easy to create and manage a web site which can be a personal site, a university site, a company site and even a middle size ecommerce web site.

WordPress, WordPress is web software which can be used to create a beautiful website or blog. WordPress is an Open Source project, which means WordPress is both free and priceless and there are hundreds of people all over the world working on it to continually improve and perfect it. As a blog web framework, WordPress has already become the NO.1 in recent years. According to the statistics of WordPress community, over 25 million people have chosen WordPress to build their web sites or blogs.

Joomla, Joomla is an award-winning content management system (CMS), which enables the users to build web sites and powerful online applications. Many aspects, including its ease-of-use and extensibility, have made Joomla the most popular Web site software available. Best of all, Joomla is an open source solution that is freely available to everyone [8].

XOOPS (eXtensible Object Oriented Portal System), XOOPS is an open source web application platform written in PHP for the MySQL database. Its object orientation makes it an ideal tool for developing small or large community websites, intra company and corporate portals, weblogs and much more. Though started as a portal system, XOOPS is in fact striving steadily on the track of Content Management System. It can serve as a web framework for use by small, medium and large sites [9].

Still there are many fine CMS we can not introduce all, such like Manbo, Frog CMS, SilverStripe, etc.

Acknowledgments. Thanks for pecuniary aid of Honghe University Bilingual Course Construction Fund (SYKC0802) and Honghe University Discipline Construction Fund (081203).

References

1. Seadle, M.: Content management systems. Library Hi Tech. 24(1), 5–7 (2006)
2. Content management system, Wikipedia, http://en.wikipedia.org/wiki/Content_Management_System
3. What is Enterprise Content Management (ECM)? AIIM. Association for Information and Image Management, http://www.aiim.org/What-is-ECM-Enterprise-Content-Management.aspx
4. Web content management system, Wikipedia, http://en.wikipedia.org/wiki/Web_Content_Management
5. Stansberry, G.: 10 Most Sought-after Skills in Web Development, http://net.tutsplus.com
6. Muin, M.: Anatomy of the Default Theme, http://underpop.free.fr/
7. Template engine (web), http://en.wikipedia.org/wiki/File:TempEngWeb017a.svg
8. Joomla! http://www.joomla.org/about-joomla.html/
9. XOOPS, http://xoops.org/modules/wfchannel/

Obtain Semi-definite Matrix Eigenvalue Based on LANCZOS Algorithm

Hong Shao[1], Zhiguo Wang[2], and Wei Xu[3]

[1] Jianghan University School of Physics and Information Wuhan, China
[2] HuBei Water Resources Research Institute Wuhan, China
[3] Yangtze River Scientific Research Institute Wuhan, China
wuhanshaohong@hotmail.com

Abstract. Orthogonal projection lanczos algorithm is the effective way to solve the complex structural vibration, vibration frequency and vibration mode. Its idea is to translate high-level vibration problem into low-level to solve the vibration problem without losing eigenvalue. In this paper, a simple and convenient method of computing is presented using Orthogonal Projection lanczos algorithm to solve semi-definite matrix generalized eigenvalue problems and re-solve the eigenvalue.

Keywords: structure, eigenvalue, orthogonalization, finite element method.

1 Heading

Suppose vibration equation as follows:

$$KX=\omega 2Mx \tag{1}$$

In the formula, K is symmetric positive semi-definite stiffness matrix, M is mass matrix, to simplify, consider: $M_{ii} > 0,\ M_{ij} = 0(i \neq j)$

Vibration in the above equation (1), K is a matrix of more than 1000 bands. According to the request to obtain just a number of previous vibration frequencies only [1]. Transform the equation (1):

$$\begin{cases} A = L^{-1}ML^{-T} \\ K + \alpha M = LL^{T} \\ Y = L^{T}X, \lambda = 1/(\omega^{2} + \alpha) \end{cases} \tag{2}$$

There: $AY = \lambda Y$

Which: L is a lower triangular matrix, α is a positive number greater than zero, called the translation factor. The choice of translational factors will be based on the condition number of stiffness matrix K to determine. Practice has proved that the finite element method to form the stiffness matrix when α is appropriate between 1 to 10^4.

D. Jin and S. Lin (Eds.): Advances in ECWAC, Vol. 1, AISC 148, pp. 253–259.
springerlink.com © Springer-Verlag Berlin Heidelberg 2012

For the formula (2), we use three recursive Lanczos algorithm, to obtain orthogonal sequence {V0}, namely, using pseudo-random numbers to select a sequence $\overline{V_1}$:

$$\begin{cases} \beta_k = (\overline{V_K}, \overline{V_K})^{\frac{1}{2}} \\ V_K = \overline{V_K} / \beta_k \\ \alpha_K = (AV_K, V_K) \\ \overline{V}_{K+1} = AV_K - \alpha_K V_K - \beta_k \overline{V}_{K-1} \end{cases} \quad (3)$$

Which $\beta n+1=0$:

With (3) produces the vector {V i} has been proved to be a set of standard orthogonal basis. However, the actual experience, found that these values rapidly degraded, the result after the (V_i) is far from being orthogonal with the earlier. The same time, K is a positive semi-definite, that is, A has a multiple root, therefore, when m (<n), then, $\beta_m = 0$. For example:

$$A= \begin{pmatrix} 2 & 0 & 0 \\ 0 & 1 & 0 \\ 0 & 0 & 1 \end{pmatrix}$$

With the eigenvalue : $\lambda 1 = \lambda 2 = 1, \lambda 3 = 2$

By (3) calculation, taking $\overline{V}_1 = (1,2,3)^T$, there are:

$$\beta_1 = \sqrt{14}$$

$$V_1 = 1/\sqrt{14}(1,2,3)^T,$$

$$\alpha_1 = 15/14, \overline{V}_2 = 1/\sqrt{14}(\frac{13}{14}, -\frac{1}{7}, -\frac{3}{14})^T,$$

$$\beta_2 = \frac{1}{14}\frac{1}{\sqrt{14}}\sqrt{182},$$

$$V_2 = 1/\sqrt{182}(13,-2,-3)^T,$$

$$AV_2 = 1/\sqrt{182}(26,-2,-3)^T,$$

$$\alpha_2 = 315/182, \overline{V_3} = (0,0,0)^T$$

So, $\beta_3 = 0$.

The cases show that in accordance with formula (3) Calculate the eigenvalue will be degraded, however, there is the calculation of the round-off error makes $\beta_3 \neq 0$, thereby losing the accuracy of the calculation.

Generally we have the following theorem:

Theorem: n-order symmetric positive definite matrix A, If m multiple roots λ_0 according to (3) type A, the orthogonal process must have $\beta_{n-m+\alpha} = 0$, namely, the initial choice of a vector, it can be uniquely expressed as

$$D=\alpha_1 V_1+\alpha_2 V_2+\ldots+\alpha_n V_n,$$

Set all the coefficients α_i not all zero, then the set of linear independent vectors.

Theorem Proving:

By reduction to absurdity, V_1, V_2, …, V_m is the feature vector of the corresponding characteristic value$\lambda 0$ about matrix A, V_{m+1}, V_{m+2}, …, V_n is the feature vector of the corresponding characteristic value λ_j (j=m+1, m+2, …, n) about matrix A,

Then:

$$\begin{cases} \overline{V} = \sum_{i=1}^{m} a_i V_i + \sum_{j=m+1}^{n} a_j V_J \\ A\overline{V} = a_0 \sum_{i=1}^{m} a_i V_i + \sum_{j=m+1}^{n} a_j \lambda_j V_J \\ \cdots \\ A^{n-m}\overline{V} = \lambda_0^{n-m} \sum_{i=1}^{m} a_i V_i + \sum_{j=m+1}^{n} a_j \lambda_j^{n-m} V_j \end{cases} \tag{4}$$

If $\overline{V}, A\overline{V}, \cdots, A^{n-m}\overline{V}$ linear correlation, then there exists a group of incomplete number of zero $\{C_i\}$,

$$C_0\overline{V} + C_1 A\overline{V} + \cdots C_{n-m} A^{n-m}\overline{V} = 0 \tag{5}$$

To formula (4) into formula (5), the merger of its various items is:

$$\begin{aligned} &(C_0 + C_1\lambda_0 + \cdots + C_{n-m}\lambda_0^{n-m})a_1V_1 \\ &+ (C_0 + C_1\lambda_0 + \cdots + C_{n-m}\lambda_0^{n-m})a_2V_2 \\ &+ \cdots + (C_0 + C_1\lambda_0 + \cdots + C_{n-m}\lambda_0^{n-m})a_mV_m \\ &+ (C_0 + C_1\lambda_{m+1} + \cdots + C_{n-m}\lambda_{m+1}^{n-m})a_{m+1}V_{m+1} \\ &+ \cdots + (C_0 + C_1\lambda_n + \cdots + C_{n-m}\lambda_0^{n-m})a_nV_n = 0 \end{aligned}$$

Because $\{V_i\}$ is linearly independent vector, so $\{V_i\}$ of the coefficients should be zero, so:

$$\begin{cases} (C_0 + C_1\lambda_0 + \cdots + C_{n-m}\lambda_0^{n-m})a_1 = 0 \\ (C_0 + C_1\lambda_0 + \cdots + C_{n-m}\lambda_0^{n-m})a_2 = 0 \\ \cdots\cdots \\ (C_0 + C_1\lambda_0 + \cdots + C_{n-m}\lambda_0^{n-m})a_m = 0 \\ (C_0 + C_1\lambda_0 + \cdots + C_{n-m}\lambda_{m+1}^{n-m})a_{m+1} = 0 \\ \cdots\cdots \\ (C_0 + C_1\lambda_n + \cdots + C_{n-m}\lambda_0^{n-m})a_n = 0 \end{cases} \tag{6}$$

Select the entire coefficients $a_i \neq 0$ base on the assumption, there:

$$\begin{cases} C_0 + C_1\lambda_0 + C_2\lambda_0^2 + \cdots + C_{n-m}\lambda_0^{n-m} = 0 \\ C_0 + C_1\lambda_{m+1} + C_2\lambda_{m+1}^2 + \cdots + C_{n-m}\lambda_{m+1}^{n-m} = 0 \\ \cdots\cdots \\ C_0 + C_1\lambda_n + C_2\lambda_n^2 + \cdots + C_{n-m}\lambda_n^{n-m} = 0 \end{cases} \tag{7}$$

The formula (7) is a homogeneous equation with n-m+1 unknowns and n-m+1 equations, Its determinant is :

$$\begin{vmatrix} 1 & \lambda_0 & \lambda_0^2 & \cdots & \lambda_0^{n-m} \\ 1 & \lambda_{m+1} & \lambda_{m+1}^2 & \cdots & \lambda_{m+1}^{n-m} \\ \cdots\cdots & & & & \\ 1 & \lambda_n & \lambda_n^2 & \cdots & \lambda_n^{n-m} \end{vmatrix}$$

Because of $\Delta \neq 0$, then the equation (7) Only the zero solution, C_i=0, this contradicts the hypothesis. Therefore, $\overline{V}, A\overline{V}, \cdots, A^{n-m}\overline{V}$ is a linearly independent vector.

2 Vector Orthogonal Decomposition Ease of Use

Now to improve the formula (3), introduce the concept of linear operator projection for vector orthogonal decomposition [2][3]. Suppose S_1 is subspace of space S, then any α vector of S can be broken down into :

$$\alpha = \alpha_1 + \alpha_2 \tag{8}$$

which : $\alpha_1 \in S_1, \alpha_2 \perp S_1$,
Then

$$\alpha_1 = \alpha - \alpha_2 \tag{9}$$

α_1 is called vector projection of α on S_1.

Can be shown: $\alpha_1 = \sum_{i=1}^{m} C_i \xi_i$

Which $\xi_1, \xi_2, \cdots, \xi_m$ is the base of S_1. It can be derived directly by rigid body translation and rotation formulas [5][6].

Where C_i is the solution of following equation (10) :

$$\begin{aligned} C_1(\xi_1,\xi_1)+C_2(\xi_1,\xi_2)+\cdots+C_m(\xi_1,\xi_m)&=(\alpha,\xi_1) \\ C_1(\xi_2,\xi_1)+C_2(\xi_2,\xi_2)+\cdots+C_m(\xi_2,\xi_m)&=(\alpha,\xi_2) \\ C_1(\xi_m,\xi_1)+C_2(\xi_m,\xi_2)+\cdots+C_m(\xi_m,\xi_m)&=(\alpha,\xi_m) \end{aligned} \tag{10}$$

Suppose rigid body vibration mode $\xi_1, \xi_2, \cdots, \xi_m$ is zero eigenvalue eigenvector that corresponds to equation (1), then $u_i = L^T \xi_i$ is eigenvector of λ_0 eigenvalue that corresponds to equation (1),

Calculation steps are as follows:

A. According to formula (3) obtain V_K, denoted by $\overline{V}_K$,

B. According to formula (10) obtain $\alpha_1 = \sum_{i=1}^{m} C_i \xi_i$, $\overline{V}_K$ place where α,

C. $\overline{V}_K = \overline{V}_K - \alpha_1$ as a new $\overline{V}_K$ iterates following formula (3) continue to.

In order to avoid the calculation error, and then introduce the concept of complex orthogonal,

Which:

$$\begin{cases} \beta_1 = (\overline{V}_1, \overline{V}_1)^{\frac{1}{2}} \\ V_K = \overline{V}_K / \beta_K \\ \alpha_K = (AV_K, V_K) \\ \overline{\overline{V}}_{K+1} = AV_K - \alpha_K V_K - \beta_K V_{K-1} \\ \overline{\overline{V}}_{K+1}^{(S+1)} = \overline{\overline{V}}_{K+1}^{(S)} - \sum_{j=1}^{K} A_j^K V_j \end{cases} \qquad (11)$$

There: $A_j^K V_j = (\overline{\overline{V}}_{K+1}^{(S)}, V_j)$,

Formula (11) is convergent, which $(\overline{\overline{V}}_{K+1}^{(P)}, V_j) < \varepsilon$.

3 Results of Calculation

Using formula (10), (11) Example calculation shows that the equation (1) three zero roots have all been destroyed. Compared with the usual Jacobi equation, the first root corresponds to the fourth root, fifth root corresponds to the second root, and so on.

Table 1. Lanczos algorithm compariseon with jacobi

Root number	Jacobi algorithm	Lanczos algorithm α=10 m=20
1	0.99986407×10^{-1}	--
2	0.99995025×10^{-1}	--
3	0.9999922×10^{-1}	--
4	0.115024×10^{-4}	0.1149935×10^{-4}
5	0.154079×10^{-5}	0.1534095×10^{-5}
6	0.408291×10^{-6}	0.4053039×10^{-6}
7	0.152411×10^{-6}	0.150744×10^{-6}
8	0.6977413×10^{-7}	0.6872529×10^{-7}
9	0.6156869×10^{-7}	0.6157158×10^{-7}
10	0.36589928×10^{-7}	0.3587629×10^{-7}
11	0.21142635×10^{-7}	--

4 Conclusion

The eigenvalues and eigenvectors is very useful as the initial vector using in synchronized iteration based on the above proposed algorithm.

As with the simultaneous generation of an iterative method of time is equivalent to using a Lanczos orthogonal cut-off at the end of computation time. Thus, It is very obvious that use these two methods alternately to make up for lack of efficiency.

References

[1] Zhang, H.X.: Automobile design. China Machine Press, Beijing (1999)

[2] Qin, X., Jiang, H.: A dynamic and reliability-driven scheduling algorithm for parallel real-time jobs executing on heterogeneous clusters. Journal of Parallel and Distributed Computing 65(8), 885–900 (2005)

[3] Huang, K.: Advanced Engineering Mathematics. People Railway Publishing House, Beijing (1999)

[4] Parsopoulos, K.E., Vrahatis, M.N.: Particle swarm optimization method in multiobjective problems. In: Proceedings of the 2002 ACM Symposium on Applied Computing, SAC 2002, pp. 603–607. ACM Press, Madrid (2002)

[5] Shi, W.Y., Guo, Y.F., Xue, X.Y.: Matrix-based Kernel Principal Component Analysis for Large-scale Data Set. In: International Joint Conference on Neural Networks (2009)

[6] Reyes-Sierra, M., Coello, C.A.C.: Fitness inheritance in multi-objective

The Management System Research Based on Information Technology

ZhongXuan Yang

Zhongyuan University of Technology, Zhengzhou, P.R. China
{yangzhongxuan111,zhenghuifan}@163.com

Abstract. Community, as a system, there are many practical problems, such as property management companies do not act, owners committees exist in name only, etc. problems arise due to community management system root causes of this paper, according to my paper, it based on information technology, and by way of the development of new systems to build new community management system, information and community to achieve flat management, which greatly improve the efficiency and level of community management.

Keywords: Community Management, Information Technology, Institution.

1 Introduction

As China's market economy is reforming, the gradual deepening of China's urban community management has made some achievements, community management system has also undergone some changes, but the new era of community management system and development requirements of the socialist market economy is not yet adapted, according to the community management of the intrinsic properties of the main problems and cluster, the analysis of survey results can be seen[1], the current institutional set up community committees are still not perfect, people can not supervise community committees, complaints; owners of the committee could not supervise or dismissal of the property management companies, property management services of higher quality issues, these problems are essentially caused by the community management system, thus building a new community management system is of very important significance.

1.1 A Community Management System Exist Problems

1) Community management, lacking of relevant laws and regulations, institution building is not perfect;
2) Government's role is unclear;
3) Community Management of Fragmentation, lack of systematic;
4) Government administration is simple, not the establishment of community-based multi-body resident self-management system of democratic participation[2~4].

D. Jin and S. Lin (Eds.): Advances in ECWAC, Vol. 1, AISC 148, pp. 261–265.
springerlink.com © Springer-Verlag Berlin Heidelberg 2012

1.2 Principle about Construction of New Community Management System

1) The Principle of Strengthening the Legal System

Community management of many problems there are due to the lack of appropriate laws and regulations, unsystematic, the establishment of new community management system, the first community management should establish and improve laws and regulations for the management of all aspects of community legal constraints, form being provided by law and community management services to the main subject of rights and obligations. Community management should follow the legal procedures, to avoid disputes event without laws, it can not be investigated, buck-passing responsibility for the situation; secondly, to strengthen law enforcement, standardized law enforcement activities should be carried out to ensure that community management of the work keep up with legal procedures and corresponding rules and regulations. Thus, in order to make community management work into the standardization, the rule of law track.

2) The Principle of Autonomy Residents

In the current community management system, the government is the lead management of community affairs, internal affairs single-handedly communities, there is nothing pipe, pipe anything bad situation, there are also functions of the government, "Offside", "absence" phenomenon, and other subjects of community management initiative not to play. Community management trend is the realization of residents of autonomy, whose essence is the transformation of government functions and community management center of gravity, making the residents have the right to self-management of community affairs, the establishment of new community management system, we must adhere to the principle of self-government community residents.

3) The Principle of Harmonious Management

According to certain principles to be built up, public organization theory is a consistent target people collaborative and social groups with a certain boundary [4]. From the public in terms of histology, is a community organization, you need to follow certain principles of the various subjects of mutual collaboration. Therefore, the establishment of a new community management system should be based on harmony and management theory, the community as an organization, using scientific management methods and means of management and coordination of community services to the main system for the community to create a harmonious environment to achieve All management of the main synergies, satisfy the needs of community residents to achieve a harmonious community management.

2 Construction of the New Community Management System

2.1 The Design of New Community Management System

After the study community management system research and analysis in the original “two levels of government, three-level management” system based on the creation of a new community management system, the original neighborhood offices to community offices, still as a government an agency with administrative functions, its role as the administration of their areas of jurisdiction, performed by government-appointed government. At the same time, strengthening the community of the original construction of the neighborhood, and enhancing its functions, making the community mass organizations of itself, for community management and settlement services. Neighborhood democratic election, except to fulfill some government (or assigned by the government and residents to develop the social and public affairs), but its main function is to serve the residents. In order to better implement the urban levels of government for community development planning, construction and administration, setting up a number of communities in each district office, a government agency, key management community committees, guidance and supervision of their work.

Based on the above analysis, to build a new community management system, divided into four levels, each level has different rights, their functions clear, responsible for different levels of community work, the framework shown in Figure 1.

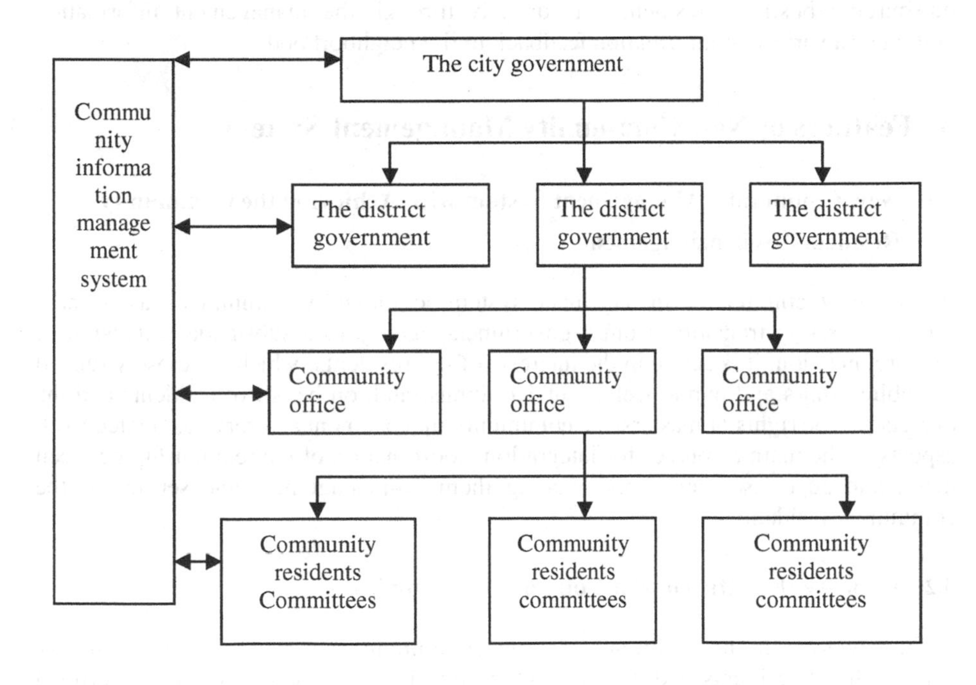

Fig. 1. The new community management system

Figure 1, the first level is the municipal government, is responsible for community development and construction within the city's unified planning and management, develop appropriate laws and regulations, the city's community development principles and policies as the basis for community management.

The second level is the district government, mainly in accordance with city planning, principles, policies and regulations in the area within the city to perform with the butt of all government functions. According to this study's definition, the district government can set the appropriate community management institutions (community offices), mainly responsible for the local community and residents in the relevant part of government, and the district government, through its management functions to guide the community, supervision and support.

The third level is the community offices, according to the ease and convenience of the residents of the community management principles, the district office following the establishment of a number of communities, is mainly responsible for household registration, security and other government and law on community building, sanitation and monitor the activities of the residents of the community living work of the Committee give support and help.

Fourth level is the community committees, community committees democratically elected by the community residents, or in the community under the guidance and assistance office, nominations, publicity, solicit the views of the residents identified. Such as community-government, finance, integrated services, information centers, and community services through the form of a contract to various professional services outsourcing business, residents can directly through the management information system or a variety of information feedback to the neighborhood.

3 Features of New Community Management System

3.1 New Community Management System with Achieving the Community Residents' Self-management

In this new community management system, community committees as a mass self-organization transition from a government agency to a substantial extension of self-organization, it is active in the interests of the residents, which are closely related to public affairs and public welfare undertakings, and, on behalf of residents and the exercise of the rights of masters, the community management and services related to all aspects of the main resources for integration, coordination of the relationship between them, and supervise their work, making them co-management and service to the community residents.

3.2 Improve the Efficiency of Community Committees

In the new system, the abolition of the owners committee, instead by the community neighborhood residents and exercise autonomy. The members of the neighborhood itself is part of the community residents or property owners, in addition, to fulfill the original owners of the Committee's functions, which is also responsible for some of the government (such as family planning, census, government-designated floating population management work), it is the owners' Commission that do not have or can not have functions. In this system, taking into account the overlapping functions and two "committees" of the actual situation, the commission revoked the owners, community residents from the community committees as autonomous agencies, representatives of community residents to exercise their rights, one of two committees, which will greatly improve the efficiency and level of community management.

3.3 Establishment of Community Information Management System to Achieving a Two-Way Communication and Exchange Based on Information Technology

Under the new system, based on information technology, the establishment of a community information management system, timely release of the community at all levels of management of the main management information, and the establishment of monitoring and complaint system. In this way, the government controls the dynamics of the community, monitoring the status of implementation of specific policy measures for the problems in a timely manner to develop appropriate policies, through the complaints system can understand resentment from the residents and suggestions to change the original one-way between the Government and the people relationship management and managed to control and be controlled relationship between government and residents to achieve a two-way interaction. Meanwhile, the Community Management can keep abreast of all the main community management position, you can also play a part in community management of the supervisory role of the interaction between the main body.

4 Summary

New community which is based on information technology management system, not only to build a two-way communication platform, and community residents through self-government, integration of community resources so that all the main achievement of the 1 +1 greater than 2 synergies, the new system can be better services for community residents to realize the harmonious management of the community.

Acknowledgments. The project is supported by the soft science research projects of Henan Province (092400420044).

References

1. Yang, Z.: Urban construction in the new Community Management Model 12, 21–25 (2009)
2. Wakefield difficult. Community Management. Sichuan People's Publishing House, Chengdu (2003)
3. Wu, L., Ping, S.: Community Management. Higher Education Press, Beijing (2003)
4. Marvin, Bing, L.: Public histology. Higher Education Press, Beijing (2003)

Construction Research of the New Community Management Model Based on Virtual Organization

ZhongXuan Yang

Zhongyuan University of Technology, Zhengzhou, China
{yangzhongxuan111,zhenghuifan}@163.com

Abstract. Community harmonious relationship to the stability of the society for community management, to effectively solve the many problems, based on information technology and virtual organization theory of traditional community management, business process analysis, the author puts forward that the community management mode, the virtual organization the model has the characteristics of agile response, can effectively integrate the community resources, solve community management of the existing problems.

Keywords: Information Technology, Virtual Organizations, Harmonious Management.

1 Introduction

The virtual enterprise is two or more independent enterprise or departments, in order to make full use of the enterprise or the resources of the department, expand the competitive advantage to the computer technology as the foundation, the integration of the enterprise resources, to realize the resource share many of the win, win cooperation organization. At present, the virtual organization as a new organizational mode by more and more enterprises and the theory attention global virtual organization structure, with flat, virtual sex, integration, flexibility, information network and other characteristics, therefore, the strengthening of the virtual enterprise in community management of research and application of the system to solve community management of the existing problems, and improve the management efficiency and level of the community has an important significance.

2 Reasons of the Community Virtual Organization Drivers

2.1 The Need of Solving the Current Community Management Problems

The current community management subject, the government administrative management as the center of gravity, community residents' committees for government affairs of the work, uploading issued and no representative residents play its role. The

D. Jin and S. Lin (Eds.): Advances in ECWAC, Vol. 1, AISC 148, pp. 267–271.
springerlink.com © Springer-Verlag Berlin Heidelberg 2012

reasons are that the dispute of the community management continuously increased, residents rights and interests are violated, and virtual organization is a quick response to community organizations, can effective integration of resources, build a two-way communication mechanism, driving the main body continuously improve their service quality.

2.2 The Need of Achieving Their Benefit of Participating in Community Management all Service Main Body

Based on the game theory, the community management between each main body form a virtual enterprise is a cooperative game. Community Duo-Ren game of virtual organization is the community participation in management participants main body, the game of strategy is the main activity involved in management, the selection of game principle is through the choice of actions to make their maximize returns.

2.3 Development and Application about Information Technology

Information network technology for enterprise management put forward higher request, but the current community management work in the main way behind, office the efficiency is low, the community resources waste, these problems and the information society does not adapt, need to change. Community participation in management, subject more fragmented, and each management subject their reform will strengthen community information management costs, and the virtual organization is with the information network technology to support, the enterprise resource integration, can save the cost, improve enterprise's office efficiency and resource allocation efficiency.

3 Construction of the Community Virtual Organization Management Mode

3.1 Principles

(1) Systematic management principle
(2) Market principle
(3) Information sharing principle

3.2 The Business Process of Traditional Community Management

In traditional community management activities, involves different management subject and service department, but its ultimate service object is the community residents. Traditional community management service flow mainly community property service flow, the community public welfare service flow, community service flow, community e-government intermediary service flow, community business service flow on several, as shown in figure 1 below. Figure 1 can see, traditional community management service process to provide related services in basically is fragmented, and between different subjects without coordination, cooperation, the community residents passively accept every subject the services provided. Traditional community service main body and the residents disputes continuously is essentially because the business flow unreasonable, no effective integration of community resources and business process, combing the main body, the relationship between the fragmented, and no effective communication mechanism of the cause.

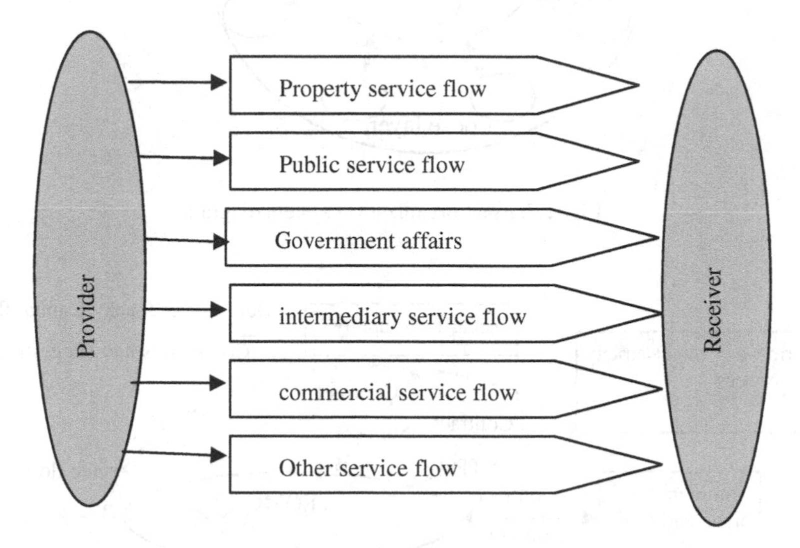

Fig. 1. The flow chart of traditional community management service

3.3 Construction of Community Virtual Organization Management Model

The virtual organization generally USES the two layers of organization system[4] is the core layer and tools, as shown in figure 2 shows, the core layer is the whole organization rules makers and managers, the member should be appropriate. Based on the virtual organization theory, the virtual community organization management model is shown in figure 2. Among them, the community residents' committees in organization system, the core layer, warm public service department, security companies, clean companies, community organizations, community service agencies subject for organization system tools.

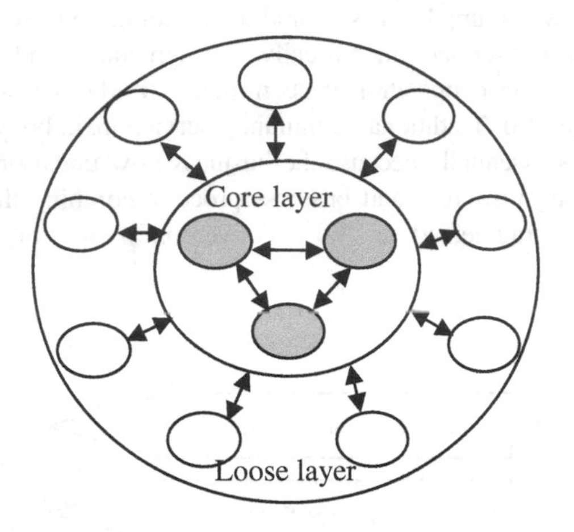

Fig. 2. Virtual organization system diagram

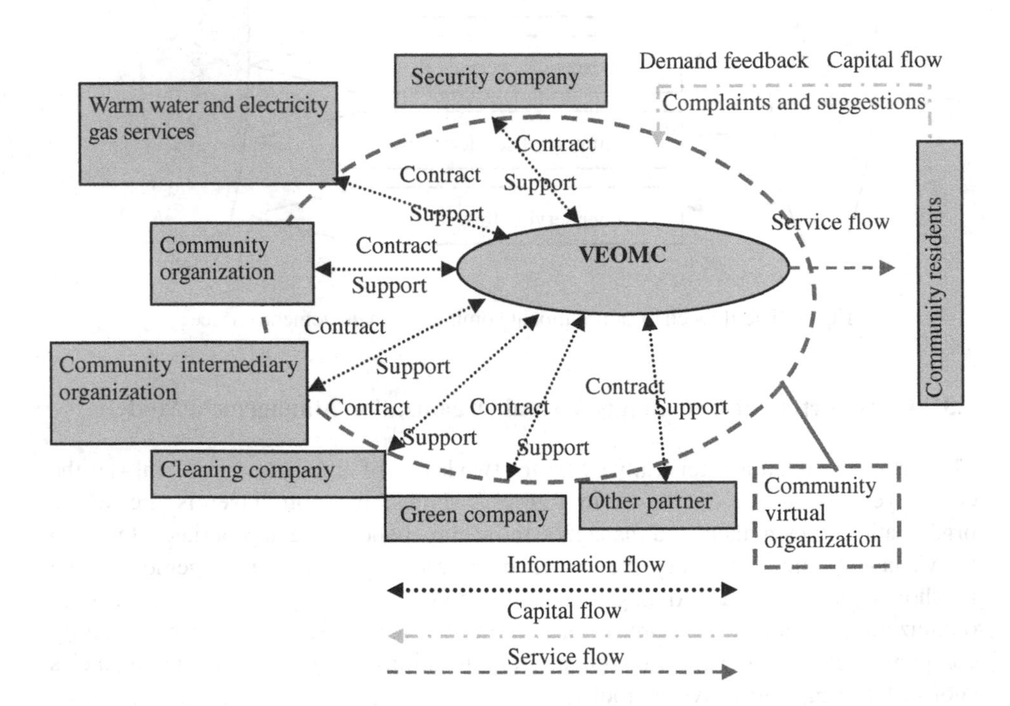

Fig. 3. Community virtual organization management mode

Community virtual organization management mode is shown in the figure 3, the community of the virtual organization management core for community residents' committees, mainly be responsible for the community affairs, community public welfare undertakings, representatives of public resources residents community management, be responsible for the community to every part of the outsourcing professional services company, and carries on the management, the community residents' committees unified coordination, distribution of community resources, the main business Community virtual organization members have equal status, the member play the professional and core ability, reflect the community management socialization and specialization principle, community virtual organizations by Duo-ge business is the different business enterprise or department composition, through the contract way to each other constraints and supervision and management and standardize the various enterprises and departments, realize community virtual organization overall interests, to realize the maximization of the harmonious community management.

The virtual community organization established based on Web technology community virtual organization management information system, and is responsible for the integration of community resources to share information, the management through the information network platform between subject to timely information and released its work to get the information, the management of subject and the community residents can through the portal between convenient communication, also facilitate the community residents of the supervision, so, the community to manage the main body participation management community work of subjective initiative got work. Through the mutual supervision and communication, reduce the community management of related problems of produce, makes community virtual organizations in a benign circulation state.

4 Summary

Community virtual organization management mode is to use systematic management thoughts and theory, emphasize the integration and win-win cooperation mechanism, relying on information technology, through the integration of the main body of the different business core ability to better for customer service, a kind of flexible, agile organization, this community virtual organization management mode can effectively solve the problems of the community management existing.

Acknowledgments. The project is supported by the soft science research projects of Henan Province (092400420044).

References

1. Bao, G., Jia, X.: The organizational structure of the virtual enterprise. Beijing: China's Industrial Economy 10, 25–28 (2005)
2. Wang, Z.: A virtual enterprise.-the age of the knowledge economy enterprise organization of the new situation. Hebei Enterprise 5, 35–37 (2006)
3. Xing, Y.: Based on the game theory the research about virtual organization theory. Tianjin university (2003)
4. Chen, J., Feng, W.: Virtual enterprise construction and management, pp. 60–70. Tsinghua university press, Beijing (2002)

Discussion on Web Development Technology Education from the Educational Philosophy View of Pragmatism

Jianhong Sun[1], Qun Cai[2], and Yingfang Li[1]

[1] Engineering College of Honghe University
[2] Science College of Honghe University
Yunnan Mengzi, 661100, China
sparkhonghe@gmail.com

Abstract. Web sites work as a major resource of obtaining information which spreads information so fast and so wide far more than any traditional media. Because of this, web sites are no longer a rarity and it is a necessity for any organization. In order to students are able to use the best technologies to develop web site, the teachers should keep catch up the development of web design technologies. From the educational philosophy view of pragmatism, we present some web design potential jobs for teaching prepare of Web Design Technology. Some teaching tips also are proposed. This work is based on a case study research.

Keywords: education, educational philosophy, pragmatism, web development.

1 Introduction

Since 1989, HTML was first proposed by physicist Tim Berners-Lee and Robert Cailliau [1], it leads the world walk into a new age with information explosion. Websites work as a major resource of obtaining information which spreads information so fast and so wide far more than any traditional media such as newspaper and television. Because of this, websites are no longer a rarity; they're a necessity for any organization. Therefore, it goes without saying that the demand for developers and maintenance technicians of web sites is on the rising each year.

In fact, building and maintenance a website is a responsibility to the information technology manager of many companies. To those students who study in computer related majors, who are the potential candidates of the position of information technology manager, it is a basic requirement to master the skill of developing website. Furthermore, the teachers who take in charge of this course bear a heavy responsibility for the leading duty. Then the focus on this issue that helping students to gain this skill and be able to use it with a sense of comfort and confidence. The teaching methods and theories as a hot topic are discussed a lot. It is well known that how fine teachers differ as their characters and styles differ. In this paper, we discuss the point of view of educational philosophy of pragmatism based on a case study on Web Development Technology teaching in Honghe University (HU).

D. Jin and S. Lin (Eds.): Advances in ECWAC, Vol. 1, AISC 148, pp. 273–277.
springerlink.com © Springer-Verlag Berlin Heidelberg 2012

2 Backgrond

HU is a developing university which from a normal specialized postsecondary college upgrade to a university since 2003. It is one of 1090 universities of China. HU serves a diverse population of full- and part-time, national and international students. In 2009~2010 academic year, more than ten thousand students enrolled in 12 undergraduate colleges with 36 undergraduate specialties [2]. Cultivating applied talents for society is the goal of HU and for this HU's administration is promoting instructional technology as a crucial part of higher education for faculty and students.

2.1 The Program Objective of Computer Science and Technology

Computer Science and Technology is one of the 36 undergraduate specialties of HU. Since 2003, with HU upgrade from a normal specialized postsecondary college to a university, the new program objectives was described as, "The Bachelor of Engineering (Engineering in Computer Science and Technology) program aims to prepare graduates for gain a professional qualification in the field of computer and entry-level positions as IT professionals. The program provides a broad and thorough study of the essential knowledge and practical skills required by graduates preparing for a career in the IT profession."

2.2 The Gap between Reality and Objective

According to survey on the diploma projects of graduates in recent years, a depressing result shows that there is a huge gap between reality and the program objective. Most of students choose developing web application system as their diploma projects, but most of them develop only an unsophisticated website. From an application point of view, their works did not achieve the purpose of any application, only a very few exceptions. So it goes without saying that the disappoint results of the employment statistics of graduates is not difficult to imagine.

3 Analysis of Factors Affecting the Effectiveness of Teaching

Face with the education can not achieve our program objective this serious problem, we have been analyzed a lot in our previous research works. In this paper, we focus on the special scope, web application techniques teaching and training to analysis. Through rigorous survey and analysis, the following crucial issues should be the causes that HU is facing.

- Some teachers lack of updating their knowledge. Web application development technologies and concepts are rapidly developing in recent years. If teachers lack knowledge of these new technologies and concepts acquisition, the impact on the effectiveness of teaching is the most serious.
- Most teachers of China have been used to teaching on textbook-directed instead of on knowledge-oriented. A textbook will narrow a student's focus on if the students lack of initiative and capability of self-study to extend the range of knowledge by the Internet or other books [3].

- Concept of student employment limited. Most of students prefer to be a state official more than to be a technician. In China, to be a public servant of state means can get stable salary and good benefits. This situation will affect the initiative of students to study technology.

4 The Web Development Technology Contributing to Potential Careers

In short, the goal of education is to enable students to master skills to earn a living. Therefore, from the point of view of educational philosophy of pragmatism, the education should focus on the potential employment of students to carry out and ensure that students master the required skills and knowledge of potential positions. Only in this way can the students easy to get a job after graduation. Contributing to this expect objective as much as possible is a responsibility to every course teaching mission.

4.1 One Point of Educational Philosophy View of Pragmatism

In the view of John Dewey, an important early developer of the philosophy of pragmatism, "I believe that the school is primarily a social institution. Education being a social process, the school is simply that form of community life in which all those agencies are concentrated that will be most effective in bringing the child to share in the inherited resources of the race, and to use his own powers for social ends. [4]" This is a very important educational thought, but only a few people realize this in China education. As a result, much of what the schools teach is out of date, especially in these specialties technology updating quickly. Therefore, it is not uncommon that the candidates are required that they must have work experience in many recruitment advertisements. Actually, this will never happen if every educator treats education being a social process, directs students to learn the skills and knowledge that society in need of.

4.2 What Are the Most Sought-after Skills in Web Development?

Based on previous point of view, for the teachers on duty, it is important not only should to know that what the most sought-after skills in web development are, but also should make every student to know clearly. Some skills in highest demand for web developers are shown as follows:

Web Framework Knowledge. A web application framework is a software framework that is designed to support the development of dynamic websites, web applications and web services. The framework aims to alleviate the overhead associated with common activities performed in web development. For example, many frameworks provide libraries for database access, template frameworks and session management, and they often promote code reuse [5]. The developers can build websites much faster with the help of web framework which helps developers cut out much of the repetitive tasks that normal custom programming would require. As reference [6] said, "Having knowledge of the top frameworks (Rails, Django, CakePHP, Symfony, and a few others), can give you a whole other dimension to your skill set."

Development Based on Content Management System (CMS) Platforms. A web content management (WCM) system is a CMS designed to simplify the publication of web content to web sites and mobile devices — in particular, allowing content creators to submit content without requiring technical knowledge of HTML or the uploading of files [7]. Develop a website based on a CMS such as Drupal, Joomla, XOOPS not only spend less time, but also do not even need to write code. It is the fist choice for these projects with low cost and short development cycle requirement. Furthermore, many customers hope to power their personal or business websites, so always there are positions open for designers and developers in custom CMS themes, CMS customizations and plug-in development.

Flash and Actionscript Knowledge. Flash technology can add a very professional dimension to any website, and large websites and corporations always pay to have their sites look professional, and often commission flash animated interfaces to showcase their products. With search engines working on ways to have flash communicated better with them, this is a skill that's sure to boom as the search technology advances [6].

MySpace/iphone/ipad/Facebook Applications Development. With the increase of Myface and Facebook users, the demand for social network apps has been huge since then. A whole new industry for web development sprang up overnight, and hundreds of applications are now added on a daily basis. The social media application platform has been found to be very viral and potentially very lucrative [6]. The same is true of iphone/ipad applications development.

5 Proposed Tips for Teaching

The most sought-after skills presented in previous section concern many technologies and knowledge. It is guidance for both teachers and students in teaching and hard be treated as a teaching objective of only one course. Here, some tips for teaching is presented for making a better use of the educational philosophy view of pragmatism.

- Select some useful reference books for students after prudent assessment. In China, it is well known that the number of textbooks for colleges and universities has been on dramatically rising in recent years, but the quality has not been improved. In fact, most textbooks are monotonous copy each other. Find a good reference book needs strong professional background.
- Knowledge-directed teaching is much better than textbook-directed teaching. Most teachers of China have been used to teaching follow a selected textbook. However, a textbook will narrow a student's focus on if he lack of initiative and capability of self-study to extend the range of his knowledge by the Internet or other books.
- Guide students how to learn, how to solve problems. As proverb says, "Give a man a fish and you feed him for a day. Teach a man to fish and you feed him for a lifetime." There are many orientation of technological development in web development as previous description. Students need spend more time than in class to continue further study, which orientation is decided by their own interest.

- Panel discussion, group projects are useful for training teamwork spirit and communication skills of students. In web development, many big projects are team work, one developer hard to finish. The teamwork spirit is important under this situation.

Practical work should be related with the practice application. Treats education being a social process, do some projects can be used in practical application.

6 Summary

In Engineering College of HU, many teachers used to focus the programming skills training and pay less attention to the use of open source web application framework, such as Wordpress, Drupal and Joomla, etc. In fact, develop website based on CMS not only save time, but also be able to meet most application requirements. For these students without interested in programming, they still have an option method to develop a website. Anyway, pay more attention to web development trends is the most important thing for teaching.

Acknowledgments. Thanks for pecuniary aid of Honghe University Bilingual Course Construction Fund (SYKC0802) and Honghe University Discipline Construction Fund (081203).

References

1. Tim, B.-L.: CERN, Information Management: A Proposal, http://www.w3.org/
2. Sun, J., Zhu, Y., Fu, J., Xu, H.: The Impact of Computer Based Education on Learning Method. In: 2010 International Conference on Education and Sport Education, vol. 1, pp. 76–79. Engineering Technology Press (China), Guang Zhou (2010)
3. Sun, J., Xu, Q., Li, Y., Li, J.: Research on Computer Education and Education Reform Based on a Case Study. In: 2011 International Conference on Computer Science, Environment and Ecoinformatics, Wuhan, China (2011) (in press)
4. Dewey, J.: My Pedagogic Creed. School Journal 54, 77–80 (1897)
5. Web application framework, http://docforge.com/wiki/Web_application_framework/
6. Stansberry, G.: 10 Most Sought-after Skills in Web Development, http://net.tutsplus.com/
7. Content management system, http://en.wikipedia.org/wiki/Content_management_system/

Analysis of Current Status of Self-defence Course in Colleges in Hubei Province from the Perspective of Course Reform

Ning Guo and Shanli Yi

Physical Education Department of Huazhong Agriculture University,
Wuhan, Hubei, China 430070

Abstract. Through the use of the research methods of documentation, questionnaire, logical analysis, this paper makes a survey of the optional course offering of self-defense in twenty universities and colleges by means of random selection. The result indicates that self-defense course is very popular among the students and suggestion is given that universities with certain qualifications should offer this course.

Keywords: course reform, colleges, self-defense, current status.

1 Introduction

Along with the deepening of the physical education (Abb. P.E.) reform, developing new sport courses has been a must. Currently, the P.E. course system adopts the competitive sport mode as it was in the past decades and could not meet the practical needs. As the uncertain insecure factors increases in society, the strengthening of college students' sense of security and ability of self-protection makes the set-up of self-defense course in colleges very necessary. Through the self-defense course, college students could not only enhance their physique but also cultivate their will and character. What's more, they could be well-informed of how to correctly react to the illegal infringement. To offer self-defense course in colleges satisfy both the demands of strengthening college students physique as well as self-defense and college P.E. course reform.

Through the survey of the current status of the optional self-defense course in colleges in Hubei Province, this paper expounds and proves the necessity and feasibility of the set-up of self-defense course in colleges in the hoping of providing theoretical and practical support for introducing this new sport course into the college P.E. system.

2 Research Object and Methodology

Through stratified random sampling, twenty colleges including Wuhan University, Huazhong University of Science and Technology, Wuhan Institute of Technology,

D. Jin and S. Lin (Eds.): Advances in ECWAC, Vol. 1, AISC 148, pp. 279–284.
springerlink.com © Springer-Verlag Berlin Heidelberg 2012

Wuhan University of Science and Technology, Hubei University, South-Central University for Nationalities, Yangtze University, China Three Gorges University, Hubei University of Technology, Hubei University of Economics etc. are chosen as objects. Through the combination of the methods of documentation, questionnaire, logical analysis and mathematical statistics, this paper conduct a survey of the set-up of self-defense course in above colleges.

Table 1. The administration of questionnaire

Items	Teacher	Students	P.E. course instructor
Distributed questionnaire (No)	88	1000	20
Collected questionnaire (No)	75	937	20
Percentage of collected questionnaire	85.2%	93.7%	100.0%
Effective questionnaire (No)	73	908	20
Percentage of effective questionnaire	97.3%	96.9%	100%

3 Result and Analysis

3.1 Forms of Self-defense Course in Colleges in Hubei

The figures in table 2 indicate that in the chosen twenty colleges, there are five colleges which have offered self-defense course, accounting for 25%; the other 15 colleges are planning to set up this course. As a whole, the rate of offered course is low.

Table 2. Forms of self-defense course in colleges in Hubei (n=20)

Forms	Optional course	Public course offered	Sport club offered	Optional course (in the plan)
Counts	2	3	5	10
Frequency	10.0%	15.0%	25.0%	50.0%

3.2 The Use of Teaching Materials in Self-defense Class in Colleges in Hubei

The investigation of the P.E. course instructors shows that the use of teaching materials in self-defense class does not comply with norms. The use of unified compiled textbook only accounts for 15%, self-compiled textbook 20%. The others (65%) make discretionary choices in teaching methods. The phenomenon of "students learn what their teachers are good at and how much they learn depends on how much the teachers master" prevails.

Table 3. The use of teaching materials in self-defense class in colleges in Hubei (n=20)

Types of teaching materials	Counts	Frequency
Self-compiled textbook	3	15.0%
Unified compiled textbook	4	20.0%
discretionary choices	13	65.0%

3.3 Venues used for Self-defense Course in Colleges in Hubei

Table 4 clearly shows that in the surveyed ten colleges which offer self-defense class, only 30% of them use indoor venues for class while the other 70% hold classes on the ground or at other places.

Table 4. Venues used for self-defense course in colleges in Hubei (n=10)

Types of venues	Counts	Frequency
Outdoor rubber ground	6	60.0%
Outdoor random ground	1	10.0%
Indoor floored ground	2	20.0%
Indoor special ground	1	10.0%

3.4 College Students' Attitude

About the question whether college students should have self-defense course or not, 32% and 57% students choose very necessary and necessary respectively; other 10.5% and 0.5% students choose not very necessary and not necessary at all. Besides, the rate between male students and female students indicates that not only female students are fond of and expecting the self-defense course, but also the male students shows a high demand of this course.

Table 5. College students' attitude (n=908)

Items	Very necessary	Necessary	Not very necessary	Not necessary at all
Counts	291	518	95	4
Frequency	32.0%	57.0%	10.5%	0.5%

3.5 College Students' Motivation to Select Self-defense Course

The top motivation of college students to select and participate self-defense course is to protect themselves from possible violation, which constitutes their main aim to learn self-defense, only 30% of the students learn for the credit; secondarily, college students select the course for the purposes of enhance their physique and cultivate their will and character. College students who choose these two items in the questionnaire learn the course for their interest and identification with the features of self-defense skills.

Table 6. College students' motivation to select self-defense course (n=908)

Items	Counts	frequency
To learn some effective self-defense skills	894	98.5%
To enhance physique and health	833	91.7%
To cultivate will and character	668	73.6%
To make friends, promote friendship	573	63.1%
To master graceful combat skills	354	39.0%
To obtain credits	272	30.0%
other	95	10.4%

3.6 College Students' Judgments of Self-defense Course

Table 7 shows that, college students generally give a positive judgment about self-defense. 27.4% of the total likes this course very much, 53.7% like it; only about 18.9% of the total makes negative judgment, which makes it clear that on the whole the offered self-defense course in colleges in Hubei province is popular.

Table 7. College students' judgments of this course (n=908)

Items	Like it very much	Like it	Does not like it very much	Does not like it very much
Counts	249	488	125	46
Frequency	27.4%	3.7%	13.8%	5.1%

4 Conclusion and Suggestion

4.1 Setting Up Self-defense Course Is in Accordance with the Practical Needs of Society

According to the current social status of our country, the aim of setting up self-defense course is not only to strengthen students' health condition and cultivate their taste, but also to help students conduct self-defense when confronted with danger and avoid being hurt, what's more, it will also help to enhance college students' legal awareness and promote their ability to safeguard their own security, which will eventually play an active role in their healthy growth and the establishment of a harmonious atmosphere of society.

4.2 Setting Up Self-defense Course Is in Accordance with College Students' Own Needs

With shallow social experience, modern college students always conduct limited lagging measures such as call the police or take legal action when confronted with emergency and infringement, their awareness of how to prevent and deal with abrupt criminal behaviors is far from enough. Self-defense can not only strengthen college students' health condition, enhance their confidence and defense ability when facing abrupt criminal behaviors, but also promote their learning interest, which corresponds with the physiological and psychological needs of modern college students.

4.3 Setting Up Self-defense Course Is in Accordance with the Needs of Complete the College Students' Volitional Quality

Exercise self-defense skills can harden people's courage, help people overcome timidity and fear and cultivate people's quality of facing danger fearlessly, it can also effectively enhance people's adaptability to changes which plays a very active role in promoting college students' abilities of dispassion, independent thinking, objective analysis and judgment making.

4.4 Setting Up Self-Defense Course Is in Accordance with the Needs of P.E. Course Reform in Colleges

By far the P.E. class has not really come down to the teaching of students' personal safety, setting up self-defense course in colleges is undoubtedly to transfuse fresh blood and introduce new elements to college P.E. class, which will adequately enrich the course content and forms. Setting up self-defense course is not only in accordance with the needs of college students to strengthen their health condition but also in line with the needs of maintaining harmonious development of society and deepening the reform in college P.E. course in general.

4.5 Suggestions for Setting Up Self-defense Course

During teaching activities, P.E. teachers could break the routine and take college students' physiological and psychological needs as well as their interests into consideration; in practice, they could also conduct teaching activities by adopting actual combat training method, case analysis method, scene simulation method and so on. As to the examination and evaluation to the students, the colleges and teachers should stick to the people- oriented educational thinking and establish a scientific examination and evaluation system based on the goal of strengthening students' health condition, cultivating students' their exercise habit, and help them gain the ability of self-defense to the full extent. All the college P.E. teachers should enhance their awareness and respect students' interest and needs. The people-oriented educational thinking should be reflected thus to promote the popularization of self-defense course in colleges in Hubei province.

References

1. Zhang, R.L.: The Guiding Ideology of "Health is the most important" and Public Physical Education Curriculum Reform. Sports Culture Guide 12, 43–45 (2004)
2. Liu, R.Q.: Feasibility Study on Implement Defensive Education of University Women. Sports Sciences Researches 9(1), 64–67 (2005)
3. Zhang, R.: Safety Education and Self-Defense: the Textbook for Optional P.E. Course of Colleges and Universities, pp. 4–6. Beijing Sport University Press, Beijing (2004)
4. Yao, Y.: Discuss the Set up of Self-Defense Course on College Sports Programs. Science and Technology Information 17, 173 (2007)
5. Sun, Y.W.: Research on Organizing the Course System of Female Self-Protection. Journal of Nanjing Institute of Physical Education (Social Science) 20(5), 92–94 (2006)

On the Development of Leisure-Oriented Mass Sport in Yangtze Delta

Jianbo Li and Yuhua Ren

Physical Education Department of Huazhong Agriculture University,
Wuhan, Hubei, China 430070

Abstract. Through the combination of methods of field research, interview and logical analysis, this paper studies the present status, features, mechanism as well as future planning of mass sport in Yangtze Delta, and tries to propose the development path and strategies for the development of mass sport in Yangtze Delta, in the hope of shedding some light on the strategical development of regional mass sport.

Keywords: leisure-oriented, Yangtze Delta, mass sport.

Introduction

"Leisure" has always been the ideal living condition pursued by human beings and playing a more and more important role along with the advancement of the human civilization. It is an significant component of human living and a decisive factor of a happy life. As a matter of fact, leisure has been the characteristic and fashion of our time. Along with the deepening of the nation fitness program, sport, as a top priority for leisure, entertainment and fit-building, has greatly promoted the human living standard and quality.

Sport has played a significant role in raising the human living standard and quality. Strategy is a momentous overall or decisive plan. The development strategy of mass sport is the high-level and overall macro plan for sport development in light with the advantages of sport activities as well as the aim and requirements of its further development. What's more, mass sport, as a basic element of the national sport, played a role that could not be underestimated in realizing the aim of national sport strategy. This paper is supposed to study the present status, features, mechanism as well as future plan of mass sport in Yangtze Delta, and propose the development path and strategies for the development of mass sport in Yangtze Delta, in the hope of shedding some light on the strategical development of regional mass sport.

1 The Environment for Mass Sport Development in Yangtze Delta

Ever since the reform and opening up, especially in the 1990s, the Yangtze Delta economic circle, which take Shanghai city as its core and Jiangsu province as well as Zhejiang province as its main body, has emerged. The economy here has been enjoying

D. Jin and S. Lin (Eds.): Advances in ECWAC, Vol. 1, AISC 148, pp. 285–291.
springerlink.com © Springer-Verlag Berlin Heidelberg 2012

a rapid development since then. Nowadays, the former core-city circle of "15+1" which formed by cities from the above mentioned two provinces and one city has been enlarged into "24+1" which almost includes all the major cites in this area.

1.1 The Social Environment for the Strategic Development Mass Sport in Yangtze Delta

The Yangtze Delta, lying in the lower reach of the Changjiang River where the Changjiang River flows into East China Sea, covers an area of nearly 10, 000 kilometers and supports more than 55 medium-sized counties of the two provinces and one city. Shanghai, Nanjing and Hangzhou are the three "pivots" of this area. Shanghai has been the most important port, financial center and comprehensive industrial center of China since modern time. Nanjing is "the capital of ten ancient dynasties" and boasts a history of more than 2400 years. Hangzhou has been crowned as "paradise under heaven" for its marvelous landscape. Since ancient time, this area has been a land of honey and milk as well as the birthplace of many outstanding people. At present time, the Yangtze Delta boasts China's most developed economy, culture and education. It is China's very first region to enter a relatively well-being society and enters in the new development phase of facilitating modernization.

1.2 The Economic Environment for the Strategic Development Mass Sport in Yangtze Delta

According to statistics, the Yangtze Delta, occupying an area of 99.6 thousand kilometers and boasting 75 million people, achieved a gross production of 56,374 billion Yuan in 2007, accounting for 22% of China's GNP. This enables it to be a city-circle of a world class. Meanwhile, it also has a radioactive and promotive effect on the economy of the surrounding area which is China's most concentrated and urbanized one with 1.5city, 6.7 counties and 148 towns upon each kilometer. Consequently, it has a relatively larger scale of health investment and sport consumption. Professional market, including sport facilities, sport wears grows rapidly with a turnover of more than 2 billion Yuan per year. Both the number and scale of the operating gymnasiums are in lead position in the whole country.

1.3 The Humane Environment for the Strategic Development Mass Sport in Yangtze Delta

The advanced culture and education in this area boosts the intensive cultural atmosphere for mass sport development. The Yangtze Delta boasts more than 100 universities and colleges, more than 1000 research institutes and densely spreaded libraries, gardens and sport facilities. In the history of reform and opening up, advanced developing styles like "Wenzhou Model", "Sunan Model" and "Putong Spirit" emerged successively.

2 Mass Sport Development Strategies in Yangtze Delta

It is known to all that Yangtze Delta enjoys comparatively well-advanced economy and culture in China. People's living standard of the most part of this area has reached a relatively high level, the public shows fairly well cultural quality, the sport facilities spread densely and mass sport activities are vigorous.

2.1 Characteristics of the Strategic Development of Mass Sport in Yangtze Delta

The most characteristic are the brand sport zones "Tai Lake Rim" and "Changjiang River Bank". Body-building is of "brand effect". On one hand, the government should popularize sport that is to lead the public into body-building activities which are of local characteristics; on the other hand, the government should promote brand activities. The two mass sport activities: "Fitness Running on New Year" and "9.29 World Walking Day" have attracted nearly one million people each year to participate and naturally become "Brand activities" of Nanjing. Also, the sport bureau of Zhejiang province has successfully organized two associative mass fitness activities in "sport circle of Yangtze Delta".

2.2 Achievements of the Mass Sport Development in Yangtze Delta

2.2.1 Mass Sport Activities Vigorously Carried Out in the Two Provinces and One City

"National Fitness Plan" is fully implemented here, the national fitness service system is gradually improved, the mass sport activities are carried out in a deeper level and massive physical quality is strengthened pervasively.

2.2.2 Mass Sport Developed in a More Scientific and Standardized Way

Mass sport activities for teenagers, workers, farmers, women and senior citizens have been carried out extensively. Meanwhile, mass sport for the handicaped people, minorities and military has been further enhanced. This area, which take community as the breakthrough point of carrying out mass sport activities in cities and focus on the community and town, thorough promoting the program of "Sending Sport to Communities and Towns", has constructed a scientific national fitness service system. Also there are more and more fitness trainers appearing.

2.2.3 Relatively Impeccable Management System, Scientific Management Mode

Shanghai, Nanjing and Suzhou has established and put into effect successively a serial of management regulations, such as fitness place management provision, fitness program (place) management and maintenance regulation, fitness place property certificate management regulation and community sport work system, etc. Meanwhile, they promise to maintain the fitness facilities under two conditions, sign third-party contract with facility-providers and –receivers, promote the establishment of four trusts through pilot projects and enhance five management modes. Therefore, they originate the scientific management mode which is of Nanjing characteristic.

2.2.4 The Network of National Physique Monitoring and Sport Organizations Gradually Improves with Widespread Physique Test Stations

"Physique test station" at districts and county level has been first set up in this area. A city -district (county) two level physique test network is established and authorized to offer scientific physique test service as well as to provide sport prescription for all people all the year round. About 30 thousand people on average could receive the test each year. In this network, districts and counties are seen as the basic elements to carry out special fitness or demonstration activities. Taking community as the breakthrough point of carrying out mass sport activities in cities and focus on the community and town, thorough promoting the program of " Sending Sport to Communities and Towns", a scientific national fitness service system has been constructed in this area. The developing mode of "mass sport in a relatively well-being society" has come into shape.

3 Planning of the Strategic Development of Mass Sport in Yangtze Delta

The mass sport in Yangtze Delta, as a result of relatively advanced economy, displays many a leading features. However, as the demonstration area, the mass sport in all the provinces(city) located in Yangtze Delta needs to be further integrated and developed, needs to form an organic whole which communicates and cooperates with each other well and needs to construct a regional sport network system. The construction of Yangtze Delta mass sport associative program and the Yangtze mass sport circle needs not only learn from the experiences and practices of Tai Lake rim sport zone as well as Eastern China collaboration zone, but also keeps paces with times and sticks to innovation and development.

3.1 Increase Sport Games

Sport games could stimulate the public's sport interests in Yangtze Delta, strengthen cultural connection and enlarge sport's social influence so as to arouse people's interests to participate in sports. Since Yangtze Delta is rich both in geographical and humane resources, each city could, according to their local circumstances, conduct fun mass sports games in order to attract more people to the field of body building. Each city should be guided to construct a national fitness project which is of high level and large scale with tranquil environment. Meanwhile, it should be coordinated with both nature and modernization as well as enjoy an advantaged position in the whole country. The project should also demonstrate a first-class service and plays a model role. The government should proactively construct the Yangtze Delta sport circle and cultivate it into an excellent model project of national fitness program and "well-being sport". The relay race along the old city wall of Nanjing for experiencing the non-material heritage and the West Lake osmanthus festival motocross are great examples.

3.2 Develop Sport Activities for Performance and Entertainment

A serial of influential, impressive and tasteful national fitness activities should be held and constructed to be brand activities. Sport performance and leisure-oriented fitness activities should be held during holidays so that more citizens have time to participate and an active atmosphere for their participation could be created. Great examples are like setting up a national fitness day (week, month) to encourage citizens to take part in activities like mountain climbing, morning exercise, aerobics, hip-hop performance or competition; organizing the public to join in the group fitness activities like drum dancing, Mulan fist, Tai Chi, ball exercise and Yangko in the city squares. The national fitness work in all respects could be promoted by drawing on experience gained in pilot project, typical fitness activities could enable the less popular ones and the brand activities could be popularized. Therefore, the national fitness activity could strive to a higher level.

3.3 Launch Public Fitness Activities Consecutively during Holidays

A set of brand activities of Guangdong characteristic should be formed and organized through the active discovering and spreading of traditional folk sport. Experiences should be gained and skills advanced in holding public brand activities. What's more, frequent foreign exchange activities should be held in this respect in order to contribute more to the promotion of Chinese traditional sport. The construction of the land of sport also should be strengthened and brandized. Games of kites flying, tug-of-war, family bowling, ping-pong, fun game for a family of three, long run of thousands of people, fitness assembly in communities, fishing, ballroom dancing competition, fitness road, public orienteering and outdoor activities are of great fun and national fitness value in that they are less competitive but suitable for the public to participate.

3.4 Construct Leisure-Oriented Tourist Sport Park

Thanks to the many a place of interest, tourism flourishes in Yangtze Delta. Jiangsu province, Zhejiang province and Shanghai city boasts many years of cooperation in this respect and will continue to give full play of their advantages to develop new sport tourism line, establish the leisure fitness park and improve their guiding service network on the basis of Tai Lake Rim sport zone. Such as the water leisure park in Qiandao Lake, Hangzhou, serial beach activities in Ningbo and Tianmu Mountain tourism festival etc. Recently Shanghai makes a "Everyone Sport" plan which proposes to establish three sport circles on weekdays, weekends and long holidays respectively. Its aim is to advocate everyone to participate in sport and bring on an organic combination of sport, tourism and health care. Meanwhile, the government will arrange national fitness park or sport theme park to help to establish a comprehensive and well structured public fitness system with appropriate scale, guide the public to communicate their thoughts on public fitness as well as carry out physique test on a regular basis and promote a scientific life style.

4 Measures for the Strategic Development of Mass Sport in Yangtze Delta

4.1 Strive to Develop Mass Sport Fitness for All Groups of People

It is a must to strengthen sport development in enterprises, rural areas, communities, and schools, enhance the development of sport societies and sport activities in all layers of government. Improve sport organizations of workers in enterprise, assist the workers in establishing sport clubs; promote sport organizations in rural areas, and setting up sport clubs and societies for farmers; strengthen the construction of sport organization at grass-root level, especially improve the sport clubs in communities so as to make them an important platform for people in the community to carry out sport activities; coordinate with the educational system to enhance the sport work in schools centering around two key points: strengthen the students physique and cultivate sport talents; organize extensively sport games for all layers of institutions and popularize fitness method for all the managers; promote the development of all kinds of sport societies especially for those sports which enjoy a high popularity among the public.

4.2 Improve the Organization and Management by the Construction of a High-Level Managerial Team

The comprehensive, coordinated and sustainable development of the city mass sport demands the construction of a highly qualified managerial team. This requires the cultivation of a group of creative leading managers with solidified professional knowledge as well as a managerial team constituted of the youth who are in the prime of their life. They should also boast both great political and professional qualities as well as forge ahead with determination. Except for strengthening the management of the places for morning and evening exercise, the government should spare no effort to set up more fixed places for mass sport, care about and support more the exercise places so as to attract more people to participate in mass sport. The management of the exercise places should be included in the construction of the city community, which may allow the street offices and neighborhood committee to be in charge of the daily work and the sport administration department above district to be in charge of the guidance and policy input.

4.3 Strengthen the Construction of Public Sport Facilities and Expand Activity Space for Residents

This could be realized through the fully open of the public sport venues and sport fields in schools to society; expanding the commercial sport stadiums to meet the different levels of consumption demands: introducing certain fitness facilities in the existing scenic spots and gardens for the residents so that the senior as well as some special fitness group may use these facilities free of charge all day long and other residents could choose what they need. Meanwhile, traditional unique sport activities could attract more visitors and live up the atmosphere. Also, the community construction needs to be normalized. More practical sport facilities should be introduced to make the frequent fitness activity more convenient for the residents.

4.4 Strengthen the Team Construction of Trainers and Improve the Incentive System

Former trainers are limited to certain group of people. To break this limitation the key point is to supply more middle-aged and young trainers, which demands to lower the standard of upgrading and break the years limitation of promotion on the basis of quality is ensured. The sportsman should give a full play of their part-time to give relative guidance to mass sport. It is not realistic to work as a social sport trainer without payment in the long run. An improved double incentive system including both spiritual and material reward may better facilitate the vigorous development of the mass sport.

References

1. Li, C.X.: The Tendency of Popular Sports of China in the 21st Century. Sport Science and Technology (01) (2005)
2. Liu, J.Y.: Discussion on the Sustainable Developement of Mass Sports in China. Fujian Sports Science and Technology (2) (2010)
3. Chen, Q., Lv, S.T.: Study of the development of mass sport in the southern coastal region from the perspective of cultural ecology. Journal of Physical Education (1) (2010)
4. Deng, Y.M.: Mass Sports Development Measures. Sports Culture Guide (1) (2010)

The Sports Dance Course in Constructing Harmonious Personality of College Students Study

Shanli Yi and Ning Guo

Physical Education Department of Huazhong Agriculture University,
Wuhan, Hubei, China 430070

Abstract. This paper unscrambles the relationship between dance-sport teaching and harmonious personality, analyses the function of dance-sport course in cultivating harmonious personality, further proposes some methods of constructing harmonious personality through dance-sport course.

Keywords: college, dance-sport, student, harmonious personality.

1 Introduction

Harmonious personality is the relatively optimal personality which could help people lead a harmonious life. It is formed through the personal consciously harmonious organization of both internal qualities and external conditions. Concretely speaking, it is a healthy and wholesome personality integrated by harmony among the quality systems constructing personality, the elements constructing quality and all the elements themselves. The aim of higher education is to cultivate highly competent people with integrated development. The prior attribute of a highly competent people is the harmonious personality. A man of harmonious personality is a spiritually healthy and wholesome people. Albert Einstein held that:" The aim of school should always be: a young man leaves school as a harmonious people but not an expert." Higher education, especially the public physical education, should take cultivating college students of harmonious personality as the fundamental pursuit.

2 The Relationship between Dance-Sport Teaching and Harmonious Personality Cultivation

Beauty symbolizes the spiritual civilization and the harmonious development of personality. All of the beautiful things are the outstanding pursuit of college students. College students' ardent love for dance-sport not only comes from their admiration of the dance-sport participants' strength, speed, excellent skills and wisdom, but also from the sense of beauty presented by dance-sport, which is colorful in content and displays unique beauty in each movement. Beauty of postures, beauty of movements, beauty of music and beauty of costumes are the basic intension of dance-sport. When they appear as concrete teaching contents, its intension of beauty would certainly help to construct a wholesome personality.

D. Jin and S. Lin (Eds.): Advances in ECWAC, Vol. 1, AISC 148, pp. 293–298.
springerlink.com © Springer-Verlag Berlin Heidelberg 2012

Table 1. Female college students before exercise, after comparison of results in SCL—90 gauge

Options	Before exercise (n=40)		After exercise(n=40)		P
	X	SD	X	SD	
Body	1.43	0.35	1.37	0.28	<0.05
Force	1.62	0.16	1.55	0.24	<0.05
Interpersonal relationship	1.63	0.22	1.50	0.39	<0.01
Depression	1.52	0.31	1.44	0.34	<0.01
Anxiety	1.57	0.38	1.39	0.45	<0.01
Hostile	1.48	0.23	1.38	0.39	<0.01
Horror	1.31	0.27	1.28	0.17	<0.05
Paranoid	1.40	0.19	1.37	0.16	<0.05

2.1 Beauty of the Human Body and Movements Are the Basic Characteristics of Dance-Sport

Dance-sport is the art of body. In dance-sport, dancing partners differ in sex, body and their postures, respectively displaying significantly different aesthetic features: man's robust, slapping and firm figures and woman's neat and shapely body both distinguishes and coordinate in their masculinity and femininity. Harmonious unity outstands from the distinction, while unique features are stressed in the unity. The combination of different aesthetic styles presents a strong and brand new aesthetic experience. Dance-sport uses the human body movements as the material technique. Through the vigorous prosodic movements of human body, it elaborates human mind and their inner feelings. Beauty is fully enjoyed when the beautiful movements act on people's visual organs.

2.2 Beauty of Music Is an Important Component of Dance-Sport

Music and dance are never separated with each other in the long history of human development. Music, which is an unique audio art using floating sound as its material technique, fully shows people's affection of beauty. It is a special form of aesthetic manifestation. Music is the soul of dance, and its beauty corresponds to the dancing styles. World classical music and light music are most frequently chosen in dance-sport, for their enchanting melody, strongly accented rhythms and colorful styles could help people to enjoy themselves freely and contribute greatly to the beauty of dance-sport. Examples are like music for tango are always full of power and grandeur, cadenced and stress syncopation; music for quick-step are at leisure and sprightly rhythmic; music for rumba are sentimental and lyric, soft and enchanting; music for paso doble are majestic and excited, energetic and powerful, etc.

2.3 Beauty of Costumes Is a Special Requirement of Dance-Sport

Along with the evolution of human being and social development, utilitarian influence of costumes decreases while its aesthetic and recreational functions increase. Costumes for dance-sport have special requirements: the dancer's costume must correspond to and be harmonious with distinct style of dance. Therefore, thanks to the different styles of dance-sport, costumes matched are also vertiginous and pleasing to the eye. In standard dance, man's costume is formal attire, in order to show the dancer's graceful figure and dignified manner; woman's evening gown like slinky dress brings a sense of elegance and luxury, demureness and good taste. As to dancers of Latin dances, man usually wears tights and loose-fitting shirts, woman usually has grass skirts, both indicate a heavy Latin flavor.

3 Function of Dance-Sport Course in Constructing College Students' Harmonious Personality

3.1 Beneficial to Cultivate Correct Aesthetic Perspective of College Students

Aesthetic perspective is the view and evaluation criteria of beauty. During the basic dance practices teaching and dance-sport accomplishment cultivation, teachers would provide students with materials as correct concept of beauty, the appreciation of perfect skins and methods of body shaping, etc. which could cultivate and raise students' perceptive ability of beautiful things. Through these materials and teacher's comment on them, students could percept the beauty of human body shape, the flexibility of healthy skin, the strength and masculinity of man, the beautiful curve of woman as well as the shapely figure of a perfect body. The cultivation of students' perceptive ability of beauty is in fact the construction of students' aesthetic perspective and taste. To conduct the education of beauty during dance-sport teaching is to cultivate student's aesthetic perspective through above mentioned teaching method of aesthetic assessment. Therefore, they could tell the truth from the false, the good from the bad, so as to abandon the mendacious, sleazy and ugly value.

3.2 Beneficial to Establish Harmonious Psychological Character

College students are always fond of following beautiful things, which in return could purify their mind and soul. Emotion nurture is one of the most important characteristics of dance-sport course, in which graceful, vivid and affecting artistic character act upon human emotion. Thus, dance-sport plays an irreplaceable role in temper controlling and raising people's spiritual character. In dance-sport, college students would liberalize their thoughts unconsciously, sparing no effort to obtain "truth, virtue and beauty" and finally form their aesthetic perspective. The beauty education in dance-sport course should not only cultivate the students' unique personal perceptive and assessment ability of beauty, but also respect students' personal creation. Under this principle, the students should be guided to model their souls by correct notions, to display lofty and grand affections through creative adoption of dance-sport skills and to express their pursuit of a happy life so that they could construct a harmonious psychological character.

3.3 Beneficial to the Coordinated and Balanced Growth of Each College Student

During dance-sport course, people can not only jump the traces of nature force, but also can be "a man of taste" by getting rational freedom, because the aesthetic objects, especially those dance-appreciation works, are all condensed with much efforts and hard work of predecessors. On one hand, dance-sport course presents colorful and specific scenes in the learners' mind to form the comprehensiveness, richness and diversity of the subjects' transcendental world; on the other hand, dance-sport course can give people the rational adjustment and subliming by adding their vitality, emotional experience and other kinds of emotional impulses to the dance in a dynamic art form, so as to realize the utopia of highly harmonious in one step. In dance-sport teaching, much attention should be paid to the coordinated and balanced growth of each college student by promoting their comprehensive quality, training their overall abilities and cultivating their active attitude to life.

3.4 Beneficial to the Communication and Mutual-Understanding between College Students and Society

The charm of dance-sport lies in the rich social factors of its own, because human beings are all living in the society and there is no one who can totally separate himself from society, so all the human activities belong to social activities. During the whole process of dance-sport teaching, people can not only enter into a brand-new living realm, but also generate the consciousness of freedom which enables the free development of each person and coordination with other person. For example, the lofty realm of harmonious coexistence, tolerance and mutual understanding between people can be reflected by the tacit coordination and whole-hearted cooperation between male and female partners during dance-sporting process. Dance-sport boasts human moral characters: the character of sincerity, kindness, beauty, rightness as well as humanism which have formed the common moral standard of human society, and can also bring stable and happy life to people. This represent the harmonious relationship between human and society.

3.5 Beneficial to Harmonious Interact between College Students and Nature

Almost every kind of dance-sport entails many creation of scenes about human and nature. For example, Jive shows the exciting image of romping amusement between people and cow. In the dancing process, college students can realize some artistic conception involved in the subtle relations between human and nature, thus to create dynamic motions to represent the artistic conception. Human beings—the important medium of providing dynamic motions, have abundant emotional experiences and behaviors. These emotions and experiences are physiological and psychological, instinctive and conscious, even rational and frantic. But, as for the attitude towards nature, emotions and instinct seem to be more wise, tolerant and far-sighted than rationalism. So, under the double guidance of sense and sensibility, the teachers should lead the learners to a natural attitude toward nature and improve the relations between human and nature, finally form a brand-new outlook on ecological development.

4 Ways of Building Harmonious Personality through Dance-Sport Teaching

4.1 Cultivate the Aesthetic Competence of College Students

The main task of aesthetic appreciation training is to cultivate college students' abilities of feeling, understanding and creating beauty. Aesthetic appreciation training can help college students to appreciate the beauty of body language in the dance directly. Though coming from daily life, these action languages have much difference with manual labors. Therefore, during the aestheticization process of dance-sport teaching, the teachers should pay attention to the guidance in rhythm and gestures to improve students' abilities in dance-sport skills and music appreciation which enables the students can feel the existence of beauty by themselves. Besides, the teachers should also give students some guidance on how to understand beauty in the teaching process, that is to say, information on the origin, background, significance and function of aesthetic object should be given to the students to stimulate the students' creation desires, fiddle their heartstrings and arouse their true sentiments.

4.2 Cultivate College Students' Innovative ability

The cultivation of college students' innovative ability is based on the innovative education to them. In the teaching process, if the college students are regarded as the real subjects of aesthetic appreciation training, they can add more sentiments to the dance and have a better understanding of dance-sport. The dance-sport teachers are supposed to start with the characteristics of aesthetic appreciation training teaching, pay attention to the beauty in dance-sport, then follow the steps of perception, understanding, and re-perception to teach students from the shallower to the deeper. By this way, the dance-sport teachers can help college students to improve their abilities of finding beauty and creating beauty. Meanwhile, the students can sublimate their aesthetic appreciation ability through beauty feeling experiences through which their personality of beauty-appreciation could be formed and they manage to have integrated and wholesome personalities.

4.3 Cultivate the Honesty of College Students

Honesty is the footstone to construct the building of personality; it is also the important premise for college students to lead a good life. The college students should be devout when appreciating the dance-sport; only in this way can they realize the beauty of dance-sport, accept this beauty, learn from it and make progress. Otherwise, any depreciation or exaggeration to the dance-sport can make terrible influences on their study and life. Dance-sport teachers should organically combine college students' direct perception and aesthetic appreciation feeling together, pay much attention to the cultivation of college students' abilities of finding and appreciating beauty, teach students to observe life and nature in the vision of aesthetic appreciation. Thus the students can be nurtured in beauty in the process of pursuing it, and their noble sentiments of loving life, lives and motherland will be stirred up in the same time.

5 Conclusion

Beauty endows everlasting vitality to dance-sport in the most harmonious artistic form. Thus dance-sport not only possesses the most charming and vivid color, but also lets the students be affected by its infinite vitality. The beauty of costume, dance, physique, music and modeling which dance-sport has shown in cultivating college students' sentiments, moistening their hearts, enriching their imagination and expanding their wisdom. The construction of college students' harmonious personality is the pursuing goal of college physical education reform; it is also the most important part of comprehensive development education to college students. Much effort should be paid to realize the unification of college students' wisdom and harmonious personality.

References

1. Sun, H.-Y.: Current Situation and Tendency of the Effects of Physical Exercises on the Personality of University Students. Bulletin of Sport Science & Technology (1) (2011)
2. Li, S.: Empirical Research on Relationship of University Students' Personality Development and Physical Education. Journal of Sports and Science (4) (2010)
3. Chen, S.-G.: Physical Activities and Mental Health of University Students. Journal of Kaifeng University (4) (2005)
4. Luo, J.: A Discussion on the Influence of Sports Activities on the Mental Health of College Students. Sport Science and Technology (4) (2005)

Design of Automobile Intelligence Security Control System Based on Microcontroller AT89C51

WeiSheng Zhong[1] and YaPing Wang[2]

[1] Nan Chang University, Nan Chang 330029, China
[2] East China Institute of Technology, Nan Chang 330013, China
ws0791jx@163.com, ypwang@ecit.cn

Abstract. AT89C51, being the main control chip, controls the auto speed, brake and clutch while the car is traveling under some situations. A long-distance displacement sensor detected the distance between the sensor itself and the obstacles in front of it, and succeeded in auto controlling the speed (brake and clutch) according to its traveling condition (including speed, maximum acceleration-the roughness of the pavement), which ensured autos to travel at the safe speed.

Keywords: AT89C51, Displacement sensor, Brake, Clutch.

1 Introduction

With the advancement of society and improvement of people's living standard, automobiles become the ordinary vehicles and play the more and more important role in our lives. It can be predicted that the car will enter into every household and become our necessary vehicle in our future life. Meanwhile, we will pay more and more attention on the driving security. The problem we should solve at first is how to ensure the security of auto traveling. Although the number of casualties in traffic accidents has the tendency of declining, approximately 100,000 people died of traffic accidents every year. The death and economic loss resulting from traffic accidents is the social issue in the field of transportation safety, which presses for solution.

2 Basic Theory

The basic idea of this design is to detect the distance between the automobile and the obstacles in front of it, making use of a long-distance displacement sensor. After the detecting result is transformed by A/D converter, it is inputted into micro-controller and then after computed with the current traveling speed and the roughness of the road (measured by sensor), whose output result succeed in auto-controlling brake and clutch after A/D transformation. Basic theory of system is shown in figure.1.

D. Jin and S. Lin (Eds.): Advances in ECWAC, Vol. 1, AISC 148, pp. 299–303.
springerlink.com © Springer-Verlag Berlin Heidelberg 2012

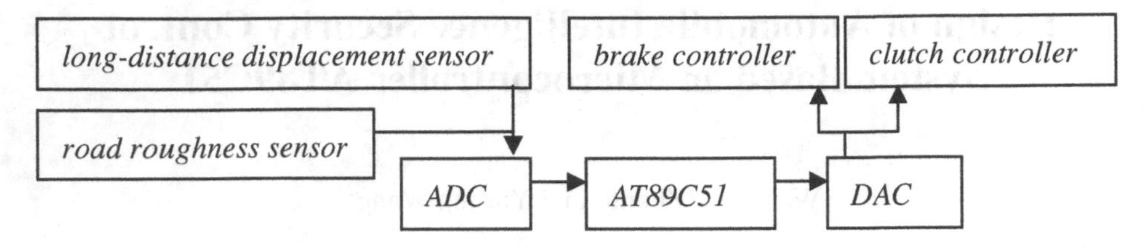

Fig. 1. Basic theory of security control system

3 Hardware Circuit Design

3.1 Section of Main Components and Chips

3.1.1 Using Microcontroller AT89C51 in MCU

AT89C51 is 8-bit microcontroller with low voltage and high performance made by American ATMEL Company, which includes 4k bytes erasable ROM and 128 bytes RAM, compatible with MCS-51 command system, and which is planted with a general 8-bit central processor and Flash memory cell [1].

Main performance parameters: Completely compatible with MCS-51 command system;4k bytes erasable Flash celerity memory;1000 erasing periods; Full static operation : 0Hz-24MHz;Three-step encrypted procedure memory 128*8 -bite interior RAM;32 programmable I/O rim lines; Two 16-bit timers;6 interrupt sources; Programmable serial UART channels; Low power dissipation free and power-failure mode.

3.1.2 Long-Distance Displacement Sensor

Based on the existing DLS-A measuring distance sensor, DLS-B improved measuring response speed. Its quickest dynamic response attains 25Hz, the longest measuring distance achieved 500m, and the highest precision was 1.5mm.

3.1.3 Selecting of A/D Converter and D/A Converter

ADC0804 is a comparative A/D conversion chip, made by CMOS integration technique, whose resolution is 8-bit, conversion period is 100μs and input voltage range is 0-5V, and which can attain 5V after adding some exterior circuits. Output data latch register is fixed in the chip, whose conversion circuit output can be connected to CPU directly while it is linked with computers.

DAC0832 is made up of an 8-bit input register, an 8-bit DAC register and an 8-bit D/A converter which uses R-2R resistor network of inverted 'T' type. Because DAC0832 has two data registers which is controlled respectively, it has much more flexibility than many others and can pick up different ways of working as needed. Without operational amplifier in DAC0832 and with current output in DAC0832, the external operational amplifier needs using while working. With Rfb set in the chips,

9 feet is connected to output point. If the operation amplifier gain is not enough, the external feedback resistor should be fixed.

By logical analysis of interior control in DAC0832, LE1 is in high voltage while ILE, CS and WR1 are effective simultaneously. Meanwhile, input data D7~D0 enter into input register. While WR2 and XFER are effective simultaneously, LE2 is in high voltage. Meantime, the data in input register enter into DAC register. 8-bit D/A converter circuit is ready to convert the data in DAC register into Analog signal (IOUT1+IOUT2) to output[2].

Besides, a multiple selector is fixed at the analog signal input point to select signals (the distance between auto and the front obstacle and the present pavement condition). Performing system should select the actuator with better control effects. With the increasing of inputting analog signal, control intensity should increase linearly or exponentially.

3.2 Connection Circuit between Main Components

Because input signals of the sensor are always weaker and have the strong interference, the voltage signals detected by sensors will first pass wave filter circuits and amplifier circuits and then input ADC. This design is required to detect two signals (the distance between auto and front obstacle and the roughness of the pavement), and both quantities must ADC and input MCU, therefore a 2-crucuit sector is fixed to connect ADC and 2-crucuit input, so that MCU can control when to input distance and when to input roughness.

3.2.1 Connections between ADC and MCU

The function of this part is to convert two detected physical quantities (the distance between auto and front obstacle and friction level between autos and pavement) into digital quantities and then to deposit them in the corresponding memory in the MCU, which is helpful to compute these data, and then achieve the appropriate control of external actuator.

AT89C51 controls the conversion of ADC0804 through P3.0, P3.1, P3.2, and P3.3. The 6th pin is connected to input point of 2-crucuit sector, while 9th pin is linked to reference voltage. And then the conversion result is inputted to internal MCU through P1.Specific circuits are as shown in Figure 2.

3.2.2 Connection between MCU and DAC and Control of Actuator

It is digital signal that MCU output, so outputted digital signal must be converted into analog signal by DAC in order to control the braking action of autos. Other circuits are shown in Figure 3.

3.2.3 Control Circuits of Brake and Clutch

As for the brake and clutch, we take different controlling methods[3].

For the clutch, we control it by simple 2-valued logic (Either completely separated or not separated). When the car maintains the safe distance from the obstacle in front of it, the clutch isn't used.

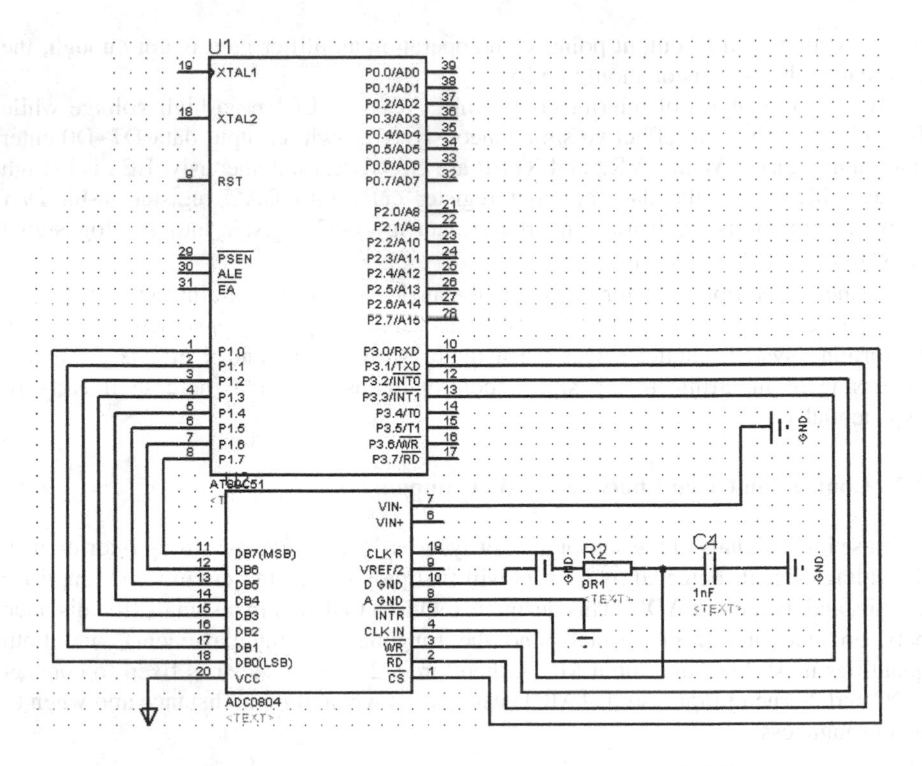

Fig. 2. Connections between ADC and MCU AT89C51

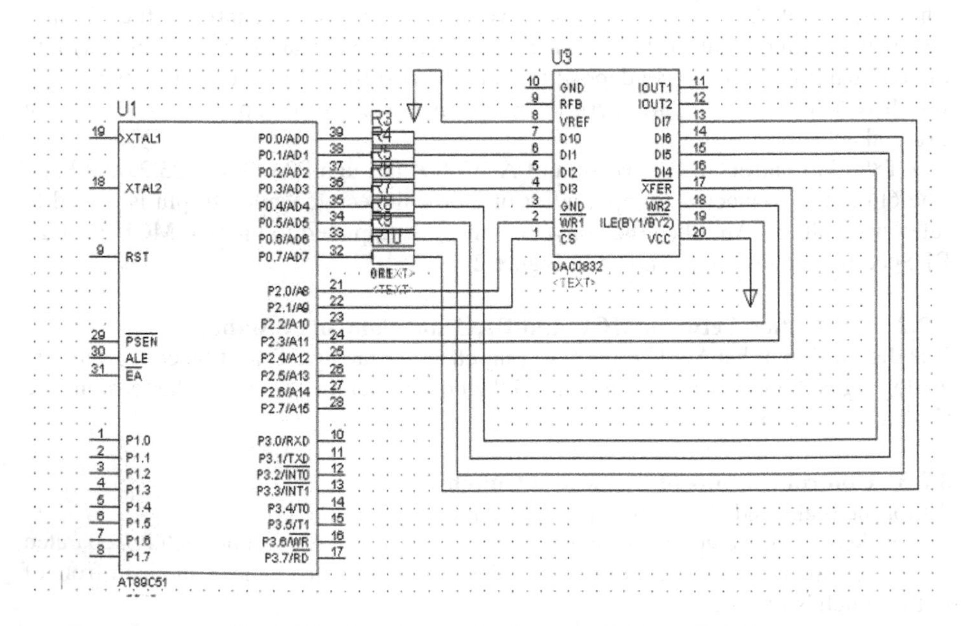

Fig. 3. Circuits of connection between MCU and DAC

When the distance between autos and obstacle is too small, we use complete clutch and it can separate the power system and auto, which is able to avoid many dangerous traffic accidents. For example, when drivers feel nervous facing with coming accidents, they would step on accelerator instead of brake. Even if drivers make a mistake in the operation and they mistake accelerator for brake, the auto will slow down by the friction between road surface and the auto itself because the auto power system is separated from wheel drive system, thus avoiding unexpected accidents as result of drivers' wrong operation by increasing the auto power.

As for the control of brakes, this design is to control braking degree through frequency transformer according to output variable of ADC. When the clutch is not separated completely, the brake doesn't implement stopping; when the clutch is separated completely, braking actuator determines whether to brake and the braking degree according to output variable of ADC, thus ensuring the safe driving of autos. Control circuit of baking is shown in figure.4.

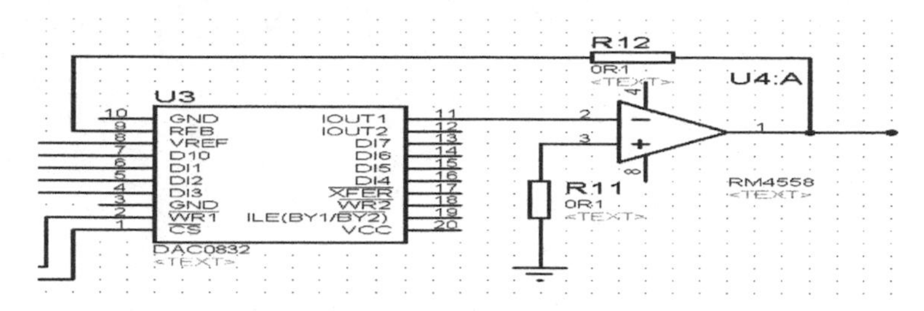

Fig. 4. Control circuit of baking

4 Software Design of Controlling Systems

Running of AT89C51 must be controlled by the corresponding program. The program is made up of the following main parts: read external distance data, read degree of road surface friction, process external input data, control auto's clutch and brake[4].

5 Conclusion

This design, most of which is founded on the basis of consumption, is restricted to the theoretical research owing to the limited conditions. For example, sensor of distance test will change with the external conditions (temperature, visibility etc.). Moreover, the corresponding control rules, in particular the stopping extent of brake, should be modified in the practical applications.

References

1. Zhang, Y.: Microcontroller Theory and Application. Higher Education Press (2006)
2. Gong, Z.: Digital Circuit. University of Electronic Science and Technology Press (1998)
3. Zhang, Q., Du, Q.: Microcontroller Application System Design Technology. Electronics Industry Press (2004)
4. Wen, Z., Guan, Z.: Design of a Data Acquisition System Based on TLC5510. Electronic Design Engineering (5), 22–25 (2008)

Blind Separation of Noisy Mixed Images Based on Neural Network and Independent Component Analysis

Hongyan Li and Xueying Zhang

College of Information Engineering of Taiyuan University of Technology,
Taiyuan 030024
lhy6018@163.com

Abstract. Blind source separation problem has recently drawn a lot of attention in unsupervised neural learning. In the current approaches, the additive noise is negligible so that it can be omitted from the consideration. To be applicable in realistic scenarios, blind source separation approaches should deal evenly with the presence of noise. In this contribution, the covariance of noise was foregone, a noisy multiple channels blind source separation algorithm was proposed based on neural network and independent component. At prewhitening, the data have no noise was used to whiten the noisy data, and the windage wipe off technique was used to correct the infection of noise, a neural network model having denoise capability was adopted to realize the multiple channels blind source separation method for mixing images corrupter with white noise. The result shows that this method may reduce the affect of noise and improve the signal-noise ratio (SNR) of separation images, accordingly renew the original images.

Keywords: Independent component analysis, Neural network, Blind source separation.

1 Introduction

Independent component analysis (ICA) is the process of extracting unknown independent source signals from sensor that are unknown combinations of the source signals. The work has been one of the most exciting topics in the fields of neural computation, advanced statistics, signal processing, and communication engineering which find independent components from observing multidimensional data based on higher order statistics.

In order to solve the problem of blind signal separation, many techniques have been proposed. Among them, the ICA methods are based on the assumption of mutual independence of the sources. Most of these methods were developed in the case of noiseless data. Some fast and efficient algorithms have been proposed such as FASTICA[1]. However, all these algorithms perform poorly when noise affects the data. Recently, some work has been done to partially overcome this limitation. In this paper, we combined neural network and ICA to separate the noisy mixed images.

D. Jin and S. Lin (Eds.): Advances in ECWAC, Vol. 1, AISC 148, pp. 305–310.
springerlink.com © Springer-Verlag Berlin Heidelberg 2012

2 ICA Recurrent Neural Network Structure

The ICA recurrent neural network structure[2] can be depicted by Fig.1.

Improve this model with a whole connected recurrent neural network, each neuron connected with whole neurons, the model can be expressed as

$$y_i(t) = x_i(t) - \sum_{j=1}^{n} w_{ij}(t) y_j(t) \tag{1}$$

With thc optimal weight w_{ij}, the output signals can be independent mutually. The independent vectors $f(x)$ and $g(x)$ have the formula

$$E[f(\boldsymbol{x})g(\boldsymbol{x})] = E[f(\boldsymbol{x})]E[g(\boldsymbol{x})] \tag{2}$$

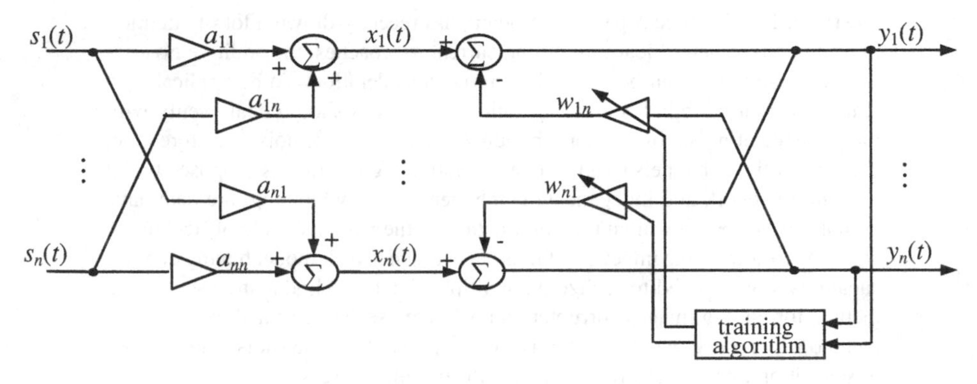

Fig. 1. ICA recurrent neural network structure

Suppose the independent signal y_i and y_j are zero mean, the generalized covariance matrix of the non-linear transform $f(y_i)$ and $f(y_j)$is a nonsingular diagonal matrix, that is

$$\begin{aligned} R_{fg} &= E\left[f(\boldsymbol{y})g^{\mathrm{T}}(\boldsymbol{y})\right] - E[f(\boldsymbol{y})]E\left[g^{\mathrm{T}}(\boldsymbol{y})\right] \\ &= \begin{bmatrix} E[f(y_1)g(y_1)] - E[f(y_1)]E[g(y_1)] & & 0 \\ & \ddots & \\ 0 & & \mathrm{E}[f(y_n)g(y_n)] - \mathrm{E}[f(y_n)]E[g(y_n)] \end{bmatrix} \end{aligned} \tag{3}$$

where the covariance $E\left[f(y_i)g(y_j)\right] - E[f(y_i)]E\left[g(y_j)\right]$ is zero when i is not equal to j, the variance $E[f(y_i)g(y_i)] - E[f(y_i)]E[g(y_i)]$ is not zero, and are different nonlinear activation function.

According to the statistic independence and the normalize condition $E[f(y_i)g(y_i)] = \lambda_i$, the real-time learning algorithm can be deduced

$$\begin{cases} \dfrac{\mathrm{d}w_{ii}(t)}{dt} = \mu(t)[\lambda_i - f[y_i(t)]g[y_i(t)]] \\ \dfrac{\mathrm{d}w_{ij}(t)}{dt} = \mu(t) f[y_i(t)]g[y_j(t)] \quad i \neq j \end{cases} \tag{4}$$

It can be expressed as matrix form

$$\frac{\mathrm{d}\boldsymbol{W}(t)}{\mathrm{d}t} = \mu(t)\left[\Lambda - f[y(t)]g^{\mathrm{T}}[y(t)]\right] \tag{5}$$

where $\Lambda = \mathrm{diag}\{\lambda_1, \lambda_2, \cdots, \lambda_n\}$

3 A Noisy Multiple Channels Blind Signal Separation Algorithm Based on Neural Network

For the noisy ICA model ***x*=*As*+*n***, the hypothesis are showed as follows

1) the noise and the independent component are independent mutually
2) the covariance of noise is foregone
3) ***n*** is Gaussian noise

3.1 The Whitening of Data

Suppose the covariance matrix of noise is Σ, the covariance matrix of noisy observation signal is $\boldsymbol{C} = E\left[\boldsymbol{x}\boldsymbol{x}^{\mathrm{T}}\right]$, whitening the observation signal with the covariance matrix $\boldsymbol{C}$-Σ, the whitening observation signal $\tilde{\mathbf{x}}$ can be expressed as

$$\tilde{\mathbf{x}} = (\mathbf{C} - \Sigma)^{-1/2}\mathbf{x} \tag{6}$$

make

$$\begin{aligned} \tilde{\mathbf{A}} &= (\mathbf{C} - \Sigma)^{-1/2}\mathbf{A} \\ \tilde{\mathbf{n}} &= (\mathbf{C} - \Sigma)^{-1/2}\mathbf{n} \end{aligned} \tag{7}$$

The whitening observation signal $\tilde{\boldsymbol{x}}$ has the ICA model

$$\tilde{\boldsymbol{x}} = \tilde{\boldsymbol{A}}\boldsymbol{s} + \tilde{\boldsymbol{n}} \tag{8}$$

where $\tilde{\boldsymbol{A}}$ is an orthogonal matrix, $\tilde{\boldsymbol{n}}$ is a linear transform of noise.

The noise covariance matrix can be transformed

$$\tilde{\Sigma} = E\left[\tilde{\mathbf{n}}\tilde{\mathbf{n}}^{\mathrm{T}}\right] = (\mathbf{C} - \Sigma)^{-1/2}\Sigma(\mathbf{C} - \Sigma)^{-1/2} \tag{9}$$

3.2 Windage Wipe Off Technique

Select G as a density function of a zero mean Gaussian random vector, the density function of vector x with variance c^2 is

$$\varphi_c(x) = \frac{1}{c}\varphi\left(\frac{x}{c}\right) = \frac{1}{\sqrt{2\pi}c}\exp\left(-\frac{x^2}{2c^2}\right) \tag{10}$$

Suppose $\varphi_c^{(k)}$ is kth order derivatives, $\varphi_c^{(-k)}$ is the kth order integral (k is a positive integer). $\varphi_c^{(-k)}(x) = \int_0^x \varphi_c^{(-k+1)}(\xi)d\xi$, a theorem can be expressed

Theorem[3]: Suppose *z* is a non-Gaussian random vector, *n* is a independent noise with variance c^2. Define Gaussian function φ with the formula (10), then for any constant $c>\sigma^2$, there is

$$E[\varphi_c(z)] = E[\varphi_d(z+n)] \tag{11}$$

where $d = \sqrt{c^2 - \sigma^2}$, to substitute $\varphi^{(k)}$ for φ, the formula still valid.

The theorem revealed that the noisy independent components can be estimated from noisy observation signals by maximization the contrast function.

3.3 The Noisy ICA Algorithm Based on Neural Network

Using the improved ICA recurrent neural network algorithm, the algorithm can be expressed

$$\frac{dw(t)}{dt} = \mu(t)[\Lambda - f[y(t)]g^{\mathrm{T}}[y(t)]] \tag{12}$$

Discrete iterative form is

$$w(n+1) = w(n) - \mu[\Lambda - f[y(n)]g^{\mathrm{T}}[y(n)]] \tag{13}$$

The algorithm has the advantage such as uncomplicated principle and handy realization. But in actual application, it has the limitations of stagnation and poor convergence, and is easy to fall in local minimizer, which are the bottlenecks of its wide application. Make further refinements to the algorithm, impose momentum term, adopt variable learning rate, the new noisy blind signal separation based on neural network is

$$w(n+1) = w(n) + \beta[w(n) - w(n-1)] - \mu(n)(I + \Sigma)[\Lambda - f[y(n)]g^{\mathrm{T}}[y(n)]] \tag{14}$$

where $0 \leqslant \beta \leqslant 1$ is a momentum factor, the affection of the weights variation for last two times passed to present weights variation through momentum factor, send weights variation along the mean direction of error nether camber surface, reduce the sensitivity of the local details in error camber surface for neural network, restrain the defect fall into local minimum. $\mu(n)$ is the variable learning rate, select the biggish learning rate in learning elementary stage for faster speed, reduce the learning rate in stage of closing to convergence for avoiding oscillation and non- convergence.

4 Simulations

Select two original images rice and cameraman shown in Fig.2. Mixed images depicted by Fig.3 are mixed by matrix A with original images, where $\boldsymbol{A} = \begin{bmatrix} 0.1712 & 0.6409 \\ 0.1070 & 0.3571 \end{bmatrix}$.

Fig. 2. The original images

Fig. 3. The mixed images

Add white Gaussian noise to mixed images for three times, the variances of noise are 10, 50 and 100, the noisy mixed images depicted by Fig.4.

The mixed images with noise once

The mixed images with much noise

The mixed images with much more noise

Fig. 4. The noisy mixed images

make $\mu(n)=\lambda^n$, λ=0.5, $\beta=0.1$, non-linear function $f(y)=y$, $g(y)=\tanh(y)$.The separated images by the algorithm can be depicted by Fig.5.

The separated images by the algorithm for the first time

The separated images by the algorithm for the second time

The separated images by the algorithm for the third time

Fig. 5. The separated images by the algorithm

We examine the signal-noise ratio of mixing images, noisy mixing images, separation images which is shown by Table 1.

Table 1. The SNR comparison of images

images		The images SNR with noise once (dB)	The images SNR with much noise(dB)	The images SNR with much more noise(dB)
mixed images	rice		-1.7964	
	cameraman		3.8416	
noisy mixing images	rice	-1.9257	-4.6801	-8.5793
	cameraman	3.5674	-0.1944	-4.7425
separation images	rice	9.7680	9.6810	9.6317
	cameraman	12.6524	11.8343	11.4182

The performance index of the algorithm can be depicted by Table 2.

Table 2. Performance index of the algorithm

simulation	iteration time（s）	PI
first time	3.5462	0.1054
second time	3.6857	0.3681
third time	3.9548	0.4471

where performance index $\mathrm{PI}=\frac{1}{2N}\sum_{i=1}^{n}\left[\left(\sum_{k=1}^{n}\frac{|g_{ik}|}{\max_j|g_{ij}|}-1\right)+\left(\sum_{k=1}^{n}\frac{|g_{ki}|}{\max_j|g_{ji}|}-1\right)\right]$, ***G*** is the transmission matrix and it is $N \times N$ matrix, and ***G=WA***, g_{ij} is the element of ***G***, $\max_j|g_{ij}|$ is the maximum absolute value of line *i*, $\max_j|g_{ji}|$ is the maximum absolute value of row *j*.

We see form above simulations that the performance of the proposed method has strong antinoise ability, it can improve the SNR of the noisy images effectively.

5 Conclusion

In this paper, a method for performing blind signal separation in the presence of additive noise is described. The method is proposed of combining neural network and independent component analysis to separate noisy mixed images. The windage wipe off technique was used to correct the infection of noise, a neural network model having denoise capability was adopted to realize the multiple channels blind source separation method for mixing images corrupter with white noise. By the computer simulation, the method can renew the original images effectively.

Acknowledgements. This work was supported by Shanxi International Science and Technology Cooperation Foundation （No. 2011081047） and Shanxi Scientific Research Foundation for the Returned Overseas（No.2011-035）.

References

1. Shi, Z., Tang, H., Tang, Y.: A new fixed-point algorithm for independent component analysis. Neural Computing 56, 467–473 (2004)
2. Zhou, Z., Dong, G., Xu, X.: Independent component analysis. Electronic Industry Press, Beijing (2007)
3. Huang, Q., Wang, S., Liu, Z.: Improved image feature extraction based on independent component analysis. Opto-Electronic 34(1), 123–125 (2007)

Improved RED Algorithm for Low-Rate DoS Attack

Li Ma[1,2,*], Jie Chen[1,2], and Bo Zhang[1,2]

[1] Jiangsu Engineering Center of Network Monitoring
[2] School of Computer & Software,
Nanjing University of Information Science & Technology, Nanjing 210044, China
mali1775088@163.com

Abstract. Low-rate DoS (LDoS) attack is very ulterior. The characteristics of LDoS attack are not very different from the characteristics of legitimate traffic. That makes prevention measures for traditional DoS attack ineffective to LDoS attacks. Although Flow Random Early Drop (FRED) can resist LDoS attack in the form of non-adaptive stream, but it can't resist LDoS attack of types of data stream. This brings great risks to the network. Therefore we propose new algorithm based on RED. NS-2 Experimental results show that LDoS can be accurately identified and be punished by improved RED algorithm. Then the network can return to normal and network security has been guaranteed.

Keywords: improved RED algorithm, low-rate DoS, FRED, NS-2.

1 Introduction

Low-rate denial of service attack (LDoS) [1-3] is different from the traditional flood-style DoS attack. It does not require the attacker to maintain a high rate of attack traffic. LDoS uses loopholes of the adaptive mechanism on the terminal or network system to send little periodic attack traffic. Because LDoS attack is hidden, the characteristics of intermittent attack are not very different from legitimate traffic. That makes prevention measures for traditional DoS attack ineffective to LDoS attack. This brings new network security threats.

LDoS attack can be divided into attack on the TCP protocol, and attack on the Active Queue Management (AQM) mechanism [4] of routers. LDoS attack on AQM mechanism impacts all links on this router. These attacks are more harmful. This paper analysis LDoS attack on AQM mechanism by experiments based on NS-2 software. We propose an improved RED [5] algorithm for LDoS attack.

2 Experiments of LDoS Attack on RED Algorithm

2.1 Analysis of LDoS Attack

RED algorithm is the only candidate algorithm of AQM. RED randomly drops packets when average queue length exceeds the maximum threshold to effectively control

[*] Corresponding author.

D. Jin and S. Lin (Eds.): Advances in ECWAC, Vol. 1, AISC 148, pp. 311–316.
springerlink.com © Springer-Verlag Berlin Heidelberg 2012

average queue length. RED uses method which is similar with low pass filter when it calculates the average queue length.

$$avg = (1\text{-}wq)*avg + q*wq \qquad (1)$$

In Equation (1), wq is weight of queue. Q is actual queue length when RED samples.

The changes of the actual queue length which is caused by short-term congestion will not make significant changes of average queue length. So this can eliminate bias to sudden flow and prevent TCP global synchronization.

LDoS attack on the RED affects all the links connected to this router through sending pulse high-intensity attacks to the router periodically.

Figure 1 is an abstract model of LDoS attack. The attack stream duration is τ. Attack period is T. Attack strength is δ. The router queue length increased sharply when the net has offensive flow. The router sends congestion control signals to the end system because of the adaptive mechanism of RED. This makes the end systems to reduce their own contract rates. The router's queue length declines to very low.

LDoS Attacks increase transmission delay, delay jitter, and reduce the throughput of the router by periodically sending data stream pulses to the router. The quality of service of network is seriously reduced.

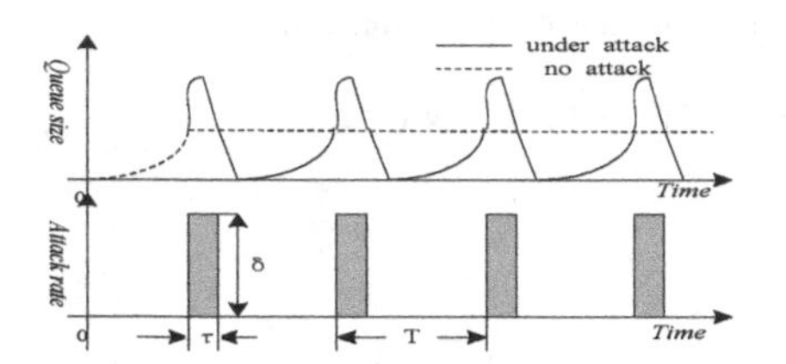

Fig. 1. The model of LDoS attack on RED mechanism

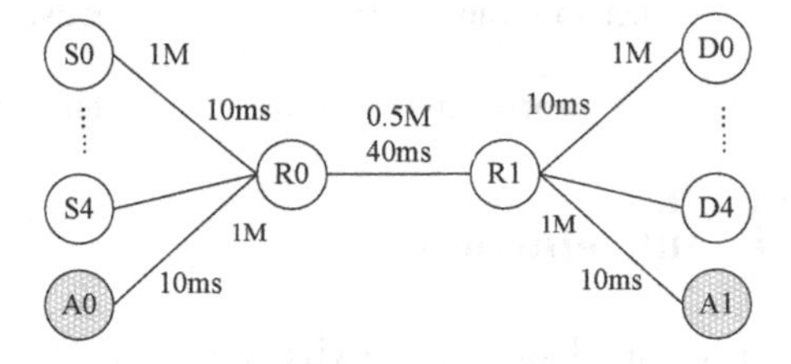

Fig. 2. The topology of LDoS attack on RED mechanism

LDoS attack algorithm is given below.

Constants:

```
    for(i=0;i<n;i++){
        at  (t0+T*i)s  attack
start;
        at (t0+T*i+τ)s attack
stop;}
```

Variables:

n: the total number of attacks;

t0: start time of attack;

T: the attack period;

τ:the duration of each attack pulse;

2.2 Experiments of LDoS Attack

This paper use NS-2 to do experiments. Figure 2 shows the experimental topology. There are two core routers R0 and R1. The link bandwidth is 0.5m/s. Propagation delay is 40ms. There are five data source nodes (S0 ~ S4) and an attack node (A0) connect to R0. There are five data receiving nodes (D0 ~ D4) and an attack receiving node (A1) connect to R1. The bandwidth between router and other nodes is 1m/s, and propagation delay is 40ms.

RED algorithm is deployed on R0. S0~S4 send data stream to D0~D4 during 0s to 30s. The window size is 128KB. The packet size is 512byte..

A0 sends UDP data stream to A1 through R0. These packets act as attack traffic. The packet size is 1000 byte. A0 sends a packet every 5ms. According to the experimental test, the effect is best when the parameters are set as follows: T = 2.3 s, τ= 0.3s. A0 starts attacking when the queue length of router restores stability. This makes queue length always shock.

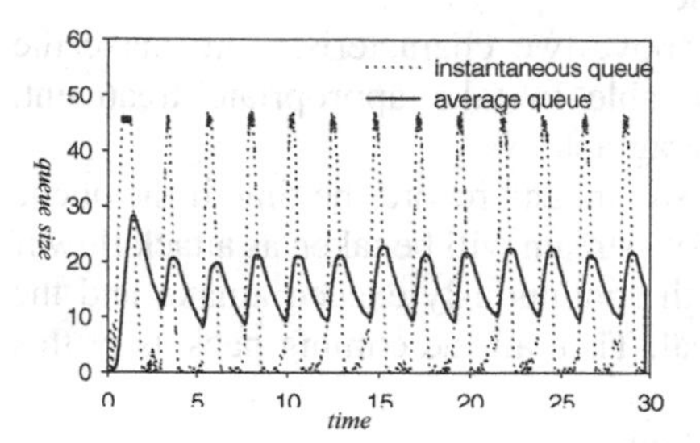

Fig. 3. Changes of the queue length when router is under LDoS attack

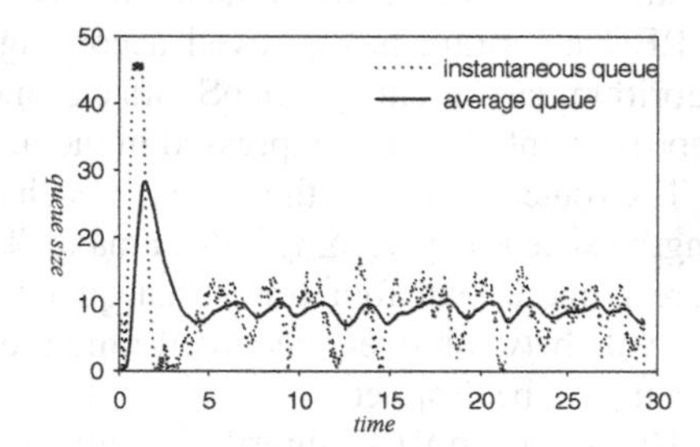

Fig. 4. Changes of the queue length when router is in normal network

Figure 3 shows the changes of average queue length when R0 is under attack. Compared with the situation of queue length in Figure 4, the queue length fluctuates. Especially the changes of the instantaneous queue length are basically the same as the attack model in figure 1.

Figure 5 shows the instantaneous throughput on R0. Solid line shows the fluctuations of throughput when R0 is under attack. Dotted lines indicate that the instantaneous throughput is stable when R0 is not under attack.

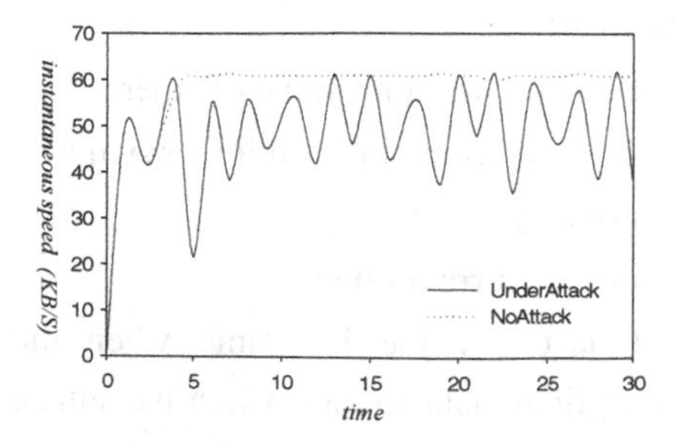

Fig. 5. Comparison of instantaneous throughput when router is under LDoS attack and in normal network

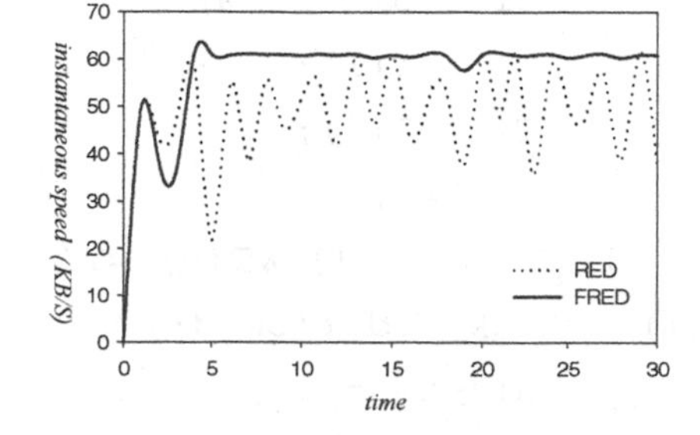

Fig. 6. Comparison of instantaneous throughput when router deploys RED and FRED

In the above experiment, the attack traffic uses UDP packets. Because the RED algorithm can not identify the type of data stream, UDP data stream can effectively implement LDoS attack.

Because UDP data stream is non-adaptive flow, FRED can prevent LDoS attack using UDP data stream. When R0 finds a non-adapted flow, it drops a large number of packets of this stream. Then the impact on the router can be reduced.

3 Improved RED Algorithm for LDoS Attack

It is found that LDoS attack stream has two characteristics. The first one is that the strength of each attack is very high. Most of the queue space is occupied by attack data stream. The second is that attack pulse has cycles.

RED algorithm is improved according to above two characteristics to make the algorithm can identify LDoS attack and be able to take appropriate treatment. Improvement ideas are expressed in the next paragraph.

The router maintains the status of each data stream and record the time if the queue length exceeds the warning value amaxq. The data stream will be taken as attack flow if there is more than four times that queue length continuously exceed amaxq and the intervals between every recorded time are equal. Then all the coming packets of this stream will be dropped.

The improved RED algorithm is given as follow:

Constants:

```
  for each arriving packet
P:{
   if(flowid==attack_fid)
       drop(P);
   if(qlen>maxq&&hit_tag==
0 && aqlen>=amaxq){
       T=time-time_old;
       time_old=time;
       T(0)=T(1);
       T(1)=T(2);
       T(2)=T;
   if(T(0)==T(1)&&T(0)==T(
2)) attack_fid=flowid;
       hit_tag==1;}
       if(qlen<maxq        &&
hit_tag==1)
       hit_tag==0;}
```

Variables:

flowid：id of every flow;

attack_fid: id of attack flow;

drop(P): drop packets;

qlen：instantaneous queue size;

maxq：warning value of qlen;

hit_tag：enable button of hit++;

aqlen：instantaneous queue size of one flow;

amaxq：warning value of aqlen;

T：the time interval of two queue abnormality;

time：current time;

time_old：the last time when the router finds data stream which has attack signatures;

T(0),T(1),T(2)：three consecutive T;

Improved RED algorithm is deployed on R0 on topology shown in Figure 2. Changes of instantaneous and average queue on R0 can be captained.

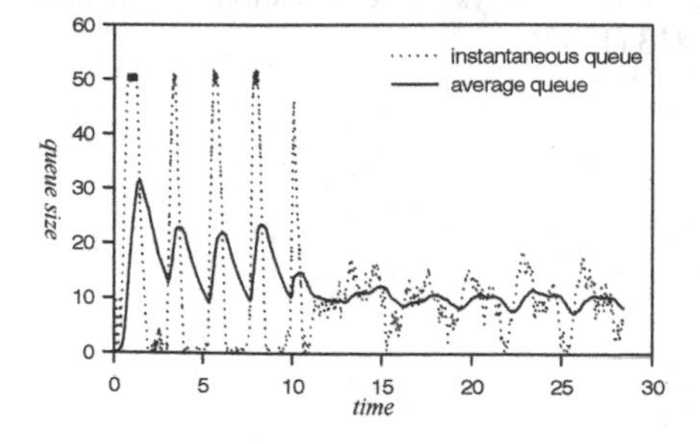

Fig. 7. Changes of instantaneous and average queue size when router deploys improved RED

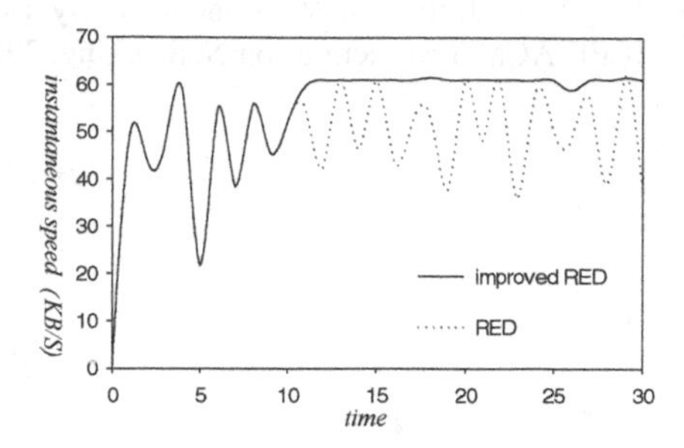

Fig. 8. Comparison of instantaneous throughput when router deploys RED and improved RED

At the beginning, there are five times (the first is not caused by attack traffic) that the instantaneous queue length exceeds the warning value, as shown in Figure 7. When the router knows that T(0) = T(1) = T(2), the attack data stream is identified. All the coming packets of attack traffic all packets are dropped. The router restores stability.

Figure 8 shows the situation of instantaneous throughput on the router R0. The contract rate of the router returns to normal and no longer fluctuate after the router identify the LDoS attacks and take appropriate measures.

4 Conclusions

LDoS is a very subtle attack and very harmful to the network. LDoS attack on the router's AQM mechanism is more harmful. It is found that the FRED algorithm can effectively resist the non-adaptive flows' LDoS attacks. This paper also improve RED algorithm based on two characteristics of LDoS attack. The router will take appropriate punitive measures when it identity the LDoS attack data stream. Network security has been guaranteed.

The research of prevention measures to the LDoS attack in this paper is just the beginning. It is necessary to do more further researches in this area.

Acknowledgments. This work was supported by A Project Funded by the Priority Academic Program Development of Jiangsu Higher Education Institutions (PAPD).We also thank Jiangsu Engineering Center of Network Monitoring for providing experimental environment.

References

1. Kuzmanovic, A., Knightly, E.W.: Low-rate TCP-targeted denial of service attacks. In: Proceedings of ACM SIGCOMM 2003, Karlsruhe, Germany, pp. 75–86 (2003)
2. Guirguis, M., Bestavros, A., Matta, I.: Exploiting the transients of adaptation for RoQ attacks on Internet resources. In: ICNP 2004, Berlin, Germany, pp. 184–195 (2004)

3. Guirguis, M.: Reduction of Quality (RoQ) attacks on Internet end- systems. In: Proceedings of the 24th IEEE INFOCOM, Miami, Florida, pp. 1362–1372 (2005)
4. Athuraliya, S., Low, S., et al.: Active Queue Management. IEEE Network, 48–53 (2001)
5. Floyd, S., Jacobson, V.: Random Early Detection Gateways for Congestion Avoidance. IEEE/ACM Transactions on Networking, 397–413 (1992)

OKN: A Presentation Method for Web Resource Based on Ontologies

Baolu Gao, Jingyu Sun*, and Xueli Yu

College of Computer Science and Technology, Taiyuan University of Technology, China
baolu_gao@163.com, whitesunpersun@163.com, xueli13287@263.net

Abstract. In order to acquire satisfying knowledge quickly in current Web and semantic Web, it is very important to find an effective method to present Web resource. We proposed a concept of Ontological Knowledge Node (OKN) as the core of semantic search model and give a design schema to organize knowledge representation from domain perspective. OKN is a kind of Web resource representation method, which consists of a 6-tuple set and can represent the intension and extension of ontologies as well as the relationship between ontologies. This method was explained in detail from the perspective of granular computing. In order to show advantages of the OKN, a Semantic Search Engine was established, which combines the OKN with Nutch to search campus resources. OKN was proved to work well on semantic search.

Keywords: Granular Computing, OKN, Semantic Search, Ontology, Knowledge Representation.

1 Introduction

Since the 1960s , the knowledge representation has undergone a long process from First-order-logic, Semantic Network, Framework, Production System, and KL-ONE etc. The KL-ONE combines Semantic Network with the features of Framework, and gradually evolves to Description Logic. The Description Logic is the theoretical foundations of the Semantic Web, and the basis of OWL (Web Ontology Language). However, OWL is a web ontology language, and earlier languages have been used to develop tools and ontologies for specific user communities, they were not defined to be compatible with the architecture of the World Wide Web in general, and the Semantic Web in particular.[1]

According to Eric Miller[2] ,the node is an important component, which is composed of data, information or knowledge. Academic Lu Ruqian said, "Knowledge is a kind of structured information" [3]. That is to say, knowledge not only includes concept and term, but also contains the relationship among concepts..

From Granular Computing [4, 5], every concept is understood as a unit of thoughts consisting of two parts, the intension and the extension. The intension of a concept is an abstract description of common features shared by elements in the extension, and the extension consists of concrete examples of the concept.

* Corresponding author.

D. Jin and S. Lin (Eds.): Advances in ECWAC, Vol. 1, AISC 148, pp. 317–322.
springerlink.com © Springer-Verlag Berlin Heidelberg 2012

We propose a concept of OKN used as the core of semantic search model and give a design schema to organize knowledge representation from domain perspective.

2 The Concept of Ontological Knowledge Node

2.1 Some Related Concepts

Before mentioning the concept of OKN, two relative notions should be introduced first.

Ontological Community
Peter Mika said, "Ontologies build upon a shared understanding within a community. This understanding represents an agreement over the concepts and their relationships that are present in a domain."[6] These communities are those ontology groups, which have the common conceptualization terminology and vocabularies. The people who have the similar background in the community can communicate each other in an ontology community covering the most of domain knowledge, so that we could organize a virtual and dynamic environment in which people can exchange their opinion or ideas.

Content Awareness
We make use of semi-automatic aware schema to detect the dynamic ontology knowledge of relative community. Content is the semantic signification of the representation of human in natural language or in words, and it includes the three aspects: content-awareness through semi-automatic interaction, getting information of personal taste, and the factor of transforming among different levels of knowledge. We call this model as the Labeled Content-Awareness Approach (LCAA), and it means that both of the content of semi-automatic interaction and the personality could be used as a knowledge label to assist content-awareness system to get user's intention [7].

2.2 Ontological Knowledge Node

The Ontological Knowledge Node (OKN) can be defined as 6-tuple set, which could be represented as **OKN (N, S, D, S', R, M)**.

- **N** stands for **node_ID**, and it is an unique tag to identify this knowledge node;
- **S** stands for **Set_of_Knowledge_Meaning**, and it represents the set in which the meaning of ontological knowledge of this node is gathered together;
- **D** stands for **Domain_Space**, and it represents entire collection of all the domains in which the ontological knowledge of this node could possibly show up;
- **S'** stands for **Set_of_Domain_Knowledge_ Meaning**, and it represents the set in which every element is the meanings of knowledge in different corresponding domain;
- **R** stands for **Relation_Set_of_Domin_ Knowledge**, and it represents the relations between the meaning of domain knowledge and the meanings of other domain knowledge;

- **M** stands for **Mapping_of_Domain_Knowledge**, and it includes two kinds of mapping rules, the first one produces S', and the second produces R.

The formal constraint is given below:

M,M':Set_of_Knowledge_Meaning, $m_i \subseteq M, m'_j \subseteq M', i = 1,2,3,\ldots,n, j = 1,2,3,\ldots,n'$;

D:Domain_SpaceDomainSpace, $d_k \subseteq D, k = 1,2,3,\ldots,l$;

S:Set_of_Domain_Knowledge_Meaning,

$e_k \subseteq S, k = 1,2,3,\ldots,l$; $e_k = \{m_i \mid m_i \in M \wedge F_{ki}(t)\}$,

$k = 1,2,3,\ldots,l, F_{ki}(t)$ is a logic function, and its value is given in **Mapping_of_Domain_Knowledge;**

S':Set_of_Domain_Knowledge_Meaning,

$e'_k \subseteq S', k = 1,2,3,\ldots,l$, $e'_k = \{m'_j \mid m'_j \in M' \wedge F'_{kj}(t)\}$,

$k = 1,2,3,\ldots,l, F'_{kj}(t)$ is a logic function, and its value is given in Mapping_of_Domain_ Knowledge;

Node_ID can be defined as segment integer value (e.g.FD37-ABCD-DFC3-345D-744F-2223);

Mapping_of_Domain_Knowledge includes the mapping rules. In fact, domain experts define those rules and set the Boolean values $F_{ki}(t)$ of $F'_{kj}(t)$ and, namely, 1 or 0; variable t represents the mapping from the element in **M** (or **M'**) to the element of **D**.

Relation_Set_of_Domin_Knowledge defines relation between one meaning of the ek and one meaning of the e'k.

2.3 An Example of OKN

The concept of OKN is quite complicated. So we will explain the concept with a familiar example in our life. For example, both "Mac" and "Format" can be defined respectively as an OKN shown in Fig.1. In order to define OKN in details, let us analyze two words as follows.

The possible implications of "Mac" are either a mackintosh (raincoat) or a brand of computer; these implications constitute the knowledge meaning space of "Mac". In the same way "Format" has two meanings: a plan or arrangement in a specified form and a command of formatting disk:

Mac's **Set_of_Knowledge_Meaning** is
{a mackintosh, a computer brand};

Format's **Set_of_Knowledge_Meaning** is
{a plan or arrangement in a specified form, a command of formatting disk}.

For the word "Mac", the possible domain space is includes **commodity** and **computer**; for the word "Format", the one possible domain space is **work** (in an office) and **computer**. So Domain_Space of "Mac" can be defined as **D** {commodity, computer}, and that of "Format" can be done as **D'** {computer, work}. Thus the domain knowledge meaning set of "Mac" has two meaning subsets. One belongs to the **commodity** field; there is only one interpretation of "Mac" in this field: a mackintosh. The other belongs to **computer** field; the only interpretation of "Mac" in this field is a computer brand. The choice of meanings is determined by the rules of domain knowledge mapping. So there are two sets.

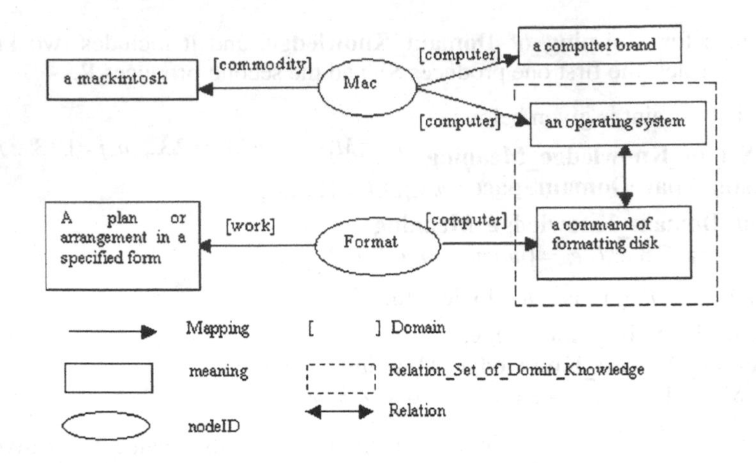

Fig. 1. An Example of OKN

Mac's **Set_of_Domain_Knowledge_Meaning**:
commodity {a mackintosh },and
computer {a computer brand};
Format's **Set_of_Domain_Knowledge_Meaning**:
work {a plan or arrangement in a specified form},and
computer {a command of formatting disk}.

However, with the development of society, it is likely that there might appear a new meaning of "Mac" in **computer** field, such as "Mac OS". At this time, "Mac" has two meanings in the subset of **computer** field; they are Mac computer and Mac OS. The knowledge meaning space of "Mac" in this way is extended to three elements: a mackintosh, a computer brand, and an operating system.

At this time, Mac's **Set_of_Knowledge_Meaning** is {a mackintosh, a computer brand, an operating system};

Accordingly, Mac's **Set_of_Domain_Knowledge_ Meaning** becomes **commodity** {a mackintosh} and **computer** {a computer brand, an operating system}.

"Format" has similar situation likewise. Suppose we are only concerned with the **computer** field, the domain knowledge meaning set of "Mac" has two meanings in the subset of **computer** field; the domain knowledge meaning set of "Format" has one meaning in the subset of computer field.

According to knowledge of human being, we only consider **computer** fields and can get **Relation_Set_of_Domin_Knowledge** of Mac and Format (←→indicates the relations):

Mac's **Relation_Set_of_Domin_Knowledge**: **[computer]** {an operating System} **←→Foramat**: [computer] {the command of formatting disk};

Format's **Relation_Set_of_Domin_Knowledge**: **[computer]** {the command of formatting disk}**←→Mac [computer]**{an operating system }.

3 Experiment: A Semantic Search Engine

In order to show advantages of the OKN, we design a Semantic Search Engine based on OKN and Nutch [8] Search Engine to search our campus resources. In the system, we use OWL to represent communities and take Jena [9] as an inference tool to manage them.

In order to simplify situation we assume community number of our system is 14. These communities can be organized with hierarchical tree and made by Protégé [10]. We represent communities as a set:(Education, Music, Movie, Computer, Software, Database, Oracle, SQL2000, OS, DOS, Windows, Unix, Internet, Management).

Suppose a user named Mr. White has completed his registration as a regular user successfully. When he logs in the system and inputs "OS" and "Windows" as his interest knowledge domain, the system will analyze and produce a CRL automatically depended on BP, CP and IKD sets [7] as follows:

Windows, Education, Music, Movie, Computer, Software, Database, Oracle, QL2000, OS, DOS, Unix, Internet, Management.

Now, White can choose "Windows" community role to log in the system, so he can search in community Windows or father community (such as OS), which is an inference of communities represented in OWL by Jena. Therefore, White can query pages about "Windows" under the backgrounds "Computer" and "OS".

Let's suppose that the user "White" wants to search some pages about "Apple Brand Computer". So he can take "Apple" as a keyword to submit, and the system can return some links at left and some recommended ones at right in browser. The process which White logs in and queries is shown as Fig.2.

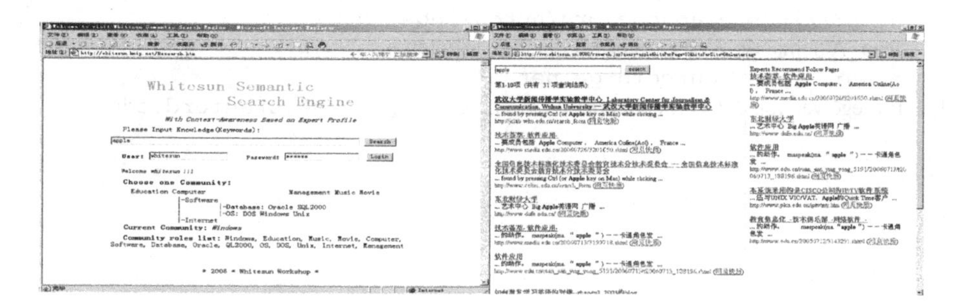

Fig. 2. Returns when White inputs "apple"

4 Conclusion

OKN was proposed based on Ontology and Granular computing, and was explained in detail to show its function. A Semantic Search Engine was developed based on OKN to show how to search knowledge quickly and effectively.

In the future, effective tools will be designed based on the method described in this paper, and an efficient algorithm is the key point of the tools and will also be the most important part of our future work.

Acknowledgement. The authors would like to acknowledge the following support: College Technology Project of Shanxi China (No. 800104-02080075);Youth Natural Science Foundation of Taiyuan University of Technology(No. 90010303010360); Natural Science Foundation of Shanxi (No. 200821024).

References

1. http://www.w3.org/2004/OWL/#ontologies (2006)
2. Miller, E.: Towards Personalized Metadata and Knowledge Systems, 2002-2003 EMMA MA-DMD Program version 0.81
3. Lu, R.: Knowledge Science and its Research Frontiers, http://ir.hit.edu.cn/cgi-bin/newbbs/topic.cgi?forum=17&topic=6&show=25
4. Yao, Y.Y.: Granular computing:basic issues and possible solutions. In: Proceedings of the 5th Joint Conference on Information Sciences, pp. 186–189 (2000)
5. Yao, Y.: A Partition Model of Granular Computing. In: Peters, J.F., Skowron, A., Grzymała-Busse, J.W., Kostek, B.z., Świniarski, R.W., Szczuka, M.S. (eds.) Transactions on Rough Sets I. LNCS, vol. 3100, pp. 232–253. Springer, Heidelberg (2004)
6. Mika, P., Akkermans, H.: Analysis of the State-of-the-Art in Ontology-based Knowledge Management, EU-IST Project IST-2001-34103 SWAP (2003)
7. Yu, X., Sun, J., Gao, B., Wang, Y., Pang, M., An, L.: Studying on the Awareness Model in Ontological Knowledge Community. In: Szczepaniak, P.S., Kacprzyk, J., Niewiadomski, A. (eds.) AWIC 2005. LNCS (LNAI), vol. 3528, pp. 458–463. Springer, Heidelberg (2005)
8. Nutch (version 0.8.1) (September 2006), http://lucene.apache.org/nutch/index.html
9. Jean- A Semantic Web Framework for Java (September 2006), http://jena.sourceforge.net
10. Protégé(version 3.2.1) (September 2006), http://protege.stanford.edu/

The Research of Network Educational Resources Managing Based on Virtual Synergistic Technology

Meifang Chen, Xiaoqiang Hu, and Liang Zeng

Jiangxi Science and Technology Normal University
Nanchang, Jiangxi, 330038, China
meif.cmf@163.com, chinavrmm@163.com,
zengliang3891@qq.com

Abstract. Nowadays the differences between different areas often result in the loss of educational equity. As the development of network, people get the resources with the diversification in the format, what relaxes the phenomenon about educational flat. However, the irregularity and insecurity of network limit effectively requiring of resources. The virtual synergistic technology can provide a distributive and cooperative platform to solve the sharing of large scale resources, and to manage the network educational resources effectively. At last, it will realize the all-round share of resources to promote the equity of education.

Keywords: flat education, virtual synergistic technology, distributive cooperation.

1 Introduction

With the popularity of modern information technology in education, network educational resources become one of the most important information sources for comprehensive utilization to teachers and students on different levels. The openness of the network promotes informatization of the area and campus network resources, which provides us abundant forms of rich, colorful diversified resources, and realizes the communication and share of the regional resources to a great extend. However, because of confused quantity, dynamic instability, possible search difficulties, and poor safety, the network resource was scattered disorderly. All these factors seriously influenced the effective searching, acquiring, sharing, and using the network information resources.

As a distributed computer environment, virtual synergistic technology can effectively organize many cooperative members and their activities which are separated by time together, distributed by space, but worked interdependently. It is to realize all-round coordination among different application systems, different data resources, different terminal equipments and different application situations. It provides a distributed cooperation platform with multiply virtual units to solve the problem of large scale resources sharing.

D. Jin and S. Lin (Eds.): Advances in ECWAC, Vol. 1, AISC 148, pp. 323–328.
springerlink.com © Springer-Verlag Berlin Heidelberg 2012

2 The Structural Form of Network Educational Resources

The large quantity of Network educational resources has brought not only mighty service function for education, but also new challenges for educational resource management. As shown in figure 1, the three-dimensional configuration classification of educational resources informatization is structured from the contents of resources organization (well-structured or bad-structured), resources presentation forms (isomorphism or isomerism), resources distribution and storage (distributive of discrete) in this article.

The difference between discreteness and distribution is that discrete resources are in disorder, while distributive resources can be sorted and combined orderly, that is, resources are distributive stored, and catalogue management is centralized. Most information resources perform as "isomerism-bad structured-discrete" form, so that data searching is inconvenient and the data security can't be guaranteed.

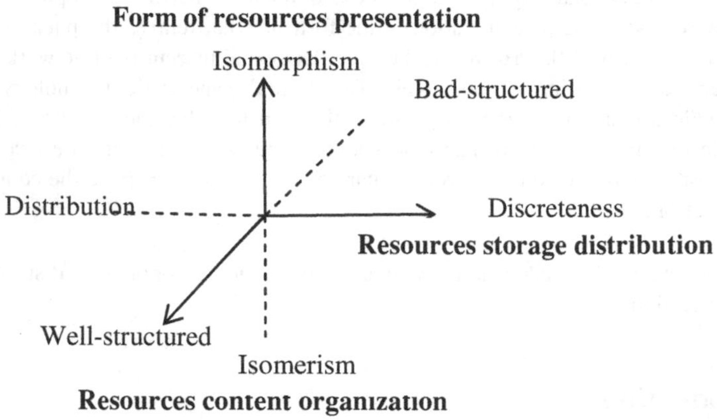

Fig. 1. The three-dimensional structural form of educational resources informatization

3 The Analysis of Network Educational Resources Management

Network educational resources management has experienced several development models such as the discrete WEB resources list, comprehensive web site, academic web site or project site, database of resource management and so on. As is shown in table 1, the information resources in most of development models have unstructured characteristics, so that it is not easy to search, to acquire, to manage and to apply the information resources.

3.1 Network Educational Resources Development Is Not Balanced

Network educational resources mainly are published by the form of web service that each of agencies uses different creating technology and database structure and the users can browse pages directly to access and share what they need. However, unstructured educational information resources are locked in a single database so that

Table 1. The examples of network educational resources classification

Resources content organization	Forms of resources presentation	Resources storage distribution	examples
well-structured	isomorphism	distributive	Project website, Net library
well-structured	isomorphism	discrete	Net work bookmarks
well-structured	isomerism	distributive	Project website
well-structured	isomerism	discrete	Blog
bad-structured	isomorphism	distributive	Online database
bad-structured	isomorphism	discrete	Forums, Online community
bad-structured	isomerism	distributive	Comprehensive website
bad-structured	isomerism	discrete	Net discrete resources

they will not be shared comprehensively, which eventually leads to the rich resources held by some of universities, while other users complain about resource scarcity.

3.2 Resources Management Modes Are Not Perfect

Fast expansion of network educational resources and the lack of effective optimization management of the resource management platform lead to great difficulties to acquire resources. The users can't effectively screen and obtain resources. Most searching engines are also unable to cover all the network information resources, especially the deep-seated searching in the database, which greatly influenced the recall ratio of searching engine.

3.3 Resource Management Platform Can't Realize Hardware Resources Sharing

Network educational resources are scattered in independent network center or laboratory, where exists expensive storage and computing devices. But it is hard to realize the sharing for that it must increase the resources management cost to some extent and go against the high utilization of hardware resources and resource sharing.

Virtual synergistic technology can provide a distributed cooperation platform with multiple units. It will get together different types of computers, external equipments, resource pools, and collaborative service support which dispersed in different places. Through the integration of resources and coordination, it can realize the full sharing of computational resources, equipment resources, service resources and information resources.

4 Distributed Resources Management System Based on the Virtual Synergistic Technology

To solve the problem of network educational resources management, virtual synergistic technology can be used to build low-cost distributed resources management system, with specific themes for task orientation, then to realize resource sharing.

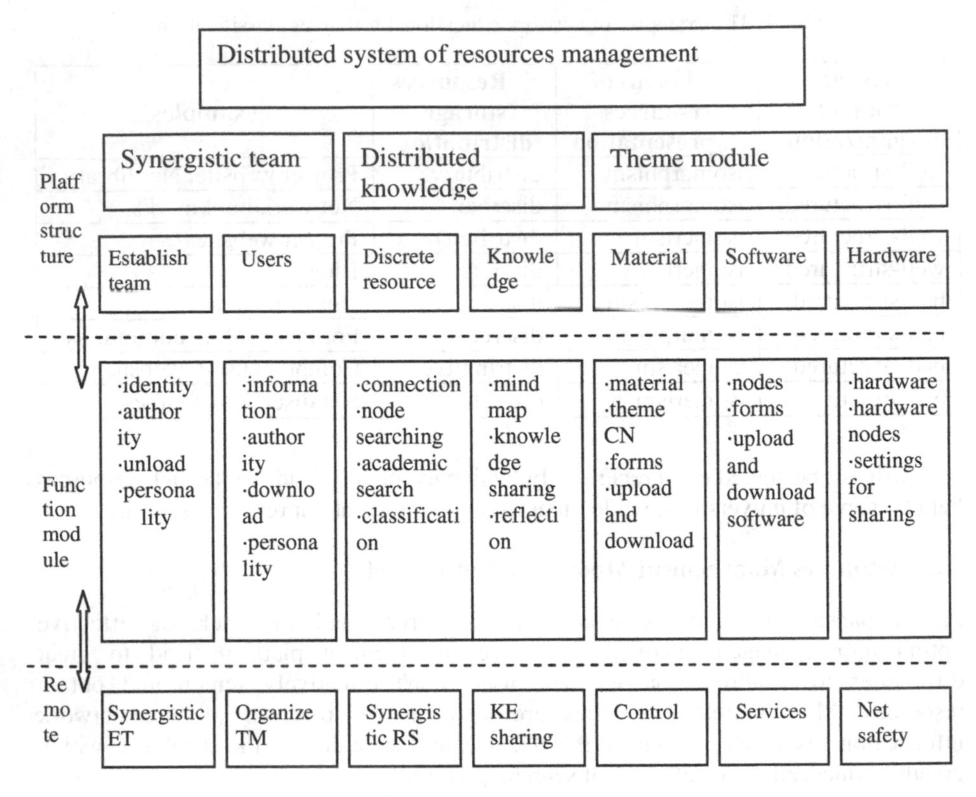

Fig. 2. Distributed system of resources management

4.1 Collaborative Team Management

Team management includes remote management for far-flung resources construction team and the resource users. Distributed resources management system allows all registered users to manage personal information. According to the individual user's need, different theme permission is distributed. With virtual cooperative technology to create collaborative environment, it provides a coordinative resources construction platform. Through the collaborative tool, resource builders can realize opened communication and promote the formation of collaborative team, at the same time, they can use remote collaboration and control technology allocation to upload resource modules.

4.2 Formulas

Distributed knowledge management includes polymerization management of discrete resources and knowledge sharing management. The system takes use of distributed collaborative tools with low cost to get together dominant resources dispersed in different place through a certain subject classification form, to manage coordinative, and to support ideas exchanging and knowledge sharing between the users.

4.2.1 Polymerization Management of Discrete Resources

Discrete resources polymerization refers to the process of resources turning from discrete to distribution. This system takes different themes as key words, and uses Web2.0 software tools to establish connection between discrete resources, thereby promotes the depth of resource searching. Not only does it save time and efforts to highlight user's personality, but also it enhances the accuracy rate and makes the resources easy to save and arrange.

4.2.2 Knowledge Sharing Management

Virtual synergistic technology provides a platform for knowledge exchanging and sharing. It also encourages the resource users to self-reflect, to construct mind map, and to form a knowledge chain centered by themes. The users will have an orderly knowledge structure by effectively managing the knowledge chain. Through personalized settings and communicative platform, the users are able to share recessive resources.

4.3 Theme Module Management

With the strongly generative and dynamic character, the half-opened theme resources mainly contain material resources, software resources and hardware resources. Through the theme classification module in collaborative management platform, resources which have strong inner connection and common theme will be disintegrated into many theme units, so that the resources are made to be structured and modular. The form of the theme resources is distributed. All the resources don't focus on local region, but through the links and mutual visits of modules between platforms, they are available for all users.

4.3.1 Material Module Management

General materials include teaching lesson plans, courseware and test questions. The platform use collaborative tools to get together material resources dispersed in different places. After classifying structural modules, managers can upload and download resources according to users' distributive authority.

4.3.2 Software Resources Coordination Management

Software resources include network curriculum resources and system resources. The resources are distributed in different website nodes. By using the relative control technology, the platform collects different software resources.

4.3.3 Distributed Management of Hardware Resources

Computer resources in different areas and storage resources can be opened up to the public as the form of web by using virtual synergistic technology. The users can use the remote distributed laboratory equipments to acquire different information resources. In the platform of distributed resources management, they cooperatively finish the same project so as to share all kinds of resources.

5 Conclusion

The application and management of educational resources are the core of educational informatization. How to improve the utilization rate of educational resources and how to realize the comprehensive resource sharing have important and practical significance for solving the flat phenomenon on educational resources.

This paper focuses on re-constructing managerial platforms for network educational resources by using the virtual synergistic technology. Through effective integration of discrete different resources, including hardware equipments and software resources, the technology improves the resource sharing, the degree of usage, and the efficiency of educational resources, sequentially to solve the practical problem of effective educational resources management.

References

1. Baidu Post Bar: Information fusion and scale-free network (DB/OL), http://post.baidu.com/f?kz=7938537
2. Chen, X.: Link analysis of a new kind evaluation method for education website. Chinese Audio-Visual Education (7), 66 (2007)
3. Vreeland, R.C.: Law libraries in hyperspace: a citation analysis of World Wide Web sites. Law library Journal 92(1), 9–25 (2002)
4. Hu, X.: Marginal utility: How to deal with educational resources? Information Edition of Chinese Electronic Education in CCE (Z6-Z7), November 11 (2006)
5. Hu, X., Zhan, B., Hu, T.: Research on regional education information resources construction and the development strategy at the present situation. Chinese Audio-Visual Education (6), 60 (2007)
6. Yang, X.: The network information age push technology and related issues. The Journal of the Library Science in Jiangxi (3), 79–80 (2005)

A Brief Study on Exploration-Internalization Mode Based on Information Technology

Xianzhi Tian

Wuchang University of Technology
Julia030712@163.com

Abstract. With the development of information technology and education, different reforms have been advocated in different fields. In the present, most Chinese reformers have realized that the core of learning is on learners and their autonomous learning abilities. Therefore, Wuchang University of Technology has promoted exploration-internalization mode to develop the autonomous learning abilities of students. During the course of reform, the author has tried many different teaching methods to stimulate learning desires of students. After one year and a half, the trained students have been improved in a great degree. But at the same time, the author has also found some problems in teaching.

Keywords: autonomous learning abilities, exploration-internalization mode, information technology.

1 Introduction

In the long run, the core of learning is not on the teaching of teachers,but on the autonomous learning abilities of learners, which has been advocated by Chomsky, Helen and some other educators in the world. Nowadays, teachers have realized the point day by day from their teaching experiences. But how can teachers dig out the inner invisible abilities? Different reformers have put forward different methods. As for the author, she has tried many different teaching methods too. But the most effective one is on exploration-internalization mode(EIM). Actually,EIM puts her focus on digging the learning potential of students out. In class, teachers should try their best to inspire learning desire and learning abilities and any other abilities. And "guide " "checking" "testing" ect. are the main means for teachers to spire different abilities of students. At the same time, students should finish many different exercises related with class teaching. In the class, teachers should try their best to control teaching time, thus they can leave so much time for students to learn by themselves and ponder about some questions by themselves. During the reform, teachers should put their main focus on designing modes on how to spire the inner potential abilities of students in class. So many different questions will be asked in class. But the questions are not the simple questions,but the deepest questions, which will be benefit for digging the potential abilities of students.

EIM is consistent with autonomous learning theory of Helen and the discovery learning theory advocated by famous psychologist J. S. Bruner also emphasizes that students' learning should be a process of actively discovering on their own, rather

D. Jin and S. Lin (Eds.): Advances in ECWAC, Vol. 1, AISC 148, pp. 329–332.
springerlink.com © Springer-Verlag Berlin Heidelberg 2012

than of receiving knowledge passively. In actual teaching and learning, students are active explorers, and the role of teachers is to set up context for problem-solving, to evoke students' interest of learning, to satisfy their cognitive needs, and to arouse the learning motivation of autonomous exploration of knowledge, rather than to pass on or to impart existing knowledge. The purpose of EIM is exactly to embody the feature of people's autonomy,which is the same with "zero-approach" teaching mode(the author has studied it in the former studies).

2 Exploration-Internalization Mode

Exploration-internalization mode has been firstly put forward by Wuchang University of Technology by the strong requirement of teaching reform. In the present, the mode has been applied in Wuchang University of Technology in China. The headmaster has already realized the importance of inspiring learning autonomous abilities and other abilities of students. Therefore, from September,2011, the university began to put forward the mode in most course to face the whole reform of the country. In the mode, the university had ruled the time of teaching or lecturing. At the same time, it also ruled the amount of exercises of students, time of questions, time of going over learning contents and so on. In order to make the mode scientifically, the university had held many conferences to discuss the effectiveness and improvement points of the mode. And some famous researchers and experts had been invited to give suggestions to improve the application of the mode. After several months, some detail measures had been put forward. At the same time, almost all leaders began to put their working emphasis on the real realization of the mode.

Exploration-internalization mode mainly includes several steps as follows:

Table 1. Detail course of Exploration-internalization mode

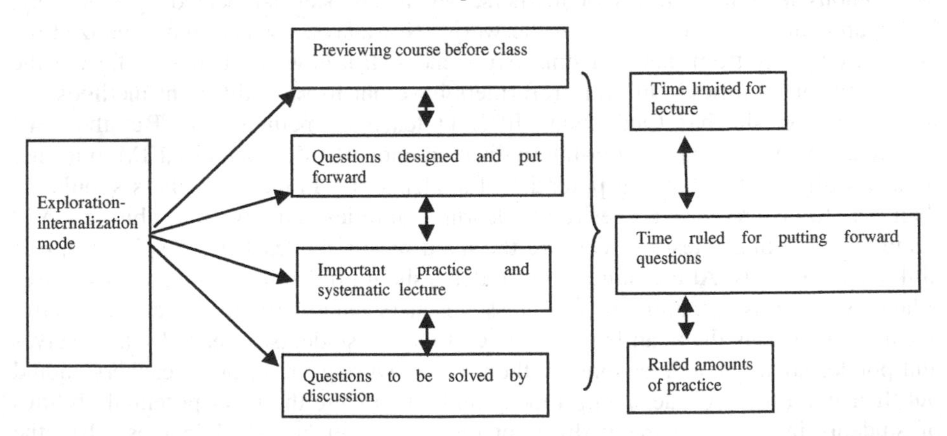

From the above table, it is clear for us to see the detail course of Exploration-internalization mode. In the course, it is mainly concerned about the main steps in class, and in fact , some steps outside of class have been omitted here. All of the above steps had been based on application of information technology. It is also combined with modern technology for the training of autonomous abilities of students. The aims of the method is on "exploration" of the inner invisible abilities by using some teaching modes to make students internalize their learning contents. "previewing course before class""questions designed and put forward""important practice and systematic lecture""questions to be solved by discussion" are the basic steps for reaching the above aims. In those steps, there are different detail measures to be adopted by different teachers or trainers. Finally,some limitations can be adopted by teachers for their controlling class and students such as "time limited for students""time ruled for putting forward questions""ruled amounts of practice". The method mentioned above is to improve teaching efficiency and to reach the real aims of teaching reform. Different teachers had different ideas on the method, but the common ideas on it is on its focus of putting students to learn by themselves and urgent students to find problems in learning and try their best to solve the problems. If students can not solve some difficult questions, then teachers can help them to analyze the problems and "stimulate" students to find the correct measures to deal with the difficult problems. During the course, students will learn how to deal with problems not only in learning course but also in social life.

3 Exploration-Internalization Mode and Information Resource

In the reform of "exploration-internalizaion mode", different information resources have been used for improving autonomous learning abilities of students. In fact, not only for students but also for teachers, they are closely connected with modern information day after day. All steps of the mode is to urge students to learn by themselves and more questions must be solved by students. In order to solve the questions, students must get help from different kinds of resources. The main in is from library, and the next main one is from web. At the same time, some autonomous learning centers are based on modern information to build resource pool, which is very important for students to get efficient information in a shorter time. As is applied by the authors, the following types often have been used by teachers and students:

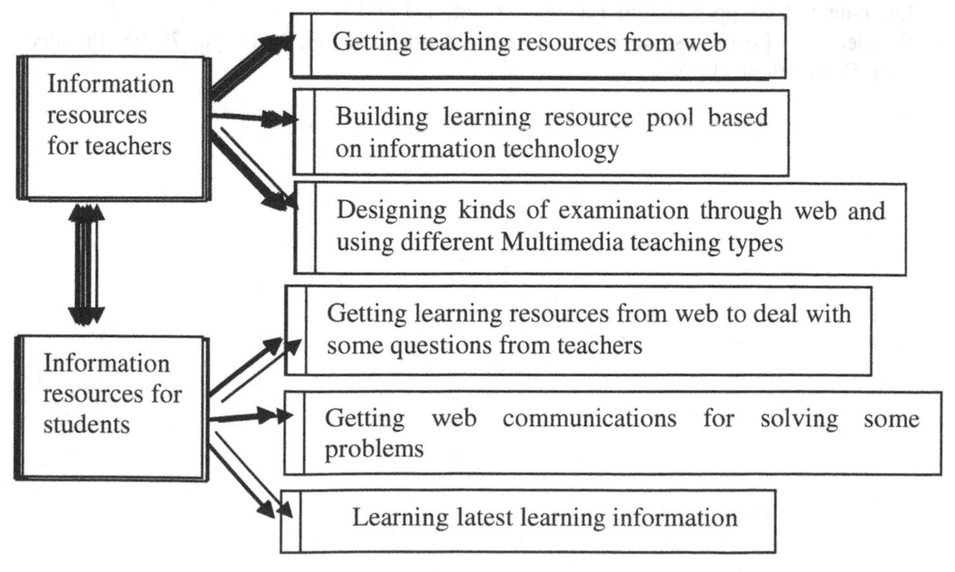

The above tables explains the different kinds of uses of information technology for teachers and students in modern information age. In fact, all of them are served for improving teaching efficiency and making students learn how to lean and how to deal with social problems for their further coming to society. So information resources play very important role in teaching and learning courses, and we must not ignore their very important role during this course. In teaching reform, different teaching methods will be used, but all of them must follow modern situation and development of information. So if learners and teachers don't realize the importance of modern information and its status, their reform will only be in form. As for the author, she has put some modern teaching mode in her teaching course, and the most important one is information-based autonomous learning mode for students. During the course, students are required to finish certain tasks based on information resources, and after a period, teachers will check them through web by autonomous checking.

4 Conclusion

In short, the "exploration-internalizaion mode" is a new mode for exploration inner abilities of students, and its aim is also very clear. Teachers will get its efficiency only through autonomous learning of students based on information technology. The exploration road is very long, and it is inevitable to meet difficulties. But the author believe that it will bring real education reform in education field. The author will study it further and try her best to find suitable teaching forms to realize the aims of it.

References

1. Qin, X.M.: Analysis of Quantitative Data from Foreign Language Teaching and Learning. Huazhong University of Science and Technology Press, Wuhan (2003)
2. Sheerin, S. A.: An Exploration of the Relationship Between Self-access and Independent Learning, vol. 63, pp. 143–146. Longman, London (1997)
3. Wenden, M.: Learner Strategies for Learner Autonomy, vol. 21(1), pp. 79–95. Prentice Hall, Great Britain (1998)

The Research of Commercial Real Estate Development Problems and Countermeasures in China

LiYuan Liu and WenKuan Chen

Economics & Management College, Sichuan Agricultural University, Chendu Campus, Chendu Sichuan 611130, China
125929979@qq.com

Abstract. With the real estate market development, commercial real estate became the type of expanding real estate market, transferred the real estate market pressures, solution of livelihood issues, adjusted the real estate market in China. This analysis Based on the rapid development of commercial real estate, combined the lack of commercial real estate development experience, financing, operated strategy, talent, China should position of items, broaden the financing channels, unified operating channels and train professionals for the development of commercial real estate .

Keywords: commercial real estate, development, countermeasure analysis.

1 Introduction

Commercial real estate included shopping centers, supermarkets, Commercial Street, shopping mall, theme malls, professional markets, wholesale markets, discount stores, factory outlets, entertainment commercial real estate, and residential and commercial real estate office of the underlying product. Commercial real estate had become the focus of real estate development, but the rapid development of the commercial real estate, faced many problems: the difficult of location, accounted for funds, long cycle, the difficult of managing and so on. Therefore, the analysis of development of commercial real estate was necessary.

2 Reason for the Rapid Development of Commercial Real Estate

2.1 State Real Estate Regulatory Policies Promoted the Development of Commercial Real Estate

In recent years, since the rigidity of the housing needs of the people, resulted in a lot of money into residential real estate, direct consequence was: the contradiction between general public demand for housing and the market supply was growing, the market price went up again, housing became a national focus on livelihood issues. In recent years, strict control related regulations of excessive growth of high prices had been put on, solved the livelihood problems that had plagued the country, promoted a large number financing existing in the residential real estate to commercial real estate,

D. Jin and S. Lin (Eds.): Advances in ECWAC, Vol. 1, AISC 148, pp. 333–337.
springerlink.com © Springer-Verlag Berlin Heidelberg 2012

tourism, pension real estate and other form of real estate investment, promoted the restructuring of the real estate market.

From the macro perspective, the state has stepped up the regulation of residential real estate market, promoted the development of commercial real estate, effectively solved the housing difficulties of low-income urban family problems, protected the livelihood of the people; promoted the diversification of the real estate market development, reduced the risk of a single investment in the real estate market and real estate bubble.

2.2 Property Developers Go after Interests

Since capital inherently seek profit, when the average profit of residential real estate declined, then real estate developers would find new profit growth point, changed investment in strategic. Since the late 1990s, the development growth of residential real estate in the amount of 20% or more, but the commercial development of structure, market positioning and housing prices and other issues in the development process made the current residential real estate investment in the overall market decline. Commercial real estate return on investment was relatively high, profitability was strong. According to the investment-benefit analysis, residential real estate investment return rate of about 6% to 8%, commercial real estate return on investment of 8% to 12%, some even up to 15% or more, long-term investment rate may be as high as 40%, the profit in the residential property declined, the commercial real estate development added a new source of profit for the developers.

2.3 The Inevitable Result of Urbanization

30 years after reform and opening up, China's urban area was expanding, the number of cities was increasing, the number of residents was expanding, income gradually increased, living standards increased, ordinary city residents also had strong demand, original business model in beginning of reform and opening up had failed to meet the residents higher demand of the community services, cultural entertainment, health care and other facilities; living and consumption patterns had undergone tremendous changes in leisure time and leisure groups gradually increased, shopping, entertainment, leisure and other social interaction activities increased, grade requirement of these commercial place had improved, so some large-scale, multi-functional, beautifully decorated and shopping, entertainment, leisure placed gradually.

2.4 Demand for Commercial Real Estate after Farmers Lived Together

With industrialization and urbanization process accelerated, industrial and urban demand for land development accelerated in many areas part of China's net agricultural area located in the outer suburbs or rural homestead areas, began demolition and reclaim under the guidance of the government, promoted large-scale farmers to live together. The basic needs driven small community commercial real estate development after large-scale farmers live together, the commercial premises located in a residential community, mainly faced residential neighborhoods, in order to meet the living, production and entertainment needs of concentrated residential areas. Convenience stores, small supermarkets, drug stores, kiosks, books, newspapers hall, a small clothing store dining, fitness facilities, beauty salon, bank, dry cleaners, kindergartens and other facilities was the main form of community shops.

3 The Problems in Rapid Development of Commercial Real Estate

3.1 Lack of Development Experience

Most of the commercial real estate investors followed the national macro-control policies in real estate market, diverted from the residential to commercial real estate development, lacked understanding of commercial real estate operation law and in-depth analysis of commercial real estate development and residential development difference, did not project site investigation and analysis directly applied the model of residential development and there was a certain blindness in investment.

3.2 Single Financing Channels

Along with the rapid development of real estate finance, real estate, stocks, trust funds, stocks, bonds and other financial products continued to emerge, but because of the commercial real estate investment was larger in general, the situation that real estate developers had been relying on bank loans had not changed, National commercial real estate depended on bank credit at the level of 70% -80%. China's current self-financing commercial real estate remained low, only 33.5% of capital in place, while the large-scale urban development depended on bank credit more than 90%, much higher than 50% of the risk discount, appeared a huge investment risk.

3.3 Shortage of Professionals

Because most commercial real estate developers changed from the residential real estate investment, they lacked practical experience for commercial real estate development field, other professional knowledge of commercial real estate site selection, functional, operational, sales and so on. Currently, the commercial real estate was still in rapid development period, the combination professionals of well-known commercial real estate and real estate was less, resulted in the current commercial real estate projects were the same pattern, impacted post-operational management. The latter part of the operations, sales, management relatively lacked of a large number of personnel, increased investment risk.

3.4 Single Operating Strategy

For large-scale commercial real estate, its business multi-used overall development, mainly collected rent as a return on investment model; commercial real estate projects could be packaged to form commercial real estate finance; for small commercial real estate, most projects were still take the way to recover the rent. At present, commercial real estate business model were: "rent not buy," "sales", "rental with sales", "sale-leaseback" four modes. A successful commercial real estate projects required length project cycle, large amount of investment, long payback period, but the developers currently hold the idea withdrawn funds quickly , sold after commercial real estate project completed, adopted sale, fragmented business model. They focused on short-term development interests, without considering long-term project operation. A direct result was loss of integrity for the large-scale commercial real estate projects, caused the later management difficulty, increased risk of post-operational management.

4 Countermeasures

4.1 Accurate Positioning of Items

Commercial real estate development and management was more complex than residential real estate, most of the commercial real estate used lease form to obtain compensation and benefits. The project site and pre-positioning decided commercial real estate later benefits, developers made market positioning should following the laws of the market, combined with field development in the case, analyzed market demand, government and other needs in the market in-depth, accurately identified the type of project to meet the market demand. Commercial real estate could attract investment first, then developed the model. This "Order Sales " items could be quickly integrated into the commercial market, achieved the direct operation of commercial real estate projects, developers realized capital flows quickly, reduced project risk .

4.2 Broaden the Financing Channels

In order to spread risk, the developer should broaden the long-term financing channels in real estate projects, changed past financing form relying on bank financing single. Through joint funding with partners to develop business, sharing profits, sharing the risk; also could make use of foreign commercial real estate development experience: developed with insurance industry and the life insurance industry; commercial real estate could also make use of financing brand. But the developers in the financing process must pay attention to the importance of reputation, improved their level of credibility, and ultimately achieved the effect of its own brand.

4.3 Unified Operating Channels

At present, the operation of commercial real estate had variety channels, but the developers often used "net sales" mode for their own interests, so the developer got profit only once, lost the opportunity to operate distribution of profits later in the project. Successful commercial real estate projects overseas business model were the "pure rent" mode, through renting shop, looked for experienced managers, then achieved its maximum benefits by rental income. The developers should considerate comprehensive, industry, background, credit and other aspects of comprehensive, when commercial real estate developers selected tenants for the project, then selected tenants for their best interest , achieved maximize long-term benefit.

4.4 Trained Professionals

Developer needed Familiar with the complex commercial and real estate real estate developer talent. Compound talents knew real estate development, operation, management, sales knowledge, provided key recommendations for a project to achieve the exact location of commercial real estate projects, a better operating fund, and maximized project profits from a global point.

In short, the rapid development of commercial real estate in the current case, the problems commercial real estate developers faced gradually increase, commercial real estate capital operating risks and competition increased, and the countermeasure for commercial real estate development problem also required constant improvement.

Acknowledgement. This work was supported by Medical and Technology Planning Project of Sichuan Province, China(NO.2010NZ0105),Research of Modern Ecological Agricultural Technology Integrated Mechanism Innovation and Its Performance Evaluation Project of Sichuan Province, China(NO.10SA065) and Rural Development Research Center Project of Sichuan Province, China(NO.CR1002).

References

1. Wu, K., Zhen, X.: Analysis of China's commercial real estate and strategic status. China Business (12) (2010) (in Chinese)
2. Guo, X.: Analysis of China's commercial real estate development Warming and characteristics. Economic Forum (12) (2005) (in Chinese)
3. Liu, X., Chen, S.: Commercial real estate risk and prevention (12) (2005) (in Chinese)
4. Wang, J.: Analysis of the commercial real estate business model in China (8) (2009) (in Chinese)

Multi-pattern Finite Automation Based Regular Expression Matching

Zhanjie Wang, Wenjuan Qiu, and Lijun Zhang

Department of Computer Science and Technology
DaLian University of Technology, DaLian 116024, China
wangzhj@dlut.edu.cn, qwjd-001@163.com, zhlijun_dlut@163.com

Abstract. Data processing and information retrieval is key technologies in information technology area. Regular expression matching plays a crucial role in information retrieval. However, in regular expression matching, the traditional DFA method has serious performance degradation with a large-scale rule set. And building a hybrid DFA will lead to a substantial increase in space requirements. To address these problems, according to the rule set's characteristics and combining with multi-pattern idea, a matching mechanism based on multi-pattern finite automation is proposed. Compared with the traditional matching algorithm, the new mechanism's time complexity can be reduced to O(M+n). Using experiment, we demonstrate that by parallel processing, it can achieve an efficient regular expression based information retrieval and matching. This provides a new technical support for regular expression in information retrieval and language processing fields, and has a wider range of applications.

Keywords: information technology, retrieval, regular expression, multi-pattern, parallel processing.

1 Introduction

Data processing and information retrieval is key technologies in information technology area. Regular expression matching plays a crucial role in information retrieval. Now the main features of regular expressions are contained in most kinds of computer application tools and software packages [1-4], making regular expression develop widely in information technology [1].

In information retrieval practice, there are two typical finite automata can complete data matching based on regular expressions: NFA and DFA. But with the explosive growth of modern information, efficiencies of information retrieval and data processing should be improved relatively. Traditional NFA and DFA is difficult to meet the increasing demands of matching efficiency: NFA must handle all states in the active state set one by one while dealing with each input character. It is not suited for parallel processing and results in very low efficiency. Although DFA only needs to access one state while dealing with a character, if each rule is compiled into a single DFA, its time complexity will increase with the number of rules. The performance degrades seriously when there are large-scale rule sets. And building a hybrid DFA will lead to a

D. Jin and S. Lin (Eds.): Advances in ECWAC, Vol. 1, AISC 148, pp. 339–344.
springerlink.com © Springer-Verlag Berlin Heidelberg 2012

substantial increase in space requirements, however, the current hardware conditions will not meet such a large memory requirement [5]. Therefore, a regular expression matching technology which can retrieve large amounts of data by parallel processing has become a priority.

This paper conducted a study for the proposed question and proposed a multi-pattern-based regular matching mechanism. It can improve matching efficiency of large scale rule sets by parallel processing many rules as well as achieve efficient regular expression matching. And experiment results proved its information matching efficiency is better than traditional NFA and DFA.

2 Defects of Traditional Algorithm

In information retrieval practice, matching based on regular expression is achieved by two typical automata: NFA and DFA. DFA construction process is as follows: Firstly, construct the regular expression's NFA according to Thompson Algorithm [6]; and then build the corresponding DFA according to NFA.

Thompson Algorithm introduces a number of ε (blank) migrations in generating automaton process, which increases the automaton states' number and takes up more space. Therefore, for regular expression rules, we can ignore the ε states and directly get into a next state to reduce the number of automaton states and space utilization.

Using the traditional DFA algorithm to match a data set T of length n is required backtracking. In matching process, if the rule sets contain M rules, data set need to be scanned again for each DFA built by every rule. Algorithm's time complexity is O(M*n), and it will increase with the number of rules or data set. The performance degrades seriously when there are large-scale rule sets. But building a hybrid DFA will lead to a substantial increase in space requirements. The current hardware conditions will not meet such a large memory requirement [5].

By studying the rule sets, we found there were a lot of rules containing similar characters. So some rules can be compiled into an automaton. Then once scanning of data set can match multiple rules. This not only reduces time complexity, but also avoids the huge space consumption when all of rules are compiled into a mixed DFA. Therefore, based on existing problems of traditional algorithms and using multi-pattern idea, a multi-pattern-based DFA matching mechanism is provided. It can improve the matching efficiency of large scale rule sets by parallel processing multiple rules at one time.

3 Multi-pattern-Based DFA

Multi-pattern-based DFA is to compile a number of rules with similar characters into an automaton and to merge the states. It can match many rules while scanning data set only once. Not only reduce DFA state space but also enhance matching speed.

3.1 Multi-pattern Automaton and Its Building

The starting point of building the multi-pattern DFA is to embody multiple rules in an automaton. By preprocessing rule sets and establishing an automaton to scan strings in matching stage, it can match a string simultaneously with a number of rules, which expands the model of building an automaton of each rule using the traditional NFA or DFA regular expression matching algorithm.

Multi-pattern automaton model and related definitions are as follows:

Definition 1: The set of regular expression rules R, x is a rule belonging to R. That is $\{xj \in R \mid 1 \leq j \leq \mid R \mid\}$, in which a rule string x = x [0 ... m-1], and its length is m.

Definition 2: The data set to be matched: y = y [0 ... n-1], and its length is n.

Definition 3: The multi-pattern tree of a rule set is a tree T with the following properties: Each edge of T has a character as label; Labels on the edges connecting to the same node are different; For each pattern $p \in P$, there is a node v satisfying L(v) = p, and L(v) is the splicing of characters on the edges from the root node to v; For each leaf node v', there is a rule $p \in P$ satisfying L(v ') = p;

In preprocessing stage, multiple rules containing similar characters form one set in accordance with regular expression syntax requirements. Then build a multi-pattern automaton for the rules in the same set. So character comparisons are transferred skillfully into automaton state transition through a multi-pattern tree.

Definition 4: The multi-pattern automaton MA is a 6-tuple array: MA= (Q, $\sum$, g, f, q0, E), and MA must satisfy following attributes: Q is a finite state set (a node in multi-pattern tree); $\sum$ is a finite input character table (all characters possibly occur in data set); g is the transfer function, whose definition is: g(s,a) is the state arrived along with the edge whose label is a starting from state s. So if edge (u, v) has a label a, then g(u, a) = v; if the root node has no edge labeled a, then g(0, a) = 0. That means if there is no match character, the state remains in initial; f (the failure function) is defined as follows: f(s): When w is the longest proper suffix of L(s) and w is the prefix of a pattern, then f(s) is the node state labeled w; $q0 \in Q$ and is the initial state (the root node, whose identifier is 0); E is final states set (node set labeled rules).

The algorithms of constructing transfer function g and failure function f are shown in Algorithm 1 and 2:

```
Algorithm 1: constructing the transfer function g
Input : rule sets X= {x1, x2, ... xk};
Output : the transfer function g and output function's intermediate results;
Algorithm description:
newstate= 0;
for i=1 to k
  enter (xi); //insert a path of xi into the transfer graph
for all a such that g (0,a)=fail
  g(0,a)=0;
```

```
Algorithm 2: constructing the failure function f
Input: Transfer function g and output function obtained from algorithm 1;
Output: the failure function f and output function;
Algorithm description :
queue=NULL;
for each a such that
```

```
g(0,a)=s0

queue=queueU{s};f(s)=0;
    while queue!=NULL
     pop();
    for each a such that
g(r,a)=s ≠fail
```

```
queue=queueU{s};state=f(r);
   while g(state,a)=fail
      state= f(state);
   f(s)=g(state,a);
   output(s)=output(s)U
output(f(s));
    End
```

The process of constructing multi-pattern DFA is to transform the rules into a multi-pattern tree which includes g, f and output function. This method of constructing multi-pattern tree can reduce the number of states for rules which contain a lot of similar characters. For example: for the rule set {dfsd_a, dfs {2} a, asd}, the constructed multi-pattern tree is as shown in Figure 1 (circles are the state nodes; the characters on the edge are labels; 0 is the root node; the gray nodes identify a final state; the string near final states are matched rules; dashed line is the state transition of failure function; unmarked nodes' failure function are all pointed to the root node).

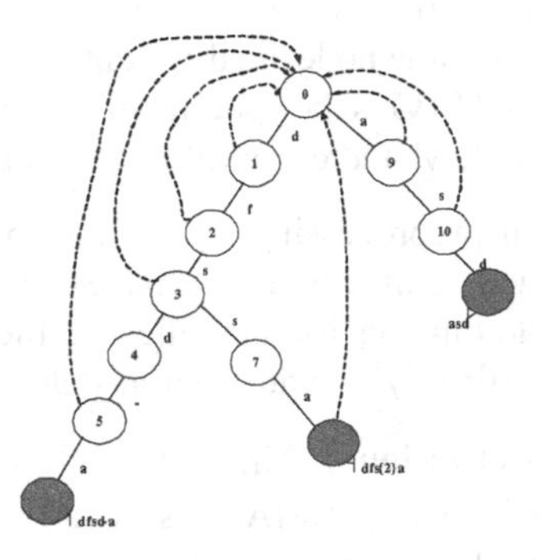

Fig. 1. Pattern tree of multiple rules

3.2 Matching Process

After building the multi-pattern automaton, the next stage is matching of data set. The detail matching process is to enter the rule strings you want to find and the automaton can find out the matched rules and their location (shown in Algorithm 3). To find rules in data set is translated to the search process in a multi-pattern tree. When scanning a data set T, start from the root node (the initial node) of a multi-pattern tree and go down along with the path labeled with characters in a rule v. If the automaton can reach a final state r, then there is a rule v in T; otherwise there are not strings described by rule v in T.

```
  Algorithm 3: Matching
algorithm
  Input: data set text and
multi-pattern DFA M
(including g, f and output);
  Output: the rules
occurring in text and their
location;
  Algorithm description:
```

```
state=0;
for i=1 to n do
while g(state, x[i])=fail
        state=f(state);
state=g(state,x[i]);
 if output(state)≠NULL
    print i;
    print output(state);
 End
```

For example: To find strings expressed by the rule set {dfsd_a, dfs {2} a, asd} in the data set adfsd-abdfsssasd according to the multi-pattern tree shown in Figure 1. Input data from left to right one by one. The specific state conversion is shown in Figure 2 according to the input character:

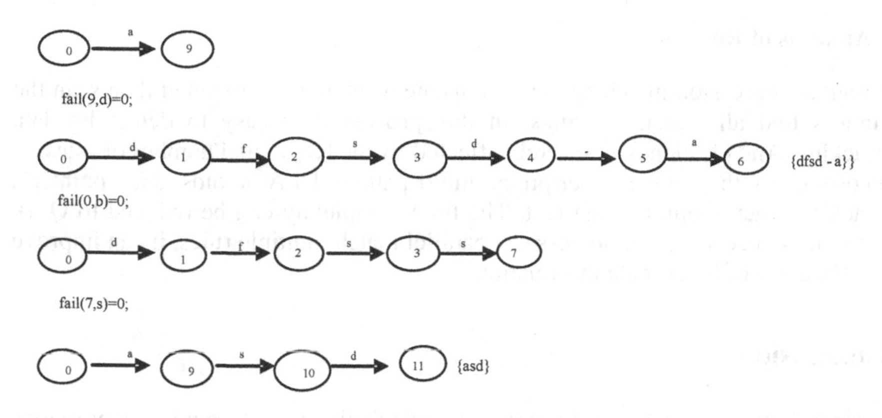

Fig. 2. State conversation

The data pointer does not need to backtrack while using multi-pattern DFA to scan a data set T of length n. The time complexity of matching algorithm is O(n), and has nothing to do with the number of rules and the length of each rule. No matter whether a rule P occurs in a target string T, each character in T must enter into the state machine. So whether it is the best case or the worst case, the time complexity of matching algorithm is O(n). Including the pretreatment time, the total time complexity is O(M+n), where M is the total length of all rules.

4 Experiments

Test the time performance of regular expression matching mechanism based on multi-pattern automaton.

4.1 Experiments Environment

200G HDD, 16G memory, two 4-core CPU; Ubuntu Linux; C language.

A data set of 1G size was used to match regular expressions. Compare the time consumption of the traditional algorithm with multi-pattern DFA using the same rules. Figure 3 shows the time performance. Compare the time occupied by the two algorithms respectively when there are 10, 30, 50 and 100 rules.

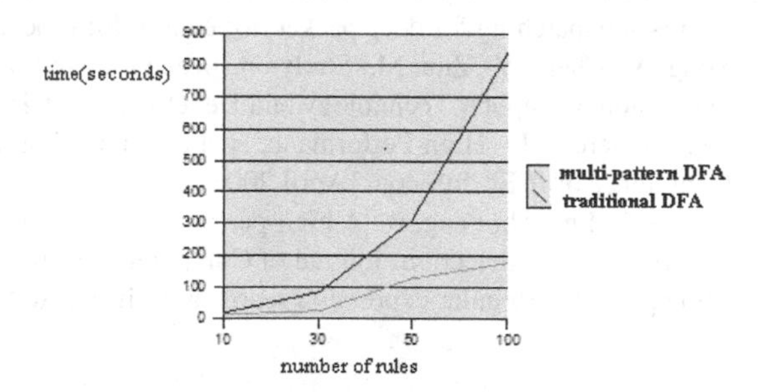

Fig. 3. Time performance comparison

Experiment results shows time performance of traditional DFA degrades seriously when the number of rules increases; and multi-pattern DFA can improve matching speed when there are larger set of rules. By comparing the four rule sets, we can find multi-pattern-based automaton matching can greatly improve the time performance.

4.2 Analysis of Results

Most regular expression matching are to translate a single rule first, and then scan the text unless find all matched strings. In this process it is easy to detect its data backtracking. And efficiency is severely affected by the length and number of rules.

According to the above description, multi-pattern DFA avoids data pointer's backtracking when scanning data text. The time complexity can be reduced to O(n). In addition, as once scanning process can parallel match multiple rules, it can improve match efficiency of large scale information.

5 Conclusions

This paper first described the importance and challenges of regular expressions matching in information technology field. Then based on traditional NFA and DFA, combining with multi-pattern mechanism, a multi-pattern-based DFA was proposed. This method first constructs a multi-pattern tree, and then matches multiple rules by parallel processing. Experiment results show our algorithm can match multiple rules in once scanning, and data pointer has no need to backtrack. Its high matching speed can achieve the requirement of large scale data retrieval with regular expressions. It will provide a new technical support for regular expression matching in information retrieval and language processing fields, and has a wider range of applications.

Acknowledgements. This work was supported in part by the National Natural Science Foundation No. 60673039.

References

1. Kumar, S., Turner, J., Williams, J.: Advanced algorithms for fast and scalable deep packet inspection. In: Proc. of ANCS 2006, pp. 81–92. ACM (2006)
2. Yu, F., Chen, Z., Diao, Y., Lakshman, T.V., Katz, R.H.: Fast and memory efficient regular expression matching for deep packet inspection. In: Proc. of ANCS 2006, pp. 93–102 (2006)
3. Song, X., Chen, W., Zhu, M.: Study on speech recognition control strategy based on regular expression. Computer Technology and Development (02) (2010) (in Chinese)
4. van Lunteren, J.: High-Performance pattern-matching for intrusion detection. In: 25th Conference of IEEE Infocom (April 2006)
5. Zhang, S., Luo, H., Fang, B.: A high-performance regular expression matching algorithm for network security detection. Journal of Computer 33(10), 1976–1986 (2010) (in Chinese)
6. Thompson, K.: Regular expression search algorithm. ACM (1968)

An Improved PID Tuning Algorithm for Mobile Robots

Wei Zhi, QingSheng Luo, and JianFeng Liu

School of Mechatronical Engineering
Beijing Institute of Technology
Beijing, China
zhiwei158@sina.com,
luoqsh@bit.edu.cn

Abstract. PID control schemes have been widely used in most mobile robot control systems for a long time. However, it is still a very important problem how to determine or tune the PID parameters, because these parameters have a great influence on the stability and the performance of the control system. In this paper, a new tuning algorithm of PID parameters for mobile robot using the shift function is proposed. PID control based on neutral network and PID control based on fuzzy are complementary each other in the whole control process, while the fast and the stability of the control system can be given consideration together.

Keywords: mobile robot, PID, shift function, neutral network, fuzzy.

1 Introduction

PID control is a generic feedback control technology and it makes up 90% of automatic controllers in industrial control system [1]. Conventional linear PID controller is not effective when the system is more complex and uncertain factors of object increase [2,3]. In the robot control field, the mathematical model of the robot is not easy to build up precisely. The time-varying influence of parameters for the robot control also can not be ignored in inaccurate model. The problems of traditional PID technology existing in the robot control mainly reflect in time-varying of parameters in inaccurate model while the PID parameters can not varying at the same time.

Therefore, the modern advanced PID technology has been developing rapidly in robot control field, infusing lots of advanced algorithm into the PID parameters tuning. For example: PID based on fuzzy technology[4,5,6]; PID based on neural network[7,8] ;PID based on the genetic algorithm[9,10] ;PID based on ant colony algorithm [11]; PID based on particle swarm algorithm[12,13,14]; PID based on simulated annealing algorithm[15], etc. All above propose the development of PID with different advanced algorithm, however, the robot control needs to be considered in the control effect and algorithm consumption. Thus, this paper puts forward to shift control based on fuzzy PID and neural network PID (NN PID). The main development of this paper is as follows:

D. Jin and S. Lin (Eds.): Advances in ECWAC, Vol. 1, AISC 148, pp. 345–353.
springerlink.com © Springer-Verlag Berlin Heidelberg 2012

i .Efficiency of the real-time control for the mobile robot is satisfied.

ii .Fast and stability can be given consideration at the same time according to different demand.

This paper is organized as follows. First, the fuzzy PID, the NN PID and the shift PID control between fuzzy PID and NN PID are described. Next, the simulink models of the experiment are illustrated. Then, the contrast effect of the different circumstance is given. Finally, reaching an conclusion of the algorithm.

2 Algorithm Statement

2.1 Shifting Control Algorithm Based on Neural Network and Fuzzy of PID Control

In order to enhance the efficiency of the system, this paper puts forward the neutral network PID and fuzzy PID control algorithm, as is shown in Figure 1.That is the choice of using fuzzy PID control algorithm in small error range and using neutral network PID control algorithm in the large error range.

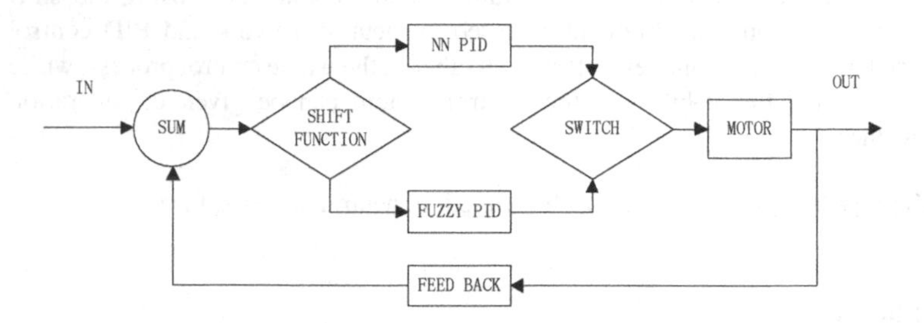

Fig. 1. Single neuron PID control principle

According to different systems and control requirements, it can be flexible to set shift points. In this paper, the parameter ***e*** is chose as the argument of shift function. In the paper, shift function f(***e***)=***e***, while ξ is the shift point argument. Setting different shift point argument can vary the control effect while the shift function is fixed.

$$\Delta u \rightarrow \left\{ \begin{matrix} \text{fuzzy-pid} & f(e) \le \xi \\ \text{nn-pid} & f(e) > \xi \end{matrix} \right\} \tag{1}$$

2.2 Neutral Network PID Control

Because single neuron PID is the simplest style of neural network, this paper chooses single neuron to be the neural network in consider of the efficiency. The single neuron self-adaption PID controller consists of single neurons which are self-learning and self-adaption, as Figure 2 shows. The single neuron self-adaption controller realize the

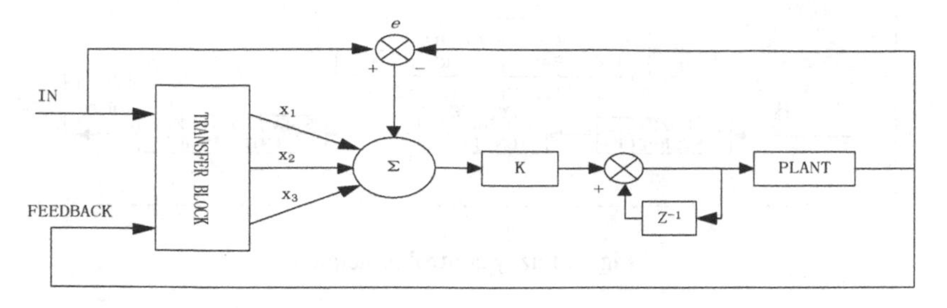

Fig. 2. Single neuron PID control principle

function of self-adaption and self-organization through adjustment of the weighted coefficient, which realizes on the basis of the supervised Hebb rule. Control algorithm and learning algorithm are:

$$u(k) = u(k-1) + K\sum_{i=1}^{3}\omega_i(k)x_i(k) \tag{2}$$

$$\omega_1(k) = \omega_1(k-1) + \eta_I\zeta(k)u(k)x_1(k) \tag{3}$$

$$\omega_2(k) = \omega_2(k-1) + \eta_P\zeta(k)u(k)x_2(k) \tag{4}$$

$$\omega_3(k) = \omega_3(k-1) + \eta_D\zeta(k)u(k)x_3(k) \tag{5}$$

Among which,

$$x_1(k) = e(k); \tag{6}$$

$$x_2(k) = e(k) - e(k-1); \tag{7}$$

$$x_3(k) = \Delta^2 e(k) = e(k) - 2e(k-1) + e(k-2); \tag{8}$$

$$\xi(k) = e(k); \tag{9}$$

η_P η_I η_D are learning rate of proportion, integration and difference separately.

$K>0$ are the proportional coefficient of neurons.

2.3 Fuzzy PID Control

Fuzzy control is computer intelligent control based on the fuzzy set theory, the fuzzy language variables and fuzzy logic. Its basic concept is first proposed by professor Zadeh from the University of California. The basic principle of fuzzy control as is shown in Figure 3.

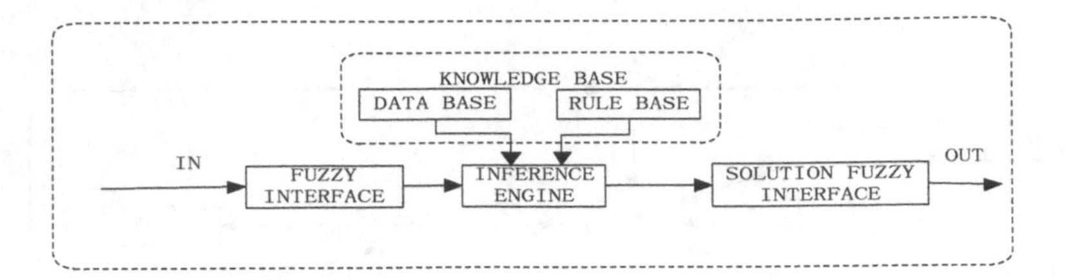

Fig. 3. Fuzzy control principle

i. Fuzzy Interface

The main function of fuzzy interface is inputting a true quantify to translate into a fuzzy vector. For example, the fuzzy input variables ***e***, the fuzzy subsets usually can be divided into:

e={NB,NS,ZO,PS,PB}
e={NB,NM,NS,ZO,PS,PM,PB}
e={NB,NM,NS,NZ,PZ,PS,PM,PB}

The higher levels of detail of the fuzzy input variables are, the better the control effect will be, but the number of fuzzy input variables and subdivision degree directly affect the doubling of fuzzy rules, besides, it has an influence on the consumption of algorithm. The fuzzy variables and refined number is few, the accuracy of the global control and adaptability is naturally affected. Taking account of the accuracy and efficiency of control, this paper takes fuzzy control when the value of error is small, that is reducing the numbers of fuzzy variables and levels of detail of the variables. To narrow down the control domain, it can weaken the problems of the control precision and poor adaptability bringing with the decrease of fuzzy variables and low levels of detail.

ii. Knowledge Base

a. Database base : membership functions of the database deposit for all input/output variables of all the fuzzy subsets provide data for reasoning machine. The membership function can illustrate the division of fuzzy subsets. Membership functions can be triangle style, trapezoidal style, Gaussian style, etc. Most of papers put forward the membership function divided with the double variables with seven divisions. Here is a single variable with three divisions of Gaussian style membership functions. Namely select ***e*** = {N, O, P}

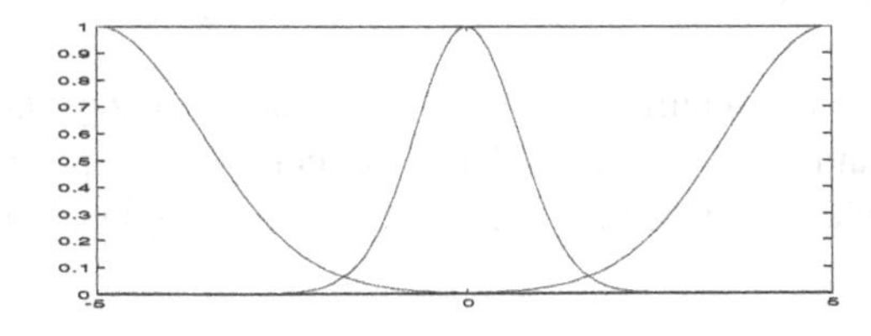

Fig. 4. Three variables domain Gaussian membership functions

b. Rule base : rules base deposits all of the fuzzy implication reasoning rules and provides effective logical result to the reasoning machine. In most paper the choice of PID rule library as is shown in table 1, including seven fuzzy variables and 49 reasoning rules. This article selects three fuzzy variables only need three rules, as is shown in table 2.It can greatly shorten the time of reasoning.

Table 1. Variable domain of fuzzy rule table

EC/E	NB	NM	NS	ZO	PS	PM	PB
NB	PB	PB	PM	PM	PS	ZO	ZO
NM	PB	PB	PM	PS	PS	ZO	NS
NS	PM	PM	PM	PS	ZO	NS	NS
ZO	PM	PS	PS	ZO	NS	NM	NM
PS	PS	PS	ZO	NS	NS	NM	NM
PM	PS	ZO	NS	NM	NM	NM	NB
PB	ZO	ZO	NM	NM	NM	NB	NB

Table 2. 3 variable domain of fuzzy rule table

E	N	O	P
ΔU	P	O	N

iii. Reasoning and defuzzy interface

Reasoning is , according to the input fuzzy quantity, solving the equation of fuzzy controller after fuzzy control rules completing the fuzzy reasoning. The fuzzy vector has solved defuzzy into controlled quantity. The method of COA to defuzzy is commonly used.

$$u_{FC}(x_k, y_k) = \frac{\sum_i u_i \bullet \mu_u(x_k, y_k, u_k)}{\sum_i \mu_u(x_k, y_k, u_k)} \tag{10}$$

$u_{FC}(x_k,y_k)$ stands for the accurate value ; $u_i \in U$; μ_u (x_k,y_k,u_i) is membership function.

3 The Experiment Simulation and Effect Analysis

3.1 Simulation Modeling Based on the SIMULINK

Simulink provides interactive graphical environment and customized module base to design, emulate, execute and test. The experiment selects DC servo motor as a controlled object, which is the most typical execution element in mobile robot control. DC servo motor parameters are: R = 3.44 Ω; L = 0.182 mH; K_T = 17.4 mNm/A; K_E = 0.0174 V / (rad/s); J = 4.45 g. As is shown in Figure 5, among them, the output is voltage signal and input is speed signal. In order to simulate the real physical model, adding to 0.02 s time delay in it.

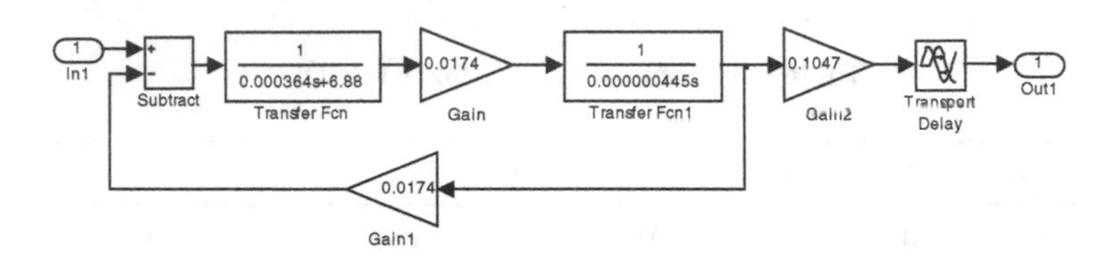

Fig. 5. SIMULINK model of dc servo motor

SIMULINK model of fuzzy PID shows in figure 6, static parameters Kp/Ki/Kd are decided with classic Z-N method. After adding output floating of fuzzy controller ΔKp/ΔKi/ΔKd to static parameters Kp/Ki/Kd respectively are the input of the PID controller. To avoid shock the system when the number of differential floating is large, it joins the differential saturated link. The model SIMULINK of single neuron of PID control as is shown in figure 7, To prevent output overshooting, it also joins a saturated link.

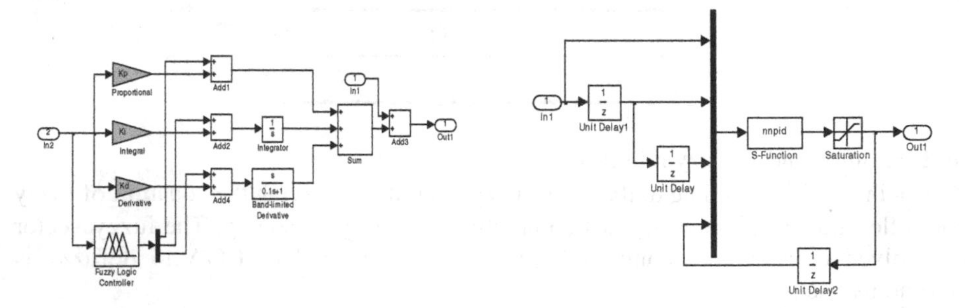

Fig. 6. SIMULINK model of fuzzy PID control

Fig. 7. SIMULINK model of the single neuron PID control

SIMULINK model of shifting system as is shown in Figure 8, according to the logic, the true value table determine input signals into which subsystems to control. Switch derives the output of the subsystem, according to the logic results above.

To encapsulate and connect each child link above , it can get switch control model based on fuzzy PID and single neuron PID, as is shown in Figure 9. In order to observe the effect of the selection of the switch function in the process of regulating parameter, a comparison model can be built by SIMULINK, as is shown in Figure 10.

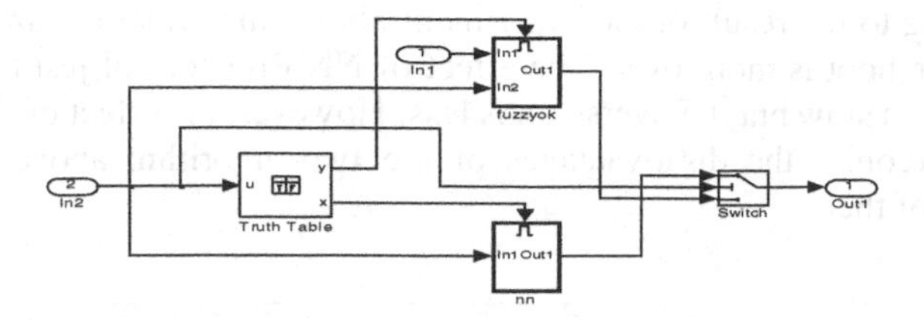

Fig. 8. SIMULINK model of switch system

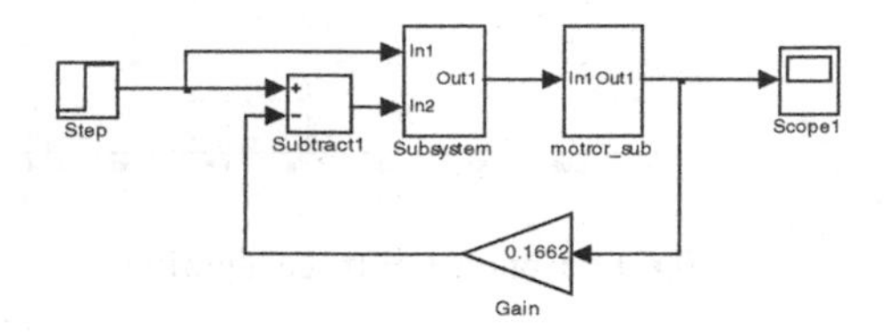

Fig. 9. Switching control model based on fuzzy PID and single neuron PID

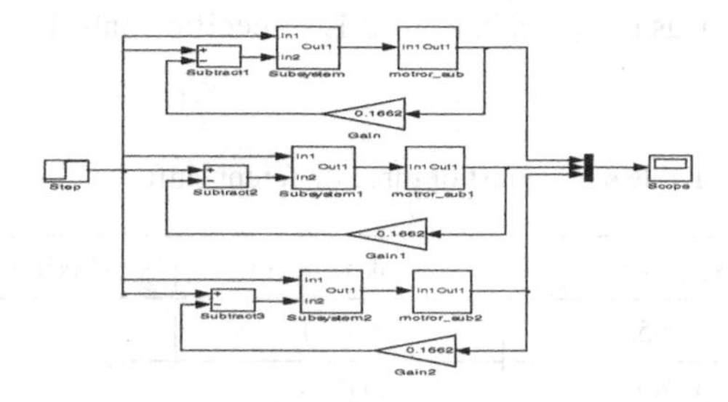

Fig. 10. SIMULINK model of contrast test

3.2 Effect Analysis of the Experiment

According to the SIMULINK model shown in the Figure 9, select factor = 5 for switching parameters, the input quantify of the voltage signal u = 10 V, theory output is speed signal n = 60.2 r/min. The simulation results are shown in Figure 11, the specific control effect parameters are shown in Table 3:

Table 3. The result of three algorithm contrast

	NN-FUZZY	FUZZY	NN
rise time(s)	0.060	0.055	0.060
peak time(s)	0.069	0.061	0.074
maximum overshoot(%)	0.532	0.674	0.437

According to the result of the experiment above, the effect of fuzzy PID control is fast but overshoot is more over. The effect of NN PID control just takes the opposite point, which is slow but the overshoot is less. However, the effect of shift function PID control overcomes the disadvantages of the two algorithm above and express the advantages of them.

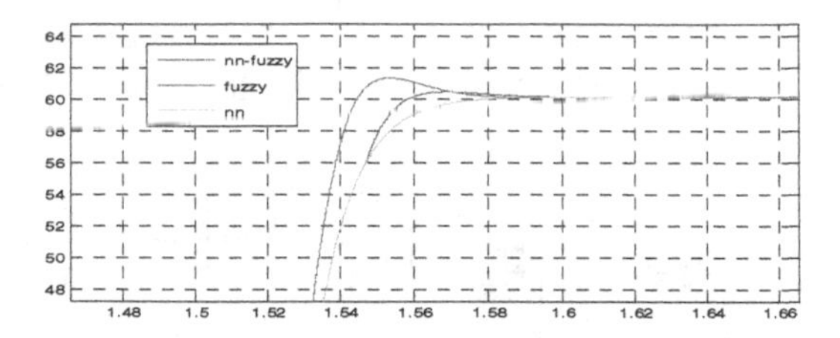

Fig. 11. Contrast of three algorithm

To investigate the influence of switch parameter ξ to control effect. Doing real-time simulation according to the SIMULINK model as is shown in Figure 10.And selecting ξ=3,4,5 to be compared, as is shown in Figure 12. Specific control effect parameters are in Table 4 :

Table 4. Contrast of three different shift point

	Rise time(s)	Peak time(s)	Maximum overshoot(%)
ξ=3	0.065	0.074	0.316
ξ=4	0.062	0.070	0.417
ξ=5	0.060	0.069	0.532

Through the contrast can come to the conclusion: the bigger the values of ξ are, the quicker of the reaction speed will be, as well as the overshoot; the smaller the values are, the slower the reaction speed will be, as well as the overshoot. According to the different control objects, it can be adjusted parameter ξ to meet the expected effect followed the demand of control.

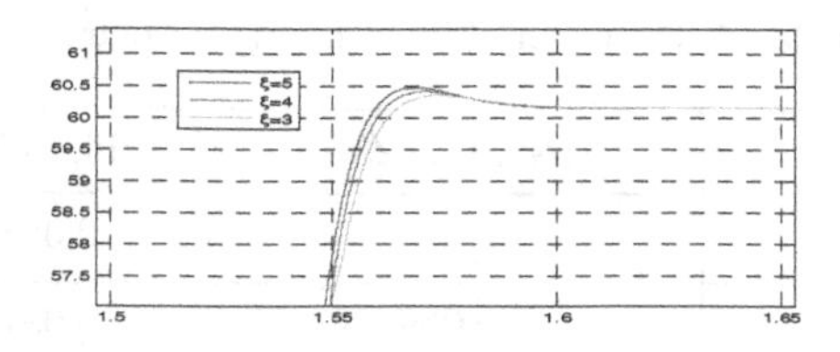

Fig. 12. Control effect contrast under different switching parameters

4 Conclusion

In this paper, shift control based on fuzzy PID and neural network PID is proposed according to the shortcoming of traditional PID for mobile robot. The contrast experiment of fuzzy PID, NN PID and shift control based on above two are illustrated to give a series conclusions about the advantage of shift control algorithm. Especially, the argument of shift point has significant influence on the control effect. The future research will be probe on the influence about the style of shift function and application in the real robots.

References

1. Knospe, C.: PID Control. IEEE Control Magazine, 30–31 (2006)
2. Li, S.-D., Zhu, J.: Optimization of MEO Regional Communication Satellite Constellation with Genetic Algorithm. Journal of System Simulation 17(6), 1366–1470 (2005)
3. Jin, J., Su, Y.: A improved adaptive genetic algorithm. Computer Engineering and Application 29(3), 64–70 (2005)
4. Precup, R.-E., Preitl, S., Faur, G.: PI predictive fuzzy controllers for electrical drive speed control:methods and software for stable development. Computers in Industry 52, 253–270 (2003)
5. Ding, Y., Ying, H., Shao, S.: Typical Takagi-Sugeno PI and PD fuzzy controllers:analytical structures and stability analysis. Information Sciences 151, 245–262 (2003)
6. Ahn, K.K., Truong, D.Q.: Online tuning fuzzy PID controller using robust extended Kalman filter. Journal of Process Control 19, 1011–1023 (2009)
7. Shan, J., Ma, C.: PID controller based on BP neural networks. Acta Photonica Sinic 34(5), 754–758 (2005)
8. Yuan, H., Li, Y.-H.: Research on neural network PID control with application to heavy-duty wheeled vehicle steering system. Journal of System Simulation 17(5), 1185–1189 (2005)
9. Gan, S.-C., Yang, P.-X.: PID self-tuning based on fuzzy genetic algorithm. Journal of North China Electric Power University 32(5), 43–46 (2005)
10. Mahony, T.O., Downing, C.J., Fatla, K.: Genetic Algorithm for PID Parameter Optimization: Minimizing Error Criteria. In: Process Control and Instrumentation, July 26-28, pp. 148–153. University of Stracthclyde (2000)
11. He, G., Tan, G.: An Optimal Nonlinear PID Controller Based on Ant Algorithm. Programmable Controller & Factory Automation, 99–105 (2007)
12. Kennedy, J., Eberhart, R.C.: Swarm Intelligence. Morgan Kaufmann Publishers (2001)
13. Gaing, Z.L.: A Particle Swarm Optimization Approach For Optimum Design of PID Controller in AVR system. IEEE Trans. on Energy Conversion 19(2), 384–391
14. Shi, Y., Eberhart, R.: A modified particle swarm optimizer. In: Proc. IEEE Int. Conf., Evolution Computer, Anchorage, AK, pp. 69–73 (1998)
15. Rutenbar, R.A.: Simulated Annealing Algorithms: An overview. IEEE Circuits and Devices Magazine 5(1), 19–26 (1989)

Study and Application of High-Speed Campus Network Model

Lianzhi Guo[1], Guo'an Zhang[2], and Guangming Han[3]

[1] Department of Computer Science and Engineering, Zhangzhou Normal University, China
glz@netease.com
[2] Department of Computer and Information Science, Fujian University of Technology, China
jvtczga@163.com
[3] Department of Mathematics and Information Science, Zhangzhou Normal University, China
hanxu9362@163.com

Abstract. With its business growing, campus network is expected to achieve real-time or quasi real-time business performance and the quality of services such as VOIP and IPTV, therefore network model suitable for rapid design and upgrading is in desperate need. Based on queuing model and self-similar model, further correlative deductions are made to the classical queuing model and self-similar model. It is shown that there are actually a few correlative parameters between the two models, especially link utilizing ratio which is the golden section number. At the critical value, multi-ports input model of campus network is established, combined with campus network business parameter testing and verification, which can meet the needs for rapid design of network engineering properties.

Keywords: Campus Network, Golden section number, Link utilization, Queuing model, Self-similar model, QoS.

1 Introduction

Campus network is of the following significant features: 1.The network's speed is becoming higher and higher, the past speed generally referred to gigabit Ethernet for backbone medium-large switched LAN, but it is 10G Ethernet nowadays, and it will be 100G Ethernet within predictable years; 2.Network scale is getting bigger and bigger, rising from hundreds of information points to tens of thousands of information points; 3.The diameter of a campus network is more and more big, already several kilometers not rare, even multiple areas of connection; 4. In addition to the traditional business, network can supply more and more services, real and half real-time services such as QQ, MSN, VoIP, IPTV, are going into the campus network service category; 5. Campus network switch tend to be of multi-layer, 100M switch to desktop has become the mainstream of campus networks. Objects of this paper's study are switching campus networks.

In recent years, the network model for research is still in the general network research, no network analysis model specific to the performance of campus networks appear. Although some new models have been put forward [1, 2], less research is

D. Jin and S. Lin (Eds.): Advances in ECWAC, Vol. 1, AISC 148, pp. 355–361.
springerlink.com © Springer-Verlag Berlin Heidelberg 2012

specific to the correlation between these models, with no much practicability. In fact, the commonly used classic continuous queuing model and discrete time queue model, and other self-similar models are of their limitations. Calculating method in classic queuing model is simple, but only for analysis of network situation with light load. The self-similar network model can better be used to analyze network layer and the application layer function than queuing model, but since calculation of Hurst self-similar index estimate is hard and its accuracy is not high, it can be influenced by distributed denial to service (DDoS) [3, 4] attack. Reference [5] gave experimental method measuring the difference of above three models, where based on fluid model used in reference [4], with the continuous time parameters instead of discrete parameters, two separated simulation tests were designed to compare. The conclusion showed that queuing performance of campus network is far more influenced by the of velocity variance than coefficient of self-similarity Hurst, and influence of Hurst coefficient on queuing performance is restricted by variance, but suitable scope is unfortunately not given.

In this paper, based on classical method for queuing model and the self-similar traffic model, mathematical proof is used to study features of the above models, and it is managed to find a simple model which can easily be applied for network engineering design. Correlation between similar model and queuing model is researched, and it is concluded that the two models have a very important correlative parameters value -- the golden section value of link utilizing ratio ρ. This important reference value is introduced into network engineering design, which enables the control and measurement of network design to be easy. Depending on the value, input model of multi-ports the campus network can be established, which can satisfy the requirements of rapid design of campus network.

2 Correlation of Two Kinds of Models

2.1 Queuing Model

In classic queuing model, packet transmission is assumed to be a Poisson distribution. The simplest and most easy calculation result occurs when standard deviation is equivalent to the average value, namely service time distributes exponentially with M/M/ 1 model. Although M/D/ 1 model has more short delay than M/M/1 model, as campus network could impossibly transmit only packets with equal length such as ATM, and exponential service time model simulates network model with the worst performance, if the situation can be calculated, the analysis result would be so conservative that no need to consider M/D/1 model application.

For M/M/1 queuing model, the relationship between the average number of packets r and utilization ratio ρ of the system is [5]:

$$r = \frac{\rho}{1-\rho} \tag{1}$$

But in actual network, packets transferring could not completely follow Poisson distribution, burst and heavy tail phenomenon exist widely, high packet loss rate will happen if sudden amount is greater than the cache of switching equipment, the above phenomenon is considered in self-similar model.

2.2 Self-similar Model

Lots of research shows that most of the network traffic, such as FTP, HTTP, VBR video flow etc, can be described using self-similar and long-range dependences. This means that in network environment, very large data burst (packet queue) and its long time spare (data arrival gap) will happen. Norros is a representative of many this sort of study, who developed a reliable analytical model for similar behavior in reference [6], based on Fractional Brownian Motion (FBM) process proposed the work load model with infinite cache under a fixed length of service time, got a simple results formula. Namely in self-similar input model, relations of storage or cache demand r and average utilization ratio ρ is: [7]

$$r = \frac{\rho^{1/2(1-H)}}{(1-\rho)^{H/(1-H)}} \tag{2}$$

When $H = 0.9$ and 0.75, curves are shown in figure 1 in detail.

2.3 The Relationship between Two Models

First, when self-similar parameters of value Hurst $H = 0.5$ (traditional short-range dependent model), equation (2) can be simplified as $r = \rho / (1 - \rho)$, this is the classic queuing formula 1 for M/M/1 system with exponential arrival interval and exponential service time.

In addition, is there any other relation between the two formulae for different models? Following is further derivation of mutual relation between the two formulae. Under similar phenomena state, namely when $H \geq 0.5$, does H parameter only have a negative effect on delay?

Make that:

$$\frac{\rho^{1/2(1-H)}}{(1-\rho)^{H/(1-H)}} \leq \frac{\rho}{1-\rho} \tag{3}$$

$$\Leftrightarrow \frac{\rho^{1/2(1-H)}}{\rho} \leq \frac{(1-\rho)^{H/(1-H)}}{1-\rho} \tag{4}$$

$$\Leftrightarrow \rho^{\frac{1}{2}\bullet\frac{2H-1}{1-H}} \leq (1-\rho)^{\frac{2H-1}{1-H}} \text{ (5)} \tag{5}$$

$$\Leftrightarrow \frac{2H-1}{1-H}\lg \rho^{\frac{1}{2}} \leq \frac{2H-1}{1-H}\lg(1-\rho) \tag{6}$$

By solution of (6):

$$1-3\rho+\rho^2 \geq 0 \tag{7}$$

The solution of ρ value is constant, and has nothing to do with H parameters. Values for ρ are 0.382 and 3.376. For 3.376 does not accord, therefore is eliminated. 0.382 is the complementary number of golden section number 0.618. Queuing model and self-similar model curves are compared in Figure 1:

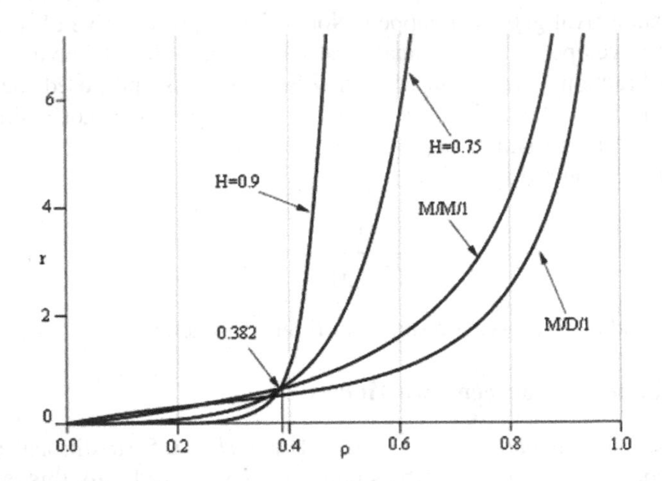

Fig. 1. Comparison between queuing model and self-similar model

3 The Engineering Significance of Golden Section Value

From table 1 and fig. 1, an interesting phenomenon can be seen, only when utilization rate ρ is greater than golden section number 0.382, self-similar values take obvious effect on the delay. When it is below 0.382, delay of self-similar model is lower than that of queuing model. H parameters is larger, the queue length from self-similar becomes smaller than the queue length of M/M/1 model. And higher than this ρ value, by contrast, such as when queue length $r = 4$, H=0.75 corresponding to ρ=0.568, and the corresponding value to H=0.9, ρ is 0.45 only. And when $\rho = 0.4$, $H = 0.9$ queue length $r = 1.0161$, also is only 1.5 times to M/M/1.

Table 1. Queue length r for different models

ρ	**0**	**0.1**	**0.2**	**0.3**	**0.35**
M/M/1	0	0.1111	0.2500	0.4286	0.5385
***H*=0.75**	0	0.0137	0.0781	0.2624	0.4461
***H*=0.9**	0	0.0000	0.0024	0.0602	0.2536

ρ	**0.382**	**0.4**	**0.45**	**0.5**	**0.6**
M/M/1	0.6181	0.6667	0.8182	1	1.500
***H*=0.75**	0.6182	0.7407	1.2171	2	5.625
***H*=0.9**	0.6186	1.0161	4.0068	16	296.63

To explain from physical phenomenon, according to the self-similarity theory, a time sequence have shown the same pattern in any resolution, by Cantor set by the standard structure rules, the longest transmission time of self-similar traffic only accounts for 2/3 Poisson traffic transmission time. In circumstances of light load, there is gap between actual sending groups; the total transmission quantity is less than the flow of Poisson distribution. Namely when packet loss rate is equal to zero, accumulated transmission time through the nodes or link traffic burst is shorter than that of continuous packet transmission time in queuing theory. Short queue length r, this is very attractive for transmission like variable bit rate (VBR) packet stream.

Usually, maximum allowable error in engineering calculation is within 20%. At the intersection of self-similar model and classic queuing model, the upper limit error of ρ value is 0.382 ×1.2 = 0.458, close to ρ=0.45 corresponding to H=0.9, where queue length is still shorter. And based on various statistical tests, while self-similar H value of Ethernet traffic is just between 0.71~0.78, variable video flow VBR can reach 0.9 [10].

In a campus network, for heavy-tailed distributions owing to transmission of different file size, priority queues can be achieved on the access layer. According to GB/T 21671-2008 "Acceptance and evaluation specification of LAN system based on Ethernet technology" [11], it is required that link health index, the average utilization rate of shared Ethernet or half-duplex Ethernet be less than 40%, the average utilization rate of full-duplex Ethernet be less than 70%. Obviously it is very close to the author's research result, if the link utilities ratio ρ is designed below 0.382, links have one times more capacity to cope with sudden quantity; the influence of heavy-tailed also can be eliminated.

As long as to control that link utilization ρ does not exceed 0.382 or at most 0.45, calculation model with classical queue to create campus network is feasible. Reasonable value ρ can simplify high-speed campus network bandwidth calculating. Because no suitable parameter are available in the past design, operating dealers usually used to consider to update equipments when average link utilization ρ reach 30%, in order to provide better service for users. The golden section value can be found, to raise the average link utilization ratio ρ more than 20%, then to consider equipment update.

4 Campus Network Reference Model

Based on the above mathematical reasoning to 0.382, calculation formulae can be established model for campus network:

$$\left.\begin{aligned} r &= \frac{\rho}{1-\rho} && \rho \le 0.382 \\ r &= \frac{\rho^{1/2(1-H)}}{(1-\rho)^{H/(1-H)}} && \rho > 0.382 \end{aligned}\right\} \tag{8}$$

When $\rho \le 0.382$ or 0.45, bandwidth of campus backbone can be calculated by M/M/1 queuing model. The campus network model is shown in figure 2:

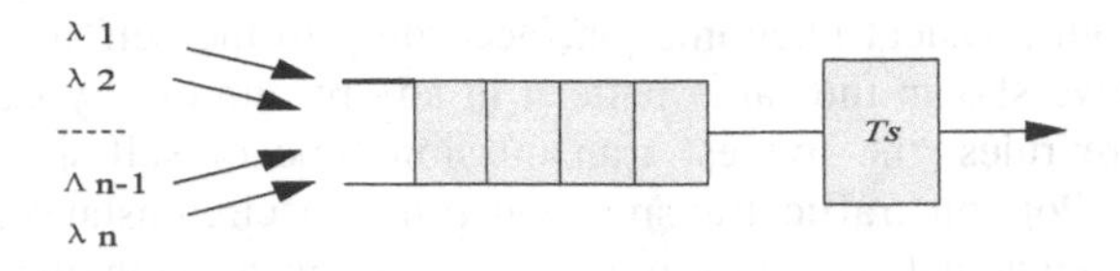

Fig. 2. Backbone model of campus network

$$\lambda = \sum_{i=1}^{n} \lambda_i \tag{9}$$

5 Conclusion

Due to the backbone of campus network must meet requirements of either doing best effort or meet the user's service quality; a suitable model is needed to estimate. In planning the network/link utilization, Mischa Schwartz [17] had assumed that at any time among the distal concentrator interfaces to work, one-third are free (in a waiting state), one-third are inputting information, and one-third are output information, but gave no theoretical proof. Usually operators began to update in average link utilization rate 30%. In this paper, mathematical formula proof shows that, when link utilization rate ρ is not more than 0.382, self-similar model is of smaller queue length than classical queuing model; when $0.382<\rho\leq 0.382\times 1.2$, self-similar parameters H influence on network still weakly. In interval $\rho\leq 0.382\times 1.2$, it is feasible to establish campus network backbone calculating model with classical queuing, and the key technologies is to control the utilization rate ρ of backbone link. Certainly through the use of the high coding efficiency video/audio algorithm and multicast technology, the user's bandwidth demand still can further be reduced. In a campus network, equipment protocol can easily achieve consistency, to enable large-scaled multicast. No matter VOD or IPTV, multicast system can be used to reduce the demand of user for bandwidth. When $\rho\geq 0.382\times 1.2$, by using simulate statistical time division multiplex technology, we can adjust the variance of traffic. It can achieve better queuing effect than by increasing cache.

Acknowledgment. This paper is supported by Fujian Natural Science Foundation of China (A0540007), Technology Key Project of Fujian Province of China (2009H01010435), Science and Technical Project of Fujian Department of Education (JA10216).

References

1. Tang, Y., Huang, S., Yun, X.: Queue length distribution of discrete time multiple vacation $Geom^X/G/1$ queuing system. Acta Electronic Sinica 37(6), 1407 (2009)
2. Su, G., Wang, H.: Self-similar input stream in the N-BurstD1 queue system performance analysis. Journal 36(5), 1011–1012 (2007)

3. Fu, L., Wang, R., Wang, H.: RS method for solving implementation and application of self-similar network traffic parameters. Journal of Aeronautics and Astronautics University of Nanjing 39(3), 361 (2007)
4. Liu, Q., Li, S., Wu, X.: Self-similar network traffic modeling and analysis. Microelectronics and Computer 26(11), 132 (2009)
5. Hong, F., Wu, Z.: The Husrst index adaptive estimation based on wavelet. Journal of Software 16(9), 1687 (2005)
6. Song, L., Chen, M., Qiu, X.: The simulation analysis and comparison of network flow characteristics of effect on queuing performance. Journal of System Simulation 17(1), 25–28 (2005)
7. Norrors, I.: On the use of fractional Brownian motion in the theory of connectionless networks. IEEE Journal on selected Areas in Communications (August 1995)
8. Norrors, I.: A storage Model with Self-Similar Input. Queueing System, 16 (1994)
9. Leland, W., Taqqu, M.: On the Self-Similar Nature of Ethernet Traffic. IEEE/ACM Transactions on Networking (February 1994)
10. Stallings, W.: High-Speed Networks and Internets: Performance and Quality of Service, 2nd edn. (2002)
11. GB/T 21671-2008, Specification for Acceptation, Testing and Evaluation of Local Area Network (LAN) Systems Based on Ethernet Technology. China standard press, p. 7 (2008)
12. Guo, L., Zhang, G.: Study of Self-Similar Traffic Control of Campus Network. In: 2010 International Conference on Multimedia Technology, pp. 2344–2347. IEEE (2010)
13. Teaare, D.: Campus Network Design Fundamentals, p. 128. Cisco press (2006)
14. Demiemichelis, C.: IP packet Delay Variation Metric for IP Performance Metrics. RFC3393 (2002)
15. Guo, L.-Z., Han, G.-M.: Research of Fast Design of Campus Network. In: International Conference of China Communication (ICCC 2010), pp. 741–745. Scientific Research Publishing, USA (2010)
16. Stallings, W.: High-Speed Networks and Internets: Performance and Quality of Service, 2nd edn. (2002)
17. Schwartz, M.: Computer Communication Network Design and Analysis, TN919 /X6. Prentice-hall (1977)

Similarity Detection Method Based on Assembly Language and String Matching

Shuqian Shan[1], Fengjuan Guo[1], and Jiaxun Ren[2]

[1] Department of Computer North China Electric Power University, Baoding, Hebei, China
shshq_2004@163.com, gfj0917@126.com
[2] BAODING LIZHONG WHEEL MANUFACTURING CO., LTD, Baoding, Hebei, China
renjiaxun@163.com

Abstract. With the rapid development of higher education, students copying work is becoming increasingly serious. The purpose of this paper is : first, converting the program into assembly language, and then using the string comparison algorithm, the improved algorithm KR, to get the similarity between programs. This method has very good results for teachers to detect plagiarism .

Keywords: assembly language, Improved Algorithm KR, Similarity.

1 Introduction

Now, accessing documents is a breeze. Particular computer program as an electronic document, is easier to copy. Students copying work is widespread, especially computer programming job. For convenience, many students submit other students' work or web documents without any modification or by a text editor after a simple transformation to teacher inorder to pass the exam.

Parker and Hamblen think[1] that if a code is a complete copy or a small number of changes of another section ,it can be believed that this code is copied from another piece of code. Faidhi and Robinson [2]describes six levels about copying the code of procedure in 1987. Jones summarized ten methods about Code plagiarism[3]. Although there is no uniform classification criteria, but the means used in the following areas: (1) program format: the increase in empty lines or spaces. (2) grammar: Add or change the comment, rename identifiers and so on. (3) semantics: replace with the equivalent structure of the original structure. The first two changes are simple and the third requires more costly, most students copied by means of the first two.

Code detection methods include property of counting method [4], structural metric method [6,8], XML modeling, and methods based on clustering and so on[5,7,9]. These methods for detecting plagiarism layout level have a good effect. But if the program is short, and has a large number of redundant statements, those methods are not effective. We propose a method to detect similarities: First, the source is converted to an assembler. Secondly, using string comparison algorithms which is Improved algorithm KR determine procedures similarity. This method can effectively eliminate the interference caused by the copying means.

D. Jin and S. Lin (Eds.): Advances in ECWAC, Vol. 1, AISC 148, pp. 363–367.
springerlink.com © Springer-Verlag Berlin Heidelberg 2012

2 Similarity Detection Based on Assembly Language, the String Comparison Algorithm

Most students do not read the source code ,and they will make the result correct.so they made changes only in thelayout grammar.For those students whichhave some knowledgeof programe ,they will use semantics means:such as change the expression ,replaced with the equivalent structure.In short program, increasing a large number of unrelated comments or statements, then similar procedures may be have very small Similarity.

Reference[10]was that converting a program to assembly language through the compiler optimization, disassemble, assembly language,and then detect the similarities about assembly language,give a threshold and clustering based on the threshold .This approach means good plagiarism detection for the semantic. This paper draws this idea that the source program will be converted into an assembly language ,then obtain similarity using string comparison algorithm not using Clustering method .

The reason converting the source program into an assembly language for comparison is that after compilation, the source of the spaces, blank lines, comments, etc. assembler will be filtered out, and using a modified copy of such grammar and means to change the comment will be identified. However, if the copy only by adjusting the code sequence, then the code which was converted into assembly language, and the difference between the two would be great.so assembly language program needs unification.

After normalization for the generated assembly language code, converting the code into token strings, then selecting a fast randomized Karp-Rabin string matching algorithm to improve the similarity between the documents obtained.

2.1 Pretreatment of Source Code

To follow-up work simple and quick, the first,make the source code normalize.

Make some of the program which is less impact on the program into a unified symbol substitution symbols, such as the use of all the constant CONST place, the identifier is replaced with a unified valume. Identifier is the final implementation of the program will not affect the results.

Simplified, unified expression is written: x <y (x <= y) into y> x (y> = x), so that the compiled expression will have the same form;

Retain only one statement per line And so on. After pretreatment, the identifier name for the change, a simple change of expression caused by noise can be eliminated.

2.2 Compiling Source Program

After pretreatment, compiling the program to form a binary code. In the compiling process, the comment such as space, blank lines, etc. are filtered out by the compiler, which will not affect the layout format; the noise brought by simple expressions' transformation in the preprocessing stage has been eliminated.

2.3 Getting Assembly Language by Disassembly Tools

Changing statements' order is a means of plagiarism in the premise ofwithout modigying the result of the program. The binary code after assembly does not have noise brought by changing the layout and grammar. Changing the order of statements will make offset diferrent, and make the assembler statements different. To eliminate the noise caused by changing statements'order, assembly code requires a unified.

First, using symbols "offsetaddr" instead of all the offset addresses. Because the position of statement written or the changes of variable declaration lead to offset dirrent, so assembly code requires a unified.Second, using symbols "funaddrr" instead of all the function addresses . Function is written in different positions, and function call address will be different. Therefore, all the function addresses are replaced with uniform symbols, inorder to eliminate the interference.

2.4 Formating TOKEN String

Converting the assenbly code into formating the TOKEN string .

2.5 KR String Comparison Algorithm and Improved

KR string comparison algorithm is a random string matching algorithm. The algorithm can find the first position of occurrence of the target string in the text ,but similarity comparison need to find all possible strings. Therefore, it is need to improve the KR algorithm: All the same hash value of the substring, one by one and compare with the target string to find the same string, and then make the count plus one, (count records the same number which has been compared with the target string) until all sub-strings are more exhausted.

Specific description of the algorithm: in the text string T of length n(T=t1t2…tn),finding the target string S of length m(S= s1s2…sm).R is a binary set, and express S by an integer and getting ASCII code for each character,then S can be expressed as Eq. (1):

$$S' = s1*R_{m-1}+s2*R_{m-2}+\ldots+s_m \,. \tag{1}$$

Finding a large prime number q, and defining the hash function of S, ¢(S')can be expressed as Eq.(2):

$$¢(S') = mod(S',q) = mod(s1*Rm-1+ s2*Rm-2+\ldots+sm ,q) \,. \tag{2}$$

Similarly, the text string of length n, is divided into n-m+1 sub-strings, and each substring expressed by Eq. (1) and Eq. (2),find the hash value for each sub-string.

k is an integer,(0<=k<n-m+1),the substrings of T can expressed:

$$T(k) = t_{k+1}t_{k+2}\ldots t_{k+m}$$

The value of T(k), using Eq. (1)

$$T'(k) = t_{k+1}*R_{m-1}+ t_{k+2}*R_{m-2}+\ldots+t_{k+m} \tag{3}$$

$$T'(k+1) = t_{k+2}*R_{m-1}+ t_{k+3}*R_{m-2}+\ldots+t_{k+m+1} = T'(k)*R+ t_{k+m+1} - t_{k+1}*R_m \tag{4}$$

Hash value of T'(k) :

$$¢(T'(k))=mod(T'(k),q)=mod(t_{k+1}*R_{m-1}+t_{k+2}*R_{m-2}+...+t_{k+m}, q) \quad (5)$$

Then contrast ¢ (T '(k)) and ¢ (S ') one by one, if not equal, then compare to ¢ (T '(k +1)),, until finding a equal value or all the hash values have been compared. If a hash value is the same as the target string, then compare the sub-string and target string.

For example:T="we are happy are ! ", target string S="are" ;

Using Eq.(1), the numerical results of S : $S'=s_1*R_{m-1}+s_2*R_{m-2}+...+s_m=10767$

Getting a prime numuber q=13, using Eq.(2), getting ¢(S')=3;

Deleting spaces in T, getting "wearehappyare ! " ;

Getting all the substrings with three characters:wea ear are reh eha hap app ppy pya yar are re!

Using Eq.(3), Eq.(4), Eq.(5) for each substring,getting hash values:

Letters	w	e	a	r	e	h	a	p	p	y	a	r	e	!
Values	1	5	3	5	5	0	3	4	4	2	3	5		

Then comparing the three substrings of the same hash values with target string,and getting results.

2.6 Calculation of Similarity

Suppose F1, F2 are the assembly language code to be compared, which have been dealed with above , then obtain the similarity through Improved algorithm KR. expressed as: sim (F1,F2) =(F1∩F2) / (F1∪F2)

F1∩F2 express the similar codes inF1and F2; F1∪F2 express the total length of two codes.

3 Experimental Results and Analysis

There are 24 programs which are getting randomly about seeking prime numbers . f0,f10,f20 are the original programs written by independently.f1-f7 is formed by means of various procedures for plagiarism as a sample to f0. f11-f17 is formed by means of various procedures for plagiarism as a sample to f10.f21-f27 is formed the same. Experimental results as follows:

f0, f10, f20, prepared three separate similarity between the original program are below 40%;

The similarity is more than 69.1% between f1-f7 and f0, up to 84.13%; f11-f17 and f10, f21-f27 and f20 also satisfy this rule;

similarity between f1 and f7 arrive 50% -60%;f11-f17, f21-f27 to meet the same results;

The similarity between f1-f7 and f10, f20 is less than 35% ,other groups of programs copied and the other two independent programs is also 35% similarity.

It can be seen from the above results, this algorithm for the detection of two similar procedures, has great practical value.

4 Conclusion

The method has great reference value and reduce the workload of teachers. However, the actual copying or not, teachers and students need to communicate further to determine . The algorithm can also be combined with the method based on attribute counting, comprehensive comparison of procedures to improve the algorithm.

Acknowledgments. This article has funds to support.The fund number is 104008.

References

1. Parker, A., Hamblen, J.O.: Computer Algorithms for Plagiarism Detection. IEEE Transactions on Education 32(2), 94–99 (1989)
2. Faidhi, J.A.W., Robinson, S.K.: An Empirical Approach for Detecting Program Similarity within a University Programming Environment. Computers and Education 11(1), 11–19 (1987)
3. Rinewalt, J.D., Elizandro, D.W., Varnell, R.C., Starks, S.A.: Development and Validation of a Plagiarism Detection Model for the Large Classroom Environment
4. Ottenstein, K.J.: An Algorithmic Approach to the Detection and Prevention of Plagiarism. CSD-TR200 103(2), 32–39 (1976)
5. Whale, G.: Plague:plagiarism detection using program structure. Dept.of Computer Science Technical Report 8805, University of NSW, Kensington, Australia (1988)
6. Schleimer, S., Wilkerson, D., Aiken, A.: Winnowing:LocalAlgorithmsforDocument Fingerprinting. In: ACM SIGMOD 2003, pp. 204–212. ACM Press, SanDiego (2003)
7. Gitchell, D., Tran, N.: Sim: A utility for detecting similarity in computer programs. In: The Proceedings of the Thirtieth SIGCSE Technical Symposium on Computer Science Education, pp. 266–270. Association for Computing Machinery, New York (1999)
8. Wise, M.J.: YAP3: improved detection of similarities in computer program and other texts. In: Proceedings of the Twenty-Seventh SIGCSE Technical Symposium on Computer Science Education, vol. 28(1), pp. 130–134. Association for Computing Machinery, New York (1996)
9. Prechelt, L., Malpohl, G., Phippsen, M.: Finding Plagiarisms among a Set of Programs with Jplag. Journal of Universal Computer Science 8(11), 1016–1038 (2002)
10. Zhao, C., Yan, H.: Based on compiler optimization and disassembly procedures for similarity detection method. Journal of Beijing University of Aeronautics and Astronautics

Design and Implementation of ADI H.264 Encoding Library on BF561 Platform

Qinghui Wang and Guian Wu

College of Information Engineering,
Shenyang University of Chemical Technology,
Shenyang, China
wangqh8008@vip.sina.com, wga1814@163.com

Abstract. Based on H.264 video coding standard and ADSP-BF561 dual-core structural features, design of the entire coding system is completed combined with H.264 encoding library provided by ADI. Image acquisition using OV7660,the use of DMA technology to achieve more efficient video capture and data format conversion. Test results show that, in the BF561 achieve H.264 video encoding, to achieve high-quality compressed video stream is completely feasible.

Keywords: H.264, Encoding library, Image capturing, BF561.

1 Introduction

As the latest third generation of H.264/AVC video coding standard, the main purpose is to develop a higher coding efficiency and better adaptability of the network. Under the same reconstructed image quality, with the H. 263 and MPEG-4 ASP standard compared to 50% of the stream section; Here, based on ADI H.264 video compression encoding library system design and implementation, use of ADI's highly efficient, complete system-level code base support, full account of the BF561 dual-core DSP in image processing in the high performance parallel code to achieve a high quality compressed video stream.

2 BF561 Introduction

ADSP-BF561 uses a symmetrical dual-core architecture, in a BF561 chip integrates two BF533 DSP cores, Two cores can be clocked up to 600MHz, supports parallel processing[1]. BF561 and low-voltage low-power design methodology, modified Harvard architecture and the hierarchical memory structure, that Level1 (L1) memory is generally run at full speed with no or only a few delays, and repeatedly calling block and frequent use of the data segment on the L1 in order to improve the dual-core coding rate. Level2 (L2) is another level of memory, located in the chip or out chip, it takes several processor cycles to access. BF561 integrates more peripheral, which provides a rich peripheral interfaces such as serial peripheral interface(SPI),

D. Jin and S. Lin (Eds.): Advances in ECWAC, Vol. 1, AISC 148, pp. 369–373.
springerlink.com © Springer-Verlag Berlin Heidelberg 2012

UART and a parallel peripheral interface (PPI)[2]. Therefore, the platform is more suitable for video systems.

3 Design and Implementation of ADI H.264 Encoder Library

The system consists of two parts of video capture and H.264 encoding. Video capture part is responsible for capturing images, and images will be captured by the PPI interface filled to the specified video frame buffer. H.264 encoding is primarily responsible for collecting the data is compressed, and compressed bit-stream output.

3.1 Video Capture Module to Achieve

Video capture part of the task is to complete the video data sample, and converted into a suitable encoder data stream .The specific flow chart shown in Fig.1.

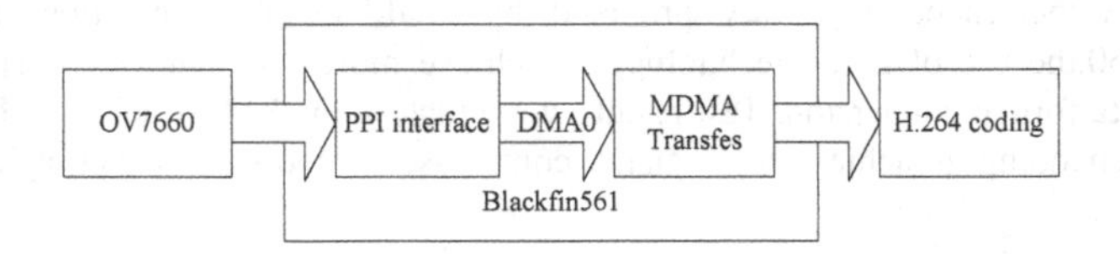

Fig. 1. Video Capture Module

Video capture is done by the camera OV7660, OV7660 OmniVision developed by the United States a CMOS color image sensor chip, support VGA, QVGA, GIF and many other resolutions [3]. Video output format Raw RGB, GRB 4:2:2 and YUV / YCbCr (4:2:2). In the design, select the CIF YVYU (4:2:2) format, need to set up the corresponding register COM1 = 0X00, CLKRC = OX80, COM7 = 0X30, TSLB = 0X05.

Video capture images are interwoven to get the YUV 4:2:2 format video data, while the H.264 video coding algorithm for 4:2:0 format video data compression [4] 4:2:0 format data brightness values and color values of the buffer is a buffer separating the use of MDMA move, the brightness can be achieved the separation buffer and color buffer. The 4:2:2 format into 4:2:0 format, to retain all of the Y, and take the line of U from 0,2,4......and line of V from 1,3,5......[6]. List of descriptors based on the "big" model MDMA, complete the Y, U and V from the mixed data of 4:2:2 format of the source area to the Y, U and V the purpose of separating the data area of the move. YUV422 and YUV420 structure shown in Fig.2:

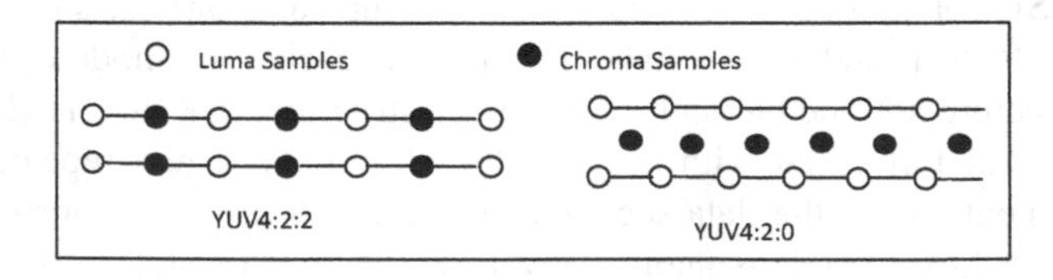

Fig. 2. YUV422 and YUV420 Structure

3.2 Implementation of ADI H.264 Video Encoding

Based on ADI H.264 encoder source development is actually a dynamic library (H.264 Encoder Library), for us to encode the actual situation. First need to write an encoder entry function file encoder.c, and call library functions to implement video coding[5].

In order to better demonstrate the encoder implementation process, a test sequence to the original file foreman.yuv as an example, the entrance from the encoder function encoder.c, details of the encoder to encode the video image compression implementation process. Coding process shown in Fig.3:

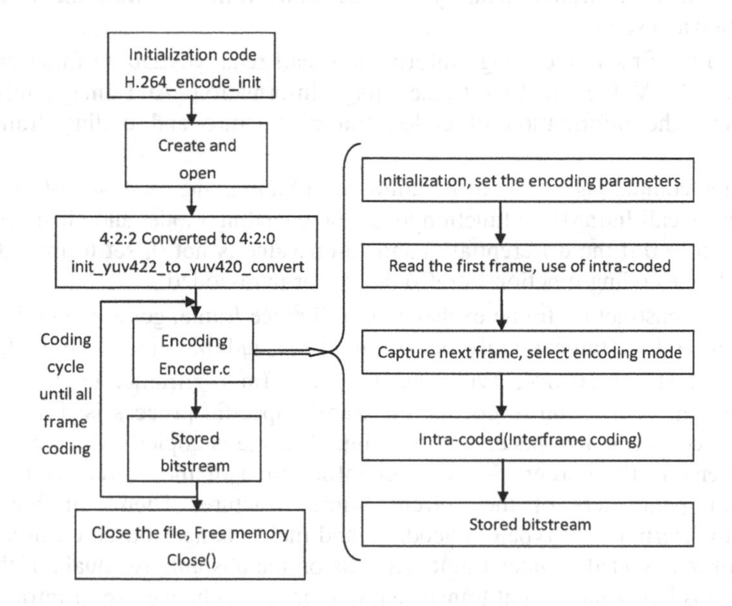

Fig. 3. Coded flow diagram

(1). Call the initialization function H.264_encode_init() to initialize and set the encoding mode parameters and quantitative parameters, and so on. Then, create a binary write and open foreman.264 file for write foreman.yuv encoded data after the completion of the file. Initialized to empty, with pointer out_file point to it, used to store the complete coding frame.

(2). As foreman.yuv format is 4:2:2, according to the H.264 coding requirements need to convert the data to 4:2:0 format, so call the function init_yuv422_to_yuv420_convert() can complete the conversion, call read_yuvdata (in_file,in_buffer) function, Read an YUV image data from foreman.yuv to in_file file, here in buffer is a pointer pointing to foreman.yuv file.

(3). To encode with the data read. Call the main encoding function encode (int argc, char *argv[ ]) To encode the input frame. Function by configuring the appropriate environment and encoder library function call to complete the coding task. encoder.c is actually the entrance into the encoder of original video file. The specific coding function to achieve the following:

① Initialization: Create the two most critical structures: one for the incoming parameters and statistical results of the coding frame encoding information structure (ifrm_t) and coding frame state structure (vidcodec_t). Called H.264 decoder init (vidcodec t *pvc, ADIMemMap *pmemblk, ubyte *bitstream, naltable_t *pnals) to initialize the encoder. Including setting encoding handle, according to passing parameters to set the value of variables, including video frame width, height, frame rate, bit rate and a variety of temporary variable, and assigned for the storage of reference frames, re-frame memory. At the same time, do well the rate control initialization and so on.

② Read the first frame image information, call read_yuvdata () function to read from a raw YUV file the first frame image information. And Image information transfer into the information of coded frame structure and coding frame status structure.

③ After completing the configuration information, the first set of intra-coded frame, can be call IntraMB () function to set the encoding mode, and all sports-related variables set to 0.If the differential quantization value is not 0, set to intra_Q. Then, call the I-frame coding function IntraCoding () for intra-coded.

④ The reconstruction frame exchange to reference frame, get a image from queue frame as the coding frame, that is, the current frame, and initialize (with the first step).

⑤ Called H.264 encoder (vidcodec t *pvc, ifrm_t *iframe, ubyte *bitstream, naltable_t *pnals) function to encode the frame. Specific process is: First, initialize the stream, convert color space, and the original image is copied to a frame of image space, extending the border. Second, get information of the current frame, set the information parameters of the current frame structure. Then call Meanalysis() function to determine the type of encoding, and make changes based on user settings. The use of intra-coded or inter-frame depends on the absolute residuals of the macro block. That is larger movement transformation, or scene change use of intra-coded, or inter-frame. The first frame is always using Intra-coded. This can greatly improve the coding efficiency.

⑥ According to the previous step to determine the encoding to encode the frame. If the frame should be encoded using intra-coded mode, then jump to the third step; If the inter-frame encoding, called P-frame coding function InterCoding() for inter-frame coding. To reduce complexity, the compression algorithm used in inter-frame coding mode for B frame coding mode, and then jump to the fourth step, save the encoded frame information.

⑦ Complete the last frame coding, end of the loop.

(4). Call fwrite () function, the frame to be encoded stream file saved in a format to the file pointed to by the pointer out_file foreman.264.

(5). Read the next frame to appropriate encoding and storage, repeat the above steps, until the coding of the last frame, can set a conditional loop statement, the test Coding total of 500 frames.

(6). Close the file, freeing up memory, the end.

4 Experimental Results

In this experiment, first select and connect to the type of target board, and in the DOS window to find the executable object file hostapp.exe to test the USB port is connected. This moment Light-emitting LED15, 16 on Debug board, and USB connection is successful. In the course of encoding light-emitting LED 18 and 19 flashing, A and B core of BF561 is being done coding, and finally the compressed stream as a file stored locally memory via a USB. After the encoding, with the H.264 decoder for decoding the compressed bit stream, Play decoded video files using YUVviewerPlus.exe. Decoded image clarity compared with the using the Image Viewer View in VisualDSP++5.0, the image quality is basically the same, and image is also clear.

5 Conclusion

Based on H.264 encoder of BF561 did not achieve the real-time encoding, because there is no optimization for H.264 open source code. As the CIF format image data is relatively large, need to be stored in external memory, the processor to access external memory rather a long time; DCT and motion estimation algorithm function is time-consuming, using C language to achieve. There are some auxiliary functions and print information in the program, thereby affecting the speed of the encoder interface. Focus of future work is to use C compiler of VisualDSP++ development environment for code optimization, Take full advantage of BF561 instruction set, implemented in assembly language; Remove the source of unnecessary auxiliary functions and print information to implement real-time encoding of H.264 encoder based on BF561.

References

1. Analog Devices. Blackfin Embedded Symmetric Multi-Processor
2. Analog Devices. ADSP-BF561 Blackfin Processor Hardware Reference (EB/OL), http://www.analog.com
3. OV7660/ OV7161 CMOS VGA (640×480) Camera Chip Implementation Guide. Rev. 1 (2004)
4. H. 264/ MPEG - 4 Part 10 White Paper (EB/OL), http://www.vcdex.com
5. Analog Devices Inc H.264 BP Encoder Developer's Guide Blackfin ADSP-BF5xx Processors. Rev. 2.1.0 (2008)
6. Bi, H.: A new generation of video compression coding standard H.264/ AVC. Beijing People's Posts and Telecommunications Press (2005)

Empirical Research on the Relationship between Chinese Inflation Level and Macroeconomic Variables

QiZhi He

School of Finance, Anhui University of Finance and Economics, Bengbu 233030, China
13956346547@163.com

Abstract. Indicators of consumer price index (CPI), gross loans of financial institutions (GL), cash in circulation (M0), narrow money (M1), money & quasi money (M2), foreign exchange reserves (FER), exchange rate (ER), nationwide interbank offered 120-day rate (R) and gross domestic product (GDP) are selected and virtual variable is introduced to measure the impact of financial crisis. The dynamic relationship between China's inflation level and relevant macroeconomic variables are empirically researched. Empirical results show that China's inflation inertia is relatively strong; Subprime mortgage crisis in United States has had a significant impact on Chinese inflation level and etc. Suggestions such as improving the transparency, credibility and independence of Chinese monetary policy, paying close attention to the international economic situation, and ensuring Chinese inflation level remain within reasonable limits are given.

Keywords: Inflation, Macroeconomic Variables; Monetary Policy.

1 Introduction

Inflation level is one of the key indicators reflecting a country's macroeconomic situation and living quality of people. Since subprime mortgage crisis in United States in 2008, to overcome the impact of the crisis, most governments take the loose macroeconomic policy, and this brings inflation pressure. Thus research on the dynamic relationship between inflation level and relevant macroeconomic variables has important theoretical and practical significance.

Researches on the relationship between inflation level and macro-economic variables focus mainly whether there is a relationship between the inflation level and money supply, as well as the Phillips curve. Some scholars, such as Ramakrishnan and Vamvakidis (2002) [1], ADEBIYI (2007) [2], Zhao and Wang (2005) [3], Yang, Chen and Wang (2008) [4] believe that there is a relationship between inflation level and money supply. But there are many scholars do not consider money growth could explain inflation, such as Durevall and Kadenge (2001) [5]. Moroney (2002) [6] summarized the three types of criticism for the quantity theory of money. De Grauwe and Polan(2005)'s empirical studies found that the relationship between inflation rate and money growth in countries with low inflation level is dim [7]. Liu and Jin (2005)'s empirical studies found that expansion of the money supply does not necessarily cause inflation in the monetary process of economy [8]. Chen, Tang and

D. Jin and S. Lin (Eds.): Advances in ECWAC, Vol. 1, AISC 148, pp. 375–382.
springerlink.com © Springer-Verlag Berlin Heidelberg 2012

Li (2009)'s empirical research shows that M0, M1 and M2 have no effect on inflation and cannot forecast inflation in China [9].

Consensus on the relationship between inflation and money supply has not yet been achieved. The same on the Phillips curve, some scholars believe that China has some form of the Phillips curve, while others hold the oppose opinion. Researches on the relationship between money supply and the inflation level, and the Phillips curve in China have important significance. Inflation is often the result of the combined effects of various factors, in addition to indicators such as monetary supply, gross domestic product, and inflation level is also affected by many other indicators.

2 The Selection of Indicators, Theoretical Analysis, and Data Characteristics

According to the customs, the paper adopts the consumer price index (CPI) to measure the inflation level. A country's inflation level is influenced by various factors, and is the result of combined effects of various factors. As for monetary factors, to reflect the effect of different monetary levels on the inflation level in China, the paper intends to select three indicators of cash in circulation (M0), narrow money (M1), and money & quasi money (M2). Money supplies are also influenced by interest rates, so nationwide interbank offered 120-day rate (r) is selected by the paper. Credit channel is one of the main transmission channels of monetary policy in China, so the paper also selects the indicator of gross loans of financial institutions (GL). Chinese increasing foreign exchange reserves force People's Bank of China to input passively the base currency to hedge the increased foreign exchange reserves, and thus there are inflation pressures and so the indicators of foreign exchange reserves (FER) and exchange rate (ER) are also selected. Inflation levels are also affected by actual production, and thus the indicator of gross domestic product (GDP) is also selected. United States subprime mortgage crisis in 2008 have enormous impact on the world economy, and Countries in the world have taken all kinds of appropriate measures to alleviate the impact of the crisis after the outbreak of the crisis. At the same time with the deepening of the process for the opening up and economic integration in the world, inter-linkages of Chinese and foreign economic are becoming more and more closely. These have an important effect on the dynamic changes of inflation level in China, and thus the paper also select the crisis variable (CRISIS). For ease of processing, the paper has used virtual variable method to express CRISIS, and thus to measure the different influencing mechanisms of the different variations, and the relevant factors on the inflation level in China before and after the crisis.

Except ER and R of the above indicators, CPI, M0, M1, M2, GL, FER and GDP have all been taken the form of year-on-year, used by the growth percentage of certain month of the year over the same period of last year to measure the variables, and the corresponding letters are still used.

Firstly, we analyze theoretically the impact of the above indicators on the inflation level. M0, M1, and M2 represent money supply. In general, when money supply increases, there are inflationary pressures, and thus the varying directions of the three indicators are the same with inflation. Increasing GL will enable the public easier to access fund, and make the social demands increasing, and thus have a catalytic role on inflation, and generally the varying directions of it are the same with inflation. When FER increases, the central bank should passively input base currency, and make the money supply expanding, and thus have a catalytic role on inflation, and generally the varying directions of it are the same with inflation. If the increasing GDP is at a reasonable range, the productions will increase and then the increasing social supply will have an inhibition effect on inflation. But if overheated economy appears, then there will be inflation pressures, and thus the varying directions of GDP are the same with inflation. In the paper, ER is represented by how much one dollar can be convertible into RMB, and thus increasing ER means the devaluation of local currency. In the case, imported goods need more local currency, and thus imports are inhibited, and at the same time the home-made products are easy to sell to foreign markets, and thus exports are promoted and there is an upward pressure on domestic inflation level. In general, increased interest rate means monetary deflation and the tight monetary policies have an inhibitory effect on inflation, and thus the varying directions of R are the opposite with the inflation level. The crisis variable (CRISIS) reflects the impact of United States subprime mortgage crisis on the level of inflation in China. After the outbreak of the subprime mortgage crisis, there is a sharp drop in the inflation level in China, and thus in theory the coefficient of CRISIS should be negative.

Time span of the study is from January 2000 to august 2009. Sources of data are respectively from state administration of foreign exchange, Dandong economic information network, the people's bank of China, information network of the State Council development research centre, China economic information network, InfoBank college financial database and genius financial terminal. Figure 1 is the dynamic changing charts of related indices. The mean, standard deviation, skewness, kurtosis, JB, Q (36) and Q^2 (36) of the related indices are also calculated. (In order to save space, the specific values are omitted). According to the JB value calculated, as to indicators of CPI, GL, M0, M2, FER and ER, the original hypothesis of normal distribution was rejected. For indices of M1, R and the GDP, the original hypothesis of normal distribution was not rejected, but their skewness and kurtosis is different from that of normal distribution. Judging from the skewness, the skewness coefficients of CPI,GL,M0,M2,R and GDP are greater than 0 and are at the right side, and the skewness coefficients of M1,FER and ER are less than 0 and are at the left side. Judging from the kurtosis, the kurtosis coefficients of CPI, GL, M0, M2 and ER are greater than 3 and thus the distribution of these variable indices are steeper than that of the normal distribution, and the kurtosis coefficients of M1, FER, R and GDP are smaller than 3 and thus the distribution of these variable indices are flatter than that of the normal distribution. In general, the distributions of China's inflation level and the related influencing factor have certain differences from the normal distribution. If using normal distribution, the corresponding indicators may not be effectively measured. The GED distribution is more flexible than the normal distribution, and the normal distribution is a special form of the GED distribution. Through the adjustment of parameters, the GED distribution can not only describe a steeper distribution than the normal distribution but also describe a flatter distribution than the normal distribution. So in the following empirical research, we will use the GED distribution. Seen from the Q(36) and Q^2(36), at 5% significance level, sequences and squares of CPI, GL, M0, M1, M2, FER, ER, R and GDP are all auto-correlation which mean that these time series are with heteroscedasticity and correlations exist among each period's variance(Li Fujun, 2006)[10]. So in the following empirical research, we will use the GARCH model.

Figure 1 reflects the dynamic changing trend of Chinese inflation level and related indices, and shows that annual growth rate of FER is the max and the hanging range of M0 is the max.These two indicators may have contained more economic information.

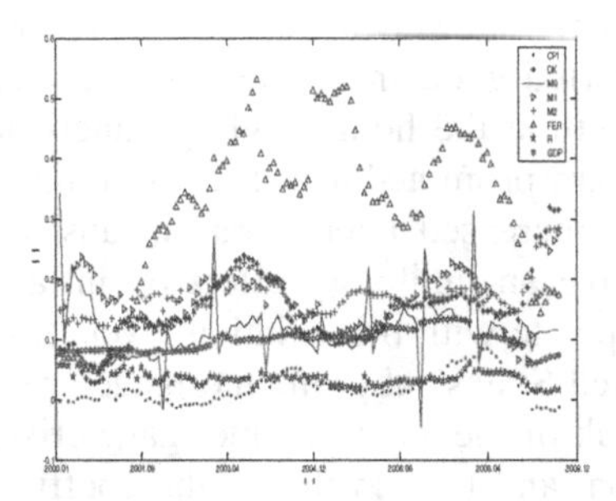

Fig. 1. The Dynamic Changing Trend of Chinese Inflation Level and the Influencing Indices

First of all, we research the effects of a single index on the inflation level in China, and then we study the effects of multiple indices on the inflation level in China.

3 Research on the Single Factor Affecting the Inflation Level in China

In the empirical research, we can determine the lag number of each index in accordance with the following formula [1, 2] :

$$CPI = a * CPI(-1) + \sum_{i=1}^{14} b_i * X(-i) + c * CRISIS + \varepsilon_t \quad (1)$$

$$\varepsilon_t / I_{t-1} \sim GED(0, \sigma_t^2, \upsilon) \;;\quad \sigma_t^2 = c_0 + c_1 \varepsilon_{t-1}^2 + c_2 \sigma_{t-1}^2 \quad (2)$$

$f(z_t) = \frac{v \exp[-|z_t / \lambda|^v / 2]}{\lambda 2^{1+v^{-1}} \Gamma(1/v)}$ is the density function of the GED distribution, where $\lambda = \sqrt{\frac{2^{-(2/v)} \Gamma(1/v)}{\Gamma(3/v)}}$, $\Gamma(*)$ is a gamma function, parameter v controls the form of the distribution[10]。

In equation (1), CPI represents the measuring factor for China's inflation level. X represents respectively the corresponding macro variable indicator such as GL, M0, M1, M2, FER, ER, R and GDP, i represents the lag number. For each indicator, the maximal lagging number is chosen as 14, and then which coefficient is significant is found out, and gradually remove the non notable lagging term. CRISIS is the dummy variable which reflects the effect of subprime crisis on the inflation level in China. a,b_i and c represent respectively the parameters which are required to be estimated.

Attached Table 1(To save space, attached tables are omitted, and if necessary, they can be obtained from the author) is the results of the regressing equation of the single indicator. The basic equation refers to the regressing equation not containing X, and the others represent respectively the equation by replacing X with GL, M0, M1, M2, FER, ER, R, and GDP.

Residuals of the equations are stationary. From Attached table 1, adjusted R^2 of various equations are over 0.9, and this shows that the fitting effect is good. Seen from the coefficient significance, indicators of M0, M1, and M2 have the better effect than the others. In the regression equation with a separate index of M0, M1 and M2, there are 7 pieces of lagging terms having significant impact on inflation. The worst is the regression model based on ER, in this model, all coefficients of the lagging ER terms are not significant, and this shows that all lagging ER terms have no significant effect on inflation. This may be because that "managed floating exchange rate system" is carried on at this stage in China, but the fluctuating degree of exchange rate is still relatively small and exchange rate still alters in a relatively narrow range. Whether according to the adjusted R^2, residual sum of squares or Log likelihood, the best is the regression model based on M0, and its adjusted R^2 is the max, and its residual sum of squares is the min and its Log likelihood is the max.

Integrated the models, we can also get: First, the first-order lagging coefficients of inflation level are all significant, and close to 1. This shows that inflation level is mainly influenced by its first-order lagging term in the single index equation based on individual indicator. Second, in the model with a separate account of GL, M0, M1, M2, FER and R, coefficients of the CRISIS are significant negative and the numerical size was relatively approximate. This shows United States "subprime" crisis had a significant impact on Chinese inflation level. The influencing mechanism of relevant factors on Chinese inflation level has a difference before and after the crisis. Many changes in levels of inflation, which can not be measured by conventional macro-economic variables, can be reflected by the crisis dummy variable. Third, there is no dynamic dependency between exchange rate (ER) and Chinese inflation level, and this is different form the research abroad, such as Ramakrishnan and Vamvakidis (2002) and ADEBIYI (2007)[1,2]. Forth, there is a significant dynamic dependency between nationwide interbank offered rate and Chinese inflation level. In the equation based on nationwide interbank offered 120-day rate (R), the R's coefficients of the 4th, the 5th, the 12th and the 14th lagging are significant. Whether according to residual sum of squares or Log likelihood criteria, the equation based on nationwide interbank offered 120-day rate (R) is better than the basic equation.

4 The Comprehensive Model of China's Inflation Level Based on Multiple Indices

The above empirical test is to study the influence of a single index on inflation. In the next we will integrate these indicators and research the influence of related index on inflation based on these indicators integrated together. The former empirical tests based on single index show those three indexes of money supply: M0, M1 and M2 have significant influences on inflation in China. Due to the M0, M1 and M2 all belong to money supply factors, in order to avoid collinearity, the paper selects only the indicator of M0 which is the best in the previous empirical tests. At the same time the previous tests indicate that there is no significant relationship between ER and inflation level in China, and thus the following comprehensive model does not include M1, M2, and ER.

First the following model is established:

$$CPI = a*CPI(-1)+\sum_{i=1}^{14} b_i*GL(-i)+\sum_{i=1}^{14} c_i*M0(-i)+ \sum_{i=1}^{14} d_i*\text{FER}(-i)+\sum_{i=1}^{14} e_i*\text{R}(-i)+\sum_{i=1}^{14} f_i*GDP(-i)+\gamma*CRISIS \quad (3)$$

In equation (5), CPI represents the measuring factor for China's inflation level. GL, M0, FER, R and GDP represent respectively the corresponding macro variable indicator, i represents the lag number. For each indicator, the maximal lagging number is chose as 14, and then which coefficient is significant of the lag term is found out, and gradually remove the non notable lagging term. CRISIS is the dummy variable which reflects the effect of subprime crisis in United States on the inflation level in China. $a, b_i, c_i, d_i, e_i, f_i$ and γ represent respectively the parameters which are required to be estimated.

Attached Table 2(To save space, attached tables are omitted, and if necessary, they can be obtained from the author) shows the regression results of comprehensive multi-index equation. In the specific empirical research, firstly we play the lagging term whose coefficient is significant of each index into the equation (3) according to the previous test results based on single indicators, and get the integrated model (1) in Attached Table 2. In the integrated model (1), the coefficients of many lagging term of macroeconomic variables have become no longer significant. Removing the lagging term whose coefficient is not significant and reestimating and we will get the integrated model (2) in Attached Table 2. In the integrated model (2), the coefficients of many lagging term of macroeconomic variables have become no longer significant. Removing the lagging term whose coefficient is not significant and re-estimating and we will get the integrated model (3). In the integrated model (3), the coefficients of many lagging term of macroeconomic variables have become no longer significant. Removing the lagging term whose coefficient is not significant and re-estimating and we will get the integrated model (4) in Attached Table 2. All coefficients of the integrated model (4) are significant, and the integrated model (4) is the integrated model of China's inflation level influencing factors what we finally gotten.

Residuals of the equations are stationary. We can know from Attached Table 1 and 2 the following results.

First, according to the Integrated Model (4), China's inflation level is mainly influenced by the first-order lagging inflation level, the money supply M0, and crisis variable.

Second, whether according to the previous test of single index, or the integrated model (1), (2), (3) or (4), the most important determining factor of inflation is the first-order lagging inflation level, and this shows that China's inflation level has a strong inertial characteristics. The following several possible reasons may partly explain this.①the transparency of monetary policy in China is not high. Xu Yaping (2006) [11] found that the present transparency of monetary policy in China is 2.5, largely smaller than full marks 8 through empirical research. ②the independence of monetary policy in China is not strong, monetary policy is also affected by many factors. For example, the implementation of monetary policy in China is easily subjected to the financial needs, and this causes China is vulnerable to financial inflation pressure. According to the view of ADEBIYI (2007)[2], when a country is vulnerable to financial inflation pressure, it will induce the creation of formal and informal indexation mechanism, and this will lead to persistent inflation.

Third, in the previous test of single index, GL, M0, M1, M2, FER and R all have relationship with China's inflation level. But after considering various factors, relationships between all lagging terms of GL and R with China's inflation level is no longer significant. This may be due to GL and R also reflecting indirectly the effects of money supply, and their impact on China's inflation level need the channel of money supply. Thus after the full account of M0 and FER, the influences of all lagging terms of GL and R on China's inflation level have been indirectly reflected by M0 and FER, and thus the lagging factor has been no longer significant.

Forth, whether the previous test of single index (except the single index model based on exchange rate) or the integrated model (1) or the integrated model (2) or the integrated model (3) or the integrated model (4), the coefficient of the crisis variable is significant and negative at all the while. Introducing crisis variable increases the fitting effect of the model and this shows that United States subprime mortgage crisis has an important impact on China's inflation level and is one of the reasons for the decline in China's inflation level after the crisis.

5 The Conclusions and Policy Recommendations

Through the previous empirical research, some of the main conclusions and policy recommendations can be drawn as follows:

First, China's inflation level is mainly influenced by the first-order lagging inflation level, the money supply M0, and crisis variable.

Second, China's inflation level has very strong inertia and the most main determining factor of level of inflation in China is the first-order lagging inflation level. Whether setting or implementing of monetary policy, the characteristic of high inertia of China's inflation level should be taken into account. In China the transparency of monetary policy should be improved, the credibility of monetary policy should be enhanced, the independence of monetary policy should be strengthened and the influence of fiscal pressure on inflation level should be reduced.

Third, In China inflation is mainly a monetary phenomenon, and in addition to the first-order lagging inflation level, money supply plays the second most important role in the determinant of inflation level in China. Thus the growth rate of money supply should be arranged reasonably. If the growth rate is arranged too low, the economic growth may be inhibited, and if the growth rate is arranged too high, there may be inflation expectation and inflationary pressure.

Forth, Subprime mortgage crisis had a significant impact on China's inflation level. With the deepening of China's opening up and process of economic integration in the world, the relationship between China and the world economy is becoming more and more close. Factors affecting inflation level are no exception, and are becoming more and more under the influences of international factors. This requires the monetary authority pay close attention to the international economic situation, and cut off the negative impact of the international factors on China. In particular, more attention should be paid to the influence of the international financial crisis and unexpected events on China's economy, and ensuring that China's inflation level is kept within a reasonable bound.

The paper researches the relationships between China's inflation level and the relevant economic variables, but the influencing mechanism of relevant financial variables on China's inflation level has not been deeply analyzed, and this will be our direction for in-depth study in the next step.

Acknowledgment. This work was supported partly by National Social Science Fund of China (11CJY080; 10CTJ008) and Postdoctoral Science Foundation (20110491399).

References

1. Ramakrishnan, U., Vamvakidis, A.: Forecasting Inflation in Indonesia. IMF Working Paper (2002)
2. Adebiyi, M.A.: Does Money Tell Us Anything About Inflation in Nigeria? The Singapore Economic Review 52(1), 117–134 (2007)
3. Zhao, L., Wang, Y.: Money stock and price level: empirical evidence in China. Economic Science (2), 26–38 (2005)
4. Yang, L., Chen, S., Wang, H.: Research on the dynamic relationship between Money supply, Bank credit and Inflation. Management World (6), 168–169 (2008)
5. Durevall, D., Ndungú, N.: A dynamic inflation model for Kenya, 1974–1996. Journal of African Economies 10(1), 92–125 (2001)
6. Moroney, J.R.: Money Growth, Output Growth and Inflation: Estimation of a Modern Quantity Theory. Southern Economic Journal 69(4), 398–418 (2002)
7. De Grauwe, P., Polan, M.: Is Inflation Always and Everywhere a Monetary Phenomenon? Scand. J. of Economics 107(2), 239–259 (2005)
8. Liu, L., Jin, Y.: Money Supply, Inflation and Economic Growth of China-The Experimental Analysis Based on Cointegration Analysis. Statistical Research (3), 14–19 (2005)
9. Chen, Y., Tang, S., Li, D.: Can Money Supply Forecast Inflation in China? Economic Theory and Business Management (2), 22–28 (2009)
10. Li, F.: Measuring VaR and ES of StockMarket Based on SV-GED Model. Systems Engineering-Theory Methodology Applications 15(1), 44–48 (2006)
11. Xu, Y.: The Effectiveness of Monetary Policy and the Flourishing of Central Bank Transparency. Economic Research Journal 8, 24–34 (2006)

On-Line Monitoring System Software Design Based on Energy-Efficient of Non-invasive Motor

Zhang Qingxin[1,2], Liu Chong[1], Li Haibin[1], and Li Jin[1]

[1] Automation Department, Shenyang Aerospace University, Shenyang, 110136, China
[2] Shenyang Institute of Automation, Chinese Academy of Sciences, Shenyang, 110016, China
zhy9712@163.com

Abstract. The motor is the main powerful output and power-consumption equipment in industrial system. Research about motor's energy conservation and on-line monitoring technology is of great significance. In this paper an energy-efficient of non-invasive motor for on-line monitoring system is introduced. The running state information of the motor is collected and analysed, and the motor's speed, torque, power factor, efficiency and other parameters can be got by the on-line monitoring. The monitoring results are analyzed to provide support for motor's energy management, monitoring interface is perfected by LabVIEW8.5. The experimental results show that the system is reliable, easy to operate, and highly safety.

Keywords: noninvasive, motor energy efficioncy, online monitoring, software design.

1 Introduction

The motor is the main powerful output in the industrial production, it plays an important role in modern industry, and it is widely used in all fields of industrial production. The motor is influenced by electricity, heat, machinery, and the surrounding environment factors in the running process, which makes the motor's performance deteriorative, not only motor's equipment can be damaged, but also the working efficiency of the motor is also reduced and power-consumption increases, the proper work of the entire system is influenced, which cause huge economic losses.

2 The System Composition

Motor's online monitoring and energy management system are composed of three parts. The first part collects information of motor state. The second part processes the data collected and monitor state, which consists of the data acquisition, the situation monitoring, the energy and energy conservation analysis, the network communication and database. The third part analyzes motor energy consumption and save energy.

2.1 On-Line Monitoring System

The on-line monitoring system for energy-efficiency of non-invasive motor consists of the on-line monitoring and the energy management. The front end has the function

D. Jin and S. Lin (Eds.): Advances in ECWAC, Vol. 1, AISC 148, pp. 383–386.
springerlink.com © Springer-Verlag Berlin Heidelberg 2012

of the signal acquisition and processing, which collects and disposes the signal of the stator currents and stator voltage, in order to get the motor running state parameter. The back end is the motor monitoring centre. It accepts motor's running parameters which the front end carries. Monitoring system can displays the instantaneous and time domain curve of these parameters in real time, meanwhile, data is stored in the database, which is the foundation of the motor's efficiency analysis and energy management. The motor speed signal is extracted with air gap torque method and the spectrum analysis technology. The high precision online monitoring system is achieved with the DC injection method.

2.2 Energy Management System

The motor's energy management is the complex system engineering in the industrial production. Firstly, all the basic condition of the motors needs to be mastered to understand the characteristics of all kinds of the motors and establish motor's resources database. Secondly on-line monitoring of the motor system is implemented effectively to get input power, output power, efficiency, loading rate, energy consumption and power factor, to establish monitoring information database. On that basis, some technologies are integrated such as signal processing technology, the condition monitoring and fault diagnosis technology, the motor health management technology, the energy consumption and energy saving analysis technology, which can make the motor save more energy and get higher efficiency. These technologies can establish the energy management of the motors and drive system, to ensure the more optimization of the motor system running.

3 Software Design

The whole system software design consists of the PC condition monitoring,which have three function,they are real-time monitoring, performance analysis and history inquiries.

3.1 PC Condition Monitoring System

LabVIEW can designs PC condition monitoring system. LabVIEW is a kind of the graphic design language, and it is accepted by the industry, the academe and the laboratory study. LabVIEW is the standard data collection and instrument control software. The designers mainly use three templates which LabVIEW provides, and they are the tool palette, the control palette and the function palette, which can design two parts of the VI floater and the flow chart.

It concludes the on-line condition monitoring, the motor performance analysis and the data history inquiries that PC condition monitoring software based on LabVIEW. The on-line monitoring interface accepts the motor's signal which each front-end device sends. These motor's signals includes the torque, the RMS voltage, the RMS current, the input power, the output power and the efficiency. the monitoring interface shows in real time by the analysis. Instrument panel, real-time wave form and logbook recording are presented in fig1. fig2 and fig3. The monitoring parameters are installed with alarm, which easier make operation staff analyze in time.

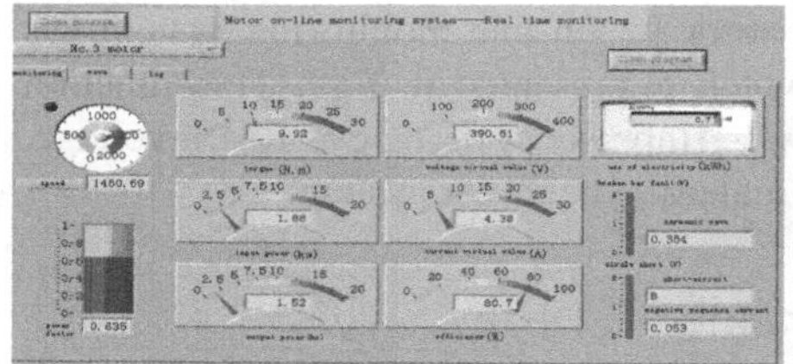
Fig. 1. The on-line monitoring of the instrument panel

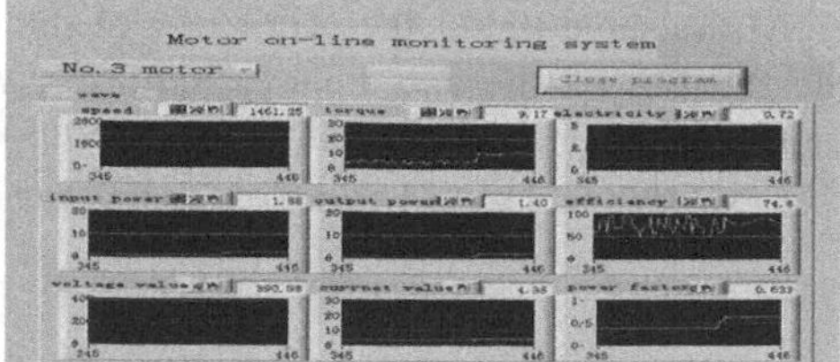

Fig. 2. The on-line monitoring of the real-time wave form

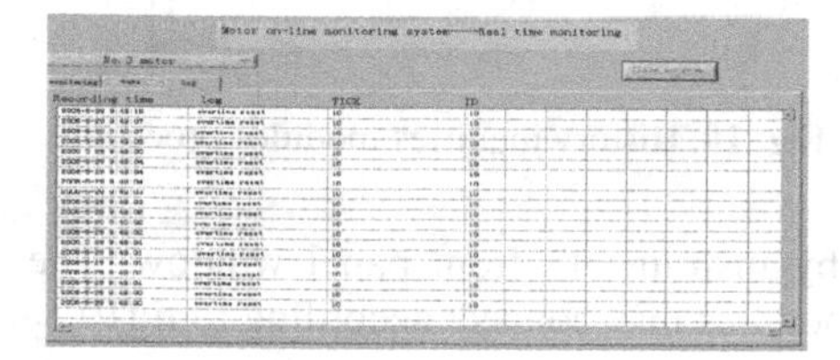
Fig. 3. Logbook recording

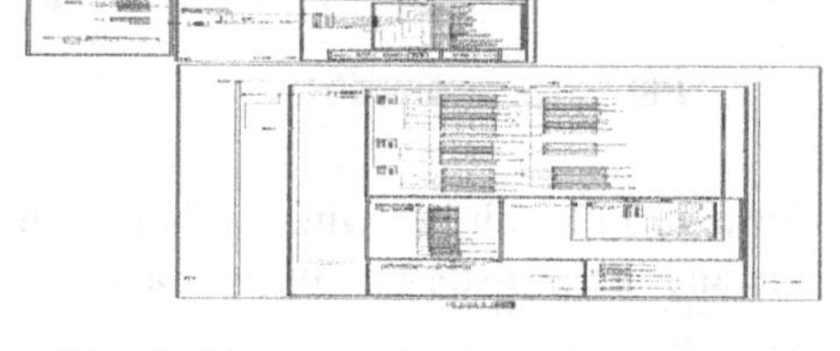
Fig. 4. Online monitoring software modules

The procedure in online monitoring software modules may connect according to fig4.

The performance analysis of the motor can includes the torque-outpower, torque-speed and input power-speed, which are presented in fig5, fig6 and fig7.

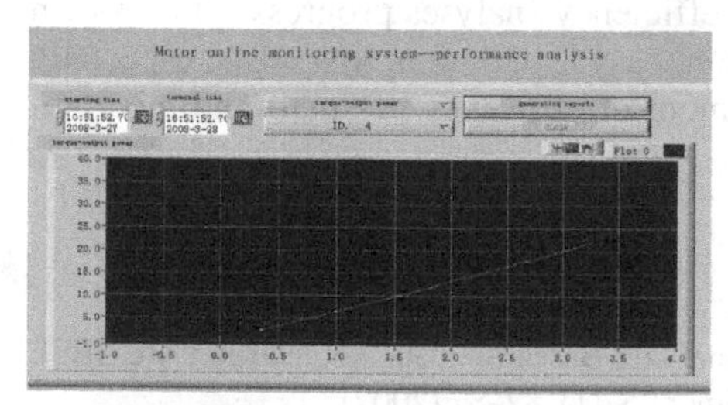

Fig. 5. Torque-outpower

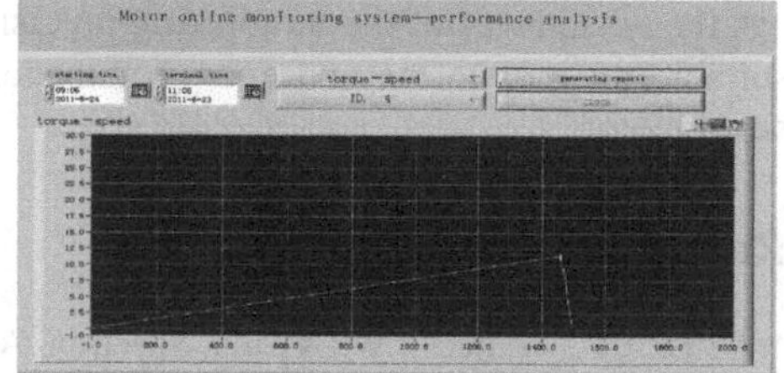

Fig. 6. Torque-speed

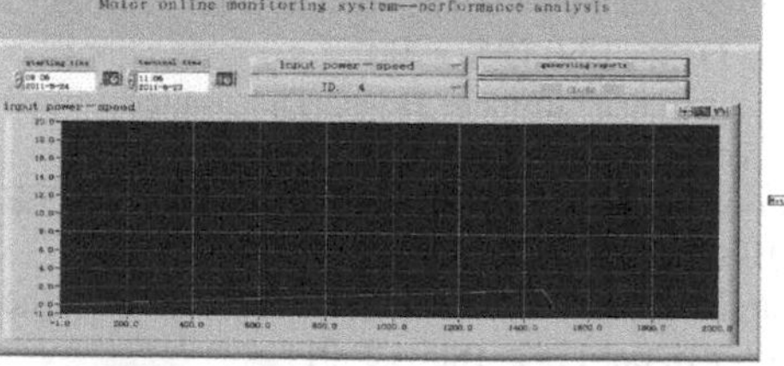

Fig. 7. Input power-speed

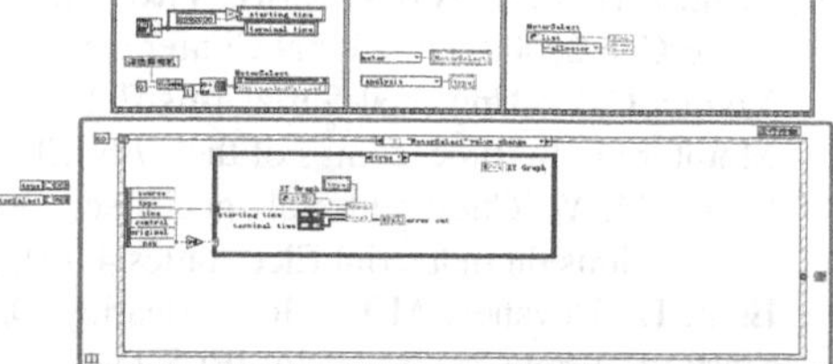
Fig. 8. Performance analysis modules procedure

Flow chart's editting window is shown by instrument's front panel window. The procedure in the performance analysis software modules may connect according to fig8.

At the same time, the efficiency of the motor for on-line monitoring system has the function of the data history inquiries, which is presented in fig9. This function inquiries the motor's speed, torque, outpower, inpower, efficiency, RMS voltage, RMS current and power factor, which easily manages the motor system.

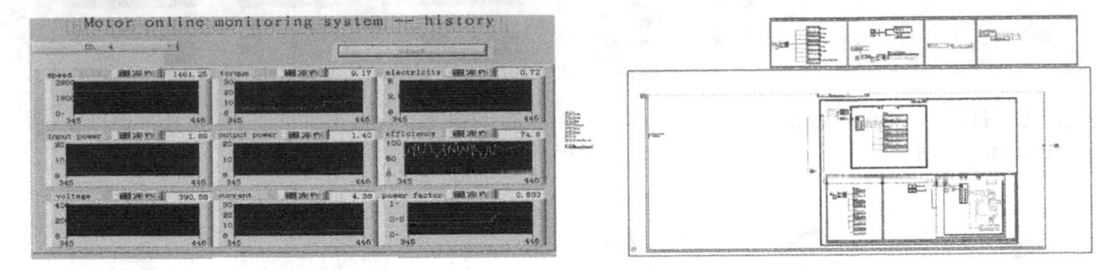

Fig. 9. History inquiries

Fig. 10. History inquiries modules procedure

Flow chart's editting window is shown by instrument's front panel window. The procedure in the history inquiries software modules may connect according to fig10.

4 Conclusion

The plan and software design of energy-efficient of non-invasive motor for on-line monitoring system are introduced. This system uses analysis technology based on the motor stator current signal feature. The on-line monitoring of the motor running state and efficiency analysis are realized. This system can supply strong analysis tool for user, to make the selection of the motor and efficiency analyses progress. This system has the advantages of the monitoring precision, easy installation, low costs, especially suitable for monitoring and energy management of efficiency of small and medium-sized motor.

References

1. Bilsky, P., Winiecki, W.: Virtual Spectrum Analyzer Based on Data Acqusition Card. IEEE Transactions on Instrumentation and Measurement 52(1), 82–87 (2002)
2. Moure, M.J.: Educational Application of Virtual Instruments Based on Reconfigurable Logic. In: IEEE International Conference, vol. (1), pp. 24–25 (1999)
3. Teng, J.-H.: A LabVIEW Based Virtual Instrument for Powe Analyzers. In: Proceedings of PowerCon 2000 International Conference, vol. (1), pp. 179–184 (2000)
4. Young, C.-P.: Universal Serial Bus Enhances Virtual Instrument Based Distributed Power Monitoring. In: Proceedings of the 17th IEEE, vol. (2), pp. 920–924 (2000)
5. Chow, M.-Y.: Guest editorial special section on motor fault detection and diagnosis. IEEE Transactions on Industrial Electronics 47(5), 1031–1041 (2000)
6. Eren, L., Devaney, M.J.: Motor Bearing Damage Detection via Wavelet Analysis of the Starting Current Transient. In: IEEE IMTC, vol. (3), pp. 1797–1800 (2001)
7. Lu, B., Habetler, T.G., Harley, R.G.: A Survey of Efficiency Estimation Methods of Condition Monitoring Requirements. In: IEEE International Conference on Electric Machines and Drives, May 15-18, pp. 1365–1372 (2005)
8. Gallegos, M.A., Alvarez, R., Nunez, C.A.: A Survey on Speed Estimation for Sensorless Control of Induction Motors. In: CIEP, Puebla, MEXICO, October 16-18 (2006)

Research on Mongolian Input Methods

S. Loglo and Sarula

College of Mongolian Studies, Inner Mongolia University, HuhHot, China, 010021
sloglo@sina.com

Abstract. In this paper, we have analyzed and studied the implementation of Mongolian input methods from technical aspects, and then proposed a training model for phrase association input methods based on large-scale corpus. When we develop an intelligent phrase association input method, training the collocation network with large-scale corpus is a desirable and efficient way.

Keywords: Mongolian, Input Method.

1 Introduction

At present, in the Mongolian office automation, publishing and printing area, the most commonly used software is Mongolian WPS-Office and Founder book publishing software, respectively developed by Mengkeli Company and Founder Cooperation. So, among the Mongolian keyboard input methods, Mengkeli input method and Founder input method occupy a comparative advantage position. In addition, in recent years, with the rapid development of Mongolian network technology, emerged many excellent input methods, such as Sain input method, Burgud input method, Oyuta input method and Mongolian smart input method, according to incomplete statistics, there already have a dozens of Mongolian input methods. Although most of the Mongolian input method is designed and developed for the realization of specific coding system, however, looking from the implementation technology, input method is independent from coding system and editing environment. Therefore, study the implementation methods and technical features of specific input methods are necessary for improving the intelligence and speed of Mongolian keyboard input. In this paper, from the two perspectives, such as programming and input methods, we have analyzed and studied the existing input methods for Mongolian language.

2 The Interface Technology for Keyboard Input

From the perspective of computer programming, we analyzed a number of Mongolian keyboard input methods, and found that these methods mainly used three types of interface technology as following.

1) Embed technology. In fact, input method that is embedded in an application is not a universal input method, because this kind of input method can only be used in applications that contain it. Under Windows environment, the keystrokes were passed to the appropriate message processing functions by WM_CHAR, WM_KEYDOWN

D. Jin and S. Lin (Eds.): Advances in ECWAC, Vol. 1, AISC 148, pp. 387–392.
springerlink.com © Springer-Verlag Berlin Heidelberg 2012

and WM_KEYUP message. In the message handler, we can map the English characters to other characters, according to our own need. This kind of input method's implementation is simple and intuitive, but can not be used for other applications. The Mongolian input method embedded in the Founder book publishing system version6.0 is a typical example.

2) Plug-in technology. Plug-in type input methods using global keyboard hook of Windows system. When a message arrives, before the handler of the target window to processing it, the hook mechanism allows the hook function to get and process the message. Whenever a particular message is sent, and while the message not arrive the destination window, hook function catch the message, in another word, the hook function first gets the control. At this time, the hook function can processed (change) the message, may it not make any deal, to pass the message. For example, Mongolian smart input method or Oyuta input method is a kind of input method that used the keyboard hook as interface. This kind of input method is a regular application, easy to implement, and not restricted by the interface. But the keyboard hook is a global hook, once enabled, each thread or application only can use the same input method, this is unlike with the IME input method, each application or thread can be there with their associated input method.

3) IMM/IME Technology. After Windows95, in the Far East version of Windows, emerged the IMM/IME (Input Method Manager / Input Method Editor) structure, this structure provides a complete and effective programming interface, therefore, the realization of input methods become even more convenient. IMM/IME is designed for developers, allowing developers to design a new input method, and added to the system. The user through using the application programming interface (IME API functions) to implement the input method. The IMM/IME's principle is shown as Fig.1.

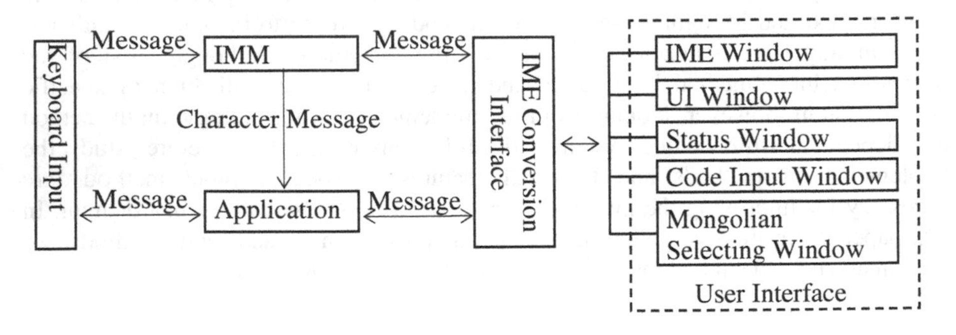

Fig. 1. The Principle of IMM/IME

After user's input is converted to keyboard input message, the system will pass the input method manager to the corresponding input method editor of current thread, firstly. Input Method Editor convert the user input actions into result string, based on the message provided by the input method manager and data recorded in the context. And then, the result string is back to the input method manager as character message forms, and is put into the application window message queue. IMM/IME is a standardized and convenient input method implementation technique, currently most of the Mongolian input method uses it as the programming interface, and is a mainstream technology.

3 The Relationship between Input Method, Code and Font

In computer system, letters, control symbols and graphical symbols are encoded into binary code and be handled. The binary code of letters and symbols are called character code. Font is the collection of graphics that are used in display and printing. In the previous dot matrix font, vector font and TrueType font, the character code and graphics only have one to one relations, however, in OpenType fonts, the situation is more complicated, there exist one to one, one to many and many to many relations. With the coding system and corresponding font library, we can use the internal code input method to input characters, but this method is slow, and need to memorize each characters code. Input method is an application software, its mission is to simplify the entry process and increase the input speed. So, easy to grasp and have a high inputting speed is the never-ending pursuit for various kinds of input methods, and there exist mutual restraint between the two sides.

Mongolian is a complex text language, so, characters have different representation forms in a word's different position. Thus the representation form's selection is the main tasks for Mongolian word processing software. Before appear Mongolian international standard code and OpenType font library, in Mongolian word processing software, the selection of representation forms are handled by input methods. In this situation, once the characters are sent to the editing environment, its shape is formed, and in the subsequent editing, the shape is not changed. Under OpenType font library that used Mongolian international standard code, the layout engine completes the selection of representation forms. When the layout engine output a passage of text, it first call the script code stored in the OpenType font library, and then according to the rules stored in the script code and context output the correct representation forms. Each time, the Mongolian characters changed in the editing environment, the layout engine re-output the passage of text according to the context, so the specific shape of a character is determined by its context. The relationship between Mongolian input method, coding system and font library shown as Fig.2.

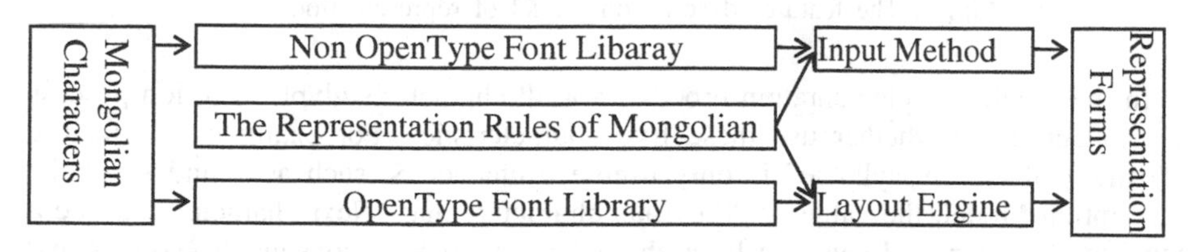

Fig. 2. The relationship between Mongolian input method, encoding system and font library

4 Comparisons of Several Mongolian Input Techniques

To use representation rules and short-code thesaurus are the main way to improve the speed of Mongolian input methods, according to these two points we can divided the Mongolian input methods into different categories.

4.1 Mongolian Input Methods without Using Any Representation Rules and Thesaurus

This kind of input method does not need thesaurus and representation rules, just need to create a simple map of keys. If the font library used TrueType technology, there can be two ways to input the characters. One way is for each different representation form map a key position, by pressing different keys to select the correct glyph, generally, called this kind of input method as character shape input method. Another way is to map one key for each character, when bit a key, pop up a window and lists the different representation forms of the current character, and manually choose the right glyph, generally, called this kind of input method as Pinyin input method. If the coding system as Mongolian international standard code, font library is OpenType font, for each nominal characters and control characters map a key position, the glyph selection process is completed by font library and layout engine automatically.

4.2 Rule Based Input Method

This kind of input method map different key for each Mongolian characters, when input, character's correct glyph is automatically selected by the representation rules, we used to call this kind of input method as pronunciation input method. Currently, this is the most common input technology in Mongolian information processing, Founder input method, MengKeLi input method, Oyuta input method, Sain input method, and Mongolian smart input method all adopted the technology. Although, encoding and programming is different in above mentioned systems, but the rules are all generated from the Mongolian orthographic rules. Rules generally described as a set of context-sensitive features. Feature description model shown as Fig.3.

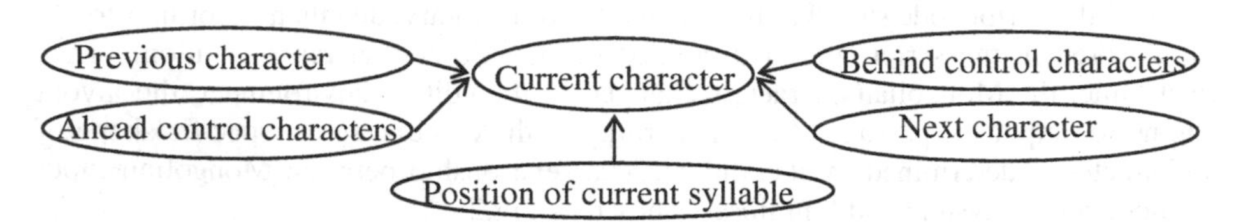

Fig. 3. The features description model of representation

In the specific implementation process, not all character's glyph selection process has to determine whether use these five characteristics. For example, the feature "Position of current syllable" is only used for characters, such as ö and ü. In the description Model, the value of "Previous character" and "Next character" can be a specific character, and can also be a character set, such as consonants, vowels and so on.

4.3 Thesaurus and Rule Based Input Method

To improve the inputting speed, some developers developed a whole word input method and phrase input method based on the pronunciation input method. Where the pronunciation input used rule-based methods, and the whole word or phrase input used corpus-based methods. Most of the whole word or phrase input methods used short-codes, but the encoding rules for short-code are different. Short-code is a encoding form, it has effectively shorten the length of a word, in theory, its length as short as possible, but in actual application, the shorter the code is the higher the repetition-rate, so the code length and repetition-rate are restricted by each other. In addition, encoding rules are needed, if short-codes have not any disciplines, the user will be unable to remember thousands of short-codes.

Here are encoding rules for some Mongolian phrase input methods:

1) Mengkeli phrase input method : Mengkeli phrase input method provides a small thesaurus and maintenance tools, short-codes are making up by first characters of words. Users can use the maintenance tool to add custom entries, and the short-codes are determined by user.

2) Sain phrase input method : For two word phrase: input the first two consonant of each word; For three word phrase: input the first two consonant of first word and second word, and then input the first two character of third word; For four word phrase: input first consonant of each word; For other situation: input first consonant of first three word and last word.

3) Mongolian smart input method : The phrase selecting mechanism of Mongolian smart input method is different from other phrase input methods, in this input method, phrases are selected by the first matching word. When user use the pronunciation input method to input the Mongolian text, the phrase association model continuously search phrases form the thesaurus with first word matching algorithm (all characters in first word+ a space + first n characters of the next word) ,if found phrases, pop-up a window and display all possible candidates. If user did not want to select phrase, do not delete the typed characters, can continually use the pronunciation input method to input the characters remaining.

5 A Corpus-Based Training Model for Phrase Input Method

Now rule-based pronunciation input method is the most commonly used Mongolian input method, in this method, user input all characters in a word, and then computer complete the glyph selection process. Although some specific pronunciation input method have set short-code for separated formalization suffixes, but the improvement of the input speed is not very obvious. To significantly improve the speed, intelligence must be added to the phrase association functions. Currently, all Mongolian phrase input methods have its thesaurus that collected by manually. However, the scale of manual collected entries is very limited, and it is very difficult to ensure that these phrases have high frequency. Thus, we proposed a large-scale corpus based training model for phrase input method. In this model morphological changes of the word can be achieved in two ways.

1) Constructs a node for all forms of the word (if appeared in corpus);

2) Forms of the words are not included in the node, relationship between two nodes are represented by different arcs. With trained by a large-scale corpus we can get a Mongolian word collocation network shown as Fig.4.

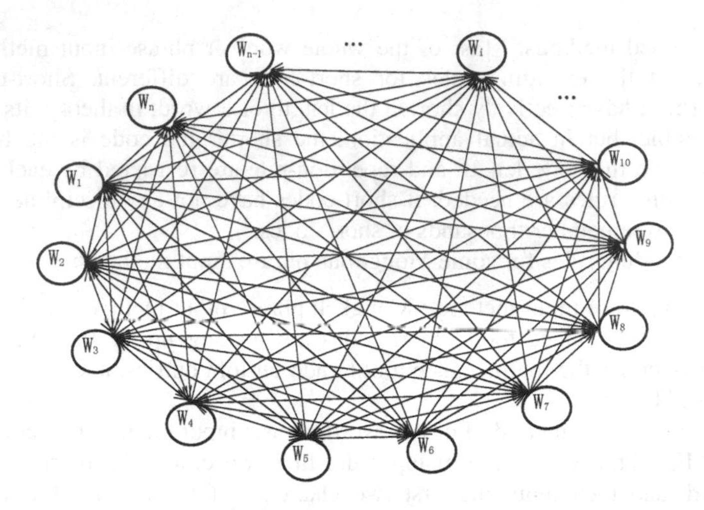

Fig. 4. The collocation network of Mongolian

In the figure, W_i indicates Mongolian word; arc between two nodes indicates that there have a collocation; figures indicates the strength of collocation. If the number is 0, it indicates that there does not exists relationship between two nodes.

When we use the network, through calculate the collocation strength of already traversed n node and next m node to determine the next step.

6 Conclusions

With Mongolian international standard code and OpenType font technology, the glyph selection process is solved by layout engine, Therefore, the focus of research and development should be placed on the improvement of intelligence. In this paper, we firstly analyzed and summarized the implementation technique of the Mongolian input method, and then proposed a training model for phrase association based on large scale corpus.

References

1. Loglo, S., Narentuya.: Mongolian Smart Input Method Based on Uniscribe and OpenType. Journal of Inner Mongolia University (Philosophy & Social Sciences in Mongolian) (5) (2007)
2. Loglo, S., Ocir: Research on Mongolian Complex Text Processing Under Windows Environment. Journal of Inner Mongolia University (Natural Science Edition) (5) (2007)
3. Hu, Y., Ma, S., Xia, Y.: Input Method Implement Based on IMM/IME. Computer Engineering and Application (1) (2002)
4. Bai, I.: The Design and Implementation of The Mongolian Whole Word Input Method. Journal of Inner Mongolia Agricultural University (9) (2008)
5. Fan, D., Bai, F., Wu, H.: Research on Mongolian Input Method in Unicode. Journal of Chinese Information Processing (6) (2010)

Influence of Introduction of Stock Index Futures on Information Efficiency of Stock Market in the Hong Kong Capital Market

Hailiang Meng

Economics & Management Institution, Beijing Information Science & Technology University
Beijing, China
zhongguoren1974@163.com

Abstract. What influence will the introduction of stock index futures exert on the information efficiency of stock market? The research on it will be of great theoretical and practical significance to both the investors and the managers of the financial market. This paper empirically researches the influence of the introduction of Stock Index futures on the information efficiency of the stock market in the Hong Kong capital market and the results show the introduction of stock index futures improves the information efficiency of stock market.

Keywords: information efficiency, Stock Index Futures, stock market.

1 Introduction

As finance has been the core of the modern economy, the financial derivatives are playing a more and more important role in the modern financial market. What influence will the introduction of stock index futures exert on the information efficiency of stock market? The research on it will be of great theoretical and practical significance to both the investors and the managers of the financial market.

The stock index futures market has its unique advantages in comparison with the stock market, such as low transaction cost, no short-sale restraint, leveraged transaction, etc, so the former is closer to the perfect competition market. The price formation in such a market can gather and convey a mass of information. The information can be conveyed to the stock market by the arbitrage mechanism between the stock index futures market and the stock market to increase the content of the price information in the stock market and guide and discover the stock price so as to improve the information efficiency of the stock market. As for the influence of the stock index futures on the information efficiency of the stock market, the empirical conclusion derived from the other developed markets has provided the evidence for that the introduction of stock index futures improves the information efficiency of the stock market.

Manaster & Rendleman (1982), based on their empirical researches, hold that the options and other derivatives markets can play the role of conveying the information to the cash market so as to improve the information efficiency of the cash market [1]. Harris (1989), Antoniou & Holmes (1995) and Gulen & Mayhew (2000)'s empirical

D. Jin and S. Lin (Eds.): Advances in ECWAC, Vol. 1, AISC 148, pp. 393–398.
springerlink.com © Springer-Verlag Berlin Heidelberg 2012

researches show that the futures market expands the information channels of the cash market and improves the information efficiency of the cash market [2, 3, 4].

However, most of the above-mentioned researches are aimed at the developed foreign markets and only few of them are about the Hong Kong markets. In view of this, the paper empirically researches the influence of the introduction of stock index futures on the information efficiency of the stock market based on the Hang Seng Index (HSI) futures in the Hong Kong capital markets.

2 Research Method

In the paper, the author follows the research ideas of Madhavan, Porter & Weaver (1999) [5] and adopts the similar event research method to analyze the changes of the information efficiency of the stock market before and after the introduction of stock index futures and, based on the analysis, concludes whether the introduction of stock index futures exerts significant influence on the information efficiency of stock market.

First, the author defines the periods to be compared before and after the event respectively as the pre-event window period and the post-event window period; then, he selects the specific indicator to be researched and calculates the values of the indicator in the pre-event window period and the post-event window period; finally, he compares the values of the indicator in the post-event window period and the pre-event window period, holds the statistical test and discovers whether the indicator changes because of the occurrence of the event. As for the statistical test, the author adopts the data-based t test method to compare the observed values of the information efficiency indicator in the post-event window period and the pre-event window period.

3 Empirical Research on the Influence of the Introduction of HSI Futures on the Information Efficiency of Its Underlying Index Constituent Stocks

In the section, the author adopts the similar event research method to compare the changes of the information efficiency of the underlying index constituent stocks before and after the introduction of HSI futures and, based on the comparison, draws the conclusion whether the information efficiency of the underling index constituent stocks are influenced by the introduction of HSI futures.

3.1 Selection of Sample Stocks

The introduction of stock index futures influences its underlying index constituent stocks most, so the author selects the underlying index of HSI futures – HSI constituent stocks – as the research samples in the section. Over the past more than 20 years from 1986, some corporations' stocks among HSI constituent stocks were delisted due to privatization, so the author cannot get their transaction data. Finally, the author selects 21 stocks the transaction data of which are available from HSI constituent stocks as the sample stocks.

3.2 Selection of Event Window Periods

Madhavan, Porter and Weaver (1999) assigned the data in a month to a window period. In the research, the author assigns the transaction data of HSI constituent stocks in two months to a window period and sets four different event window periods: one pre-event window period and three post-event window periods. Because HSI futures was introduced in Hong Kong on May 6, 1986, the author assigns the data in March 1986 and April 1986 to the pre-event window period, the data in June 1986 and July 1986 to the first post-event window period, the data in August 1986 and September 1986 to the second post-event window period and the data in October 1986 and November 1986 to the third post-event window period, then compares the performance of the information efficiency indicator respectively in the three post-event window periods with that in the pre-event window period to discover whether the information efficiency of the stock market changes because of the introduction of the stock index futures.

3.3 Selection of Information Efficiency Indicator

In the previous researches, the scholars (Boehmer, Saar & Yu, 2005) usually observed the information efficiency by the absolute value of the first-order autocorrelation coefficient of return series. According to the efficient market theory, the higher the information efficiency of market is, the random walk the stock prices approach to and the lower the autocorrelation degree of return series is. In the paper, the author takes the autocorrelation coefficient of daily return series as the first-order autocorrelation coefficient of return series. Because the research focuses on the changes of the autocorrelation degree of return series and has nothing to do with the positive or negative nature of the autocorrelation coefficient, the author observes the information efficiency by the absolute value of the first-order autocorrelation coefficient of return series in the section.

First, the author calculates the absolute values of the autocorrelation coefficients of the daily return series of the 21 sample stocks in each window period and gets 21 observed values from each window period; then he adopts the statistical test method to compare the observed values in the post-event window periods with those in the pre-event window periods; finally, he, based on the comparison results, deduces whether the introduction of HSI futures has significant influence on the information efficiency of HSI constituent stocks.

In the section, the author gets the transaction data of the sample stocks from the financial terminal of Wind Information Co., Ltd., which are from March 3, 1986 to November 28, 1986 excluding those in May 1986.

3.4 Data Processing and Results

Table 1. The test results of first-order autocorrelation coefficients absolute value（I）

statistical indicators	t-test: paired two sample mean analysis	
	the absolute values of the first-order autocorrelation coefficients in the pre-event window period	the absolute values of the first-order autocorrelation coefficients in the first post-event window period
average	0.158571	0.144062
variance	0.006909	0.008997
observations	21	21
poisson coefficient	0.0245	
hypothesized mean difference	0	
freedom	20	
the absolute value of t statistics	0.15804	
P(T<=t) (one-tailed)	0.438003	
t (one-tailed critical value)	1.724718	
P(T<=t)(two-tailed)	0.876007	
t (two-tailed critical value)	2.085963	

Table 2. The test results of first-order autocorrelation coefficients absolute value（II）

statistical indicators	t-test: paired two sample mean analysis	
	the absolute values of the first-order autocorrelation coefficients in the pre-event window period	the absolute values of the first-order autocorrelation coefficients in the second post-event window period
average	0.158571	0.125338
variance	0.006909	0.003964
observations	21	21
poisson coefficient	-0.19989	
hypothesized mean difference	0	
freedom	20	
the absolute value of t statistics	2.142396	
P(T<=t) (one-tailed)	0.022325	
t (one-tailed critical value)	1.724718	
P(T<=t)(two-tailed)	0.044649	
t (two-tailed critical value)	2.085963	

Table 3. The test results of first-order autocorrelation coefficients absolute value (III)

statistical indicators	t-test: paired two sample mean analysis	
	the absolute values of the first-order autocorrelation coefficients in the pre-event window period	the absolute values of the first-order autocorrelation coefficients in the pre-event window period
average	0.158571	0.101895
variance	0.006909	0.002918
observations	21	21
poisson coefficient	-0.26701	
hypothesized mean difference	0	
freedom	20	
the absolute value of t statistics	4.892103	
P(T<=t) (one-tailed)	0.191473	
t (one-tailed critical value)	1.724718	
P(T<=t)(two-tailed)	0.382946	
t (two-tailed critical value)	2.085963	

First, the author turns the price data of the sample stocks from the financial terminal of Wind Information Co., Ltd. into the return rate data by the function “price2ret” in Matlab; then he finds the first-order autocorrelation coefficients of the return rates by the function “autocorr”; finally, he gets the test results by the paired two-sample analysis tool of the function “average” in Excel, respectively listed in Table 1, Table 2 and Table 3.

In Table 1, Table 2 and Table 3 are listed the average absolute values of the first-order autocorrelation coefficients in the pre-event window period and the post-event window periods as well as the statistic t and the value p in the t test, of which, the statistic t is the post-event data minus the pre-event data. The significance level is 0.05.

3.5 Analysis of Test Results

The test results in Table 1 show that the absolute value of the statistic t is less than the two-tailed critical value 2.085963. The test results in Table 2 and Table 3 show that the absolute value of the statistic t is more than the two-tailed critical value 2.085963. That is to say, there is no significant difference between the average absolute value of the first-order autocorrelation efficient in the first post-event window period and the corresponding value in the pre-event window period while there is significant difference between the average absolute value of the first-order autocorrelation efficient in the second/third post-event window period and the corresponding value in the pre-event window period and the significant difference dramatically increases with the time. It is evident that the information efficiency of the underlying index constituent stocks does not obviously change within the first two months after the introduction of HSI futures but that dramatically improves within the later four months.

4 Conclusion

In the paper, the author adopts the similar event research method to hold the empirical research on the influence of the introduction of HSI futures on the information efficiency of its underlying index constituent stocks. The research results show that the information efficiency of the underlying index constituent stocks does not obviously change within the first two months after the introduction of HSI ex futures but that dramatically improves within the later four months. In conclusion, the introduction of stock index futures improves the information efficiency of stock market in general.

Acknowledgments. This paper was supported by Funding Project for Academic Human Resources Development in Institutions of Higher Learning Under the Jurisdiction of Beijing Municipality (PHR20110514) and Project for Construction of Disciplines and Specialties of Beijing Municipality (71M1110823).

References

1. Manaster, S., Rendleman, R.J.: Option Prices as Predictors of Equilibrium Stock Prices. Journal of Finance 37, 1043–1057 (1982)
2. Harris, L.: S&P 500 Cash Stock Price Volatilities. Journal of Finance 44, 1155–1176 (1989)
3. Antoniou, A., Holmes, P.: Futures Trading, Information and Spot Price Volatility: Evidence for the FTSE-100 Stock Index Futures Contract Using GARCH. Journal of Banking and Finance 19, 117–129 (1995)
4. Gulen, H., Mayhew, S.: Stock Index Futures Trading and Volatility in International Equity Markets. Journal of Futures Markets 20, 661–685 (2000)
5. Madhavan, A., Porter, D., Weaver, D.: Should Securities Markets be Transparency. Working Paper, Baruch College (1999)

Research and Application of License Plate Recognition System

Tao Zhang and Yong qi Qi

School of Mechanical Engineering
North China Institute of Water Conservancy and Hydroelectric Power
450011 Zhengzhou, China
osama1980@126.com, qiyongqi@ncwu.edu.cn

Abstract. License plate recognition (LPR) is the technique to draw the character information from the vehicle image and confirm the identity of the vehicle through using image manipulation, pattern recognition, statistical analysis, and so on. LPR consists of five steps: image gathering, image pre-processing, license plate location, character segmentation and character recognition. A method is proposed to realize the LPR based on MATLAB, comprehensively uses a number of ways to increase effective recognition of the system. The method solves the complex problem of license plate location, license plate tilt, and character extraction in the bad case of the noise and light conditions. The result shows that this system can work effectively, which help to orientate and identify license plate number.

Keywords: license plate recognition, image manipulation, license plate location, character recognition.

1 Introduction

The number of vehicles on the road continues to increase dramatically, which has led to heavy traffic congestion that can be blamed for the growing number of car accidents. The license plate number has unique information of a vehicle, so image-based license plate recognition (LPR) technology came out to be an important means of automatic vehicle identification. Now the LPR systems are widely and successfully used in highway surveillance, electronic toll collection, red-light violation enforcement, secure-access control at parking lots and identifying vehicles stolen, or those registered to fugitives, criminals and smugglers. So the key problem [1] for LPR is to improve performance of recognition and real-time.

MATLAB, the language of technical computing, is a programming environment for algorithm development, data analysis, visualization, and numeric computation. The paper use MATLAB image processing toolbox to implement the LPR algorithm based on the machine vision, and get good effects.

D. Jin and S. Lin (Eds.): Advances in ECWAC, Vol. 1, AISC 148, pp. 399–404.
springerlink.com © Springer-Verlag Berlin Heidelberg 2012

2 Design of LPR

LPR [2] is consisting of five parts: image gathering, image pre-processing, license plate location, character segmentation and character recognition. Image gathering module detects the arrival of vehicle and triggers the capture of vehicle image through CCD camera. Captured image is pre-processed by methods, such as binarization, gray-scale, median filter, edge detection, and so on. The study of system is the license plate, not the entire car image acquired from the previous step, so the license plate will be positioned out from the image. Character segmentation is to split seven characters in the located license plate one by one. At last, it is the step to recognize the segmented characters. The flow of LPR is pictured as Fig.1.

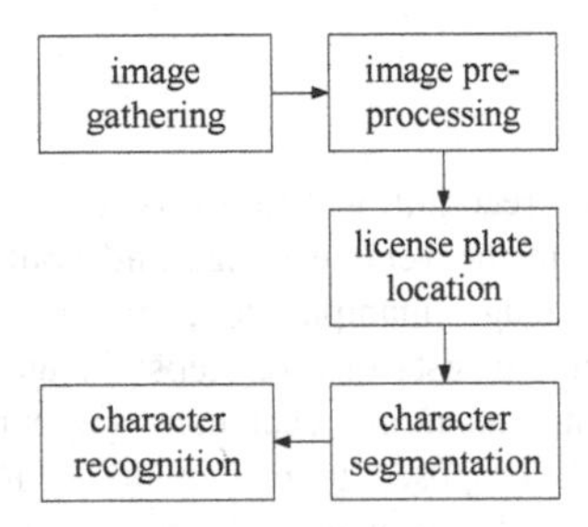

Fig. 1. Flow of LPR

3 License Plate Recognition Algorithm

3.1 Image Gathering

Image gathering is the first step of LPR. Extracted photo quality is directly related to accuracy of system identification, so it is the key to design appropriate vehicle image detection. Background difference [3], which can detect vehicles in the video and extract the trigger better image, is a widely used vehicle detection method. The method calculates the difference between real-time image and saved image to extract the difference image.

3.2 Image Pre-processing

In the process of form, transport or transform, it produces certain difference [4], which is affected by many factors, such as optical distortion, system noise, inadequate or excessive exposure, relative motion, etc., between collected image and original image. Therefore, in the image recognition must be preceded by pre-treatment, including gray-scale processing, image binarization, edge detection and filtering, and so on.

(1) gray-scale processing and binarization
Transforming color image to gray-scale image not only improves image processing speed, but also harmonizes the various license plate. Binarization algorithm is the key step for image recognition to get distinctly target figure and background by threshold.

Threshold is generated by experience or algorithm. The specific procedures are as follows.

```
I=imread('car.jpg');
I2 = rgb2gray(I);
fmax1=double(max(max(I2)));
fmin1=double(min(min(I2)));
level=(fmax1-(fmax1-fmin1)/3)/255;
I3=im2bw(I2,level);
```

(2)Median filter
In the image obtained from traffic environment, a large number of point-like or spike-like noises are appeared mostly due to weather. Furthermore, quickly shooting speed is liable to appear motion blur. Median filter is a sliding window from inside the gray value pixel sequencing, using a window center pixel value to replace the original gray value. Median filtering method can effectively remove the noise in the image point, while protecting the details of image edge.

(3) Edge detection
Edge detection [5] is an important step of image processing algorithms. Edges often occur at points where there is a large variation in the luminance values in the image, and consequently they often indicate the edges, or occluding boundaries, of the objects in a scene. Common edge detection operator: Roberts operator, Prewitt operator, Sobel operator, Canny operator. In the paper, canny operator is selected as a license plate image edge detection operator. Program format is "`I5=edge(I4, 'canny',[0.1,0.5])`". The effect of edge detection is shown as Fig.2.

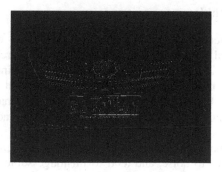

Fig. 2. Effect of edge detection

3.3 License Plate Location

Determination of license plate area is the key to the process of automatic identification. Due to the national standards, the license plate has a uniform specific size. After morphological processing, there are many connected areas in the vehicle image. The areas are marked by bwlabel function, and are calculated to obtain the information of length, width, height, area and aspect ratio. The marked area is filter by experience data to get license plate. The located license plate is shown as Fig.3.

Fig. 3. Located license plate

3.4 Character Segmentation

Character segmentation is to connect the module of the license plate location and character recognition. It split the license plate based on the location result, and provide to the next recognition module. Seen by PRC public safety industry standard GA36—2007.

The location of license plate is fixed, even the distance between characters, frame and rivet. So the plate templates are acquired by ratio algorithm. The image is scanned, upwards and downwards, from the middle of it and can be found separately the rows without character pixel point according to the digital region's ongoing connectivity. The determined rows are defined as the upper and lower bounds of each font. Then the sum of pixel value at the column direction is calculated to record the number of transition column, which is the left and right margins. Fig.4 illustrates the result of character segmentation.

Fig. 4. Character segmentation

3.5 Character Recognition

The methods of character recognition [6] mainly are based on template matching algorithms and artificial neural network. The former is to search for the most similar image pattern in the image with a template image obtained from prearranged information. The latter faces the problem of selection of input data and optimization of network parameters. Experiments show that the template matching, which has the features of high recognition ratio, works well for damage character. So the paper uses template matching method for character recognition.

Common template matching method based on binary recognition has a more quick speed, while is sensitive to noise, and tend to misjudge some similar character. For solve the problem, the template matching method with four gray-scale weighting function is provided in the paper. The method can ensure the accuracy and stability of character recognition, and has well recognition speed. Four-gray-scale is to normalize template and license plate and to transform image includes four gray-scales. Through the analysis for the actual collected license plate, there is obvious boundary in character gray value, background gray value, gray value between character and background. Suppose the template is a matrix of $n \times m$, the gray values of pixel corresponding to the region in template and match plate are described as $gm\times n$ and $fm\times n$, the similar matrix between template and match plate is defined as $sm\times n$, the above matrix are as followed:

$$G_{m\times n}=\begin{pmatrix} g(1,1) & \dots & g(1,n) \\ \vdots & \ddots & \vdots \\ g(m,1) & \cdots & g(m,n) \end{pmatrix}, g(i,j)\in\{A,B,C,D\} \tag{1}$$

$$F_{m\times n}=\begin{pmatrix} f(1,1) & \dots & f(1,n) \\ \vdots & \ddots & \vdots \\ f(m,1) & \cdots & f(m,n) \end{pmatrix}, f(i,j)\in\{A,B,C,D\} \tag{2}$$

$$S_{m\times n}=\begin{pmatrix} s(1,1) & \dots & s(1,n) \\ \vdots & \ddots & \vdots \\ s(m,1) & \cdots & s(m,n) \end{pmatrix}=G\oplus F=\begin{pmatrix} g(1,1)\oplus f(1,1) & \dots & g(1,n)\oplus f(1,n) \\ \vdots & \ddots & \vdots \\ g(m,1)\oplus f(m,1) & \cdots & g(m,n)\oplus f(m,n) \end{pmatrix} \tag{3}$$

The similar function is adjusted by formula (4):

$$M=\frac{\sum_{i=1}^{m}\sum_{j=1}^{n}s(i,j)}{9\times m\times n} \tag{4}$$

The operation " $\oplus$ " is practically weight value. Weight value is a reasonable reflection to the relation between character, background, character border and background border. Fig.5 shows the final results of license plate recognition.

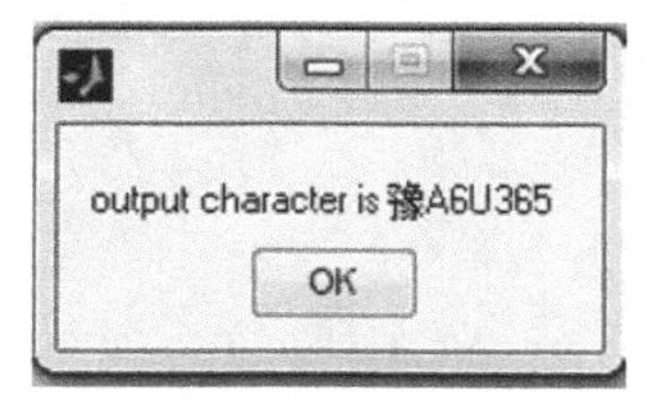

Fig. 5. Output of character

4 Conclusion

Experimental results show that the system can completes well plate position, character segmentation and character recognition. With traditional use of C++ languages, the method reduces much workload and development cycle. In practical application, the recognition rate of LPR is closely related to the license plate quality and shooting quality. Now, we are still conducting extensive research and developing more widely adaptable algorithm to improve the performance of our LPR system.

References

1. Zhang, B.M., Yu, W.B., Yu, X.P.: A Survey of License Plate Location Technology. Journal of Dalian University 23, 6–12 (2002)
2. Lin, W.T.: Image Treatment and License Plates Identification. Journal of Shangdong University of Technology 11, 46–48 (2007)

3. Zhang, H.L., Li, K.P., Deng, T.M.: Study on Threshold Segmentation Algorithm in Video Vehicle Measurement. Journal of Highway and Transportation Research and Development 26, 116–119 (2009)
4. Jiang, S.M.: The Research on Image Processing of Vehicle License Plate. China Science and Technology Information 5, 264–265 (2009)
5. Ji, H., Sun, J.X., Shao, X.F.: The Algorithm for Image Edge Detection and Prospect. Computer Engineering and Applications 14, 70–73 (2004)
6. Gu, C.Q., Ge, W.C.: Character Recognition Based on Template Matching Method. Communications Technology 42, 220–222 (2009)

Applying BCH Error Correcting Code in Digital Watermark

Wei Wang

Engineering University of Chinese Armed Police Force, Xi'an, 710086, P.R. China

Abstract. This paper presents a watermarking algorithm with BCH error correcting code. Compared with the traditional digital watermarking algorithm, the new algorithm has more stronger ability of error correction for malice attack after embedding error correction in robust watermark.Thus it sharply enhances the solidity of the robust property of the watermark.

Keywords: BCH code, Digital Watermark.

1 Introduction

The rapid growth of digital technologies has made the security of internet increasingly important. Information hiding is a novel technology for embedding secret content in multimedia or text [3, 5–9]. Digital watermarking is an important branch of information hiding [2, 4]. It is an effective method for protecting copyright and tracing pirates. Essentially, watermarking is defined as the process of embedding secret information directly into digital multimedia (such as stationary image, audio, or video signal), and this secret information can not be perceptual by human perception system[11]. Moreover, it is difficult to remove or spoil the digital watermarking. In copyright protection application field, there are three requirements to digital watermarking system: the imperceptibility of the watermark within the host signal, the security of the watermark against attacks and the robustness to signal processing. Besides these basic requirements, the watermarking capacity should be as high as possible[1,10]. In this paper, we exploit the possibility of applying BCH error correction codes in watermarking. And as a result, a multipurpose image watermarking scheme based on BCH error correction has been designed. First it chooses a BCH generator polynomial g(x), then use g(x) and system code to get generator matrix G and check matrix H. Finally it gets the new watermark by original watermark multiplying generator matrix G. The result shows that, the new multipurpose image has stronger ability of error correction for malice attack after embedding error correction in robust watermark. Thus it sharply enhances the solidity of the robust property of the watermark.

2 BCH Error Correcting Code

The code error correction technology took coding theory as its foundation, developed along with the information society's development and the semiconductor Technology

D. Jin and S. Lin (Eds.): Advances in ECWAC, Vol. 1, AISC 148, pp. 405–410.
springerlink.com © Springer-Verlag Berlin Heidelberg 2012

popularization. It has been developed vigorously as one kind of coding technique; its main purpose is to enhance number system's reliability. Even if it could not complete the error correction work in some situations, it might around the reference factor estimate information and so on relations by reason of the code relevance. It also carries on certain wrong revision work. The code error correction technology's major technique thought is: the length of information is k, the information has been increased m wrong verification position or investigates the dislocation, and then it constitutes (k add m equals n) coded word to carry on the transmission and the use again in the data transmission or in the information record system.

2.1 Binary BCH Code

The BCH code is corrects many random errors about the cyclic codes, Its definition is as follows: Given any Galois field GF(q) and its expands territory GF(q^m), q is the prime number or the prime number power, m is some positive integer. If the element which is from GF(q) is a cyclic code, and the R is set of roots of its production multinomial g(x) .If R includes a continual root: $R \supseteq \{a^{m_0}, a^{m_0+1}, \ldots, a^{m_0+\delta-2}\}$, the cyclic code which is made by the g(x) production is called q BCH code. The BCH code's production multinomial and the code length respectively are:

$$\begin{aligned} g(x) &= LCM\,(m_0(x), m_1(x), \ldots, m_{\delta-2}(x)) \\ n &= LCM\,(e_0, e_1, \ldots, e_{\delta-2}) \end{aligned} \qquad (1\text{-}1)$$

$m_i(x)$ And e_i respectively are $a^{m_0+i} (i = 0,1,\ldots,\delta-2)$ the element smallest multinomial and the level. When q=2, it is the binary BCH code (the dual BCH code). BCH code's parameter is (n, k, t), n is the symbol length, which refers to code generally: $n = 2^m - 1$. K is the length of bit of code information before encoding, n-k is the length of the encoder adding on check bit's the length, t is error correction ability. The BCH code is subclass of the cyclic code, which product by the production multinomial g(x). A code's error correction ability decided completely by its minimum Hamming distance d_m , but d_m of BCH code is completely decided by g(x)'s root. g(x) is polynomial with the time is n-k in GF(2) territory. The BCH code's parameter is (n, k, t) which is not selects willfully. Different (n, k, t) corresponds different production multinomial g(x) separately.

It can show that the ability of BCH code error correction is t which affected by n of the code length after encoding, and by the k of the code information bit's length which hasn't encoding. When code length n is bigger, it may select t's scope of the error correction ability is bigger. When n determines, k is bigger and the t of error correction ability is smaller, otherwise is stronger. It will obtain check matrix G by obtaining production multinomial g(x) which completes the code, W(x)= m(x)g(x), so the transmission character W(x) has been obtained with error correction function.

Table 1. The parameter (n,k,t) and g(x) of BCH code

n	k	t	g(x)
7	4	1	13
15	11	1	23
15	7	2	721
15	5	3	2467
31	26	1	45
31	21	2	3551
31	16	3	107657
31	11	5	5423325
31	6	7	313365047
63	45	3	1701317
63	39	4	166623567
63	36	5	1033500423
63	30	6	157464165547
127	106	3	11554743
…	…	…	…

2.2 BCH Error Correcting Code

The basic philosophy of decoding about correcting the cyclic code of burst error is follow: it supposed that g(x) produce C which is cyclic code of (n, k). It supposed that C's symbol C(x) has been transmitted, the receive is R(x),

$$R(x) = C(x) + E(x) \tag{1-2}$$

E(x) is the wrong design. The decoder's definition:

$$S(x) = R(x) \bmod g(x) \tag{1-3}$$

To calculate surplus follows S(x), the vector which obtains by R(x) subtracts S(x) is

$$\hat{C}(x) = R(x) - S(x) \tag{1-4}$$

it is a symbol Certainly. This is because $\hat{C}(x) \bmod g(x) = R(x) \bmod g(x)$ $-S(x) \bmod g(x) = 0$. Therefore, if code C can correct all mistakes in the set, and S(x) belongs to the set, the decoder can certain reliably which is the actual transmission symbol, the reason is that the difference of other symbols and the R(x) hasn't exist in ε . This point establishes to any cyclic code.

3 The Design for BCH Error Correcting Code in Digital Watermark

In order to cause the robust watermark to be steadier in the digital watermarks, applying error correcting code in robust watermark only. When robust watermark product, the size of two value watermark image is N1×N2, it will been rearranged into 36×N3 .Uses the BCH code to reset the codes' the size to 63×N3 for two value sequence, the size of each line is 63, it is been expended to 64. It become must insert finally. We design a dual source BCH code which consists of a code which length is 63 and report which length is 36. It may correct error for 5 bits.

Reference table 1, The production multinomial g(x) is correspondence with the octal numeral which is 1033500423, polynomial form is: $g(x) = x^{27} + x^{22} + x^{21} + x^{19} + x^{18} + x^{17} + x^{15} + x^{8} + x^{4} + x + 1$. The systematic code generator matrix $G = [I_k p]$ may obtain by production multinomial g(x) according to the systematic code structure method; left side is $k \times k$ step unit matrix.

$$G = \begin{bmatrix} 1\ 0 & \cdots & \cdots & 0 & r_1(x) \\ 0\ 1 & \cdots & \cdots & 0 & r_2(x) \\ \cdots & \cdots & \cdots & & \cdots \\ 0\ 0 & \cdots & \cdots & 1 & r_k(x) \end{bmatrix}$$

$r_i \equiv x^{n-k} x^{k-i} \equiv x^{n-i} (\bmod g(x))$ And $i = 1, 2, \cdots, k$. It is equal that the information bit which coefficient from n-1 time to an n-k time, but other positions is verification position.

The robust watermark restores. Firstly, each line of 64 bits' watermark information intercept to each line of 63 bits. The watermark information will been decoded by its correction ability, the procedure is more complex opposition the code. The decode procedure divides into 4 steps:

(1). It is supposed that watermark is $C(x) = m(x)g(x)$ before inserting the watermark, the withdraws watermark is $R(x) = C(x) + E(x)$. It is Supposed that the wrong pattern is $E(x) = e_{n-1}x^{n-1} + e_{n-2}x^{n-2} + \cdots + e_1 x + e_0$,if the number of mistake is t, Then $E(x) = Y_t x^{l_t} + Y_{t-1} x^{l_{t-1}} + \cdots + Y_1 x^{l_1} = \sum_{i=1}^{t} Y_i x^{l_i}$,$Y_i \in GF(q)$, x^{l_i} is called the wrong position in the formula. It explained that the error occurs in the position $n - l_i$ of R(x) ,the wrong value is Y_i which value is 1 or 0 because it is binary BCH. According to withdraw's watermark information $R(x) = r_{n-1}x^{n-1} + r_{n-2}x^{n-2} + \cdots + r_1 x + r_0$, $n = 63$. S is follow type, if S is equal to 0, it explained that the watermark have not the wrong occurrence. If S is not equal to 0, using the type 4-5 method to solve.

According to theorem: to any positive integer m and t, they have certainly a binary BCH code, its root is $\alpha, \alpha^3, \cdots, \alpha^{2t-1}$, its code length $n = 2^m - 1$ or perhaps it is the factor of $2^m - 1$. It can correct t random error which length is t; the number of verification position is $\partial^\circ g(x) = mt$ at most. Therefore the root of BCH (63, 36, 5) is $\alpha, \alpha^3, \alpha^5, \alpha^7, \alpha^9$ on $GF(2)$. Get the follows type S by formula 4-5.

(2) By follows type S, it will extract the mistake position multinomial $\sigma(x) = 1 + \sigma_1 x + \sigma_2 x^2 + \cdots + \sigma_t x^t$ 。

(3) The root of the mistake position multinomial will be getting by researching method (the wrong position is E (X));

(4) According to $C(x) = R(x) - E(x)$ obtaining the correct symbol, because it is the systematic code, the front bit by interception figure is the information, so the decoding is been accomplished.

It has the slight attack; the capability of information which can resist attacking is improved greatly by using the error correction code, when the ability of BCH (63, 36, 5) which can correct the error code is in its scope.

$$S^T = H \bullet R^T = H \bullet E^T = \begin{bmatrix} (\alpha^{m_0})^{n-1} & (\alpha^{m_0})^{n-2} & \cdots & \alpha^{m_0} & 1 \\ (\alpha^{m_0+1})^{n-1} & (\alpha^{m_0+1})^{n-1} & \cdots & \alpha^{m_0+1} & 1 \\ \vdots & \vdots & & \vdots & \vdots \\ (\alpha^{m_0+2t-1})^{n-1} & (\alpha^{m_0+2t-1})^{n-2} & \cdots & \alpha^{m_0+2t-1} & 1 \end{bmatrix} \begin{bmatrix} 0 \\ \vdots \\ 0 \\ Y_t \\ \vdots \\ \vdots \\ Y_1 \\ 0 \\ \vdots \\ 0 \end{bmatrix} \tag{4-5}$$

4 Conclusion

Image watermarking scheme which imports BCH error correction has been designed. It uses the BCH to get generator polynomial g(x). Use the g(x) and system code to get generator matrix G and checkout matrix H. So it gets the watermark of error correction coding by original watermark multiplying generator matrix G. After study, the multipurpose image has the ability of error correction by malice attack after embedding error correction in robust watermark. So it enhances solidity of robust watermark.

Acknowledgement. This work is supported by the Natural Science Foundation of Engineering College of Chinese Armed Police Force under contract no. WJY201116and WJY201120.

References

1. Shen, M., Huang, J., Beadle, P.J.: Application of ICA to the Digital Image Watermarking, 1485–1488 (2003)
2. Tirkel, A.Z., Rankin, G.A., van Schyndel, R.M., Ho, W.J., Mee, N.R.A., Osborne, C.F.: Electronic Water Mark, DICTA 1993, Macquarie University, Sydney, pp. 666–672 (December 1993)
3. Anderson, R. (ed.): IH 1996. LNCS, vol. 1174. Springer, Heidelberg (1996)

4. van Schyndel, R.G., Tirkel, A.Z., Osborne, C.F.: A Digital Watermark. In: International Conference on Image Processing, vol. 2, pp. 86–90 (1994)
5. Aucsmith, D. (ed.): Proc. of the 2nd International Workshop on Information Hiding, IH 1998. LNCS, vol. 1525. Springer, Heidelberg (1998)
6. Pfitzmann, A. (ed.): Proc. of the 3rd International Workshop on Information Hiding, IH 1999. LNCS, vol. 1768. Springer, Heidelberg (2000)
7. Moskowitz, I.S. (ed.): Proc. of the 4th International Workshop on Information Hiding, IH 2001. LNCS, vol. 2137. Springer, Heidelberg (2001)
8. Petitcolas, F.A.P. (ed.): Proc. of the 5th International Workshop on Information Hiding, IH 2002. LNCS, vol. 2578. Springer, Heidelberg (2003)
9. Fridrich, J. (ed.): Proc. of the 6th International Workshop on Information Hiding, IH 2004. LNCS, vol. 3200. Springer, Heidelberg (2004)
10. Barni, M., Herrera-Joancomartí, J., Katzenbeisser, S., Pérez-González, F. (eds.): Proc. of the 7th International Workshop on Information Hiding, IH 2005. LNCS, vol. 3727. Springer, Heidelberg (2005)
11. Mese, M., Vaidyanathan, P.P.: Recent advance in digital halftoning and inverse halftoning methods. IEEE Trans. on CAS 49(6), 790–805 (2001)
12. Podilchuk, C.I., Zeng, W.: Image-adaptive watermarking using visual models. IEEE Journal on Selected Areas in communications 16(4), 525–539 (1998)

Extraction of Interest Association Rule in Web-Based Education

Yang Shen, Shangqin Yang, Kuanmin Lu, and Yang Chang

School of Humanities, Economics and Law, Northwestern Polytechnical University, Xi'an, P.R. China
miniqueen@126.com

Abstract. As the rapid development of information and network technology, the Web-based education is playing an important role in universities of China. The object of Web-based education is human-being. Both the uncertainty and virtuality of web behavior make it a hard task to analyze the rules in education with a quantitative scientific approach. This paper presents an approach to extract interested association rules using Apriori algorithm. Based on data pretreatment of website visiting, this approach can generate association rules with the confidence of different interest points of the educatee from the frequent itemsets. Through a case study of the Web-based education in Northwestern Polytechnical University (NWPU), some convincing recommendations for innovations of the Web-based education in universities are also discussed in the end.

Keywords: Web-based Education, Data Mining, Interest Point, Association Rules, Apriori algorithm.

1 Introduction

As the progress of the network technology, the Internet has become a very important part of daily life, and the Web-based information has become a main source acquired by people. Since the 1990s, both the faculty and the researchers of education in universities have carried out various ways of innovations for education under the Web-based environment. It is known to all that the internet provides an effective communication platform for students and teachers, and makes the traditional education mode revolutionarily changed. Currently, there is increasing number of new types of social network popular with students, such as BBS, Blog and E-Mail Box etc. It is impossible for the educators to use all the new approaches to cater to all the students' interest. The educators should find out the association rules of students' interest, and provide more interesting information by internet in order to make the education more effectively.

How to find out the association rules of the students' interest? The famous "diapers and beer" case of Wal-mart inspires the writer: diapers and beer are irrelevant products, if not using data mining technology, but Wal-mart supermarket makes all beer and

D. Jin and S. Lin (Eds.): Advances in ECWAC, Vol. 1, AISC 148, pp. 411–416.
springerlink.com © Springer-Verlag Berlin Heidelberg 2012

diapers placed nearby, according to the association rules, and greatly improves the sales.

Compare the Web-based education in universities as a supermarket: when students log onto the internet and visit the website, we recognize that a "product" has been sold, and record it. In order to obtain the association rules of the students' interest, data mining provides a reference with decision-making. The main idea of this paper is to calculate the website visiting quantity at first. Then, qualify the relationship among these websites with the Apriori algorithm. Finally, create the dynamic database with the association rules to support the development of the Web-based education.

2 A Brief Introduction of the Apriori Algorithm

The Apriori Algorithm is the most well known association rule algorithm. The basic idea of the Apriori algorithm is to generate candidate itemsets of a particular size and then scan the database to count these to see if they are large. During scan i, candidates of size i, C_i are counted. Only those candidates that are large are used to generate candidate for the next pass. That is L_i are used to generate C_{i+1} .An itemset is considered as a candidate if all its subsets also are large. To generate candidates of size $i+1$, joins are made of large itemsets found in the precious pass.

3 A Case Study on Data Mining of University Web-Based Education in China

Interest is one of the irrational factors of human-beings, which reflects people's inner tendency to actively explore or to understand or to be interested in something. In Web-based education system, educators, educatee and the Internet will rely and effect to each other. We should make full use of the interest as an irrational factor. So it is important for the educators to capture the students' interests from the massive information on the internet.

Here, we study on a case of data mining, and find out the association rules of the potential interest points of the educatee in Web-based education. In this empirical study, we choose the most typical examples of the 20 candidate interest points as the samples, shown in Table 1.

Table 1. Samples of the interest points

	Internet column	Meaning and content
I1	Experts' Viewpoint	Viewpoint and analysis of experts on social issues and the current events.
I2	National Defense Education	Military situation, patriotism education, and the international situation.
I3	Law Online	Popularize an elementary knowledge of law.
I4	Campus News	Current affairs in university.
I5	Character Interview	Interview of Famous characters, such as the Nobel Prize winners, Academicians and sociologists etc.
I6	Celebrity Reports	Video or records of famous characters' public lecture.
I7	Wonderful Association	News and activities of students' associations, such as donations, volunteer services etc.
I8	Popularization of Science	Knowledge of animals, plants, information technology etc.
I9	Youth Literature	Description of the reality in the perspective of the youth.
I10	Golden Campus	Photos or videos of the campus or other Tourist attractions.
I11	Brightness of aeronautics, astronautics, and navigation	Aeronautics, astronautics, and marine technology education.
I12	Job Hunting Guide	Guidance for students to improve themselves in interviewing competition.
I13	Study Abroad Consultant	Guidance for students to study abroad.
I14	Academic Science	Academic knowledge on science, engineering etc.
I15	the Magpies Bridge	The platform for express the feeling of love of students.
I16	Video Graphic	Wonderful video and graphic records of good memories of students' various life.
I17	Psychological Communication	Platform to communication specially on the psychological field.
I18	Party School	Knowledge training and history education of the Communist Party.
I19	President' Email-box	Platform for students to communicate with the presidents about any issues of university.
I20	Good Campaign	Column of excellent students who win the Prizes and encourage others.

We first get the visiting volume of each Internet column in a fixed period (one week in the case). Using the Apriori method to generate candidate itemsets and frequent itemsets, and finally get the association rules for the visitors' interest, with respective confidence.And the flow chart of the data mining is shown in Figure1.

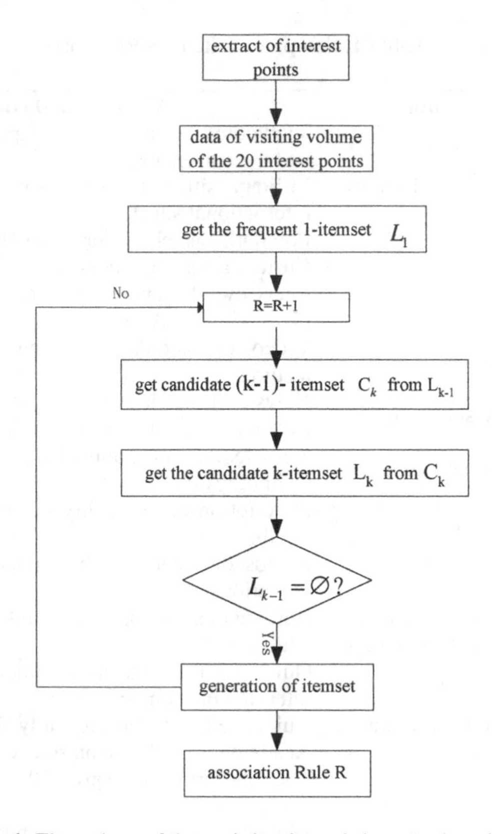

Fig. 1. Flow chart of data mining in web-based education

Setting the minimum $confidence = 60\%$, we get the association rules R as follows:

$$I1 \wedge I15 \rightarrow I6 \quad confidence = 100\%$$
$$I6 \wedge I15 \rightarrow I1 \quad confidence = 100\%$$
$$I15 \rightarrow I1 \wedge I6 \quad confidence = 100\%$$

Applying the data mining method into the education system, a manual process of result assessment will be necessary, because each of the results from the Apriori algorithm will not be valid or truthful to the real situation. The database of association rules shown as Fig.2 should be established to record all the valid rules after deleting the invalid ones. In the case, the administrators and teachers manually judge the output rules and screen the invalid ones according to their experiences. Meanwhile they study the different cases with Ethics and Behavior Science, in order to master the development rules of students' interest, and finally cultivate the healthy mentality and human dignity for them.

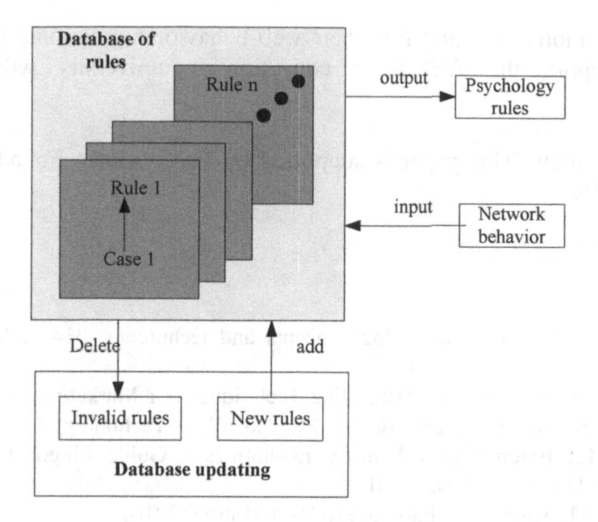

Fig. 2. Database of association rules in Web-based education

4 Analysis and Discussion

In the above case, by the mined association rules R with its confidence 100%, we can deduce a Web-based behavior track like this: when the students log onto the Web-based system of education, if they visit the Expert Viewpoint (I1), they must visit the Magpies Bridge (I15) and the Celebrity Reports (I6).Thus we try to figure out the inner world of the students through the above track. Desiring for a true love, looking forward to listening to the celebrity's reports and getting the experts' viewpoints and guidance for their whole life would be the most important activities for the youth.

In the case above, this is just one of the many association rules by the mining method. If data mining is populated in the Web-based system in universities of China, as well as the database is well constructed, we can record all the association rules by which we can get the dynamic behavior rules, analyzes the students' thoughts and minds by the internet-based activities. That will be quite beneficial for the Web-based education.

5 Conclusion and Evaluation

There is close relationship between the interest-point and education. It is very helpful to find the interest and its intrinsic association rules for the education. We should pay more attention to explore students' interest, and cultivate students' favorable interest through the Web-based so as to give full play of the positive role of the interest as an irrational factor in the education. This paper introduced the idea of data mining to the empirical research on the Web-based education with the case analysis of NWPU. The case proves that the data mining technology application in Web-based education system can provide several useful invisible rules of the students' interest and behavior on the internet, and is very helpful for the educators to understand the students' Web psychological motivation and find their web-behavior regulations. With the reliable technical support, the Web-based education in university will be optimized dynamically.

Acknowledgement. This paper is supported by Postgraduate Foundation of NWPU (No. Z2011128).

References

[1] Han, J., Kamber, M.: Data Ming concepts and techniques, 314 p. Morgan Kaufmann Publishers
[2] Berry, M.J.A., Linoff, G.S.: Data Ming Techniques For Marketing, Sales, and Customer Relationship Manaement, 2nd edn., pp. 87–120. Wiley Technology Publishing
[3] Vogel, H.L.: Entertainment Industry Economics:A Guide Fingor Financial Analysis. Cambridge University Press (2001)
[4] Stainbach, M., Kumar, V.: Introduce to Data Mining (2010)
[5] Dunham, M.H.: Southern Methodist University Data Mining Introductory and Advanced Topics. Pearson Education Inc.
[6] Fayyad, U., Grinstein, G.G., Wierse, A.: Information Visualization in Data Mining and Knowledge Discovery. Mogan Kaufmann Publishers
[7] Locatis, C.: PhD, National Library of Medicine,Designing Internet and Web-Based Education
[8] Macormack, C., Jones, D.: Building a Web-based Education System. Robert Ipsen Publisher

A Routing Algorithm in Wireless Sensor Network for Monitoring System of Machine Tools*

Qingqing Yang and Chunguang Han

School of Electronic and Information, Ningbo Dahongying University, Ningbo, China
yqq201@163.com, hancg66@163.com

Abstract. To decrease the delay of monitoring information on machining environment, a routing algorithm for the wireless monitoring system of machine tools is presented. The architecture of the monitoring system is designed and the details of mixed tree routing (MTR) are discussed according to the characteristic of the monitoring system of machine tools. Simulation results reveal that MTR not only outperforms tree routing(TR) in terms of hop-counts, but also is more energy-efficient than TR.

Keywords: Routing; Zigbee, Sensor network, Monitoring system.

1 Introduction

Recently, the application of the information technology into manufacturing fields gets more and more importance. Many studies reflect the effectiveness of monitoring and control system related to the exchange and management of various manufacturing information via Internet[1-3]. Using advanced multi-sensor monitoring and automation technology is an effective way to improving the quality of manufacture[4-5]. Over the years, various methods have been proposed to achieve machine tool condition monitoring, and recently wireless sensor networks have become highly popular in monitoring system. In this paper, a routing algorithm in wireless monitoring system of machine tools is presents according to the machining requirement.

2 The Architecture of Monitoring System

As shown in Fig.1, the monitoring system has been designed to provide users with a wireless and sensor-driven intuitive environment where distributed process planning, real-time monitoring and remote monitoring are undertaken. The system accomplishes machining information sharing using Intranet and Internet. Within the architecture, a wireless node device, connecting with sensors and machine tools, is used to monitor, track, compare, and analyze machining state parameters. The servers in shop floors enable remote real-time monitoring on the Zigbee wireless networks by acquiring and dealing with the process data. Utilizing Zigbee wireless network, the influence of

* Supported by the 2010 Sci-Tech Projects of Ningbo City(No. 2010A610124).

D. Jin and S. Lin (Eds.): Advances in ECWAC, Vol. 1, AISC 148, pp. 417–421.
springerlink.com © Springer-Verlag Berlin Heidelberg 2012

communication devices on machining environment is decreased enormously. Simultaneity, each server shares information and the processing data with a factory server by one-to-one socket communication on intranet.

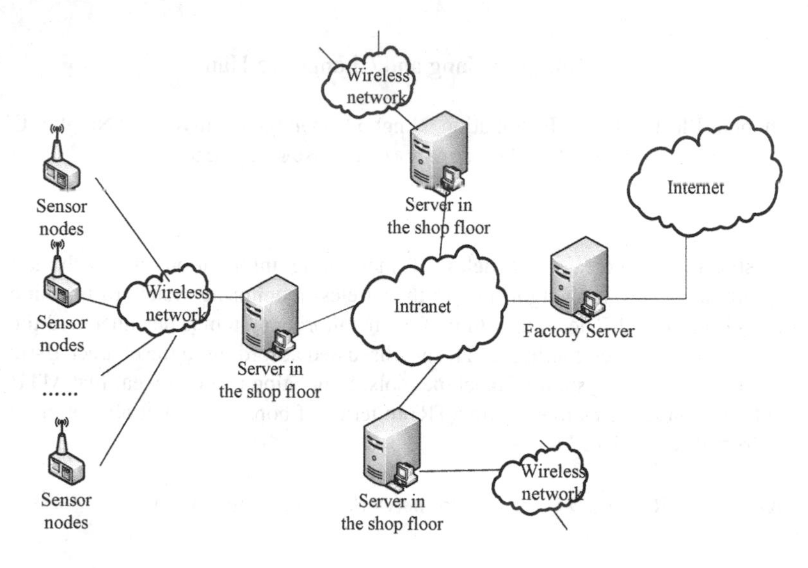

Fig. 1. System architecture

3 Routing Structure of Network

In a Zigbee network of tree structure, network addresses are assigned to devices by a distributed address assignment scheme. After forming a network, the Zigbee coordinator determines the maximum number of children (Cm) of a Zigbee router, the maximum number of child routers (Rm) of a parent node, and the depth of the network (Lm). Note that Cm≥Rm and a parent can have (Cm-Rm) end devices as its children. In this algorithm, their parents assign addresses of devices. For the coordinator, the whole address space is logically partitioned into Rm+1 blocks. The first Rm blocks are to be assigned to the coordinator's child routers and the last block is reversed for the coordinator's own child end devices. In this scheme, a parent device utilizes Cm, Rm, and Lm to compute a parameter called Cskip, which is used to compute the starting addresses of its children's address pools. The Cskip for the Zigbee coordinator or a router in depth d is defined as:

$$Cskip(d)=\begin{cases} 1+Cm\cdot(Lm-d-1) & \text{if } Rm=1 \\ \dfrac{1+Cm-Rm-Cm\cdot Rm^{Lm-d-1})}{1-Rm} & \text{Otherwise} \end{cases} \tag{1}$$

To begin with we will provide a brief background on the Tree route introduction. Tree routing (TR) is a simplified routing algorithm proposed for such networks. In TR, internode communication is restricted to parent-child links only. That is, while the network's physical topology is quite complex, the logical tree topology is used for data forwarding. By relying solely on the parent-child links, tree routing eliminates path searching and updating and, therefore, avoids extensive message exchanges associated with those procedures. TR is most suited for networks consisting of small-memory, low-power and low-complexity lightweight nodes. TR could also be used by a node at some operation stages such as when its battery supply is below certain threshold. The main drawback of TR is the increased hop-counts as compared with more sophisticated path search protocols.

TR is an extremely simple routing algorithm where a node only forwards packets to its parent or child nodes. TR avoids intensive message exchanges of path search/update processes and the overhead of storing routing tables or other information that is expensive to update.

4 Routing Algorithm

Enhanced Tree Routing. To facilitate peer-to-peer communication among the nodes, each node must have a unique identity called its network address or identification number. Due to the ad-hoc nature of the network topology, an address assignment scheme has to be in place to assign addresses to nodes when they join the network. In Zigbee network, each node must have a unique identity called its network address. An address assignment scheme is built to assign addresses to nodes whey they join the network. Each node in the network has neighbor table which records nodes' assignment address in its RF connection scope. An enhanced TR algorithm is designed utilizing information of the neighbor table and the structured address assignment schemes.

As shown in Fig.2, each node has a network depth in TR. The node's network depth equals its parent's network depth plus 1. For example, the network depth of root node is 0 and its child node's network depth is 1. In TR, A common parent node of any two nodes exists. The first same ancestor (FSA) node is actually the first node where the two node-to-root TR routes meet. In other words, For two nodes, the FSA node is their common ancestor node with the highest network depth[6].

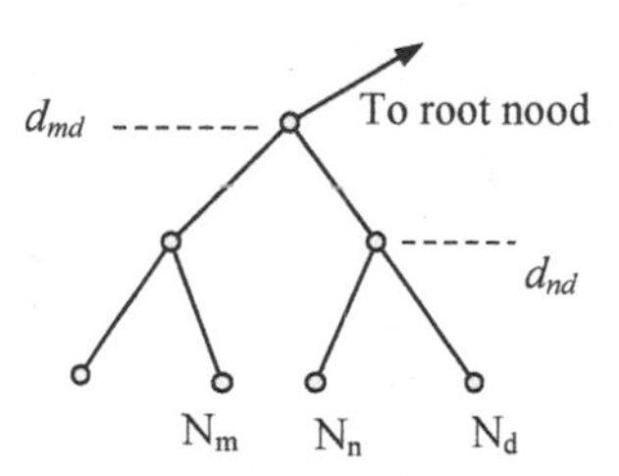

Fig. 2. Nodes relationship

As shown in Fig.2, For a node Nm, Nn is the neighbor node of Nm and Nd is the destination node of Nm. The networks depth of node Nm is dm. the networks depth of node Nd is dd. The networks depth of FSA node of Nm and Nd is dmd. When a packet from a node travels upwards along the tree to the root, it will reach all its ancestors.

In TR, the hop-count from Nm to Nd is as follows:

$$Cstep_{tr}=d_m-d_{md}+d_d-d_{md} \tag{2}$$

While the route via Nn has a hop-count

$$Cstep_{etr}=1+d_n-d_{nd}+d_d-d_{nd} \tag{3}$$

The difference of Csteptr and Cstepetr is

$$\Delta_c=C_{TR}-C_{ETR}=d_m-d_n+2(d_{nd}-d_{md})-1 \tag{4}$$

If the condition

$$\Delta_c>0$$

holds, is means that

$$C_{TR}< C_{ETR}.$$

Therefore, the hop-count is decreaced if the Nn is next hop node.

Mixed Tree Routing. Although the hop-counts are decreased in ETR algorithm, the calculation cost is increased in each node. Therefore, the routing time is increased. To achieve the faster routing speed, a mixed approach to routing is designed. The routing algorithm is used according to the neighbor table of routing node. The detail is as follows:

The number of interlaced tree is calculated by the neighbor table. If it greater than a given threshold, the ETR algorithm is adopted. Otherwise, the TR algorithm is adopted.

5 Simulation

A wireless network of temperature monitoring system of machine tools is simulated by Matlab program. The sensor node transmits the temperature information to IPC

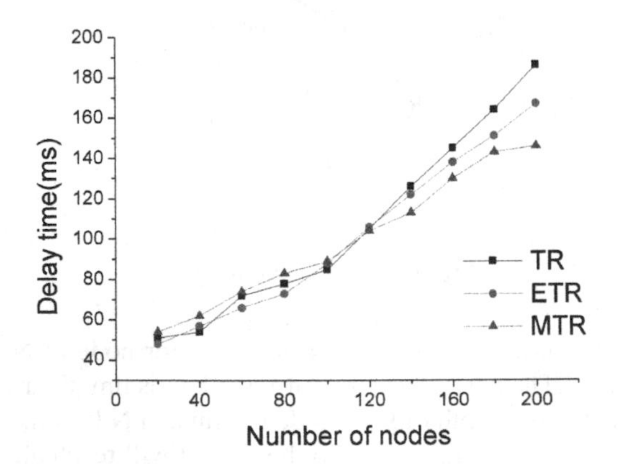

Fig. 3. Simulation result

(Industrial Personal Computer) via wireless sensor network. The routing algorithms including TR, ETR and MTR are adopted to the wireless network, respectively. The comparisons of routing algorithms are shown in Fig.3. While the number of nodes is less than about 120, the delay time of three algorithms is almost no difference. The delay of TR algorithm is greater than ETR and MTR algorithm while the number of nodes reaches 120. With the increasing of number of nodes, the delay of ETR algorithm is greater than that of MTR. The MTR algorithm has a good performance on the network delay because it makes a balance between the cost of calculation and hop-counts.

6 Conclusions

This study develops a mixed tree routing for wireless monitoring system of machine tools. Utilizing the mixed tree routing, the delay of monitoring information on machining environment is decreased enormously. The experiment results indicate that the mix routing algorithm is an effective approach in the wireless monitoring system of machine tools.

References

1. Xiaojun, G., Taiyong, W., Xianchang, R., et al.: Study on Technology of Network Monitoring in NC Manufacturing Process. Modular Machine Tool & Automatic Manufacturing Technique 9, 56–62 (2008)
2. Bong-cheol, S., Gun-hee, K., Jin-hwa, C., et al.: A Web-based Machining Process Monitoring System for E-manufacturing Implementation. Journal of Zhejiang University Science 7(9), 1467–1473 (2006)
3. Steven, Y.L., Rogelio, L.H., Robert, G.L.: Machining Process Monitoring and Control: The State-of-the-Art. Journal of Manufacturing Science and Engineering 126(5), 297–310 (2004)
4. Lihui, W., Weiming, S., Sherman, L.: Wise-ShopFloor: A Web-based and Sensor-driven E-shop Floor. Journal of Computing and Information Science in Engineering 4(1), 56–60 (2004)
5. Kang, Y.G., Wang, Z., Li, R., Jiang, C.: A Fixture Design System for Networked Manufacturing. International Journal of Computer Integrated Manufacturing 20(2), 143–159 (2007)
6. Wanzhi, Q., Skafidas, E., Peng, H.: Enhanced Tree Routing for Wireless Sensor Networks. Ad Hoc Networks 7(3), 638–650 (2009)

The Design of Remote Automatic Meter-Reading System's Communicational Reliability

BingYun Qian[1] and Song Zhu

[1] Huaibei Vocational and Technical College, Huaibei, Anhui, China.
hbqianbingyun@163.com

Abstract. RS-485 communication mode is widely used in remote automatic meter-reading system. In practical application, no reliance can be placed on the transmitting-receiving communication data. What's more, under the mode of multi-machine communication, the fault of one panel point may lead to the breakdown of the whole system of the communication structure. Thus there may be difficulties in identifying the faults. Therefore, effective measures should be taken in the software and hardware design for RS-485 bus interface in meter-reading system. In this way, the reliability and stability of the network system can be improved.

Keywords: AMR, RS-485 bus, reliability.

1 Remote Automatic Meter-Reading System

ARM is a system to automatically collect and record the electricity-using conditions of clients of the electric power system, and to automatically conduct the data and information back to the power management center. The system processes the data and realizes the function of power-charging and management[2]. AMR is a typical collecting and distributing control system, which involves the following parts: center computer, meter-reading concentrator, collecting terminal, and electricity meter of the clients (electronic meter or pulse meter). It carries out the communication through the data port of the clients' meters, the collection of terminal, and concentrator. Then, it receives the orders issued from the center computer. Finally, thc data is to be retransmitted to the remote center computer through the concentrator[3].

The remote automatic meter-reading system has the separate client-processing devices. Different clients work separately, so there are no interference among them, therefore no data lost or disorder will be caused. Thus, prompt meter-reading and data collection and accuracy can be ensured. This has a strong network control function.

2 The Communication Mode of the Remote Meter-Reading System

The selection and design of the communication mode is the key link of automatic meter-reading system. It is directly related to the performance of the system. Many aspects should be comprehensively considered: the target of the system, the

D. Jin and S. Lin (Eds.): Advances in ECWAC, Vol. 1, AISC 148, pp. 423–428.
springerlink.com © Springer-Verlag Berlin Heidelberg 2012

distribution of clients, the amount of clients, geographical conditions, the aim expected to achieve, the expansion and update of the system and its compatibility with other networks, etc.

At present, the communication mode of bus communication system is most widely adopted in the remote meter-reading system in China. This mode is represented by RS-485 bus. RS-485 is standardized CCITT. V.11/X.27compatible balanced electrical characteristic standard. This standard adopts integrated electric circuit. Differential signals are transmitted in a pair of balanced interconnection cables, and at the receiving end, differential receptor is used to judge the signals. This interface has the ability of controlling common-mode interference, so the performance of anti-interference is quite good. The frequency of signal-sending can reach as high as 10 MHz. When using the diagonal axis, the signal rate is smaller than 100 Kbps, and the transmission range can reach 1200m. The RS-485 interface can carry out half-duplex communication in one channel, so only two wires are needed to achieve two-way communication, and the intercommunication network of one spot-to-multispot or multispot-to-multispot can be easily formed. The amount of panel point varies based on the interface drive chip. At most there can be 128 panel points, with high data transmission quality and fast transmission speed[6]. The communication network structure is to be shown in figure 1.

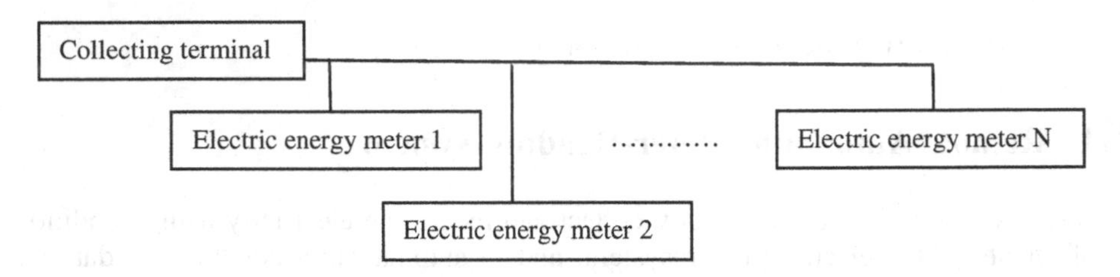

Fig. 1. RS-485 Bus Network Structure Diagram

The collection terminal internal chip adopts the storage manufacturing technology of high density, and it is not easily lost. The technology is based on 8051 with CMOS 8 microcontroller SST89C58. The data collected by collection terminal is transmitted to the concentrator through RS-485. The concentrator collects terminal according to the address resolution, and stores into the working chip AT89C51 in the concentrator. ROM and RAM in AT89C51 are separately used for storing program and data. There are maintenance-free powers in collecting equipment in order to make sure of the work after the power-off, protect data, accept and executive orders from position machine. AT89C51 realizes the wireless communication with position machine through the connection between serial port and the module of the radio transmitting-receiving[1].

However, if the interface circuit of RS-485 bus is designed simply according to the routine pattern, some problems may be caused by some reasons: there may be too many load equipments in practical application; wire and diameter are too small; impedance does not meet a criterion; and the quality of the converter is poor. The problems are as follows: the transmitting and receiving of the communication data is not reliable; under the mode of multicomputer communication, the fault of one panel point may lead to the breakdown of the communication structure of the whole system, and thus may bring difficulty in investigating faults. Therefore, in meter-reading system, effective measures are taken to improve the software and hardware design in RS-485 bus interface. Thus the reliability and stability of the network system can be improved.

3 The Reliable Design of RS-485 Bus

3.1 The Design of RS-485 Bus Interface Circuit

3.1.1 Choosing SN75LBC184 Interface Chip[4,5]

To choose SN75LBC184 as the interface chip of RS-485, and the wiring principle is to be shown in figure 2. It adopts the single power Vcc, and it can work normally with the voltage range of +3～+5.5V. It can not only fight against surge from the thunder and lightning, but also endure electrostatic discharge surge as high as 8KV. In the chip, there are 4 transient overload protecting tube integrated, and it can endure transient pulse voltage as high as 400V. When the input end opens circuit, the output is high level. This can ensure that when there are open-circuit faults in receptor input-end cable, normal work of the system will not be affected. In addition, its input impedance is the twice of the standard input impedance of RS-485(≧24kΩ), so 64 transceivers can be connected in the bus. Setting a linear slope drive in the chip may help to prevent the output signal edge from being too steep, and there will not be too much high-frequency component in the transmission line. Thus, electromagnetic interference (EMI) can be effectively stopped/held back. When designing circuit, we should try to make sure that under the circumstance of power-on reset of the system, the electric potential of end DE of 75LBC184 is 0.

3.1.2 Setting Up Electric Isolation

AT89C51 monolithic processor asynchronous communication interface and outward chip 75LBC184 realize the electric isolation through 3 photoelectric couplers TLP521, which can help improve the reliability of the system. When the P1.6 end of 89C51 is 0, the photoelectric coupler outputs high level (+5V), so the DE end of 75LBC184 chip should be selected, and transmission should be permitted; when the P1.6 end of 89C15 is 1, photoelectric coupler outputs low level, and so the RE end of chip 75LBC184 should be selected, and reception should be permitted. In actual circuit design, the adoption of high-speed optical coupling can also be taken into consideration. For example chip 6N137, 6N136, etc. Or we may optimize the circuit parameter of TLP521 (the specific value can be fixed through experiments) to make it work in optimum state.

3.1.3 The Design of 485 Bus Output Circuit

In the design of output circuit, besides considering the driver design of 75LBC184 as the way of linear slope output, we should also pay attention that it can fight against the surge from thunder and lightning and electrostatic discharge. So it has the strong ability of anti-jamming and anti-overvoltage. Also we set signal limiter diode VD1～VD4, the voltage-regulation value of VD1 and VD3 is 12V, VD2 and VD4 is 7V.

This is to make sure that the signal amplitude is limited between -7~+12V, and the ability of anti-overvoltage can be further heightened. In addition, in order to protect the chip RS-48 of a panel point in the system from being broken down, and thus affect the communication of other panel points, two 20Ω resistance R1 and R2 are installed in series in the signal output end of 75LBC184; considering that the characteristic impedance of unshielded twisted pair is about 120Ω, at the top and end of network transmission line RS-485 are separately connected with a matching resistance of 120Ω in order to reduce the reflection of transmission signals in the circuit; if pull-up resistor R4 and pull-down resistor R5 are added to output end A and B in circuit 485, the electric potential of end A is always higher than that of end B, the electrical level of RXD is kept to show high level when the bus is not in transmission, so the monolithic processor will not be interrupted or broken off.

3.1.4 Setting Up Watchdog Circuit

Watchdog circuit MAX813L is to be set up when drop-dead halt or other faults occur in panel points. It can help automatically reset the program. At this time, it stops controlling the bus, which may make sure that the whole system will not be broken down because of one panel point fault.

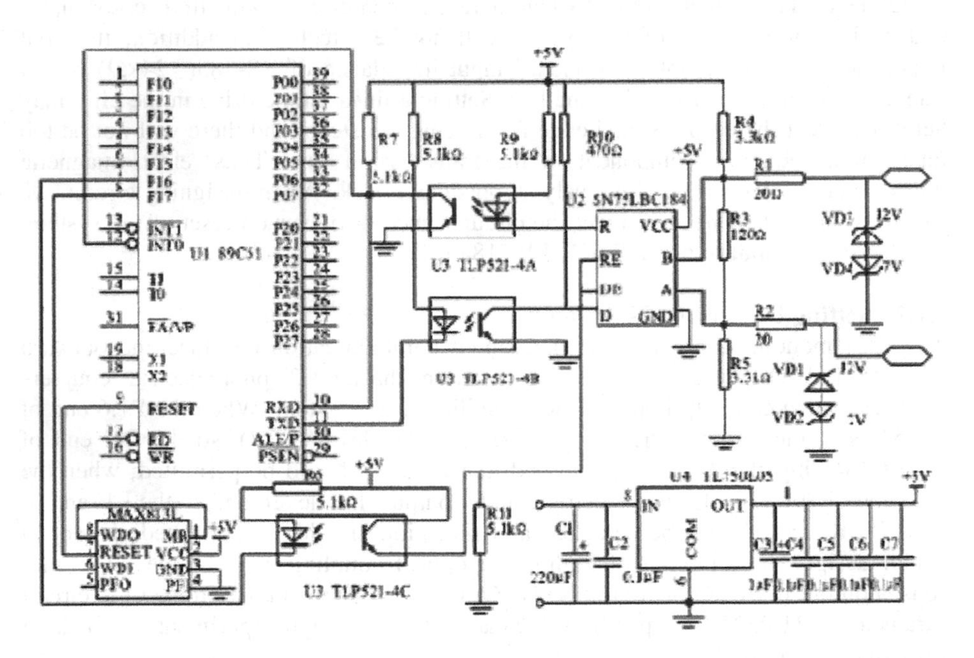

Fig. 2. Schematic Diagram of Improved RS-485 communication Interface

3.2 The Software Design of Bus Interface RS-485

The software design of interface RS-485 has great influence on the reliability of the system's communication. Because bus 485 is an asynchronous half-duplex communication bus, that is to say, at a certain time the bus can only present one condition. It generally applies to the communication of the inquiry mode from the host machine to the panel point, so a set of reasonable communication agreement to coordinate the share of time of the bus is needed. The system adopts data package communication mode, which means communication data send in frame and package. Each package is composed of guidance code, length code, address code, command code, content code and check code. The guidance code is used to synchronize the homing head of the data in each package; the length code is the total length of this package of data; the command code is the controlling order from the host computer to the panel point; the address code is the location number of the panel point; the content is the data information in this package of data; the check code is the efficacy symbol of this package of data. Parity check, "AND" check, and CRC check can be adopted[7].

In designing communication software RS-485, the programming of 485controlling end DE should be paid special attention to. In order to ensure the reliability of data transmitting and receiving, data should be transmitted and received once more when proper time-delay is added to state switching of bus 485. The specific method is to set DE to 1 under the state of data transmitting; after the time-delay of about 1 ms, effective data is to be transmitted. When one data packet is transmitted, delay another 1 ms, and then clear DE to 0. In this way, the bus is in a stable working process when switching the state.

4 Conclusion

After remote automatic meter-reading system has gone through the application of the above designs on communication software and hardware interface, great improvements have taken place in the reliability and stability of the system's communication. It can meet the real demands of the meter-reading system, and give the whole system the advantages of simple structure, convenient maintenance, low operating cost, wide covering range, etc. As it should be, the actual project must go through strict material selecting, standard construction and reasonable debugging. For example, multistrand shielded twisted pair should be adopted in wiring; when the communication distance is too long or load quantity is too much, repeater or 485HUB should be adopted, and the equipment interface should be ensured to be correct. All of these can guarantee the optimum performance of the system.

References

1. Xu, C.: Design of Communication Interface for Remote Meter-Reading System. Joumal of East China Jiaotong University 22(2) (2005)
2. Wang, X., Liu, W., Wang, H.: Research on Handshaking Srial Communication Protocol for Electrical Meter Rrading System. Measurement & Control Technology 25(3), 70–72 (2006)
3. Xu, F., Geng, X., Jia, Z.: Long-distance Meter Reading System Embedded GPRS Infinite Communication Module. The Design and Implementation of Electrical Automation (3) (2010)

4. Yang, B.: GPRS Technology Application in Power System Analysis. Telecommunications Engineering Technology and Standardization (2010)
5. Ren, Z., Wang, Y., Li, L.: GPRS-based Smart Meter Designs. Micro-computer Information (2007)
6. Zhang, Y.: GPRS-based Automatic Meter Reading System Design and Development. Ocean University of China (2010)
7. Liu, X.: Automatic Meter Reading Systems Research And Design. Huazhong University of Science (2008)

How to Balance the Motivators and Inhibitors for Consumers to Continue Using New Mobile Services

Mian Zhang

School of Management, Huazhong University of Science and Technology;
School of Business Administration, Hubei University of Economics, Wuhan 430205, China
zcmian@163.com

Abstract. Nowadays, we are in 3G (third generation) mobile network days, kinds of new mobile services and appications come into forth. Many customers pay close attention to mobile services, but only a part of them would choose to keep using mobile services. This study used the updated Delone and McLean IS success model to find the motivators and inhibitors to mobile services. We used information quality, system quality, and service quality as motivators, and information compultion and privacy concern as inhibitors, we find the relationship between these factors and intention to keep using new mobile services, and the the balance between these motivators and inhibitors.

Keywords: new mobile services, motivator, inhibitor, intention to keep using.

1 Introduction

Compared to all kinds of traditional Internet services, mobile services can provide special functionality regardless of users' access time and access position, and the strong relationship between an user and his or her mobile terminal makes spatial positioning and identification much easier in the mobile context than in the fixed environment of e-commerce[1].

In the earlier years, the core function of mobile phone is making telephone calls, but now, many mobile phone users considers the most important function of their mobile phone is not calling friends, but using kinds of mobile services and applications. Users perceive that they get more benefit in the process of using these new and personalized new mobile services, and lots of users intend to keep using these services.

But on the other hand, after a period of these new mobile services'using, some customers find there are some factors inhibit them to keep using mobile services. These causes the actual increment speed of mobile services customer amounts is lower than what mobile service providers estimated.

Most IS researches suggested that people should try to do their best to improve the motivating factors in IS, but we assume in new mobile context, there exists the balance between these motivating factors and inhibiting factors. So, the research question in this paper is: What are the balance between motivators and inhibitors for consumers to continue using new mobile services?

D. Jin and S. Lin (Eds.): Advances in ECWAC, Vol. 1, AISC 148, pp. 429–434.
springerlink.com © Springer-Verlag Berlin Heidelberg 2012

2 Literature Review and Hypothesis

The Delone and McLean IS success model is a widely cited framework in the IS literature that provides a comprehensive view of "IS success"..[2,3,4]

In this paper, we use the variable: information quality, system quality, and service quality as the motivators for customer's continue using new mobile services. As in the updated Delone and McLean IS success model, we can hypothesize:

$H_{1a\backslash b\backslash c}$: *Information quality\ system quality\ service quality of mobile service is positively related to an individual's intention to keep using mobile service.*

$H_{2a\backslash b\backslash c}$: *Information quality\ system quality\ service quality of mobile service is positively related to an individual's satisfaction toward mobile service.*

H_3: *The level of satisfaction with mobile service is positively associated with intention to continue using that mobile service.*

But, more sufficient, more timely, and more direct information the mobile service users obtain, they feel higher degree of information compulsion, which means the users are obliged to gain information when and where the users need not that information. Now new mobile service will also suffer from this problem, we can hypothesize:

H_4: *Information quality of mobile service is positively related to the information compulsion of mobile service.*

$H_{8a\backslash 9a}$: *Information compulsion of mobile service is negatively related to an individual's intention to keep using mobile service\ satisfaction toward mobile service.*

Similar as above, it is assumed that personalization is critically dependent on two factors: companies' ability to acquire and process consumer information, and consumers' willingness to share information and use personalization services[5]. So, despite the above benefits personalized services can provide to the customers, personalization requires the users to give up some of their personal information to their service provider, which raises privacy concerns, and creates an interesting tradeoff between personalization and privacy[6]. Thus, we can hypothesize:

H_5: *Service quality of mobile service is positively related to the privacy concern of mobile service.*

$H_{8b\backslash 9b}$: *Privacy concern of mobile service is negatively related to an individual's intention to keep using mobile service\ satisfaction toward mobile service.*

And if the user think mobile service is easy to use, and the technology is trustworthy, he is more likely to send his personal information to get various mobile services. So, after a period of new mobile services using, the service provider collect more and more personal information about himself, and the service provider may use the personal information in many ways, including sending information the user unwanted more frequently, then, the user would perceive higher level of information compultion in the mobile services keep using stage. Thus, we can hypothesize:

$H_{6\backslash 7}$: *System quality of mobile service is positively related to the information compultion\ privacy concern of mobile service.*

We proposed the research model in Figure 1.

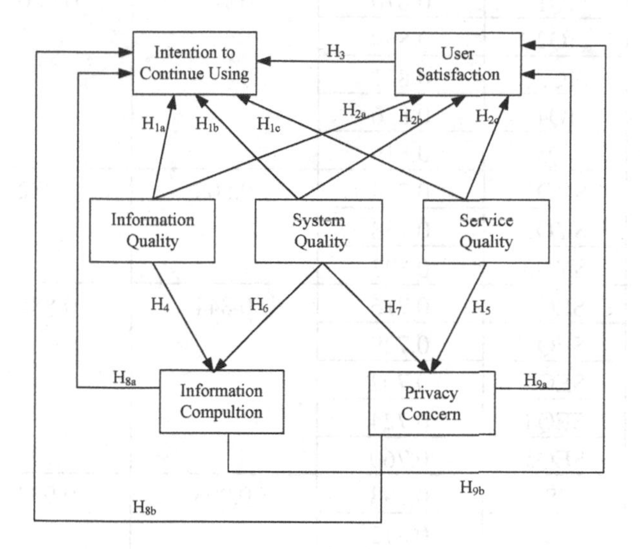

Fig. 1. Research model

3 Research Method and Results

The research model was empirically tested using data collected with a survey that included items for the constructs specified in the research model. The questionnaires use 7-point Likert scales, with responses choices ranging from one (strongly disagree) to seven (strongly agree). Participants needed to be familiar and have experience with typical mobile personalized services. We collect our empirical data for this study via a web-based survey. Of the 400 surveys sent, we received a total of 350 usable responses, the receive rate was 87.5%.

Partial Least Squares (PLS) was used to analyze the data in this paper. Table 1 provides the confirmatory factor analysis results. All the items exhibit high factor loadings (all above 0.70 criterion). And the average variance extracted (AVE) for each construct were above the 0.50 threshold. All items are significantly related to their specified constructs, the data support the convergent validity of the confirmatory factor analysis model.Discriminant validity was assessed by testing whether the correlations between pairs of construct items were significantly different from unity, where the results were shown in Table 2. And we could find all AVEs exceeded the scores of squared correlations, which confirmed discriminant validity.

Table 1. Confirmatory factor analysis results

Construct	Item	Factor	Cronbach's α	Construct	AVE
Information Quality	IQ1	0.870	0.948	0.960	0.829
	IQ2	0.830			
	IQ3	0.821			
	IQ4	0.856			
	IQ5	0.816			
System Quality	SYQ1	0.744	0.808	0.882	0.714
	SYQ2	0.738			
	SYQ3	0.823			
Service Quality	SEQ1	0.716	0.844	0.888	0.614
	SEQ2	0.729			
	SEQ3	0.702			
	SEQ4	0.724			
	SEQ5	0.760			
User Satisfaction	US1	0.798	0.905	0.934	0.779
	US2	0.812			
	US3	0.809			
	US4	0.748			
Intention to Keep Using	IN1	0.793	0.844	0.907	0.765
	IN2	0.767			
	IN3	0.765			
Information Compulsion	IC1	0.747	0.824	0.888	0.726
	IC2	0.755			
	IC3	0.747			
Privacy Concern	PC1	0.716	0.816	0.878	0.643
	PC2	0.760			
	PC3	0.738			
	PC4	0.709			

Table 2. The squared correlation scores of each construct and AVE

	IQ	SYQ	SEQ	US	IN	IC	PC
IQ	**0.829**						
SYQ	0.187	**0.714**					
SEQ	0.046	0.154	**0.614**				
US	0.092	0.066	0.205	**0.779**			
IN	0.122	0.069	0.198	0.226	**0.765**		
IC	0.048	0.133	0.182	0.059	0.078	**0.726**	
PC	0.010	0.175	0.042	0.239	0.339	0.090	**0.643**

Figure 2 and Table 4 presents the standardized paths and results for each hypothesis (* means $p<0.1$; ** means $p<0.05$; *** means $p<0.01$).

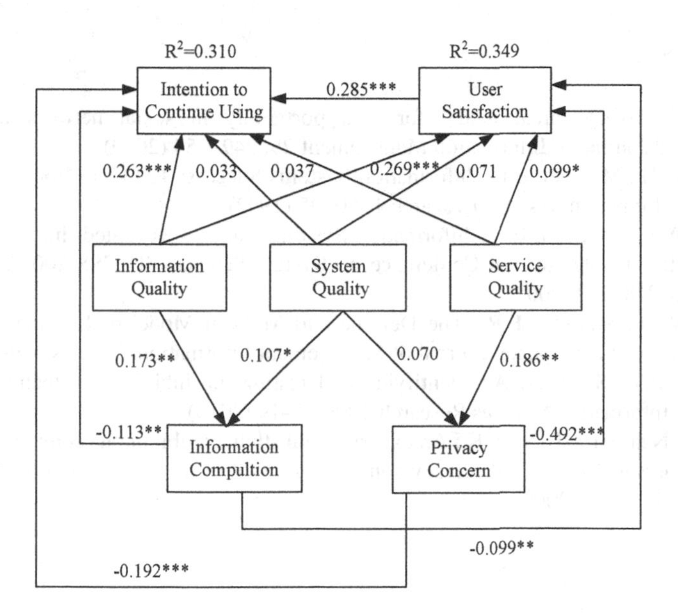

Fig. 2. Research results

4 Conclusion

This study used the updated Delone and McLean IS success model to find the motivaors and inhibitors to new mobile services. We found information quality do positively affect the mobile service users' satisfaction and intention to keep using that new mobile service, but system quality has no significant relationship with that, and service quality positively affect the mobile service users' satisfaction, but has no significant relationship with the users' intention to keep using that new mobile service, which is different with many results in many other research using IS success model.

Besides these motivaors, we found in the mobile service context, there is positive relationship between information quality and information compulsion, service quality and privacy concern, system quality and information compultion, but no significant relation ship between system quality and privacy concern. Information compulsion and privacy concern were used as inhibitors to new mobile service, we found the negative relationship between these inhibiting factors and user's satisfaction about new mobile service and the user's intention to keep using that mobile service.

Besides above implications, this research has limitations. First is potential sample bias, the majority of the participants lives in big cities in China, and weather there is differences in the mobile adoption process of other areas is our future research. Second is the R^2 value of user's satisfaction is 34.9%, and of user's intention to keep using is 31%, other variables are needed to explain part of the remaining variance.

Acknowledgement. This work was partially supported by the grants from the NSFC (70971049, 70731001) and a grant from NSFC/RGC (71061160505).

References

1. Weiss, S.: Privacy threat model for data portability in social network applications. International Journal of Information Management 29, 249–254 (2009)
2. DeLone, W.H., McLean, E.R.: Information Systems Success: The Quest for the Dependent Variable. Information Systems Research 1, 60–95 (1992)
3. DeLone, W.H., McLean, E.R.: Information Systems Success Revisited. In: Proceedings of the 35th Hawaii International Conference on System Sciences (HICSS 2002), Big Island, Hawaii, pp. 238–249 (2002)
4. DeLone, W.H., McLean, E.R.: The DeLone and McLean Model of Information Systems Success: A Ten-Year Update. Journal of Management Information Systems 3, 9–30 (2003)
5. Cenfetelli, R.T., Schwarz, A.: Identifying and testing the inhibitors of technology usage intentions. Information Systems Research 20(8), 1–19 (2010)
6. Sheng, H., Nah, F.F.-H., Siau, K.: An experimental study on ubiquitous commerce adoption: impact of personalization and privacy concerns. Journal of the Association for Information Systems 9, 344–376 (2008)

Public Sentiment Monitoring Method Based on the Combination of Semantic Network and Apriori Algorism

ChengQi Liu[1], Sui Luo[2], and JianFeng Xu[3,*]

[1] Network Center of Nanchang University, NanChang, Jiangxi 330031, China
[2] The Academy of Information Technology of Nanchang University, Nanchang 330031, China
[3] School of Software, NanChang University, NanChang, Jiangxi, 330047, China

Abstract. Apriori algorithm is a typical data mining method of association rules which can be used in data mining of the public sentiment monitoring. This paper brings out a new data mining algorithm, with combining semantic network's select function and Apriori Algorism's check function to quickly identify the events we are concerned with. We are going to use certain bbs's public opinion monitor as the example to prove this method.

Keywords: data mining, semantic network, Apriori Algorism, Sensitive Word-stock.

1 Introduction

Public sentiment monitoring is a convenient way for government departments, universities and large enterprises to get important information in time. Public sentiment monitoring aims at getting sensitive data inside the control unit itself or related events. This purpose requires two things: 1. Given the large amount of data to be processed, high processing efficiency is necessary; 2.To extract only the information we concerned. Traditional Apriori algorithm uses network data processing has its obvious shortcomings: 1. One has to scan the database for too many times which downlows the data processing efficiency; 2. Generated association rules may be an irrevalent one.Semantic network is characterized by the structural relationships of certain data item, which can be relatively fast to delete those we do not care about, while at the same time, it can not find the association rules among different groups. From this we can see that Apriori algorithm used alone or semantic networks are not fully qualified to monitor the public sentiment, but if we combine these two approaches, we can get a very good effect.

2 An Introduction to Apriori Algorithm

Apriori algorithm is one of the most influential method to dig into Boolean association rules of frequent itemsets. The association rules is divided into two parts. The first step is to retrieve a transaction database of all frequent itemsets through the

* Corresponding author.

D. Jin and S. Lin (Eds.): Advances in ECWAC, Vol. 1, AISC 148, pp. 435–440.
springerlink.com © Springer-Verlag Berlin Heidelberg 2012

iteration, and the support is not less than the threshold set by the user; the second step uses frequent itemsets to construct confidence to meet the minimum rules for the user. The specific approach is to find the frequent 1 - itemsets as L1; then use L1 to generate candidate set C2, the C2 can be used to determine the tap L2(the frequent 2 - item set); constantly so the cycle continues, until one can not find more frequent k-itemsets so far. Each excavation layer Lk needs to scan the entire database [1].

3 Semantic Network and Semantic Network Introduction

Semantic network is a knowledge representing system that imitates the human brain semantic memory, it its widely used in the expression of complex concepts and relationships between them so as to form a semantic network graph [2] composed by nodes and arcs. The specific performance of the methods can be divided into the following categories:

Wide contact is used to represent a class of generalized nodes and a more abstract class links between nodes, that is AKO (a kind of) . For example, "the sun belongs to the Milky Way."

Physical contact with the respective instance of the node used to represent the links between nodes, usually marked as ISA. For example, "Tom is a teacher."

Characterized Contact gathered for a certain individual, characteristics and the link between its components. For example, the relationship between the eye and the human body.

Property is used to represent individual links, the links between attributes and values. Usually directed arcs represent the property, with the arc pointing to the node of respective values. For example, "Jack, 25, male."

4 Data Mining Method Combined Apriori Algorithm and Semantic Networks

4.1 Semantic Web Modeling

4.1.1 The Introduction of Sensitive Word-Stock

Sensitive Word-stock is the lexicon of public safety, sex, violence, personal privacy etc that based on the current social situation. Sensitive Word-stock is usually used to shield the user's improper forum opinion, and in turn can be used to analyze whether an article or have a bad attempt to or worthy of our concern. As the word-stock is the one-dimensional storage structure, which is unable to demonstrate the relationship between each word, so in order to achieve a better analysis of results, we can establish the relationship between the words.in this paper, we use the approach of semantic networks to organize these sensitive words, to have a hierarchical and classificated effect.

4.1.2 Designed to Monitor Public Opinion, the Semantic Network Structure

We can organize a group of sensitive words into a hierarchical tree structure, marking each level a number to facilitate the positioning of these sensitive word. For example, in security incidents, suicide, violent crime, and many other sensitive words, suicide was divided into poison, run away, jumping, etc, which can be used to describe the organizational relationship to the diagram, Figure 1 shows This hierarchy.

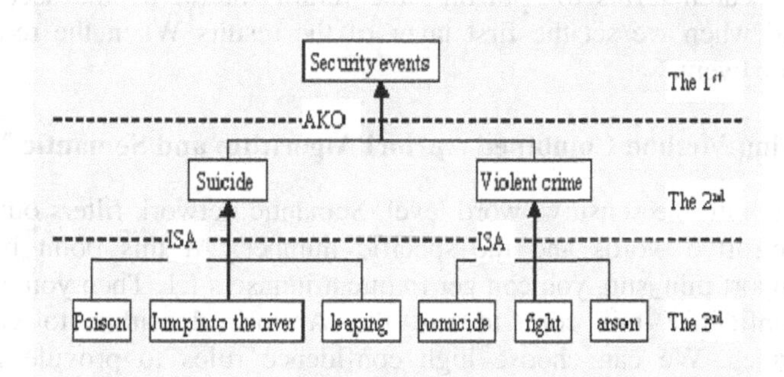

Fig. 1. Directed graph marking of security incidents sensitive words

4.1.3 The Storage Method of Semantic Network

There are many semantic network storage method, this paper uses the database storage. We store each sensitive word as a node to marked out its position in the tree structure and its relationship with the upper node. For a unit dedicated to monitoring systems of public opinion, it should be concerned about the influence of these large public influenced unit events, we can consider some words that raised servious results. For example, in the research of a company, we can put its name, organization, leadership as sensitive words. Some nodes marked as shown in Table 1 below:

Table 1. Storage method of sematic network

id	name	kind	level	logic	parent	flag
0	root	0	0	null	null	0
1	Security events	1	1	null	0	0
2	suicide	1	2	AKO	1	0

As the table shows, "root" represents root; "name" represents a new node and his name; "level" said node level; "kind" said node categories, for example, all the "security incidents" is a category; "logic" represents the logical node and its parent relationship; "parent" represents node's parent; "flag" represents the one which had been generated in association rules. If a rules item set contains the node, then this rule is a preserved one, if the rules here does not contain any sign of a flag node, we will discard this rule.

4.2 Semantic Web Search

Each single forum post, e-mail, articles has a complete content, those with obvious public opinions are set as a processing unit. In dealing with them, the first word needs to divide words, to remove repeat words (keep repeating the word count) and other unnecessary candidate word. Then, use the word processed to retrieve the semantic network. Taking into account the results to be retrieved for later use, so we need to category search results according to level, so you can check large about the number of candidates. For example, if we retrieve the final result set to the second level, when we enter the search terms to "poison", the return should be the second layer of "suicide"; and when we set the first layer of the results When the returned result should be "safe event."

4.3 Processing Method Combined Apriori Algorithm and Semantic Networks

By comparing with the sensitive word level, Semantic network filters out the articles containing sensitive words and the specific number. At this point by input the minimum support min_sup, you can get frequent itemsets L1. Then you can input the minimum confidence min_conf to use the Apriori algorithm to calculate the association rules. We can choose high confidence rules to provide for network monitoring to determine the sensitivity. Specific steps can be divided into three parts:

```
Step 1: using semantic net work to get the candidate option D
Input: Word after word to the complex set of C
Output: data set D
(1)  for each word wd ∈ C {//scan word set C
(2)  w=searchWord(wd,level);
(3)  if(w≠Φ) add w to D;
(4)  }
Function: searchWord
Input: Retrieval layer L; search term W;
Output: date recording word
(1)  Select w from database;
(2)  if exist word.name =w{
(3)  if (word.level≤L) return word;
(4)  else{
(5)  do while (word.level>L) {word=word.parent;}
(6)  Return word;
(7)  }}
(8)  if (word≠φ) return word;
(9)  else return null;
Step2: generate frequent itemsets
Input: data sets D, minimum support :min_sup
Output: all the frequent itemsets
(1)  L1 = find_frequent_1-itemsets(D);
(2)  for (k=2;Lk-1 ≠Φ ;k++) {
(3)  Ck = apriori_gen(Lk-1 ,min_sup);
(4)  for each transaction t ∈ D {//scan D for counts
(5)  Ct = subset(Ck,t);//get the subsets of t that are candidates
(6)  for each candidate c ∈ Ct
(7)  c.count++;
(8)  }
(9)  Lk ={c ∈ Ck|c.count≥min_sup}
```

(10) }
(11) return L= kU Lk;
Among which:

Function find_frequent_1-itemsets (D) is a database which does its work according to a given minimum support to find all frequent 1 item sets; function apriori_gen (Lk-1, min_sup) is to get candidate k Lk-1 project set Ck.
Step3: According to frequent itemsets generate association rules
Input: frequent itemsets L , the minimum confidence coefficient min_conf
Output: association rules
(1) Each frequent itemset l in L, generates l nonempty set;
(2) Each non-empty set s if l s of the support divided by the result of greater support equal to the minimum confidence, writing support (l) / support (s) ≥ min_conf, the output rule as "s => (ls)".
(3) For the output of the rules, keeping rules contains the rules of special symbolic items, that is the relevant rules and monitoring unit, independent from rules and monitoring unit discarded.

5 Analysis of Experiment and Performance

5.1 Experimental Environment and Data

In this paper, we take an enterprise for example to analysie the business-related news pages and forums. and 1000 news pages, forum posts and e-mails as sample. Because it is devoted to do the public opinion analysis, we need to establish common vocabulary sensitive to the semantic network by adding a dedicated corporate network. For example, the organization category contains the various departments of the enterprise and branch structure of the property; business leaders inside the company requires us to add the names of leaders at all levels; The sentive enterprise's name, category leadership which is set to 1 to improve the importance of rules generated.

5.2 Experimental Analysis

Experiment 1, using the Apriori algorithm to deal directly with the sample, due to the large set of frequent dimensions, the test operation is very large, and the sample has not been classified, the identification to sensitive topics is not high.

Experiment 2, by defining the specified sensitivity level, we use the semantic network to classify and reduce dimensionality, and then handled the results with the Apriori algorithm. This operation decreased the working time, the recognition rate of sensitive topics is largely improved. Table 2 lists the performance comparison of two methods.

Statistics proves that, in 1000 news pages, forum posts and email composition of the samples tested, there are 100 that we should pay attention to. By direct use Apriori algorithm to calculate the number we get 80 articles, and 68 of them are right; By the combination of Apriori algorithm and semantic network , we select the first three layers to get 91 sensitive samples, among which 86 are the right; By selecting layer 2, the calculated number comes to 93 , among which 90 of them are right.

Table 2. The comparsion of two experimental analysis

	Apriori algorithm	Combined Semantic Apriori algorithm level=3	The combintion of semantic network and Apriori algorithm level=2
The sample has been identificated as sensitive topics	80	91	93
The right number	68	86	90
The recognition rate of sensitive topics	85%	95%	96%

6 Conclusion

This paper proposes a data mining method which is combination of semantic network and Apriori algorithm. The method can generate more efficient and accuracte association rules by reducing frequent itemsets, and the Apriori computation. Experimental results show that this method can greatly improve the efficiency of monitoring public opinion.

References

1. Meng, S.: Research on Improving Apriori Algorithm for Mining Association Rules. Tianjin University, TianJin (2003)
2. Qiu, S., Li, Z., Wang, D.: Study on Semantic Web and Its Information Retrieval Mechanism on Web. Computer Engineering (23), 118–120 (2004)
3. Wu, C., Zuo, C.: The Research of Apriori Algorism on Association Rules Mining. Value Engineering (02), 194–195 (2010)
4. Carminati, B., Ferrari, E., Heatherly, R., Kantarcioglu, M., Thuraisingham, B.: Semantic web-based social network access control Source. Computers and Security 30(2-3), 108–115 (2011)

An Efficient Relocation Algorithm in Mobile Sensor Network Based on Improved Artificial Bee Colony

Wei Zheng* and Jian Shu

School of Software, NanChang HangKong University, NanChang 330063, China
zhengwei_nchu@126.com

Abstract. The effectiveness of distributed mobile sensor network depends on the coverage provided by the mobile sensor deployment. In order to maximize the effective coverage area and minimum the movement distance, an improved artificial bee colony (IABC) approach is proposed to relocate mobile sensors. New schemes of onlooker bees and scouts are presented to accelerate the convergence of the algorithm and avoid local optimization. Simulation results show that compared with other algorithms, the proposed relocation algorithm has better performances on effective coverage area and has a faster convergence. And the IABC also obtain the shortest movement distance.

Keywords: Mobile sensor network, Relocation algorithm, Artificial bee colony.

1 Introduction

Wireless sensor networks have recently emerged as an important research area [1,2]. Sensor networks are distributed networks composed of sensor nodes that have the ability of computation, memory and wireless communication. The effectiveness of sensor network is determined to a large extent by the coverage provided by the sensor deployment. The position of sensors affects coverage, communication cost, and resource management. So far there are two schemes to optimize the network topology, including deploying additional sensor nodes and adopting mobile sensors [3,4]. Mobile sensors have the capability of self-deploying and self-repairing, so it received more and more attentions [5].

Recently, researchers focus on using intelligent algorithms to improve effective coverage area in sensor network. In paper [6], a genetic algorithm-based approach is proposed to optimize mobile node relocation. The approach wants to find a balance between maximizing the effective coverage area and minimizing node number, and can get a good performance. A differential evolution-based relocation algorithm is presented in paper [7], and the algorithm can get a better performance than genetic algorithm -based approach.

Although the above intelligent algorithms can obtain good performances on mobile nodes relocation, these intelligent algorithms have some restrictions. Some control

* Wei Zheng, doctor, the lecturer in Nanchang hangkong university, his researches focus on network optimization and sensor network.

D. Jin and S. Lin (Eds.): Advances in ECWAC, Vol. 1, AISC 148, pp. 441–448.
springerlink.com © Springer-Verlag Berlin Heidelberg 2012

parameters have to be adjusted in all these intelligent algorithms, and these parameters depend on different application scenarios. And the above algorithms do not consider the movement distance of all mobile sensor nodes. In order to save sensor node energy, an efficient algorithm should be proposed. So we adopt artificial bee colony algorithm which is better than to these intelligent algorithms and we can save sensor energy [8,9].

Dervis Karaboga developed a novel model for artificial bee colony with three types of bees: employed bees, onlooker bees and scouts. The latter algorithm has a better or similar performance than other population-based intelligent algorithms and it employs fewer control parameters. In this paper, we propose a new improved artificial bee colony algorithm to solve the problem of relocating in mobile sensor network. In our approach, a new scheme is proposed for onlooker bees to select food resource based on employed bees' information. The onlooker bee adopts pseudo-random proportional rule to choose food resource, so that the algorithm can converge quicker. Meantime, we take a new updating scheme for scouts. In each round of iteration, artificial bee colony eliminates the bee with worst solution and sends a scout to find new food resource, so that the algorithm can avoid local optimization.

2 Preliminaries

We consider the sensor field as a m by n field grid, and assume that there are N sensors deployed in the random deployment stage. Each sensor has a detection rang r. Assume sensor s_i is deployed at point (x_i,y_i). For any grid point p at (x,y), we denote the distance between s_i and p as $d(s_i, p)$, i.e. $d(s_i,p)=\sqrt{(x_i-x)^2+(y_i-y)^2}$. Eq. 1 shows the binary model that expresses the coverage $C(s_i, p)$ of a grid point p by sensor s_i.

$$C(s_i,p)=\begin{cases}1, & if\ \ d(s_i,p)<r\\ 0, & otherwise\end{cases}. \tag{1}$$

The binary sensor model assumes that sensor readings have no associated uncertainty. And we can define the non-coverage:

$$C(\overline{s_i},p)=1-C(s_i,p). \tag{2}$$

Eq. 2 expresses a gird point p is not covered by sensor s_i. If s_i is independent of s_j, we obtain

$$C(s_i\cup s_j,p)=1-C(\overline{s_i}\cap\overline{s_j},p)=1-C(\overline{s_i},p)C(\overline{s_j},p). \tag{3}$$

Assume $S=\{s_1,s_2,\ldots,s_N\}$ is the sensor node set, and we can get the following expression:

$$C(S,p)=C(\bigcup_{i=1}^{N}s_i,p)=1-C(\bigcap_{i=1}^{N}\overline{s_i},p)=1-\prod_{i=1}^{N}C(\overline{s_i},p)=1-\prod_{i=1}^{N}(1-C(s_i,p)) . \tag{4}$$

Based on the above expressions, we can propose the definition of coverage rate:

$$R_{area}(S)=\frac{A(S)}{A_{total}}=\frac{\sum_{p=1}^{m\times n}C(S,p)}{m\times n}. \tag{5}$$

3 Improved Artificial Bee Colony-Based Relocation Algorithm

Artificial Bee Colony algorithm was proposed by Karaboga for optimizing numerical problems. The algorithm simulates the intelligent foraging behavior of honey bee swarms. The artificial bee colony algorithm is a population-based stochastic optimization algorithm and it is very simple and robust. Compared with those of other well-known modern heuristic algorithms such as genetic algorithm, differential evolution, particle swarm optimization, the performance of artificial bee colony algorithm is better or similar with the advantage of employing fewer control parameters.

In artificial bee colony algorithm, each cycle of the search consists of three steps: firstly, sending the employed bees onto their food sources and evaluating their nectar amounts; then, after sharing the nectar information of food sources, the selection of food source regions by the onlooker and evaluating the nectar amount of the food sources; finally, determining the scout bees and sending them randomly onto possible new food sources.

Based on the above descriptions, we propose a improved artificial bee colony algorithm to maximize the effective coverage area rate and minimum the movement distance. So the fitness value function is defined: $f=\varsigma e^{R_{area}(S)-1}+(1-\varsigma)e^{-\sigma D}$. Where $0\le\varsigma\le 1$, D is the sum of movement distance and σ is a control parameter. At the first step, the algorithm generates a randomly distributed initial population P of Q solutions, where Q denotes the size of employed bees or onlooker bees. Each solution is a $2N$-dimensional vector: $v_i=(x^i_1,y^i_1,x^i_2,y^i_2,\ldots,x^i_N,y^i_N)$, $i=1,2,\ldots,Q$, where N is the number of sensor nodes, and we define $v_{ij}=(x^i_j,y^i_j)$ as the position of node j. After initialization, the population of the positions is subject to repeated cycles of the search processes of the employed bees, the onlooker bees and the scout bees.

An employed bee produces a modification on the position in her memory depending on the local information. In order to produce a candidate position from the old one in memory, our algorithm uses the following Eq. 6:

$$v_{ij}=v_{ij}+\varphi_{ij}(v_{ij}-v_{kj}). \tag{6}$$

Where $k\in\{1,2,\ldots,Q\}$ and $j\in\{1,2,\ldots,N\}$ are randomly chosen indexes and $k\neq j$, φ_{ij} is a random number between[-1, 1].

An artificial onlooker bee chooses position (solution) depending on the probability value associated with it, and it will do local search near the position. The probability value p_i of each position is calculated by the following Eq. 7:

$$p_i = \frac{f_i}{\sum_{n=1}^{Q} f_n}. \tag{7}$$

In our algorithm, each onlooker bee prefers the position with bigger p_i so it chooses the position v_j based on the following pseudo-random proportional rule:

$$j = \begin{cases} \arg\max_{i \in \{1,2,\cdots,Q\}} \{f_i\}, \ q \le q_0 \\ J, \ otherwise \end{cases}. \tag{8}$$

Where q is a random variable uniform distributing between the region [0,1]; $q_0(0 \le q_0 \le 1)$ is parameter and J is a random variable obtained from Eq. 7. The pseudo-random proportional rule can assure that the onlooker bee selects the position with bigger p_i with larger possibility and can avoid precocious.

In our algorithm, if a position cannot be improved further through a predetermined number of cycles, then it is assumed to be abandoned. The value of predetermined number of cycles is an important control parameter of the algorithm, which is called "*LimitCycle*" for abandonment. Assume that the abandoned position is v_l and the scout discovers a new position to be replaced with v_l. This operation can be defined as in Eq. 9:

$$v_l = v_{lb} + rand[0,1](v_{up} - v_{lb}). \tag{9}$$

Where v_{lb} is the lower boundary of the position and v_{ub} is the upper boundary of the position. In our algorithm, besides abandoning the position with repeat cycles bigger than *LimitCycle*, the worst position v_{worst} with minimal fitness value in each cycle will be abandoned. The operation is same as Eq. 9:

$$v_{worst} = v_{lb} + rand[0,1](v_{up} - v_{lb}). \tag{10}$$

With the updating of v_{worst}, our algorithm can avoid local optimization and speed convergence. Based on the above description, detailed pseudo-code of improved ABC algorithm is given below:

1: Initialize the population of solution v_i, i=1,2,…,Q
2: Evaluate the population
3: cycle=1
4: *repeat*
5: produce new solutions v_i for the employed bees by using Eq. 6 and evaluate them.
6: calculate the probability values p_i by Eq. 7
7: apply pseudo-random proportional rule for each onlooker bee selecting solution and produce the new solutions with Eq. 6
8: abandon the worst solution v_{worst} and replace it with a new randomly produced solution by Eq. 10

9: if a solution cannot be improved further through *LimitCycle*, replace it with a new randomly produced solution by Eq. 9

10: memorize the best solution achieved so far

11: cycle=cycle+1

12: *until* cycle=MaxCycle

4 Simulation

In this section, we do some simulations to evaluate the performance of our algorithm. It is supposed that mobile sensor nodes are random deployed in a square area of 100×100m. Each sensor has a detection radius as 12m and has a communication radius as 36m.

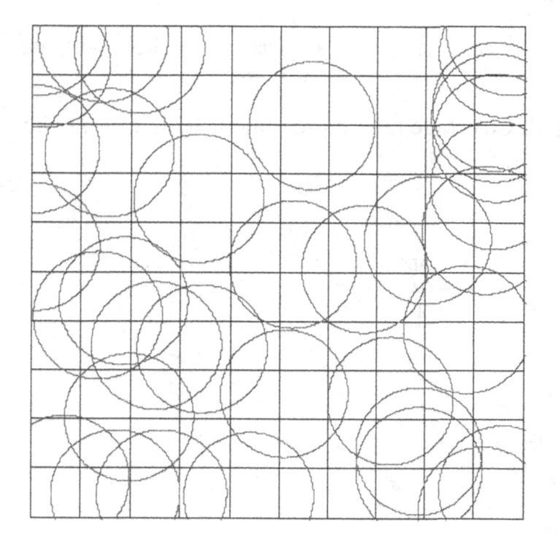

Fig. 1. Initial nodes distribution

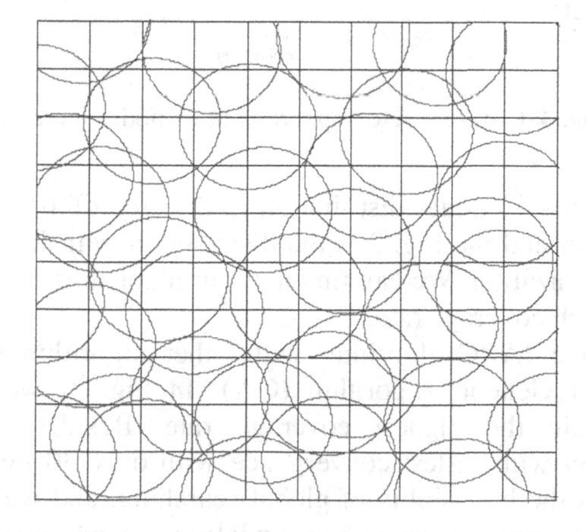

Fig. 2. Nodes distribution with IABC optimization

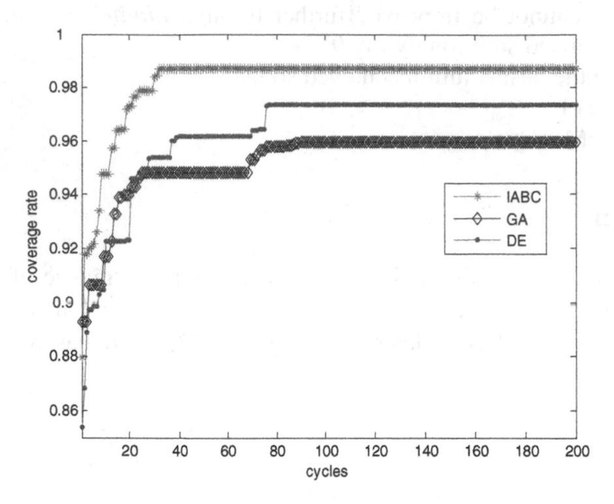

Fig. 3. Coverage rate improvement by iterations increasing

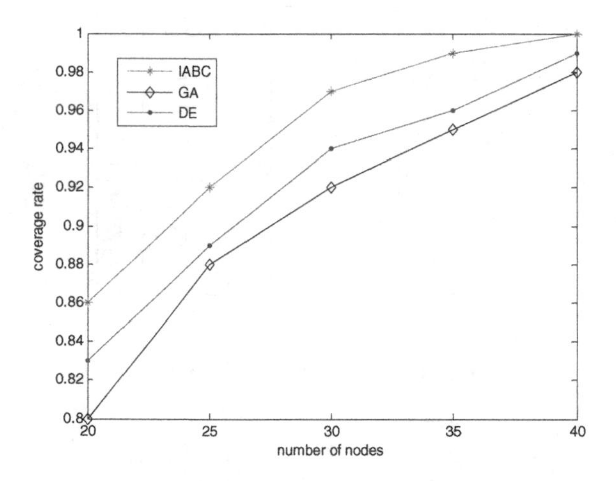

Fig. 4. Coverage rate improvement by nodes increasing

Firstly, we show the 35 nodes distribution in the area of 100×100m. The Fig. 1 is the initial distribution and the Fig. 2 is the distribution with Improved Artificial Bee Colony (IABC) optimization. We can find that our algorithm can obtain a good nodes distribution and a high coverage rate.

Then, we compare IABC algorithm with other algorithm such as Differential Evolution (DE) and Genetic Algorithm (GA). In Fig. 3, we can see the IABC algorithm can obtain the highest coverage rate. Besides, IABC can get the optimization solution with fastest convergence with only 30 iterations. Because the IABC algorithm has the best ability of global searching, and we adopt new onlooker bees selection scheme and new scout bees updating scheme to speed convergence. In Fig. 4, we test the performance of three algorithms with nodes increasing from 20 to

40 and the iteration number for all is 200. We can also find IABC algorithm has the best performance. In Fig. 5, we can see that the IABC can obtain the shortest movement distance.

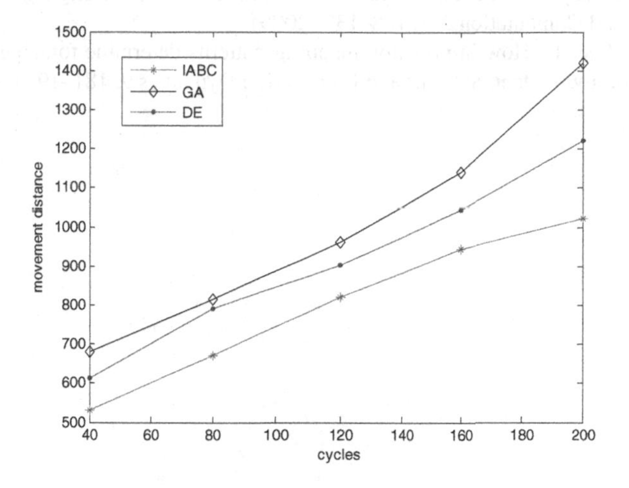

Fig. 5. Movement distance

5 Summary

In this paper, we present an improved artificial bee colony algorithm to solve the mobile sensors relocations problem. The advantages of our approach are: (1) the performance of our algorithm is better than other intelligent algorithm. (2) We design a new onlooker selection scheme to speed algorithm convergence. (3) A new scout bee updating scheme is proposed to avoid local optimization.

References

1. Estrin, D., Girod, L., Pottie, G., Srivastava, M.: Instrumenting the World with Wireless Sensor Networks. In: International Conference on Acoustics, Speech, and Signal Processing, Salt Lake City, Utah (May 2001)
2. Akyildiz, I.F., Su, W., Sankarasubramaniam, Y., Cayirci, E.: wireless sensor network: A survey. Computer Networks 38(4), 393–422 (2002)
3. Chelappan, S., Bai, X., Ma, B., Xuan, D.: Sensor networks deployment using flip-based sensors. In: IEEE International Conference on MASS, Washington DC, US, pp. 298–305 (2005)
4. Chellappan, S., Bai, X., Ma, B., Xuan, D., Xu, C.: Mobility limited flip-based sensor networks deployment. IEEE Transactions on Parallel Distributed Systems 18(2), 199–211 (2007)
5. Pei, Z.Q., Xu, C.Q., Teng, J.: Relocation algorithm for non-uniform distribution in mobile sensor network. Journal of Electronics 26(2), 222–228 (2009)

6. Zhang, S., Bao, X.R., Chen, J.: Optimal distribution of mobile nodes in wireless sensor network. Journal of Northeastern University 28(4), 489–492 (2007)
7. Jin, L.-Z., Chang, G.-R., Jia, J.: Node distribution optimization in mobile sensor networks based on differential evolution algorithm. Control and Decision 25(12), 1857–1860 (2010)
8. Karaboga, D., Akay, B.: A comparative study of Artificial Bee Colony algorithm. Applied Mathematics and Computation 214, 108–132 (2009)
9. Tereshko, V., Lee, T.: How information mapping patterns determine foraging behaviour of a honeybee colony. J. Open Systems and Information Dynamics 9, 181–193 (2002)

Research and Implementation of Information Systems about Digital Underground Engineering Based on 3D GIS

Yuliang Qiu[1,2], Man Huang[2], Zhu Wang[2], Xiaoyong Hu[2], and Mingyi Zhou[2]

[1] School of Civil Engineering, Beijing Jiaotong University, Beijing 100044, China
[2] CCCC First Highway Consultants Co Ltd, Xi'an 710075, China
vetec@163.com, 309757091@qq.com, wz123456789@sohu.com, zjygy@qq.com, 284933558@qq.com

Abstract. With underground space era coming, it has become the research hotspot to seek for scientific planning, design, management methods and means of underground engineering. Based on 3D GIS, breaking through the traditional method, this study developed a set of edit, query and analysis, 3D roam and display in one of digital information system about the underground engineering. In this study, 3D visualization and spatial analysis are achieved in the underground space by 2D and 3D integration technology. It improved the condition that underground engineering information system less involved 3D information on the ground in the past, and completed uniform information integration above and below ground. The Research results can effectively assist underground space to plan, develop and utilize. The systems development experience can also provide a reference for other similar software development.

Keywords: The geo-digital underground space, 3D GIS, The 2d and 3d integration, The ground and underground integration, Spatial analysis.

1 Introduction

Today's world, with information technology as a symbol, the science technology advances rapidly. Since U.S.Vice President Al Gore proposed "Digital Earth" ,until today, it has become one of the main topics concerned about and studied by many scholars to process amounts of geographic data in digital means and use more scientific, effective and intuitive way to manage, express and analyze spatial data [1] .

Mengshu Wang academician pointed out that "The 21st century is the age that urban underground space is exploited and used"[2]. But the underground engineering is often affected by the surrounding environment, resulting in it is difficult to plan, develop and recover, therefore seeking scientific and effective method is imperative or establishing a underground engineering digital information system [3,4]. Based on 3D GIS thinking, this study developed an underground engineering digital information system based on ArcEngine and c #. net. The system can assist us to plan and manage underground space, and also derive more valuable information; at the same time, the system can provide a wealth of spatial analysis functions, and give a reference and decision support for underground engineering.

D. Jin and S. Lin (Eds.): Advances in ECWAC, Vol. 1, AISC 148, pp. 449–454.
springerlink.com © Springer-Verlag Berlin Heidelberg 2012

2 System General Design

We select the structural system design patterns to design the various modules and link them, reducing "volatility effect." .Each function module is both separate and linked. Overall structure of the design is shown in Figure 1. Each module is interconnected through GIS standard data formats, to ensure data management, analysis, maintaining order. The system design is divided into six structural modules, as Fig 1 shown, including data management module, edit and modeling module, 3D display module, the query analysis module, output module.

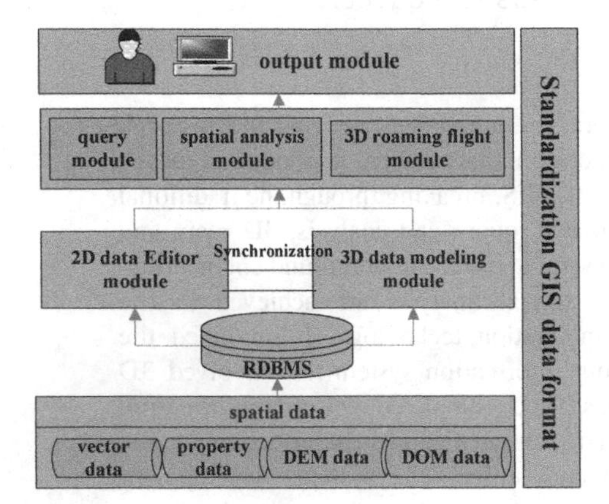

Fig. 1. General structure design of the system

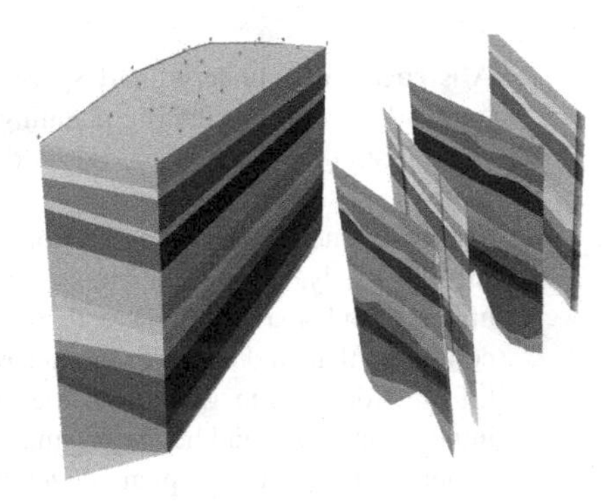

Fig. 2. Geologic body and geological cross section

3 3D Modeling

The System achieved 3D digital modeling of the visible ground and the invisible underground space. The data models on the ground and under the underground are uniformed to the same coordinate system space, to achieve expression integrated information above and below ground.

3.1 Modeling 3D Terrain

That is to build ground digital elevation model (DEM) whose data source is usually elevation points, contour line data and InSAR data. There are two ways to convert elevation points to grid DEM. Firstly, converting directly, in this way, every gird value is calculated with a variety of interpolation methods; the other is the indirect calculation, producing TIN firstly and then converting it into a grid with elevation information [5]. In contrast, the latter method can better handle the terrain features so that the system uses this method of linear interpolation that has clear physical meaning and can facilitate the calculation.

3.2 Modeling 3D Terrain

That is based on geological stratification model of drilling data [6]. We can use of drilling to Simulate geological stratification, with the appropriate interpolation method according to the characteristics of the regional geological environment. The main interpolation methods are: IDW, kriging and spline, etc. Among them, the kriging interpolation method is widely used in geostatistics. Fig2 is graphics of 3D geoscience body model..

Kriging interpolation method is as follow: $Z\ (x)$ is for the second-order stationary random function, its values is $Z\ (x_1)$, ... , $Z\ (x_n)$ in n locations. At the point x_0, the estimated value is:

$$Z^*(x_0) = \sum_{i=1}^{n} \lambda_i Z(x_i) \tag{1}$$

In Eq.1, λ_i is the weight factor, which represents contribution level of the observations value $Z\ (x_i)$ to the estimated value $Z^*(x_0)$. Kriging algorithm's main process is calculating the weight coefficient λi. Ordinary Kriging equations are as follow:

$$\begin{cases} \sum_{i=0}^{n} \lambda_i \gamma(x_i, x_j) + \mu = \gamma(x_j, x_0), i = 1, 2...n \\ \sum_{i=0}^{n} \lambda_i = 1 \end{cases} \tag{2}$$

In Eq.2, $\gamma(x_{i,}\ x_j)$is variogram value between the sample point x_i and $x_{j,}$ $\mu$is lagrange constant. We can obtain weighting coefficients λ_i by Eq.2, and take it into Eq.1 to calculate estimated value of $Z^*(x_0)$ in x_0.

Viewing from the Eq.2, it becomes a key issue to determine the best formula for variance function $\gamma(x_{i,}\ x_j)$.Currently, the main variogram is spherical model, index model, linear model, Gaussian curve model. Generally, the fitting result of spherical function model is satisfactory. Spherical model is as follow:

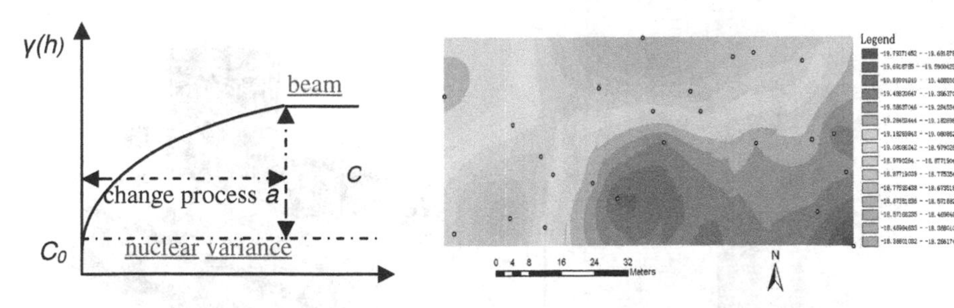

Fig. 3. Variogram spherical models

Fig. 4. Kriging Spherical models interpolation

In Fig3, a is called change process, reflecting the variable's sphere of influence.C_1 is called arch height, $C_0 + C_1$ is as the base platform value. C_0 is nugget constant, describing the discontinuity of the variogram at the origin, reflecting the variable's continuity of space. Spherical model formula is as follow:

$$\gamma(h)=\begin{cases} c_0, & h=0 \\ c_0+c_1\left[\dfrac{3h}{2a}-(h/a)^3/2\right], & 0<h<a \\ c_0+c_1, & h>=a \end{cases} \tag{3}$$

Fig 4 shows an example kriging spherical function model interpolation in system.

The prerequisite of geological cross-section analysis is geological stratification 3D model. Users can designate any of the vector line in the study area, calculate the geological cross-section between the upper and lower geological stratification, and give the corresponding attribute to the profiles, until all geological stratifications are calculated.

3.3 Modeling 3D Structures

Underground structures include underground building facilities, transport facilities and pipe network facilities. It is very important to build 3D underground structures models and effectively query their attribute information. It is also the ultimate goal of underground activities of human. There are two main ways about modeling 3D underground structures: GIS parametric modeling and CAD, 3DS Max professional modeling, both with different features for different areas.

(1) GIS parametric modeling. In this method, the standard GIS point, line, surface vector data is for the geometry reference, and the attribute data is for the parameters. This method is simple and efficient for mass production of simple 3D model, but it can not meet the requirements of complex 3D model.

(2) CAD, 3DS Max professional modeling. This method is suitable for professional complex 3d model. In this way, 3D CAD models and 3DSMax models are created and managed in the GIS database. Users can import the models into the system according to the actual real geographic coordinates, to achieve synergy analysis with the surrounding environment. As shown in Fig 5, we can view the attribute information of interest location after the tunnel model was imported.

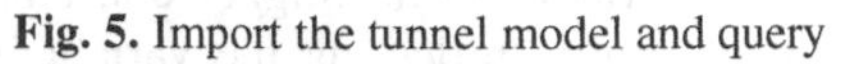
Fig. 5. Import the tunnel model and query

Fig. 6. Ground and underground integrated pattern

3.4 Ground and Underground Space Integration Expression

After modeling the terrain, the buildings and the underground structures, it comes true to achieve the integration expression of the ground and underground, automatically intergrading the spatial data through a unified coordinate system, as shown in Fig 6. It is important planning of underground space to master the topography, buildings, land utilization, can provide valuable information for supporting a safe design.

4 Spatial Query and Analysis

As professional geographic information processing systems, is able to rationally organize spatial data and attribute data, handle and analyze efficiently massive data. Space query in it can help users to extract useful information from much complex data. As GIS a core function, spatial analysis helps us to get the hidden value information, using geometric logic operations, algebra and other mathematical tools to support decision-making [7].

4.1 Spatial Query

Spatial characteristics query needs to determine the type of mask graphic as a selection tool, then determine the topology relationship between mask area graphics and the vector graphics in view. If the former includes the latter, they can be judged as selected objects, as shown in Fig 7. Non-spatial attribute query is by the SQL model to query property information in GIS database .For example, the tunnels longer than 1000m.in a road project are queried.

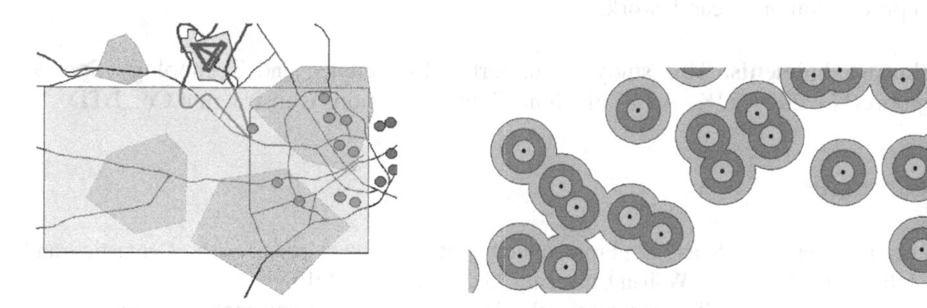

Fig. 7. Select the geography objects by rectangle

Fig. 8. Drill holes buffer

4.2 Spatial Analysis

Spatial analysis mainly includes space measurement and calculation, buffer analysis, slope and aspect analysis, excavation backfill analysis and so on.

(1) Buffer analysis based on point, line, polygon entities establishes a certain distance of the strip area, to express the main object's radiation scope and impact to nearby objects. For example, as shown in Fig 8, the buffer zone established under the

drilling plane coordinates can effectively reflect the reasonableness on the drillings position, gives the feedback and helps us to correct exploration points of layout, makes them more representative and gives a detailed description of regional engineering geology feature.

(2) Slope and aspect analysis can reflect the tilt and orientation of the space surface, as an important part of understanding the actual terrain, provide a reference basis for good underground space design.

(3) Excavation backfill analysis is to determine the size of the area and volume of excavation and backfill, by calculating the difference volume between the surface before excavation and the surface excavation.

5 Conclusion

As city progressed, the development and utilization of underground space is imminent. It has become an important issue to explore effective methods to plan and develop underground space. The system based on GIS-thinking, combining the characteristics of underground engineering, has been developed into a underground engineering digital design system with 2D and 3D integration. It improves the situation that few underground projects involved in ground 3D information in the past, integrates of ground information and the underground engineering information in one. It standardizes multi-source heterogeneous data based on GIS standards, establishes 3D models about terrain, buildings, geological stratification, underground structures, and provides a rich query methods and spatial analysis tools. This study result builds a digital information platform for the development and utilization of underground space, can provide a reference and decision support for users, but also directly serve the design and production or research work.

Acknowledgements. This study is supported by Science and Technology Special Support (No. 2007-ZJKJ-A-08) of China Communications Construction Co., LTD.

References

1. Li, D., Gong, J., Shao, Z.: From digital globe to the wisdom earth. Geomatics and Information Science of Wuhan University 35(2), 127–132 (2010)
2. Wang, M., Tan, Z.: The construct technology of tunnel and underground engineering in china. Engineering Sciences 12(12), 4–10 (2010)
3. Liu, Y., Pan, M., Peng, B.: Bulletin of Surveying and Mapping, vol. 2, pp. 45–47 (2011)
4. Cong, W., Pan, M., Zhang, L.: D igitalization ofurban underground space based on 3DGIS. Chinese Journal of Geotechnical Engineering 31(5), 789–792 (2009)
5. Tang, G., Liu, X., Lv, G.: The principles and methods of digital elevation model and geological analysis, pp. 82–141. Science Press, Beijing (2005)
6. Zhu, H., Li, X.: Digital underground space and engineering. Chinese Journal of Rock Mechanics and Engineering 26(11), 2278–2287 (2007)
7. Wu, L., Liu, Y., Zhang, J.: GIS principles methods and application, pp. 161–178. Science Press, Beijing (2001)

Research on Human Resources Efficiency of Project Department Based on PCA-DEA Complex Model

WanQing Li[1,*], Yang Liu[1], and WenQing Meng[2]

[1] College of Economics and Management, Hebei University of Engineering. 056038, Handan, China

[2] College of Civil Engineering, Hebei University of Engineering. 056038, Handan, China

noblely@163.com

Abstract. For the problem of that the evaluation process of the project contractor involves a strong subjective sense and human judgment, the paper will quantify the efficiency of human resources in analysis and evaluation by the method of data envelopment analysis. At the same time, in order to meet the limit to the number of input and output indicators, the paper will take dimensionality reduction of indicators by the method of principal component analysis which can add data envelopment analysis effectively. In the other words, to establish PCA-DEA complex model. Finally, it will avoid the subjective human impact to the evaluation of the efficiency of human resources.

Keywords: EPC, Human Resources Efficiency, DEA, PCA.

1 Introduction

With rapid development of the EPC model, it has become an important model for the construction market[1]. At the same time, as the market participants, the competition in all project contractors becomes more and more fierce. Therefore, the human resource is the most basic, important and creative resources of construction projects, which is the determining factor of success of the project. It is a key factor in winning competition to improve the efficiency of human resources of the General Contractor projects department.

The academia domestic and abroad made some theoretical research and analysis to the problem of efficiency of human resources. However, the processing of evaluation of these studies involves artificial strong subjective judgments, in addition, the index system they established is often too simple, which can not make a comprehensive evaluation of the sample. To solve the above problem, it was proposed the principal component analysis and utilization of Data Envelopment Analysis, based on previous research.

---

[*] The author: WanQing Li (1954-), male, Ph.d., professor, Hebei University of Engineering, engaged in the research of engineering project management.

D. Jin and S. Lin (Eds.): Advances in ECWAC, Vol. 1, AISC 148, pp. 455–460.
springerlink.com © Springer-Verlag Berlin Heidelberg 2012

2 Compound PCA-DEA Model

2.1 Principal Components Analysis

Principal Components Analysis is a multivariate statistical method t[2], which converses the multi-index into a few composite indicators under the premise of little loss of information, just by the opinion of the ideological dimension reduction. Based on research involving a thing of the p indicators, the indicator can be respected by $X_1, X_2, \ldots, X_p$, therefore, the p indicators constitute p-dimensional vector: $X=(X_1, X_2, \ldots, X_p)$. Let mean of random vector X μ, standard deviation matrix Σ. It can form a new integrated variable by linear transformation of X, which can be respected by Y. In other words, new integrated variables can be expressed by the linear variable, just to satisfy the following formula:

$$\begin{cases} Y_1 = u_{11}X_1 + u_{12}X_2 + \cdots + u_{1p}X_p \\ Y_2 = u_{21}X_1 + u_{22}X_2 + \cdots + u_{2p}X_P \\ \cdots\cdots\cdots\cdots \\ Y_p = u_{p1}X_1 + u_{p2}X_2 + \cdots u_{pp}X_p \end{cases}$$

2.2 Data Envelopment Analysis

Data Envelopment Analysis is a method of relative efficiency of non-parametric evaluation.Data Eenvelopment Analysis used a variety of planning models t[3], its most basic models are CCR model and BCC model. In CCR model, if the decision-making unit is valid, then it is effective in the technical and scale.Technical efficiency includes pure technical efficiency and scale efficiency of two parts. They introduced convex assumptions and increase the constraints on the optimal solution λ, therefore, excluded the impact of scale efficiency θ and gained BCC model of measuring pure technical efficiency θ.

$$\begin{cases} \text{Min}[\theta - \varepsilon(e_1^T S^- + e_2^T S^+)] \\ \text{s. t.} \sum_{j=1}^{n} \lambda_j x_j + S^- = \theta x_0 \\ \sum_{j=1}^{n} \lambda_j y_j - S^+ = y_0 \\ S^- \ge 0, \quad S^+ \ge 0 \\ \lambda_j \ge 0, \quad j = 1,2,\cdots n \end{cases} \qquad \begin{cases} \text{Min}[\theta - \varepsilon(e_1^T S^- + e_2^T S^+)] \\ \text{s. t.} \sum_{j=1}^{n} \lambda_j x_j + S^- = \theta x_0 \\ \sum_{j=1}^{n} \lambda_j y_j - S^+ = y_0 \\ S^- \ge 0, \quad S^+ \ge 0 \\ \sum_{j=1}^{n} \lambda_j = 1, \quad j = 1,2,\cdots n \end{cases}$$

Based on total technical efficiency from CCR model and pure technical efficiency from BCC model, it can obtain scale efficiency by dividing the former to the latter. Scale efficiency is a measure to evaluate whether company is in the optimal size of state. In the economic sense, the so-called optimal size is the average cost curve refers to the lowest point in the production of the state. Enterprise is able to achieve profits or optimal levels of performance of efficiency in size. When the efficiency of the evaluated value of the company's size is equal to 1, it indicates that the relative size of the company is in active state or the state of constant returns to scale. If the scale efficiency is less than 1, then the size of the company is non-effective.

2.3 PCA-DEA Composite Model

Model features of Data Envelopment Analysis depend that when the index increased to a certain extent, not only the workload of data processing will increase but also universal validity of all decision-making unit is close to 1 for related effects between each input and output indicators, finally impact results of assessment.

The input-output index of the problem of the human resource efficiency of projects department studied by this paper is a complex system, the evaluation of which will involve a lot of evaluation indexes. To avoid the effect of the above problems and ensure the accuracy of the evaluation results, this paper chooses PCA-DEA composite evaluation model, reduce the interconnections between each index by using the advantage of the principal component analysis in the index of feature extraction on the basis of guarantying the index information, and fully reflect the characteristics which evaluate the relative effectiveness in every decision unit and analysis the direction ineffective optimization and adjustment.

3 The Struction of Index System

DEA model is a relative efficiency method of analyzing many decision making units of input and output, it needs to select both aspects of the index of input and output of the human resource efficiency of projects department. As the determination of input-output index is the key of the evaluation efficiency of the DEA model, different index selection is likely to lead to different analysis results.

The evaluation of the human resource efficiency of projects department is complex system problems which is often hard to measure by a single or a few indexes overall and objectively, so it should be setting up the index system to describe. On the side of input of human resources, This paper are copied from more mature human resource accounting theory, the related human resource accounting theory are introduced as input index of evaluation of the human resource efficiency. On the side of output of human resources, this paper consider more about business efficiency of projects department.

Table 1. Input and Output indicators

Input indicators	Preliminary cost X_1	Recruitments of cost X_{11}
		Selection of cost X_{12}
		Eemployment of cost X_{13}
		Before the work education fee X_{14}
		Post regularly training fee X_{15}
	Use cost X_2	Salary X_{21}
		Bonus X_{22}
		Endowment insurance X_{23}
		Medical insurance X_{24}
		Unemployment insurance X_{25}
		Industrials injury insurance X_{26}
	Recessive cost X_3	Education background X_{31}
		Professional skills X_{32}
		Work experience X_{33}
	Item cost X_4	lime limit X_{41}
		Be number of managers X_{42}
		Business and support number of employees X_{43}
Output indicators	Management scale Y_1	General contracting turnover Y_{11}
		Main business profit Y_{12}
	Profit ability Y_2	Total profit Y_{21}
		Net profit Y_{22}
		Property profit margin Y_{23}
		Average per profit Y_{24}
	Project delivers Y_3	Period in advance Y_{31}
		First party's satisfaction Y_{32}

4 Case Analyses

4.1 Using the Principal Component Analysis Method to Input and Output Index

It may see from the principal components expression $I_1=0.276X_{11}+0.271X_{12}+0.242X_{13}+0.275X_{14}+0.273X_{15}+0.276X_{21}+0.274X_{22}+0.258X_{23}+0.279X_{24}+0.274X_{25}+0.274X_{26}+0.032X_{31}+0.063X_{32}+0.049X_{33}+0.244X_{41}+0.258X_{42}+0.255X_{43}$ and $I_2=-0.021X_{11}-0.035X_{12}-0.076X_{13}-0.035X_{14}-0.034X_{15}-0.025X_{21}-0.011X_{22}-0.002X_{23}-0.005X_{24}-0.045X_{25}-0.003X_{26}+0.610X_{31}+0.549X_{32}+0.559X_{33}-0.021X_{41}-0.042X_{42}+0.021X_{43}$ that explains the first principal components reflection currency, the time and the population aspect investment. Likewise obtains, it explained that the second principal components reflect the non-currency aspect the investment.

It may see from the principal components expression $O_1=0.442Y_{11}+0.460Y_{12}+0.460Y_{21}+0.460Y_{22}+0.122Y_{23}+0.352Y_{24}+0.126Y_{31}+0.123Y_{32}$ and $O_2=0.129Y_{11}-0.082Y_{12}-0.083Y_{21}-0.082Y_{22}+0.631Y_{23}-0.381Y_{24}+0.414Y_{31}+0.499Y_{32}$ that explained the first principal components reflect the project department finance management effect aspect output. Likewise obtains,It explained that the second principal components reflect the project appraisal effect aspect output..

4. 2 Carries on the Human Resources Efficiency Rating Using the Data Envelope Analysis

This paper chooses quite mature DEA computation software DEAP2.1 to carry on the computation, when computation chooses the input orientation the CCR model and the BCC model to the sample.

Table 2. DMU TE. PTE. SE

DMU	TE	PTE	SE	Scale stage
Department 01	1.000	1.000	1.000	-
Department 02	0.557	0.658	0.846	irs
Department 03	0.583	0.806	0.723	drs
Department 04	1.000	1.000	1.000	-
Department 05	0.467	0.485	0.963	irs
Department 06	1.000	1.000	1.000	-
Department 07	0.689	0.910	0.757	drs
Department 08	0.547	0.981	0.558	drs
Department 09	0.415	0.465	0.893	irs
Department 10	0.568	0.983	0.578	drs
Department 11	1.000	1.000	1.000	-
Department 12	0.883	1.000	0.883	drs
Department 13	0.804	0.979	0.822	drs
Department 14	0.811	1.000	0.811	drs
Department 15	0.741	0.767	0.966	irs

4.3 Analysis of the Results

Technology efficiency analysis,Table 2 shows technology efficiency (TE) of the human resource efficiency of projects department.From the table 5, there are 4 project department DEA effective(TE=1): accounting for 27% of the total. And there are 11 project department DEA invaild(TE <1): accounting for 73% of the total. Pure technical efficiency analysis,Table 2 shows pure techology efficiency (PTE) of the human resource efficiency of project department.Form the table 4, there are 5 project depariments DEA effective(PTE = 1): accounting for 30% of the total.The human resources efficiency of project department is efficient of pure technical efficiency.There are 10 project depariments DEA ineffective(PTE <1): paccounting for 70% of the total.Scale efficiency analysis,In the selected 15 projects, There are 4 project departments by constant returns to scale in the phase: accounting for 27% of the total.This 4 projects department have been in the best scale. There are 7 project departments by decreasing returns to scale in the phase: accounting for 47% of the total. This 7 projects departments should not increase investment, that even if the increase is also only a certain proportion of inputs to bring less than the proportion of output.

5 Conclusion

The paper analyze human resource efficiency of 15 project department of the general contracting company by PCA-DEA complex model.Of the 11 invalid technical efficiency project department, 6 projects have reached the Ministry of pure technical efficiency. The void of DEA is due entirely to scale efficiency. Though the other 5 project department do not meet the pure technical efficiency, its scale efficiency is less than pure technical efficiency. This indicates that the scale efficiency is a key factor which effect the the efficiency of human resources of the project department. However, most of the project department is at the stage of scale of decreasing, in the words, the investment to human resources of project department is in larger redundancy. Therefore, PCA-DEA complex model can be more objectively reflect the efficiency of the actual situation of human resources of the project department, so as to provide the scientific basis for decision making of further improving the efficiency of human resources of general contractor.

References

1. Liu, L., Du, P., Zhao, S.: DEA-based Research on Service Efficiency of Human Resource in Science and Technology in China. Science and Technology and Economy (6), 73–76 (2010)
2. Liu, B., Wang, X., Cao, J.L.: DEA Model Design and Application with Principal Component Cone. System Engineering 8, 101–105 (2009)
3. Wei, Q.: Data Envelopment Analysis. Science press, Beijing (2004)

Management of Empty Nester's Caring System

QingYing Zhang, MengYa Zhang, and Yan Chen

School of Logistics Engineering, Wuhan University of Technology
Wuhan, Hubei, P.R.China
kathy8899@126.com, kmno40311@163.com, chenyan@whut.edu.cn

Abstract. With the increasing aging population and small family structure occupying the mainstream, more and more families become empty nest. However family pension and institutional pension can not meet the current development of aging society, in order to meet the diverse needs of empty nesters, a three-dimensional empty nesters care system, which is combined with community cadres, neighbors, homemaking, volunteers, community hospitals, general hospitals and the governments at different levels, is proposed. Then the advices to implement the system are added.

Keywords: empty nester, multi-dimensional, caring system.

1 Basic Requirements of Empty Nester's Caring

The 6th population census shows that about 13.26% of the general population, i.e. 177 million people of China, is elder than 60 years old. According to internationally accepted standards, more than 7% of the population is 65 years old, or 10% of the total number is 60, implies that the country or the region has gone into aging society. At the very beginning of the new millennium, China was adversely affected by the silver hair wave.

Four features are coming into notice, which are: big magnitude of the aged, the acceleration of aging, prominent aging population, and the high proportion living alone rather than with their posterity [1], called empty nester. Because of physical decline and psychological depression, the aged become weaker and sicklier, and need help from others. Meeting the elder's various requirements, and serving them, have become the main trend currently. It is urgent to establish a full dimensional caring system to care the aged persons, especially the empty nester, so as to build a harmonious society.

2 Current Situation about Caring for the Aged

Over the past few years, a series of measure s have been put forward to support the project on the elderly care in many parts of our country. While in spite of the great development, there are still problems existing in the "Empty Nester" caring system in China compared with that of developed country, mainly in following aspects, such as only part of the elders in the city being able to enjoy social security system, the information infrastructures are not advanced, the system is poor in comprehensiveness and flexibility [2], and without an sound operation mechanism.

D. Jin and S. Lin (Eds.): Advances in ECWAC, Vol. 1, AISC 148, pp. 461–466.
springerlink.com © Springer-Verlag Berlin Heidelberg 2012

Summarizing those problems, the top 5 are listed as follows: (1) Unable to totally fulfill the requirement of empty nester's aged service; (2) The service business for the empty nester is immature; (3) Empty nesters feel unsafe popularly; (4) Low informationization of elderly care system; (5) The spiritual and cultural life of elderly is not rich.

3 Two Basic Ways for the Old-Age Caring

In China, there are mainly two traditionally basic ways for the old-age caring, i.e. home-based care and institution endowment. Of them, the former one is usually considered to be the best, and typical traditional pattern for the most aged people, living with their kids together. But, now, it is difficult to be kept. The young people have to earn life, and seek personal development. They leave their parents to pursue their professional carrier, and most of them are under a lot of pressure, while unable to take care of their parents' heart and soul. Lots of empty nester then appeared.

Besides, since our family planning, or, birth control policy operated from 1982, only one child is permitted to be born for every couple, especially in cities, which brings a series problem. 4-2-1 family pattern is the result.

So, the traditional home-based care is getting increasingly tough to satisfy the requirement.

As for the institution endowment, it must be a good choice. It is popularly said that the needs of elder service institutions are derived from reduce of healthy condition of the elder and the care function of family. But, the narrow covering (people usually have to wait for a long time, even years, for a good institution) and the high cost (a good nursing house cost a lot), as well as negative effects by news (people frequently read this kind of news form newspaper, such as aged person are badly treated in the nursing house), make the rest home hard to be accepted by a normal family. At the same time, high pay of a good nursing house is absolutely a big burden, for a common family.

Combining two traditional ways, home-based care and institution endowment, community endowment patterns is considered as the best way for the aged people, esp. the empty nesters, which resolves conflicts and helps the elderly physically and psychologically, while reduces the burden of the family on both economics and labor power. "Aged-care at home" fit into the basic state condition in China as well as complies with the international tide of social parents-supporting.

4 A Multi-dimensional Caring System

Caring the elderly comprehensively is a difficult issue, which is capable to be accomplished with the help of a multi-dimensional caring system which makes an effective combination of community endowment and family endowment [3], integrate the power of the whole society, to care the aged persons, especially the empty nester, and to build a harmonious society. The goals of the caring system are to drop incidence of disease of the elderly, make them a peace and contentment living environment, and to cut down the nursing expenditure, so as to contribute the society. A multi-dimensional caring system is designed as Fig.1.

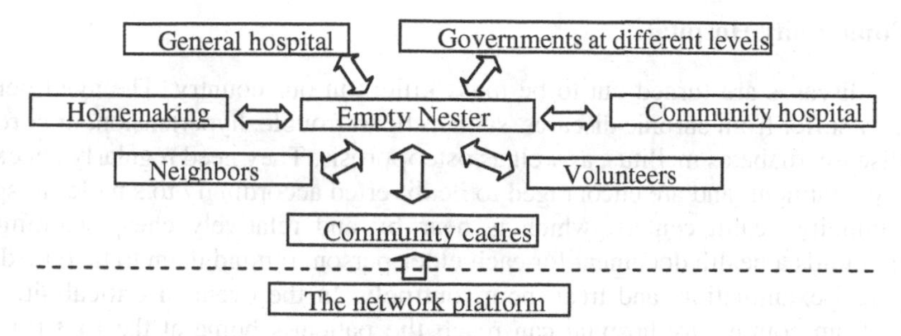

Fig. 1. A multi-dimensional caring system

4.1 Community Cadres

The caring system is community based. The cadres are in charge of the management. Their duties include arranging the system, settling the detailed affairs of empty nesters' caring, visiting the elderly, and communicating with different parts. Community hotline or COM care Call connect the cadres and the empty nesters. The community center provides all sorts of leisure activities. Community amenities, access to community facilities help the residents, including the aged, do physical exercise.

4.2 Neighbors

Neighborhood attention is now considered as a useful way to secure the elderly, especially the empty nester. An old saying goes that a near neighbor is better than a distant cousin. But previously, separate inhabitant living in the same residential area has no effective communication with each other, which brings the safety problems, esp. for the weak empty nester. The system set up a good relation with neighbors, which encourage the younger persons to take care their aged neighborhood, by frequent visiting daily or the emergency alarming bell urgently.

4.3 Homemaking

Domestic service or hourly employment is helpful for the empty nesters, which offer the homemaking, including cleaning, cooking, and other chore. As for the payment, there are different ways. It can be paid totally by the aged people themselves, if they are able to support. Or, it could be assisted by the government partly, while the aged paid half.

4.4 Volunteers

Important factors for the elderly to spend their remaining years are economic support, life care and spiritual comfort. A new trouble called empty nest syndrome is a general feeling of loneliness that parents or other guardian relatives may feel when their children leave home. They need psychological counseling to get spiritual wellness. Currently, a mass campaign encouraging kindness to disadvantaged groups is now widespread and volunteers are numerous, which become the main force of empty nester's mental caring or psychological comforting to remove loneliness and disorientation of the old persons.

4.5 Community Hospital

Chronic diseases are turned out to be mass killers in our country. The aged persons popularly suffer from chronic diseases, such as heart trouble, hypertension, or coronary heart disease, diabetes mellitus, as well as osteoporosis. They need regularly check and ordinary treatment, and are encouraged to be diverted accordingly to smaller hospitals or community health centers, which is near by and relatively cheap. Community hospital builds a health document for each elder person, remind them to take medicine of physical examination, and treat them routinely. In the event of critical situation, doctors from community hospital can reach the patient's home at the first time the rescue the sick person.

4.6 General Hospital

Polyclinic or general hospital has better hygienic conditions and experienced doctors and is able to treat the patients preferably and availably. The system requires the general hospital open a green channel to give the first aid to the elderly to ensure them to receive the fastest ambulance at the critical time. Not only chronic diseases, but acute trouble could be cured there.

4.7 Governments at Different Levels

Chinese government consistently attaches importance to proving the well-being of the people, including the providing for the aged. A series of policies are made on the subjects about people's life, such as social security, social assistance, social welfare, basic medical insurance, minimum living standard security, as well as the encouragement of charitable donations, etc. It is obvious that a powerful support by the government is evidently vital to the establishment of the system. Governments at all levels are responsible for the system's operation, a comprehensive coordination and managing between various communities, experience sharing and improve the whole system.

5 Implementation of the System

The full-dimensional caring system framework offers the old persons a friendly living environment.

The empty nesters are cared by the community organizer, neighbor, volunteer, housekeeper, and the community hospital. Among them, day-to-day affaires of elderly caring are done by the community cadre, neighbor, and housekeeper. Volunteers mainly offer the elderly with psychological assistance. Community hospital is in charge of regularly checking and chronic disease treatment. Polyclinic hospital offers green channel for the aged empty nester to ensure them to receive the fastest ambulance at the critical time. General hospital and government at different levels are all in this caring system.

All of the caring jobs are completed on the base of information network platform.

The aged helping mechanism is led by the government, organized by the community, provided with the commercialized operation, the participation of the enterprises, and a combination of public welfare and low pay service.

The implementation of the multi-dimensional caring system depends on 5 aspects.

5.1 Government Support

The constructing of the full-dimensional caring system depends much on the support of government. The result and effect of elderly caring is suggested to be scheduled in the evaluation index of government branches [4]. Not only the policy, but the financial and taxation support, as well as organizations and management are encouraged.

5.2 Standardizing Service Market

Standardizing service market and creating more jobs are thought to be meaningful to both the accepter of the service, i.e. the aged people, and the provider, e.g. nursing personnel.

The reward of service workers consists of two parts. One is from the government in the form of service bill, and the second part is cash payment given by the aged people themselves.

5.3 Improving the Volunteer Mechanism

Volunteer management institutions should be encouraged and guided to enhance the empty nester elderly caring and the scope and impact of volunteer activities is then to be expanded. Volunteers go into community, join the care team, and confabulate with the "empty nester", from the mental health of the elderly, emotional support, to the terminal care, to help the elderly actually and pressingly.

5.4 Providing Elderly Medicine Assistance

Basic medicine assistance is to be provided to the empty nesters [5]. It including: (1) Organizing community hospital to visit and check the elderly regularly; (2) After receiving the information asked for help, community hospital must arrive the site of the incident at the first moment and treat the elderly timely; (3) General hospital opens a green channel for the elderly assistance when receives the call for help.

5.5 Accelerating the Informationization in Communities

In a sense, information technology is the bottleneck, or kernel of the service system, including the elderly caring platform.

The objective of the network system services contains two aspects: (1) the fast and reliable communication between "empty nester" and the neighbors, domestic staff, community, community hospitals, general hospitals and their families; (2) location checking of the elderly, and help information sending to the relevant department. It is clear that the informationization in communities is positively crucial.

6 Conclusion

Developing an urban community fashion for the aged is the best way to deal with the problem of aging. The mode respects the society tradition and the aged custom. With the help of the multi-dimensional caring system, the problems caused by the aged society will be reduced to the smallest extent.

References

1. Li, H., Ding, J.: The Analyzing on the Character of the Nest Families in the City. Science of Social Psychology, Hangzhou Teachers College, Hangzhou China 5(21), 91–95 (2006)
2. Scottish Qualifications Authority, Pension provision: an introduction. Chinese Modern Economic Publishing House, Beijing (2006)
3. Zhang, Q., Yu, Y., Zhang, P.: Research on the "Empty Nester" Caring System. In: The Third International Conference on Management and Service Science (MASS 2009) (September 2009) ISBN 978-1-4244-4639-1
4. Lin, M., Liu, Y., Zhan, G.: Empirical study on the urban elderly people of 'empty nest' and their loneliness. Modern Preventive Medicine, Shangdong Institution of Business and Technology, Shangdong China 36, 77–80 (2003)
5. Wang, Y., He, G.: Probe into health problems of empty nest aged people in city community and its strategies. Chinese Nursing Research, Nursing College of Zhongnan University, Hunan China 22(1), 100–102 (2008)

Internet of Things Applied in the Home-Based Caring System for the Aged

Qingying Zhang, Zhimin Chen, and Peng Zhang

School of Logistics Engineering, Wuhan University of Technology
Wuhan, Hubei, P.R.China
kathy8899@126.com, chenzhimin19880210@126.com,
wh.zhangpeng@qq.com

Abstract. Aging society brings a lot of pressure and problem which is urgent to be solved. Based on the requirement of aged caring, combined with the current status of home-based caring, A multi-dimensional caring system is built. The technology of Internet of Things is used to make the home-based caring valid and effective. The living place becomes a smart home, which can help the youngsters monitor their parents and grand parents remotely, so do the cadres, neighbors, volunteers, and the doctors.

Keywords: aged caring, Internet of Things, caring system.

China has already entered an aging society since the very beginning of the new millennium. Faced up to the great wave of white hair, it becomes necessary and inevitable to reform the mode of caring service for the elderly [1]. The demand for the aging society service is growing increasingly and quickly.

1 Ways of Aged Caring

There are several ways of aged caring nowadays. The most traditional one is home-based. Parents and grand parents live together with the youngsters. But it is not popular anymore. The main reason is about the life pressure of the young people especially in a 4-2-1 family [2]. The second mode of aged caring is institution endowment. The obstacles are primarily the narrow covering, the high pay, and the negative effects by news. The third way is then produced by combining the two ways, which is called community endowment. The aged people live at home where they used to live, while are cared not only by the kids, but the community, involving cadres, neighbors, volunteers, household service, and the medical personnel on the base of internet of things and various technologies.

2 Internet of Things

2.1 IOT

The Internet of Things (IOT), also called Web of Things, is a network of Internet-enabled objects, which is the vital part of new information technology. IOT is,

D. Jin and S. Lin (Eds.): Advances in ECWAC, Vol. 1, AISC 148, pp. 467–471.
springerlink.com © Springer-Verlag Berlin Heidelberg 2012

in fact, a web which connects things to each other. It has two meanings. First, the core and foundation of IOT is still the web, a broadening of the internet; second, its clients extend to among any things to exchange and communicate information. IOT works under specific agreements through separate devices, such as RFID, infrared ray sensors, GPS, laser scanner etc [3]. To bond anything to the internet to exchange information between the web and the equipment so as to recognize, locate, trace, monitor and manage things intelligently. IOT is an extension of the internet, with the innovation as the key element, and customer experience as the soul.

2.2 Basic Features of IOT

IOT refers to ubiquitous devices and facilities, including infrared ray sensors, laser scanners, gas sensors, mobile terminals, family intelligent facilities, digital videos etc. to collect status and procedures data to get supervisory signal in real time. The signal could be all-embracing such as sound, light, heat, electricity, mechanics, chemist, biology, location and other kind according to the given purpose. The devices are bond with internet to form a huge web. The objective of IOT is to connect thing to thing, things to people, all the things to the web in order to accomplish the recognition, management and control intelligently.

2.3 Prominent Characteristics

IOT is quite different compared to the tradition internet. It's a wide application of various sensor technologies, and has a large number of multiple sensors as the information source, to collect information with different content and format. Instantaneity implies that the collection is made and refreshed momentarily. While, the foundation and core of IOT technology are still internet. It integrates various wired and wireless networks, and outputs information of objects precisely and instantaneously. Besides, IOT has a strong ability to calculate, called cloud computing, so as to get useful information for the massive data by analyzing, processing and manipulation.

2.4 Architecture

In term of the architecture, or technological structures, IOT has three layers: sense layer, web layer and application layer: (1) Sensor layer is constituted by many sensors, like people's eyes, ears, noses and nerve endings. It is the source of recognition of things and collecting of data in IOT. Its main function is things' recognition and information collection; (2)Web layer is composed of private net, the internet, wired and wireless net, web management system, cloud computing etc. It is similar to people's nerve center and brain. It transports and processes information collected from sensor layer; (3) Application layer contains an interface of IOT and clients (including people, organization and other systems). It accommodates to the needs of trades, and makes IOT's intelligent application available.

3 Aged Caring Method and System

3.1 Basic Requirement

Because of physical decline and psychological depression, the aged become weaker and sicklier, and have to be cared by others. Most aged people are suffer from different diseases, especially the chronic ones, such as heart disease, diabetes mellitus, hypertensive or high pressure of blood, stroke, or apoplectic sequelae, senile dementia, arthralgia and osteoporosis. They definitely need help from others.

3.2 Aged-Care at Home

"Aged-care at home" adheres to the basic state condition in China as well as complies with the international tide of social parents-supporting. But even for those elderly people who can look after themselves in their daily life, getting help and caring form others is still necessary. The traditional way, home-based and looked after by their kids, has to be changed. The community caring, not live together with other elderly, but at home and cared by different persons is then thought to be effectual.

3.3 Multi-dimensional Caring System

A multi-dimensional caring system is designed, shown in Fig.1. Caring and assistance is provided from many aspects, including housekeeper, volunteer, community cadre, neighbor, and community hospital, general hospital, and government at different levels [4].

4 IOT- Aided Aged Caring

The application of IOT is to be able to make the home-based caring valid and effective. With the help of IOT, an all-around surveying system is absolutely serviceable. It makes the living place to be a smart home.

4.1 A Smart Home

In such a smart home, various sensors are fixed at the given places. For example, camera on the wall catches the pictures or makes video surveillance, impact force transducer set on the floor raises an alarm when the elderly falls down, sphygmomanometer tied on the wrist of the aged people gets the blood-pressure and heartbeat, glycosometer fixed on the in-wall of flush toilet give the result of urinalysis or urine sugar. Beside, some annunciators established at separate places of the rooms help the people to send the alerting signal conveniently when needed by pressing the button [5]. Conductor arrangement provides a sound connection of all those sensors to the web. The youngsters could monitor their parents and grand parents remotely, so do the cadres, neighbors, volunteers, and the doctors.

4.2 Records and Alarm

Smart home care is an advanced way of pension. Various physical signs of elderly are measured at home and then upload to the background system. The measured data is directly transmitted into their electronic health records at the health care centers. Once the data is unusual, intelligent system launches a caution and gives a remote medical treatment automatically, on-the-spot medical service is made by the doctor and nurse if necessary.

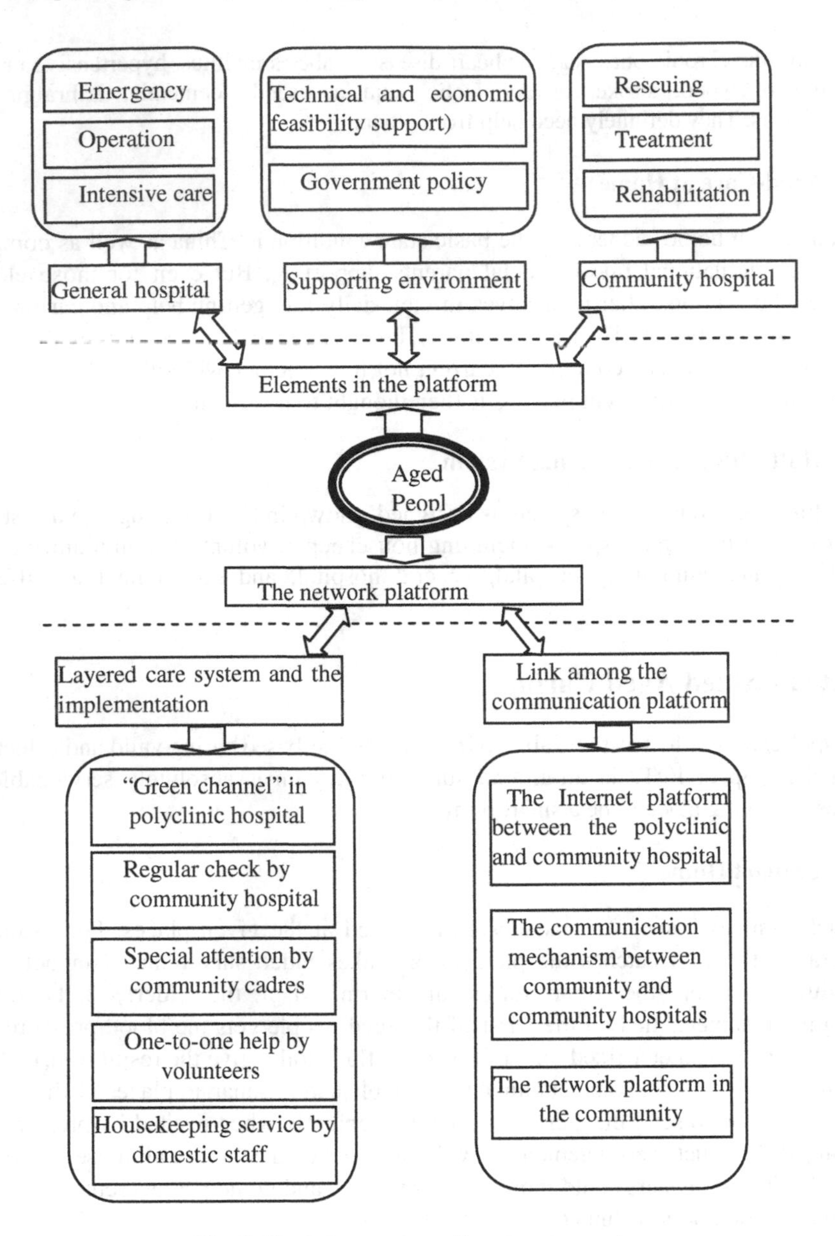

Fig. 1. The infrastructure of aged caring system

4.3 On the Way

If the aged person goes outside, a locator carried tells the position instantly in case of any accident. The operating principle of the locator is like GPS. It communicates with the control centre in the same manner that a mobile works.

5 Conclusion

In an aging society, meeting the elder's various requirements and serving them have become the main trend currently, which is able to be accomplished by a social caring system. To establish a multi-dimensional caring system to care the aged persons, and to use IOT to help people to accomplish the caring are proved to be evidently significant and available.

References

1. Zhang, Q., Yu, Y., Zhang, P.: Research on the 'Empty Nester' Caring System. In: The Third International Conference on Management and Service Science (MASS 2009) (September 2009) ISBN 978-1-4244-4639-1
2. Qi, F.: A New Model of Social Aged-care: Aged-care at Home. Journal of Liaoning Normal University (Social Science Edition) 28(3), 16–18
3. Hancke, G.P., Markantonakis, K., Mayes, K.E.: Security Challenges for User-Oriented RFID Applications within the 'Internet of Things. Journal of Internet Technology 11(3), 307–313 (2010)
4. Scottish Qualifications Authority, Pension provision: an introduction. Chinese Modern Economic Publishing House, Beijing (2006)
5. Yang, G., Li, F., Mu, X., Yu, T.: Design of security and defense system for home based on Internet of things. Journal of Computer Applications 30 (2010)

Text Clustering Method Based on K-medoids Social Evolutionary Programming

ZhanGang Hao

Shandong Institute of Business and Technology, Shandong Yantai 264005, China

Abstract. This article presents a improved social evolutionary programming. The algorithm is the k-medoids algorithm as the main cognitive reasoning algorithm, and improved to learning of Paradigm、Optimal paradigm strengthening and attenuation and Cognitive agent betrayal of paradigm.

Keywords: Text clustering, K-medoids algorithm, social evolutionary programming.

1 Introduction

Social evolutionary programming (SEP) is an algorithm based on paradigm conversion into global search algorithm [1-2], has been used to solve the problem of clustering [3-4], But not solved the problem of isolated points. This article presents a improved social evolutionary programming (K-medoids Social Evolutionary Programming,KSEP), this algorithm will increase the diversity of species group and enhance the optimization capability of social evolutionary programming.

2 Text Clustering Method Based on KSEP

2.1 K-medoids-Based Body of Cognitive Reasoning Algorithm

k-medoids algorithm operating principle is shown below: The primary idea of the k-medoids algorithm is that it firstly needs to set a random representative object for each clustering to form k clustering of n data. Then according to the principle of minimum distance, other data will be distributed to corresponding clustering according to the distance from the representative objects. The old clustering representative object will be replaced with a new one if the replacement can improve the clustering quality. A cost function is used to evaluate if the clustering quality has been improved [5]. The function is as follows:

$$\Delta E = E_2 - E_1 \tag{1}$$

where ΔE denotes the change of mean square error; E_2 denotes the sum of mean square error after the old representative object is replaced with new one; E_1 denotes the sum of mean square error before the old representative object is replaced with new one.

D. Jin and S. Lin (Eds.): Advances in ECWAC, Vol. 1, AISC 148, pp. 473–477.
springerlink.com © Springer-Verlag Berlin Heidelberg 2012

2.2 Evolved Self-optimization Process of Paradigm-Based Learning and Updating

A good paradigm is a good viable solution record. Here, F stands for paradigm, M for the number of paradigms, $F[i]$ for $NO.i$ paradigm ($i = 1,2 \cdots M$). M number of paradigms are arranged according to object function value $f(F[i])$ in ascending sequence, as shown below: $f(F[1]) \leq f(F[2]) \leq \cdots f(F[M-1]) \leq f(F[M])$ .We can obtain series of individuals through application of K-means algorithm. Once a new individual $F[l]$ is obtained, it is inserted in the proper position of M number of paradigms arranged in object function value ascending manner if its object function is smaller than a certain object function value already having individuals, i.e. for $j \in (1, M)$, if $f(F[j-1]) < f(F[l]) < f(F[j])$, then $F[j] = F[l]$, $F[j+1] = F[j], \cdots, F[M] = F[M-1]$. Suc-h as, in the entire evolving process, M number of paradigms are constantly in a dynamic up-dating status.

2.3 Learning Paradigm Form of Cognitive Agent in Cluster

A new paradigm produced in $NO.k$ generation cognitive agent should refer to $NO.(k-1)$ generation paradigm. In the k generations, all sorted in the selected paradigm after 1 / 3 of the paradigm(Because the function value in accordance with small to large order, after the 1 / 3 of the paradigm is the paradigm of the best part of all). If there are M paradigm, that is, select the M / 3 (rounded up) a paradigm. In this M / 3 个 paradigm paradigm in a randomly selected. In this paradigm, the category $h(h \in (1, c), c$ is the number of clusters)of the number of datum $i(i \in (1, n), n$ is the numberof data to be clustered) to undergo can be explicitly displayed. It is believable that some data near clustering center p (p is the mean value of category h) under category h embody better cognitive behavior of the agents so that the heritage of such behavior should proceed from $NO.k$ individual, i.e. each categories under category h reserves certain data (to the preset ratio, e.g. ratio=0.45 as assigned in this article and taken round numbers upward). The reserved data are still of c number of categories and the rest will be allocated to these categories according to the similarity (Euclidean distance is used in this article) to the clustering center (means of all data under the category) as to complete heritage and generate a new paradigm.

2.4 Optimal Paradigm Strengthening and Attenuation

To strengthen SEP local self-optimizing ability, learning probability p_1 of "currently most optimal paradigm" $F[1]$ may be artificially enlarged. Mean-while, p_1 value should also be attenuated step by step in order to prevent entire social population convergence toward most optimal paradigm to lead to reduction of global self-optimization capability. The specifics are illustrated as follows:

If a new "currently most optimal paradigm" $F[1]$ generated in k generation, in the process of $k+1$ generation clusters, the probability of learning paradigm $F[1]$ is designated as p_1, $p_1 \in (0,1)$, and the probability $p_i (i = 2,3,\cdots,M)$ learned by other paradigms as

$$p_i = [1/f(F[i])](1-p_1)/\sum_{i=2}^{M} 1/f(F[i]) \tag{2}$$

In the process of each generation clusters between $k+2$ generation and $k+t$ generation(the supposition is renewed once more in $k+t$ generation of "currently most optimal paradigm"), the probability p_1 of paradigm $F[1]$ learned by other paradigms in turn as:

$$p_1^{k+i} = \begin{cases} p_1^k \times (100 - \mu^{(i-1)})/100 & \text{if} \quad p_1^k \times (100 - \mu^{(i-1)})/100 > 1/n \\ 1/n & \text{if} \quad p_1^k \times (100 - \mu^{(i-1)})/100 \le 1/n \end{cases} \tag{3}$$

In which, $i \in 2,3,\cdots,t)$, parameter μ controls attenuation rate, its right shoulder mark $(i-1)$ is the times of power. The less the μ, the slower the attenuation. In general, $\mu \in (1,3)$ is a proper setting. Other "paradigm" genetic rate $p_i, i \in (2,3,\cdots,M)$ will stil luse Eq. (2) for computation.

In general, the more the algorithm is more close to the evolution of post-its optimal solution, In order to not destroy the optimal solution as much as possible,the learning probability p_1 of Optimal paradigm, In the period (such as the first half of the cycle) and given its relatively small value, Later re-assigned a higher value. By contrast, Parameters μ in the larger pre-, post less. This is to keep a good paradigm, but also can increase the diversity of population.

2.5 Cognitive Agent Betrayal of Paradigm

①. Assume cognitive agent mutation probability threshold α, which is used to determine whether a certain cognitive agent bears the nature of betrayal, whereas behavioral mutation probability threshold β is used to determine on which time or times of specific behavior in the entire process the cognitive agent inclining to betray fall into betrayal.

②. Prior to cognition of each cognitive agent, a random number is given by an evenly distributed generator. If it is not greater than mutation threshold α, it is considered that it does not have betrayal nature and its behavioral process rigorously follows "cognitive agent learning paradigm" form to complete genetic process as mentioned above; otherwise, this agent has the nature of betrayal and is continued instep ③.

③. If the cognitive agent is identified bearing betrayal nature, a random number is assigned by the evenly distributed generator. If the random number is not greater than behavioral mutation rate β, the behavior does not belong to betrayal behavior and follow existing paradigm genetic form as described in cognitive agent learning paradigm; otherwise, this agent has the nature of betrayal and chaotic mutation operator will be applied to produce a new individual.

Similar to previous p_1、 μ, the value of α and β,in the early to give its larger value, given its relatively small in the latter part of the value.

3 Experimental Analysis

This paper picks up 505 articles in 6 categories from CQVIP as experiment data. The first 5 categories contain 100 articles each and the last category contains 5, which form the isolated points. The first 500 articles for the experiment are sourced from http://www.lib.tju.edu.cn. The last category is current affair and news sourced from http://www.baidu.com/.

3.1 Experiment 1

k-medoids algorithm is used for clustering analysis. The results are shown in Table 1.

Table 1. Results from k-medoids algorithm

	CM	Aviation	ES	MI	Finance	CAN
Wrong articles	59	56	49	56	53	2
Correct articles	45	41	51	47	42	4
Percentage of correct ones	43	42	51	46	44	67
Time(second)	35.3					

3.2 Experiment 2

KSEP algorithm is used for clustering analysis. The results are shown in Table 2.

Table 2. Results from GA-K algorithm

	CM	Aviation	ES	MI	Finance	CAN
Wrong articles	10	11	9	7	8	0
Correct articles	90	89	91	93	92	5
Percentage of correct ones	90	89	91	93	92	100
Time(second)	2978					

4 Summary

This paper embeds k-medoids algorithm into Social Evolutionary Programming, and Improved to learning of Paradigm、Optimal paradigm strengthening and attenuation and Cognitive agent betrayal of paradigm. This algorithm will increase the diversity of species group and enhance the optimization capability of social evolutionary programming, thus improve the accuracy of clustering and the capacity of acquiring isolated points.

Acknowledgements. This paper is supported by the National Natural Science Foundation of China （Grant No.70971077）、Shandong Province Doctoral Foundation (2008BS01028)、Natural Science Foundation of Shandong Province(Grant No.ZR2009HQ005、ZR2009HM008).

References

1. Yu, Y., Zhang, H.: A social cognition model applied to general combination optimization problem. In: Proceedings of the First International Conference on Machine Learning and Cybernetics, Beijing China, November 4-5, pp. 1208–1213 (2002)
2. Picault, S., Collinot, A.: Designing Social Cognition Models for Multi-Agent Systems through Simulating Primate Societies. In: Proceedings of ICMAS 1998 (3rd International Conference on Multi-Agent Systems), pp. 238–245 (1998)
3. Hao, Z.: Building Text Knowledge Map for Product Development based on CSEP Method. In: 2009 International Conference on Computer Network and Multimedia Technology, vol. 12, pp. 1081–1085 (2009)
4. Hao, Z., Yang, J.: Building Knowledge Map for Product Development based on GAKME Method. In: The Second International Workshop on Education Technology and Computer, vol. 3, pp. 696–699 (2010)
5. Zhu, M., He, F.: Data Ming, pp. 129–164. China Science and Technology University Press (2002)

E-commerce Recommendation Algorithm Based on Multi-level Association Rules

ShuiYuan Huang and LongZhen Duan

School of Information, Nanchang University, Nanchang, 330031, China
huangshuiyuan@163.com, dlz@163.com

Abstract. With the rapid development of the E-commerce, recommendation system and algorithm has become the relevant research hotspot.In this paper, we proposed a new recommendation algorithm,based on Multi-level Association Rules (MAR).The algorithm improves the precision and individuation degree of the recommendation,and significantly reduces the time needed for the recommendation. It mines the rule of the customer's choice of commodity,by using multi-level association rule, builds a model for choice prediction.

Keywords: Recommendation algorithm, Multi-level association, E-commerce, Mining.

1 Introduction

Recommendation algorithm is the important constituent of E-commerce recommendation system, using E-commerce sites to provide customer commodity information and advice,to help customers to decide what products to buy and analog the sale staff to help customers to complete the purchasing process.The most successful recommendation technique is the collaborative filtering recommendation, it often uses "user"and "item" to handle recommended transaction."User" represents customer,"item" represents product/item.Specific recommendation process begins from an initial rating-data matrix of "user"×"item",The rating-data matrix records the "user's" subjective rating-data of "item". Matrix "user"×"item" is identified either by the user, or implicitly inferred by the system.Once the rating-data matrix is identified, the recommendation system will be able to make the "user's" rating-data of other "item" through the function of rating-data,and then generate recommendation.As the number of customers and goods increasing,"user"×"item" will become more larger and more sparser.The sparse rating-data matrix will make the recommendation's quality greatly reduced;The huge rating-data matrix makes the real-time quality of recommendation reduced. This is the problem about typical sparseness and scalability of the current recommendation algorithm.

To solve the above problem, we propose a recommendation algorithm based on model,it solves the problem about sparseness and scalability of the current recommendation algorithm, it mines the rule of the customer's preference for commodity and then builds a model for choice prediction,by using multi-level association rule.

D. Jin and S. Lin (Eds.): Advances in ECWAC, Vol. 1, AISC 148, pp. 479–485.
springerlink.com © Springer-Verlag Berlin Heidelberg 2012

2 Relevant Research

A. Commodity Catalogy

In the virtual commercial buildings of E-commerce,all the commodities are stored according to a certain classification standard,meanwhile,the classification standard could be obtained.The classification standard presents as a tree structure.such as shown in figure 1, the leave of bottom layer is 0 layer, on behalf of a specific commodity,othes are hierarchy classification.Such as "Ferrari California SportyCar" is a specific commodity,and then "car" is general commodity category.

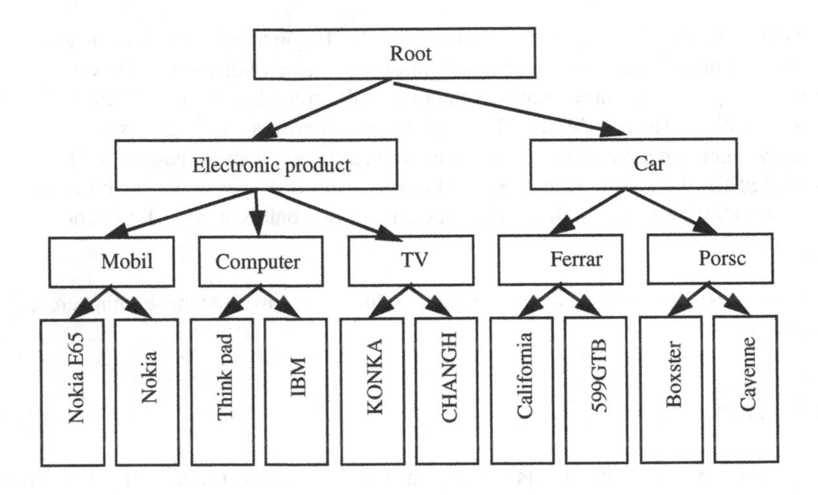

Fig. 1. Items classification hierarchy figure

B. Recommendation Agorithm Based on Multi-level Association Rules

Association rules refer to the statistics in a transaction database of trading transactions of commodity set X in the purchase of commodity set Y, which intuitively shows user's inclination to buy something else when purchasing certain commodity.The recommendation algorithm based on association rules creates association rules by using the user data,builds recommended models,and then brings recommendation to customer according to recommended models and their buying behaviors.The recommendation model based on association rules can be built offline,so it can ensure the effective recommendation algorithm for real-time requirements. Data set which was mined by association rules was stored in the transaction database, each record in the database presents a transaction. Formal definition of association rules mining is as follows: Suppose I = { i1, i2,..., im } is the set of items. Suppose the task-related data D is a set of database transactions,and each transaction T is a set of items,make $T \subseteq I$. each transaction has an identifier, called T ID. Suppose A is a set of items,transaction T contains A only if $A \subseteq T$. Asociation rules is an implicative form such as A , B,and $A \subset I, B \subset I$,besides,$A \cap B = \varphi$.The ruls A B is established in transaction set D,with support s ,and s is the percentage of the transactions which include $A \cup B$ in D,that is probability P(A ∪ B). The rule have the confidence c in transaction set D, If D contains A transaction also includes the percentage of B is c.This is the conditional probability P (B|A). That is :

$$\text{support}(A\ \ B) = P(A \cup B). \tag{1}$$

$$\text{confidence}(A\ \ B) = P(B \mid A) \tag{2}$$

The rule which can satisfy the minimum support threshold(min_sup) and the minimum confidence threshold(min_conf) at the same time is called strong rule.The set of item is called item-set, The item-set contains k items is called k_item-set . The item-set's frequency of appearance is the number of transactions that contain item-set, we shortly call it item-set's frequency.If the item-set satisfy min-sup,we call it frequent item-set. The set of frequent k_item-set is usually denoted by Lk . Mining of association rules is a two-steps process:

1) Find out the all frequent item-set:by definition,the frequency of these item-set's appearing was the same as the predefined minimum support count at least.

2) Strong association rules generated by frequent item-set:by definition,these rules must satisfy minimum support and minimum confidence.

Recomendation algorithm based on association rules described can be described as follow.Suppose P is the matrix of user-n's preference for m items of commodities .If user-i is interested in commodity-j, Pij is 1, otherwise Pij is 0.Suppose A is a association matrix ,contains confidence association rules of m items of commodities.Obviously, A can be calculated from the P,aij is the confidence of association rules i,j.So,recommendation vector r of the target user can be calculated by using association matrix A and preference vector u,according the formula (3). Therefore the user can select the most preferred top - N items of commodities as recommendation results,according to r.

$$r = u \times A \tag{3}$$

Recommendation algorithms Based on association rules x association rules can be divided into offline building phase of recommendation model based on association rules and online application phase of recommendation model based on association rules.we can use a variety of mining algorithms based on association rules to establish recommendation model of association rules in offline phase, this step is time-consuming, but it can be happened off-line;we we can provide users with real-time recommendation service, according to recommendation models of association rules which have been established and user's purchase behavior to in line phase. The steps of algorithm which uses recommendation algorithm based on association rules to generate top-N recommendation are as follows:

1) Use mining algorithms based on association rules to mine the transaction database,carry out all the association rules which can satisfy the minimum support threshold(min_sup) and the minimum confidence threshold(min_conf),we denote it association rules set R;

2) Search association rules set R,find out the all association rules set Ru, those the user supports, generate candidate recommendation set;

3) Remove those goods that the user-u have already purchased from the candidate recommendation set Pu,and select the top N of the highest confidence items as recommendation results to return to the current user-u.

3 Recommendation System Based on Multi-level Association Rules

A. Core Algorithm

In the recommendation algorithm based on association rules, if the amount of available preference information is small, strong association rules are relatively small,therefore correlation matrix becomes very sparse.In fact,if only use the data of purchase to predict user preference,the number of association rules will be very small,because each customer will usually just buy a small portion of a large number of goods sets. In this case, predict customer preferences for many commodities is impossible,meanwihle, performance and quality of recommendation will be very low. To overcome this problem, We propose to use additional information to produce recommendation,that is,according to commodity organizations,and the level types of storage, use the association rules between high-level categories of goods to recommend.

$$u_k = u_0 \times C_1 \times C_2 \times \ldots \times C_{k-1} \times C_k \quad (4)$$

Recommendation algorithm based on multi-level association rules can be described as follow:suppose C_k is a relational matrix between categories of goods, it expresses inclusion relation betwee below class level(k-1)and above class level(k), level(0) expresses goods.In the matrix, If class-i in level (k-1) belongs to class-j in level(k),$C_k(i,j)=1$,otherwise, $C_k(i,j)=0$.Use matrix C_k,Target user's preference matrix P_k and preference vector u_k for class level(k) can be calculated by the following formula:

$$P_k = P_0 \times C_1 \times C_2 \times \ldots \times C_{k-1} \times C_k \quad (5)$$

$$u_k = u_0 \times C_1 \times C_2 \times \ldots \times C_{k-1} \times C_k \quad (6)$$

In the formula (5),element value of $k \geq 1$ in P_k is not only.If user prefer a number of products in a class, the value may be greater than 1. At this time,correlation matrix A_k of level(k) can be calculated by the P_k,It consists of confidence association rules between different classes . Note that all the diagonal elements of correlation matrix A_k should be 1,other goods of the same class users interested in. In other words,for the class level $k \geq 1$,the confidence association rules c_i , c_i is always 1. According correlation matrix based on the expected count and class relationship matrix,recommendation r can be calculated,the formula is as follow:

$$r = a_0 \times u_0 \times A_0 + a_1 \times u_1 \times A_1 \times C^T_1 + \ldots + a_k \times u_k \times A_k \times C^T_k (\Sigma a_k = 1) \quad (7)$$

In the formula (7),a_k is weight,through association rules level(k), the sum is equal to 1. Better recommendation weight is determined according to experience,Each predicted preference of all class layers is regular to the interval[0.5,1],effected by S-shaped function.

B. Architecture

According to the theoretical analysis above, this paper designed a complete e-commerce recommendation system based on association rules.The recommendation system is based on XML, The user's recommendation request and the results which are returned by recommendation engine are supported in the mode of XML. The system is not only fit for C/S (Client / Server) two-tiers structure, but also B/W/S (Browser/Webserver/database server) three-tiers structure.In figure 2,we can see clearly that the data from the original transaction form to recommendation results output is the flowing process between the client, WWWserver, database server and recommendation engines.

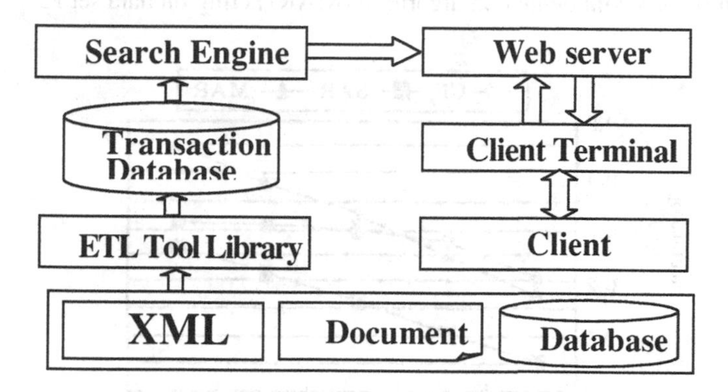

Fig. 2. Architecture of E-commerce recommendation system

From the function module, the system mainly consists of two core modules--the rules mining module (recommendation server) and the Web recommendation module (WWW server).

1) Rules mining module: extract loading data sets from a variety of external data sources by ETL tools (extract、 transform and load), standard data files generated is the basic object of mining,do the above multi-level association rules model as the basic algorithm, through configuring parameter of support, confidence and interest, complete the extraction of association rules and written the rules exrtacted into the rules library in form of relationship data.

2) Web recommendation module:directly face to buyers, has the following functions:1)Record the goods items,in the customer's shopping cart,as explicit/implicit input of the module. Explicit input is inputted their own evaluation of goods by user,according to their interests; Implicit input means input with artificial intelligence, machine learning and other means,the degree of user's preference for goods is predicted by the computer,the customer don't know completely in the circumstances,that do not need customer to specifically support some personal information for recommendation, this also reflects the system's characteristic--intelligent.2) Provide the application program interface of recommendation services to user;according to the needs of recommendation engine, collect user's access logs,

analyze user's navigation patterns and behavior patterns;use BP neural network and other artificial intelligence technology to analyze user's preferences, upload to the recommendation system database;meanwhile,translate the recommendation request submitted in form of XML and send to recommendation engine;receive the recommendation results provided by recommendation engine, and return the recommendation results to the e-commerce system in specific form.

C. *Performance Analysis*

Figure 3 describes the results of collaborative filtering recommendation algorithm (CF),single-level association rules recommendation algorithm (SAR) and multi-level association rules recommendation algorithm (MAR) acting on data set K.

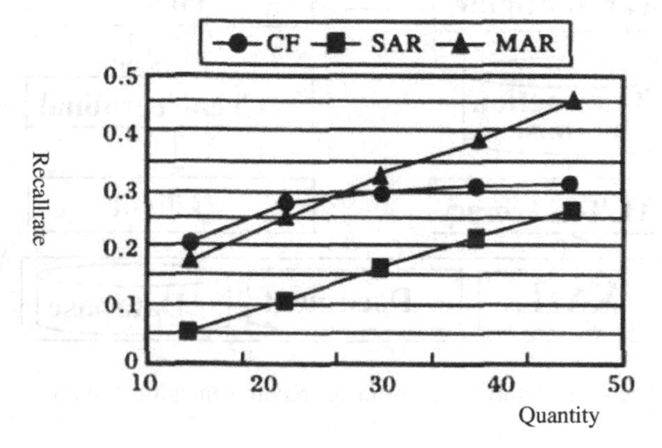

Fig. 3. Recommendation algorithm accuacy

Dataset K is very sparse, which inevitably makes strong association rules less. Suppose a_0=0.6, a_1= 0.3, a_2= 0.1,use recall rate (Recall) to measure accuracy.We can find out that MAR's performance is the higthest . For CF,when the recommendation number is much larger than 30,the performance is nearly the same, which reveals the impact of scalability and sparseness on the CF.

4 Conclusion

Realize personalized service through recommendation algorithm has become an emerging technology e-commerce applications, which also helps to implement content of site and adaptive structure.At the same time, people can apply the experience and implementation methods of system's humanity and intelligence which have been researched for a long time to recommendation system.We believe that the recommendation system technology will enable E-commerce Web site to become more individual, more personal,and can fit the preference of each user bitterly.

References

[1] Li, N., Li, C.: Zero-Sum Reward and Punishment Collaborative Filtering Recommendation Algorithm. IEEE Computer Society (September 2009)
[2] Wang, Q., Wang, J.: Improved Collaborative Filtering Recommendation Algorithm. Computer Science (2006)
[3] Feyyad, U.M.: Data mining and knowledge discovery: making sense out of data. In: IEEE Expert (October 1996)
[4] Li, Q., Khosla, R.: An Adaptive Algorithm for Improving Recommendation Quality of E-recommendation Systems. In: Computational Intelligence for Measurement Systems and Applications (July 2003)
[5] Muthukumar, A., Nadarajan, R.: TDML: A Data Mining Language for Transaction Databases. In: Fuzzy Systems and Knowledge Discovery (August 2007)
[6] Shaw, G., Xu, Y., Geva, S.: Utilizing Non-redundant Association Rules from Multi-level Datasets. In: Web Intelligence and Intelligent Agent Technology (December 2008)

Electroencephalograph Automatic Diagnosis Based on Kernel Principal Angle

Haikuan Liu[1], Xijuan Wang[2], and Liangzhi Gan[1,*]

[1] School of Electrical Engineering and Automation, Xuzhou Normal University, Xuzhou 221116, China
[2] College of Physics and Electronic Information, Luoyang Normal University, Luoyang 471022, China
well178@sina.com, fjxwxj@126.com, lzh_box@163.com

Abstract. Electroencephalograph (EEG) is the most important method in diagnosis of brain diseases. Unfortunately, the doctor interprets EEG according to personal experiences, which results in over interpretation or incorrect judgment. In this paper, we derive a sparse kernel principal angle (SKPA). The SKPA is applied to disease diagnosis in examples: we select Electroencephalograph data of different diseases, calculate the principal angles between the datasets. Reading the principal angles, one can find diseases or even diagnose disease type. The electroencephalograph automatic diagnosis system based on kernel principal angle will be helpful to doctors.

Keywords: Principal Angle, Kernel Method, Electroencephalograph.

1 Introduction

Principal angle (PA) is the minimal angle between two subspaces. It had been applied to various fields [1, 2]. Traditional principal angle has two shortcomings. Firstly it solves only linear problems. Secondly, if we calculate the principal angle in low dimensional space and the samples are huge, it is sure that the principal angle is zero because two data sets with large samples in low dimensional space are surely linear dependent. To overcome the shortcomings of the traditional PA, several nonlinear kernel method were reported [3]. But kernel principal angle (KPA) may lead to problem of over learning. According to statistic learning theory [4], some restrictions must be imposed. In this paper, we developed a sparse method of kernel principal angles, SKPA. The basic idea of SKPA is to first map the input space into Reproduced Hilbert Feature Space via nonlinear mapping and then to compute the basis or approximate basis of the feature space. The basis or approximate basis can represent all the samples by the way of linear combination. Then one can compute the principal angles in that feature space. By this way, the principal angles can be represented by the basis or approximate basis, so the solution is sparse. According to Vapnik's theory, SKPA can get better generalization.

* Corresponding author.

D. Jin and S. Lin (Eds.): Advances in ECWAC, Vol. 1, AISC 148, pp. 487–492.
springerlink.com © Springer-Verlag Berlin Heidelberg 2012

2 Principal Angle

Two random variables X and Y with zero mean. $\{a_1;\cdots;a_N\}$ and $\{b_1;\cdots;b_M\}$ are samples from X and Y. Subspaces $U_A=\mathrm{span}\{a_1;\cdots;a_N\}$, $U_B=\mathrm{span}\{b_1;\cdots;b_M\}$ are spanned by matrix $A=[a_1;\cdots;a_N]$ and $B=[b_1;\cdots;b_M]$. Principal angles between the two subspaces are defined as

$$\cos(\theta)=\max_{u\in U_A}\max_{v\in U_B} u^T v, \tag{1}$$

$$\text{subject to } u^T u = v^T v = 1. \tag{2}$$

Problem (1) can be written as:

$$\cos\theta=\max_{x,y} y^T B^T Ax, \tag{3}$$

$$\text{subject to } \|Ax\|_2=1,\|By\|_2=1. \tag{4}$$

To overcome the shortcoming of the traditional linear principal angles, several nonlinear extensions of kernel method were reported [4].

Suppose in the feature space $\Phi_A=[\varphi(a_1);\cdots;\varphi(a_N)]$ and $\Phi_B=[\varphi(b_1);\cdots;\varphi(b_M)]$, problem (3) can be written as:

$$\cos\theta=\max_{x,y} y^T {\Phi_B}^T \Phi_A x, \tag{5}$$

subject to

$$\|\Phi_A x\|_2=1,\|\Phi_B y\|_2=1. \tag{6}$$

Kernel principal angles can be written as an eigenproblem:

$$\begin{bmatrix} 0 & \mathbf{\Phi_B}^T\mathbf{\Phi_A} \\ \mathbf{\Phi_A}^T\mathbf{\Phi_B} & 0 \end{bmatrix}\begin{bmatrix} y \\ x \end{bmatrix}=\lambda\begin{bmatrix} \mathbf{\Phi_B}^T\mathbf{\Phi_B} & 0 \\ 0 & \mathbf{\Phi_A}^T\mathbf{\Phi_A} \end{bmatrix}\begin{bmatrix} y \\ x \end{bmatrix}. \tag{7}$$

In equation (7), the explicit expression of $\varphi(a)$ is not necessary. $\mathbf{\Phi_A}^T\mathbf{\Phi_B}$, $\mathbf{\Phi_B}^T\mathbf{\Phi_B}$ and $\Phi_A^T\Phi_A$ can be represented by kernel functions. So the eigenproblem is still linear eigenproblem, the dimension of the equation is M+N, where M is the column of A and N is the column of B.

3 Online Sparse Kernel Principal Angle for EEG Automatic Diagnosis System

Observing equation (3), we could find that Ax and $y^T B^T$ are the linear combination of all samples in feature space. If we could get the basis of sampling data sets A and B, vectors Ax and $y^T B^T$ can be simply represented by the basis. Without losing generality, suppose $\varphi(a_k)$ is an element of sampling data set. By solving the following minimizing problem, we know whether $\varphi(a_k)$ is one of the basis or not.

$$\min \; f(\lambda)=(\varphi(a_k)-\sum_{i\neq k}\lambda_i\varphi(a_i))^T(\varphi(a_k)-\sum_{i\neq k}\lambda_i\varphi(a_i)) \tag{8}$$

It is easy to know that $f(\lambda)\geq 0$. If $f(\lambda)=0$, $\varphi(a_k)$ is not one element of the basis because it can be linearly represented by other elements. But if $f(\lambda)>0$, $\varphi(a_k)$ is one of the basis. According to

The following steps are used to find the basis of sampling data set $\{\varphi(a_k):k=1,\cdots,N\}$ in feature space, which we called *Basis Selecting Method*:

Step1. Found two new sets $A_i = \{\varphi(a_1)\}$ and $A_d=\phi$, ϕ is an empty set;

Step2. To every $\varphi(a_k)$, $k=2,\cdots,N$, get value of $\min(\varphi(a_k)-\sum_{\varphi(a_i)\in A_d}\lambda_i\varphi(a_i))^T(\varphi(a_k)-\sum_{\varphi(a_i)\in A_d}\lambda_i\varphi(a_i))$;

Step3. If the value is less than ε (a very small positive number that approximate to zero, which is called *sparse sensitive parameter*), $\varphi(x_k)$ is added to set A_d, else $\varphi(x_k)$ is added to set A_i;

Step4. Go back to step 2 until all the vectors are checked.

It is easy to know that there must be an x' satisfying $\Phi_A x = \mathbf{A}_i x'$ and a y' satisfying $\Phi_B y = \mathbf{B}_i y'$. We could rewrite formula (7) as:

$$\begin{bmatrix} 0 & \mathbf{B}_i^T\mathbf{A}_i \\ \mathbf{A}_i^T\mathbf{B}_i & 0 \end{bmatrix}\begin{bmatrix} y \\ x \end{bmatrix}=\lambda\begin{bmatrix} \mathbf{B}_i^T\mathbf{B}_i & 0 \\ 0 & \mathbf{A}_i^T\mathbf{A}_i \end{bmatrix}\begin{bmatrix} y \\ x \end{bmatrix}. \tag{9}$$

This is the SKPA. The dimension of the equation (9) is the sum of columns of A_i and B_i, which is much smaller than M+N.

4 Experiments

4.1 Testing Samples

Five kinds of EEG data sets are selected. They are: virus meningitis (VM), cerebral arteriosclerosis (CA), infraction of the brain (IB), epilepsy (EP) and normal (NR). Every type of disease has two groups of data, which is called training and testing data. We can calculate the angles between the normal data (NR) and the training data. Because the training data is partly different from normal data, some segments of the angles may be big.

4.2 Experiment Schema

As shown in fig.1, to diagnose the type of disease by electroencephalograph data sets is very difficult. But it is theoretically feasible if the data sets sampled from patients are different from normal person. The following steps are used to find whether the data sets sampled from patients are different from normal person:

Step1. Acquire normal data set, get the basis of the normal operating data set in the feature space by the *Basis Selecting Method* introduced before, and the basis is expressed as vector A_i;

Step2. Carry out centering in the feature space: $\tilde{A}_i = A_i - \frac{1}{N}\mathbf{1}_{1\times N} A_i \mathbf{1}_{N\times 1}$, where N is the columns of vector A_i, $\mathbf{1}_{1\times N} = [1,\cdots,1]$ and $\mathbf{1}_{N\times 1} = [1,\cdots,1]^T$;

Step3. Acquire the testing data set B (for example, data set sampled from an epilepsy), carry out the basis B_i in the feature space, and $\tilde{B}_i = B_i - \frac{1}{M}\mathbf{1}_{1\times M} B_i \mathbf{1}_{M\times 1}$, where M is the columns of vector B_i;

Step4. Solve the eigenvalue-system problem

$$\begin{bmatrix} 0 & \tilde{\mathbf{B}}_i^{\mathrm{T}}\tilde{\mathbf{A}}_i \\ \tilde{\mathbf{A}}_i^{\mathrm{T}}\tilde{\mathbf{B}}_i & 0 \end{bmatrix}\begin{bmatrix} y \\ x \end{bmatrix} = \lambda \begin{bmatrix} \tilde{\mathbf{B}}_i^{\mathrm{T}}\tilde{\mathbf{B}}_i & 0 \\ 0 & \tilde{\mathbf{A}}_i^{\mathrm{T}}\tilde{\mathbf{A}}_i \end{bmatrix}\begin{bmatrix} y \\ x \end{bmatrix};$$

Step5. Determine the control limits of the principal angle charts.

As an online monitoring method B_i can be constructed as time-serial variable: $B_i = [b_{t-L+1}, b_{t-L+2}, \cdots, b_t]$.

The Detecting results of epilepsy (EP) and cerebral arteriosclerosis (CA) are shown in fig.1.

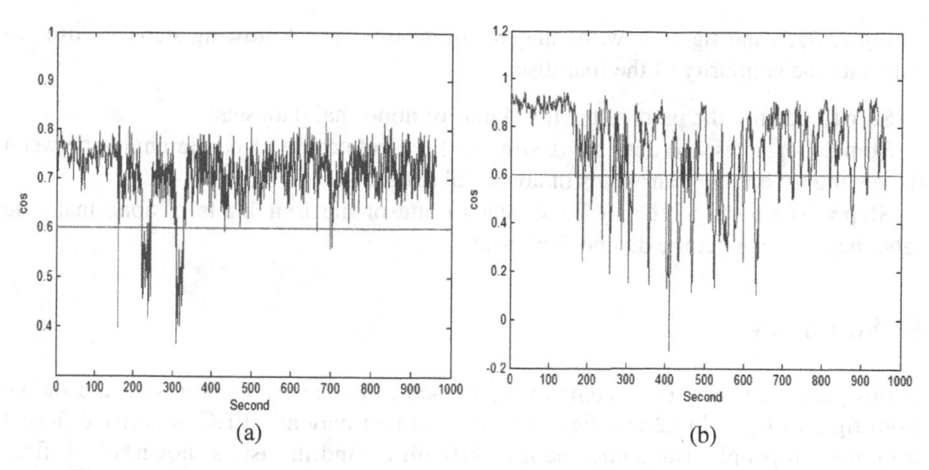

Fig. 1. (a) Detecting Epilepsy (EP); (b) Detecting Cerebral Arteriosclerosis (CA)

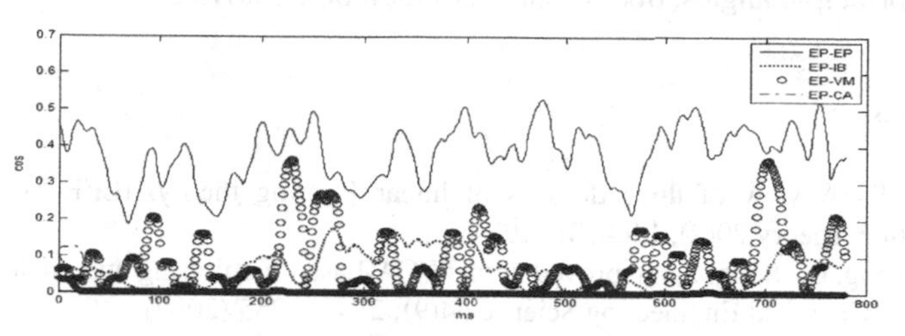

Fig. 2. Diagnosing Epilepsy

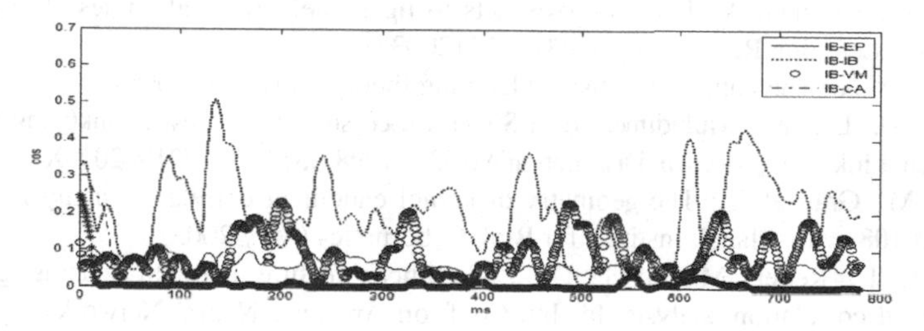

Fig. 3. Diagnosing Infraction of Brain

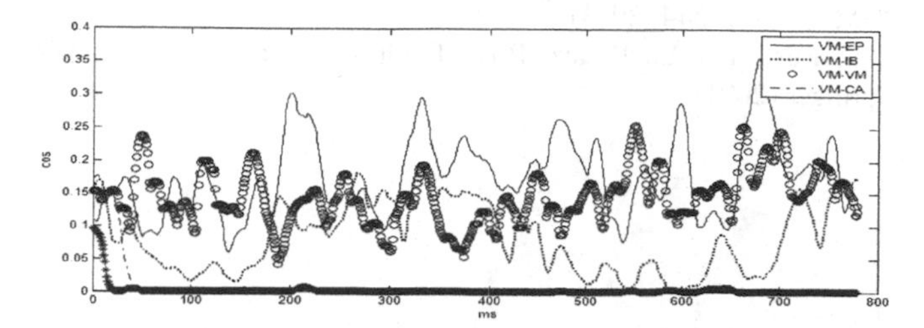

Fig. 4. Diagnosing Virus Meningitis

Fig.2, fig.3 and fig.4 show the diagnosing results. The following steps are used to calculate the similarity of the four diseases:

Step1. Acquire the parts of the four kinds of abnormal data sets;

Step2. For a disease data set (testing set), calculate the principle angles between the testing set and the four kinds of abnormal data sets;

Step3. The testing set may be similar to one of the four kinds of abnormal data sets, then one can get the diagnosing result.

5 Summary

In this paper, we carry out a complex experiment. The experimental results are shown from fig.2 to fig.6. Fig.2 and fig.3 demonstrate that patient's EEG is partly different from normal people. But sometime it is difficult to find it, just as shown. Fig.4, fig.5 and fig.6 demonstrate that patients with the same disease have similar EEGs. By calculating principle angles, doctors can get much better advice.

References

1. Kailath, T.: A view of three decades of linear filtering theory. IEEE Transactions on Information Theory 20(2), 146–181 (1974)
2. Ge, Z., Yang, C., Song, Z.: Improved kernel PCA-based monitoring approach for nonlinear processes. Chemical Engineering Science 64(9), 2245–2255 (2009)
3. Gittins, R.: Canonical analysis; a review with applications in ecology. Acta Biotheoretica 35(1), 135–136 (1986)
4. Wolf, L., Shashua, A.: Learning over sets using kernel principal angles. The Journal of Machine Learning Research 6(4), 931–937 (2003)
5. Vapnik, V.N.: The nature of statistical learning theory. Springer (2000)
6. Borga, M.: Learning Multidimensional Signal Processing. PhD thesis, Linkoping University, Sweden. Linkoping, Sweden, Dissertation No 531 (1998) ISBN 91–7219–202–X
7. Kuss, M., Graepel, T.: The geometry of kernel canonical correlation analysis. Technical Report 108, Max-Planck Institute for Biol. Cybernetics (May 2003)
8. Melzer, T., Reiter, M., Bischof, H.: Nonlinear feature extraction using generalized canonical correlation analysis. In: Int. Conf. on Artificial Neural Networks, pp. 353–360 (2001)
9. Hotho, A., Staab, S., Stumme, G.: Ontology Improves Text Documents Clustering. In: Proc.of ICDM, pp. 541–544 (2003)
10. Chiang, L.H., Russell, E.L., Braatz, R.D.: Fault Detection and Diagnosis in Industrial Systems. Springer (2001)

The Category of the 4-Valued Fuzzy Sets

Xiaoshen Li[1], Hongjian Zhang[2], and Xuehai Yuan[3]

[1] Henan University of Science and Technology, School of Mathematics and Statistics, Xiyuanroad 48, 471003 Luoyang, China
[2] Xuchang Technical and Economical College, Department of Electrical Engineering, Jiansheroad S.S., 461500 Changge, China
[3] Dalian University of Technology, School of Electronic and Information Engineering, Linggongroad 2, 116024 Dalian, China
hnykli@163.com

Abstract. This paper considers the connection of the 4-valued fuzzy sets with topos theory. The authors constructed the category **QFuz** of the 4-valued fuzzy sets. The category **QFuz** has a middle object and all topos properties except Subobject Classifier, consequently forms a weak topos.

Keywords: 4-valued fuzzy sets, category, topos, weak topos.

1 Introduction

Although the theory of fuzzy sets and the theory of topos are developed separately, they should be connected since both of them are used to deal with vagueness through logic. Recently this has been verified by various investigators[1-12]. It is well known that the category **Set** of classical sets is a topos. By the use of topos **Set**, the logic operators of classical sets such as negation, conjunction, implication and disjunction, are obtained both logically and naturally[13]. In [9-11], the category **Fuz** of fuzzy sets is redefined and the concepts of middle object and weak topos are introduced. By the use of weak topos, the logic operators of fuzzy sets as defined by Zadeh are described naturally, and consequently the category **Fuz** has a fimilar function to topos **Set**. In [12], the category **FuzFuz** of morphisms in category **Fuz** is considered and proved to be a weak topos. In [4], the category of Y-N sets is considered and is proved to be a weak topos.

It is well known that some special L-fuzzy sets such as the interval-valued fuzzy sets, the intuitionistic fuzzy sets, the interval-valued intuitionistic fuzzy sets, and the type 2 fuzzy sets et al. play an important role in fuzzy systems. In [14], the concept of the three-dimensional fuzzy set, a special L-fuzzy set, is proposed. For the three-dimensional fuzzy sets, definition of cut sets is given, decomposition theorems and representation theorems are developed. Those results have shown that the three-dimensional fuzzy sets have the intimate connection with 4-valued fuzzy sets. However, the relations between the 4-valued fuzzy sets and topos theory have not been discussed yet. In this paper the category **QFuz** of the 4-valued fuzzy sets is constructed and proved to form a weak topos.

D. Jin and S. Lin (Eds.): Advances in ECWAC, Vol. 1, AISC 148, pp. 493–500.
springerlink.com © Springer-Verlag Berlin Heidelberg 2012

The rest of this paper is organized as follows. In section 2, some preliminaries are provided. In section 3 , the category **QFuz** of the 4-valued fuzzy sets is constructed and proved to be weak topos. In section 4, the conclusions are given.

2 Preliminary

Definition 1[14]. Let X be a set. The mapping $A: X \to \{0,1/3,2/3,1\}$ is called a 4-valued fuzzy set.
Definition 2[13]. A topos is a category $\mathscr{C}$ satisfying the five conditions:
(1) Finite products exist in $\mathscr{C}$, i.e., for any objects A, B in $\mathscr{C}$, there exists an object C and morphisms $p_1: C \to A$, $p_2: C \to B$ satisfying for any morphisms $f: D \to A$ and $g: D \to B$, there exists a unique morphism $h: D \to C$ such that $p_1 \circ h = f$ and $p_2 \circ h = g$, i.e., The Fig. 1 is commutative. C is denoted as $A \times B$.

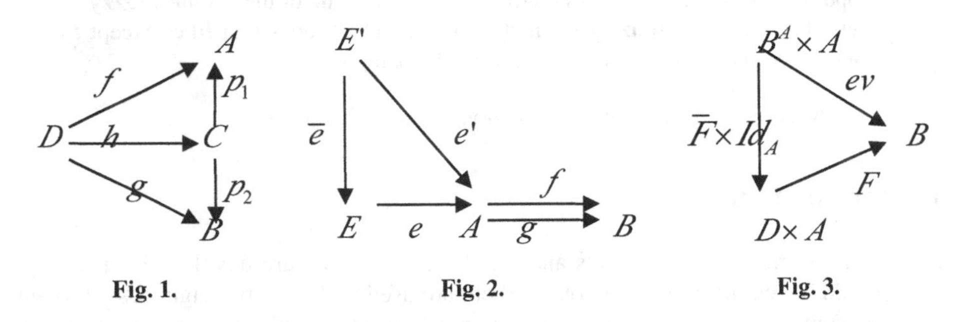

Fig. 1. **Fig. 2.** **Fig. 3.**

(2) Equalizer exists in $\mathscr{C}$, i.e., for any morphisms f, g : $A \to B$, there exists an object E and a morphism $e: E \to A$ satisfying
(a) $f \circ e = g \circ e$;
(b) For any $e': E' \to A$ satisfying $f \circ e' = g \circ e'$, there exists a unique morphism $\overline{e}: E' \to E$ such that $e \circ \overline{e} = e'$, i.e., Fig. 2 is commutative.
(3) There is a terminal object U in $\mathscr{C}$. This means that for each object A, there is one and only one morphism from A to U , which is denoted as !.
(4) Exponentials exist in $\mathscr{C}$, i.e., for any objects A,B in $\mathscr{C}$, there exists an object B^A in $\mathscr{C}$ and a morphism $ev: B^A \times A \to B$ satisfying, for any morphism $F: D \times A \to B$, there exists a unique morphiam $\overline{F}: D \to B^A$ such that $ev \circ (\overline{F} \times Id_A) = F$, i.e., Fig.3 is commutative.
(5) There is a subobject classifier in $\mathscr{C}$, i.e., there is an object Ω and a morphism T from U to Ω such that for each monomorphism f from A' to A there exists a unique morphism χ_f from A to Ω such that the Fig. 4 is a pullback. This means

(a) The Figure 4 is commutative, i.e. $\chi_f \circ f = \mathrm{T} \circ !$.

(b) For each object B and a morphism g from B to A such that $\chi_f \circ g = \mathrm{T} \circ !$, then there exists a unique morphism $\overline{g}$ from B to A' such that $g = f \circ \overline{g}$ (see Figure 5). where χ_f is called CH(characteristic morphism) of (A', f).

For example, the category **Set** of classical sets is a topos, $U = \{0\}$ is a terminal object; $\Omega = 2 = \{0,1\}$, $\mathrm{T}: U \to \Omega$, $\mathrm{T}(0)=1$ is a subobject classifier. Then for any monomorphism $f: A' \to A$,

$$\chi_f = \begin{cases} 1, a \in f(A'); \\ 0, a \notin f(A'). \end{cases}$$

The χ_f can be seen as the CH(characteristic function) of A'.

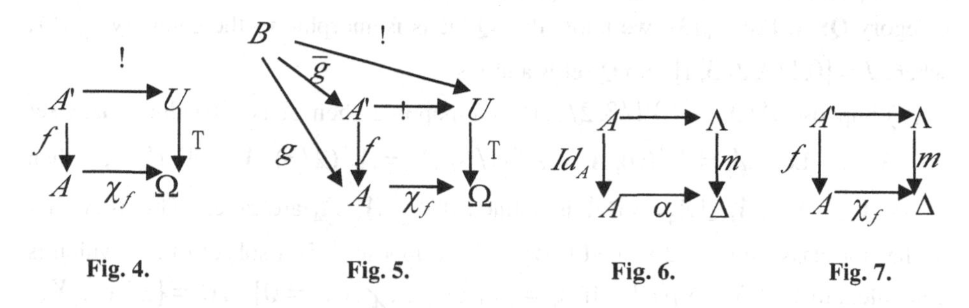

Fig. 4. **Fig. 5.** **Fig. 6.** **Fig. 7.**

Definition 3[10]. A weak topos is a category $\mathscr{C}$ satisfying the five conditions:

(1) Finite products exist in $\mathscr{C}$.
(2) Equalizers exist in $\mathscr{C}$.
(3) There is a terminal object in $\mathscr{C}$.
(4) Exponentials exist in $\mathscr{C}$.
(5) There is a middle object in $\mathscr{C}$. i.e., there exists a monomorphism $m: \Lambda \to \Delta$ such that

(a) Hom$(A, \Delta) = \{ f \mid f: A \to \Delta$ is a morphismg$\}$ is partially ordered for all object A in $\mathscr{C}$.

(b) There is a unique smallest morphism α so that the square (Fig. 6) is a pullback.

(c) For any monomorphism $f: A' \to A$ there is a unique morphism $\chi_f: A \to \Delta$ such that $\chi_f \leq \alpha$ and the square (Fig. 7) is a pullback. χ_f is called a CH of f.

For example, the category **Fuz** of fuzzy sets is a weak topos, where

$$m: (0,1] \to [0,1], x \mapsto x$$

and for any monomorphism $f:(A',\alpha')\to(A,\alpha)$,

$$\chi_f(a)=\begin{cases}\alpha'(a'), & if\ a=f(a')\in f(A');\\ 0, & else.\end{cases}$$

The χ_f can be seen as the membership function of fuzzy set (A',α').

3 The Category QFuz of the 4-Valued Fuzzy Sets

In this section, We construct categories of the 4-valued fuzzy sets in two ways.

(I) Suppose A is a 4-valued fuzzy set on X. Let (X,A) be an object; a mapping $f:X\to Y$ be a morphism from (X,A) to (Y,B), which satisfies $B(f(x))=A(x), \forall x\in X$; the composition of the mappings be the composition of the morphisms; and the identical mapping be the identical morphism. We construct a category **QSet**. From [13], we know that **QSet** is isomorphic to the category $B_n(J)$, where $J=\{0,1/3,2/3,1\}$. So **QSet** is a topos.

(II) Suppose $A:X\to\{0,1/3,2/3,1\}$ is a mapping, then A is a 4-valued fuzzy set on X. Let $A_1=A^{-1}(0), A_2=A^{-1}(1/3), A_3=A^{-1}(2/3), A_4=A^{-1}(1)$, then $X=A_1\cup A_2\cup A_3\cup A_4$. So A is defined if A_2, A_3, A_4 are given. This is similar to that for classical sets (2-valued fuzzy sets). Suppose A is a subset of X, which is equivalent to $\chi_A:X\to\{0,1\}$. If $A_1=\{x\mid x\in X,\chi_A(x)=0\}$, $A_2=\{x\mid x\in X, \chi_A(x)=1\}$,then $A=A_2$. So to consider A, ,it is enough to consider A_2. Thus we can construct another category **QFuz** similarly.

Suppose $A:X\to\{1/3,2/3,1\}$ is a mapping, Let (X,A) be an object. A mapping $f:X\to Y$ be a morphism from (X,A) to (Y,B), which satisfies $B(f(x))\geq A(x)$. The composition of the mappings be the composition of the morphisms; and the identical mapping be the identical morphism. We construct the category **QFuz**.

Theorem 1. The category QFuz is not a topos but a weak topos.
Proof (1) There exists an Equalizer for any two objects in **QFuz**.

Suppose $(X,A)\overset{f}{\underset{g}{\rightrightarrows}}(Y,B)$ are two morphisms. Let

$$Z=\{x\mid x\in X,f(x)=g(x)\}, C(x)=A(x), \forall x\in Z, e:Z\to X, x\mapsto x$$

Then $\{(Z,C),e\}$ is an Equalizer of f and g.

(2) There exists a Finite Product for any two objects in **QFuz**.

Suppose (X, A) and (Y, B) are two objects, then (Z, C) is an object in **QFuz** where $Z = X \times Y$; $C(x, y) = \min\{A(x), B(y)\}$. Let $p_1 : Z \to Y, (x, y) \mapsto x$; $p_2 : Z \to Y, (x, y) \mapsto y$. Then $\{(Z, C), p_1, p_2\}$ is a Finite Product of (X, A) and (Y, B).

(3) There exists a terminal object in **QFuz**.

Let $I = (\{0\}, M), M(0) = 1$. Then I is a terminal object in **QFuz.**

(4) There exists an Exponential for any two objects in **QFuz**.

Suppose (X, A) and (Y, B) arc two objects in **QFuz**, $Z = Y^X = \{f \mid f : X \to Y$ is a mapping$\}$. Let

$C(f) = \max\{\lambda \mid \lambda \in \{1/3, 2/3, 1\}\}, \min\{\lambda, A(x)\} \leq B(f(x)), \forall x \in X\}$.

Then $C(f) \geq 1/3$. So (Z, C) is an object in **QFuz**.

Let $ev : Z \times X \to Y, (f, x) \mapsto f(x)$. Obviously, $\min\{C(f), A(x)\} \leq B(f(x)), \forall x \in X$. So ev is a morphism.

Suppose $F : (D, H) \times (X, A) \to (Y, B)$ is a morphism. Then $F : D \times X \to Y$ is a mapping satisfying $\min\{H(d), A(x)\} \leq B(F(d, x))$. Let $\overline{F} : D \to Z$, $d \mapsto \overline{F}(d)$, where $\overline{F}(d)(x) = F(d, x)$. Then F is a mapping. Due to

$$\min\{H(d), A(x)\} \leq B(F(d, x)) = B(\overline{F}(d)(x)), \forall x \in X,$$

$H(d) \leq C(\overline{F}(d))$. Consequently, $\overline{F}$ is a morphism. Obviously, $ev \circ (\overline{F} \times Id_X) = F$ and $\overline{F}$ is unique.

Hence $\{(Z, C), ev\}$ is an Exponential of (X, A) and (Y, B).

(5) There does not exist any Subobject Classifier in **QFuz**.

Assume there exists an Ω and a morphism T satisfying Ω-axiom. Then Fig. 8 is a commutative pullback square for $X = \{0\}, A(0) = 1/3, Y = \{0\}, B(0) = 1$, where $f(0) = 0$. So $\chi_f \circ Id_Y = \mathsf{T} \circ !$. Consequently, there exists a morphism $\overline{f} : Y \to X$ such that $A(\overline{f}(0)) \geq B(0)$. It follows that $1/3 \geq 1$, which is a contradiction. Hence there does not exist any Subobject Classifier in **QFuz**.

(6) There t exists a middle object in **QFuz**.

Let $\Omega = \{0, 1/3, 2/3, 1\}, \omega(a) = 1, \forall a \in \Omega; M = \{1/3, 2/3, 1\}, m(a) = a$, $\forall a \in M$. Then (Ω, ω) and (M, m) are two objects in **QFuz.** Let $\mathsf{T} : M \to \Omega$, $a \mapsto a$. Then T is a morphism.

(a) Suppose (X, A) is an object in **QFuz**. $\hom((X, A), (\Omega, \omega)) = \{f \mid f : X \to \Omega$ is a mapping$\}$. In $\hom((X, A), (\Omega, \omega))$ we define

$$f \le g \Leftrightarrow f(x) \le g(x), \forall x \in X.$$

$\hom((X, A), (\Omega, \omega))$ is a poset.

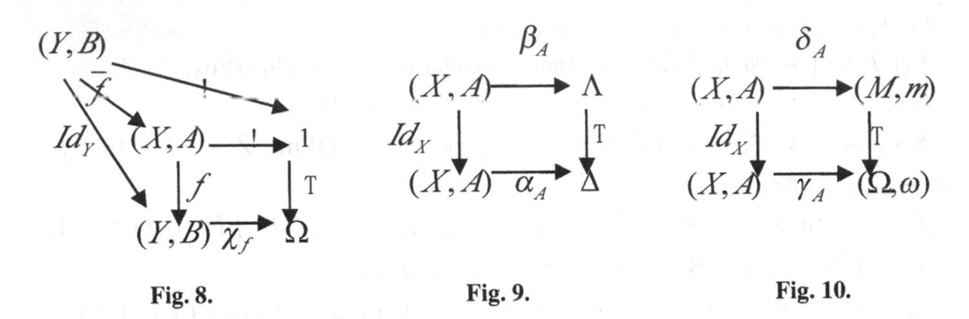

Fig. 8. **Fig. 9.** **Fig. 10.**

(b) Suppose (X, A) is an object in **QFuz**. Then Fig. 9 is a pullback square where $\alpha_A(x) = A(x), \beta_A(x) = A(x)$.

(c) Suppose A is a morphism such that Fig. 10 is a pullback square.

Because $\gamma_A \circ Id_X = \top \circ \delta_A$ and $m(\delta_A(x)) \ge A(x), \delta_A(x) \ge A(x)$. Hence $\gamma_A(x) = \gamma_A(Id_X(x)) = \top(\delta_A(x)) = \delta_A(x) \ge A(x) = \alpha_A(x), \forall x \in X$.

(d) Suppose $f : (X, A) \to (Y, B)$ is a monomorphism, then f is an injection satisfying $B(f(x)) \ge A(x)$. Let

$$\chi_f(y)(x) = \begin{cases} A(x), & y = f(x) \in f(X); \\ 0, & else. \end{cases}$$

Then $\chi_f(y) \le B(y), \forall y \in Y$. Consider Fig. 11, where $g(x) = A(x)$.

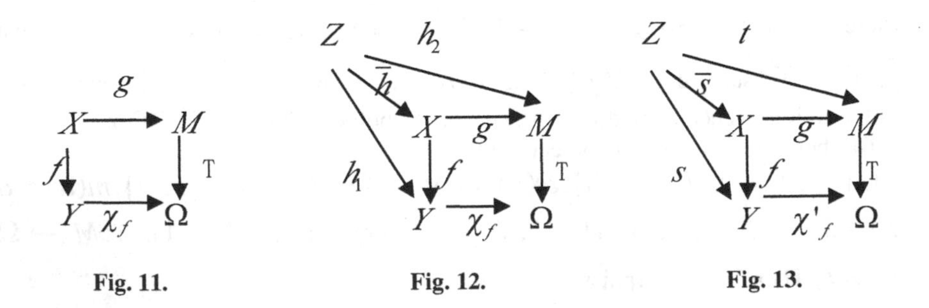

Fig. 11. **Fig. 12.** **Fig. 13.**

We have that $\chi_f \circ f = \top \circ g$. Let (Z, C) is an object in QFuz. Next consider Fig. 12.

$\chi_f(h_1(z)) = \mathsf{T}(h_2(z)) = h_2(z) > 0$, where $\chi_f \circ h_1 = \mathsf{T} \circ h_2$. So there exists a $x \in X$. Such that $h_1(z) = f(x)$. Let $\bar{h}: Z \to X, z \mapsto x, if\ h_1(z) = f(x)$.

$\bar{h}$ is a mapping since f is an injection. Because $C(z) \le m(h_2(z)) = h_2(z) = \mathsf{T}(h_2(z)) = \chi_f(h_1(z)) = \chi_f(f(x)) = A(x) = A(\bar{h}(z))$, $\bar{h}$ is a morphism. Obviously, Fig. 12 is commutative and $\bar{h}$ is unique. Fig. 11 is a pullback square.

(e) χ_f is unique.

Suppose $\chi_f': Y \to \Omega$ satisfying $\chi_f'(y) \le B(y)$, and substituting for χ_f, Fig.12 is still a pullback square. Then $\chi_f'(f(x)) = \mathsf{T}(g(x)) = g(x) = A(x)$.

Consequently, $\chi_f'(y) = \chi_f(y), \forall y \in f(X)$.

Assume there exists a $y \in Y \setminus f(X)$ satisfying $\chi_f'(y) > 0$. Let $Z = \{y\}$, $C(y) = \chi_f'(y)$. Consider Fig. 13. Due to $C(y) = \chi_f'(y) \le B(y) = B(s(y))$, $s: s(y) = y$ is a morphism. Let $t(y) = \chi_f'(y)$, then $\chi_f' \circ s = \mathsf{T} \circ t$. There exists a unique morphism $\bar{s}: Z \to X$ such that Fig.13 is commutative. So $y = s(y) = (f \circ \bar{s})(y) = f(\bar{s}(y)) \in f(X)$. This contradicts $y \notin f(X)$. Hence $\chi_f'(y) = 0, \forall y \in Y \setminus f(X)$. Consequently,

$$\chi_f'(y) = \begin{cases} A(x), & y = f(x) \in f(X) \\ 0, & y \notin f(X) \end{cases} = \chi_f(y).$$

Hence $\mathsf{T}: (M, m) \to (\Omega, \omega)$ is a middle object in **QFuz**.

From (1) -(5) we conclude that the category **QFuz** does not form a topos. From (1) -(4) and (6) we conclude that the category QFuz form a weak topos.

4 Conclusions

The category QFuz of the 4-valued fuzzy sets is constructed. It is shown that the category has the following properties.

(1) It has the topos properties such as Equalizer, Finite Product, Terminal object and Exponential.
(2) It has no Subobject Classifier.
(3) It has Middle object

Hence the category is not topos but a weak topos.

References

1. Atanassov, K.: Intuitionistic fuzzy sets. Fuzzy Sets and Systems 20(1), 87–96 (1986)
2. Atanassov, K., Gargov, G.: Interval-valued intuitionistic fuzzy sets. Fuzzy Sets and Systems 31, 343–349 (1989)
3. Barr, M.: Fuzzy sets and topos theory. Canada Math. Bull. 24, 501–508 (1986)
4. Carrega, C.: The categories SetH and FuzH. Fuzzy Sets and Systems 9, 322–327 (1983)
5. Eytan, M.: Fuzzy sets: a topos-logical point of view. Fuzzy Sets and Systems 5, 47–67 (1981)
6. Goguen, J.: L-fuzzy Sets. J. Math. Anal. Appl. 18, 145–174 (1967)
7. Goldblatt, R.: Topoi: The categorical analysis of logic. North-Holland, Amsterdam (1979)
8. Höhle, U., Stout, L.N.: Foundation of fuzzy set. Fuzzy Sets and Systems 40, 257–296 (1991)
9. Li, X.-S., Yuan, X.-H.: The category RSC of I rough sets. In: Fifth International Conference on Fuzzy Systems and Knowledge Discovery, vol. 1, pp. 448–452. IEEE (2008)
10. Li, X.-S., Yuan, X.-H.: Y-N sets and category of Y-N sets. In: Sixth International Conference on Fuzzy Systems and Knowledge Discovery, vol. 6, pp. 494–497. IEEE (2009)
11. Li, X.-S., Yuan, X.-H., Lee, E.S.: The three-dimensional fuzzy sets and their cut sets. Computers and Mathematics with Applications 58(7), 1349–1359 (2009)
12. Mac Lane, S.: Categories for the Working Mathematician. Springer, New York (1971)
13. Pitts, A.M.: Fuzzy sets do not form a topos. Fuzzy Sets and Systems 8, 101–104 (1982)
14. Ponasse, D.: Some remarks on the category FuzH of M. Eytan. Fuzzy Sets and Systems 9, 199–204 (1983)

About a System of Periodic Signal Analysis

JingBo Xu

Information Engineering University, Zhengzhou 450002, China
15937101761@139.com

Abstract. This articlc aims to discuss a few problems of the "Signals and Systems" course. as an example, it analyzed the periodic signal after a system response characteristics with sine signal . The response is weighted by the system function (signal frequency ω0). The mode of response signal is weighted by the mode of the same frequency system function to the mode of excitation signal . The response signal phase Angle is changed by the phase Angle of the same frequency system function and the excitation signal. That response is effected by system characteristic and excitation.

Keywords: System, periodic, response, excitation.

The Fourier transform applied to the communication system has a long history and a wide range. The development of modern communication system with the application of the Fourier transform such as modulation, filtering, distortion, sampling, frequency division multiplexing, etc. Generally the research mainly focuses on the relationship between the excitation and response of the LTI system. That is when an excitation signal through a LTI system and how about the response? It can explain the difference between response and excitation signal with the Fourier transform analysis system. The physical meanings are definite.

1 The Response of LTI System Is Acted by Basic Plural Exponential ejω0t

Because the domain of excitation signal e(t)= ejω0t is (-∞, ∞) , And t =-∞,the system state can always think of 0 . So the response can be thought to be the zero-state response. Suppose system function H(ω) is shown as figure 1 ,response is y(t),so:

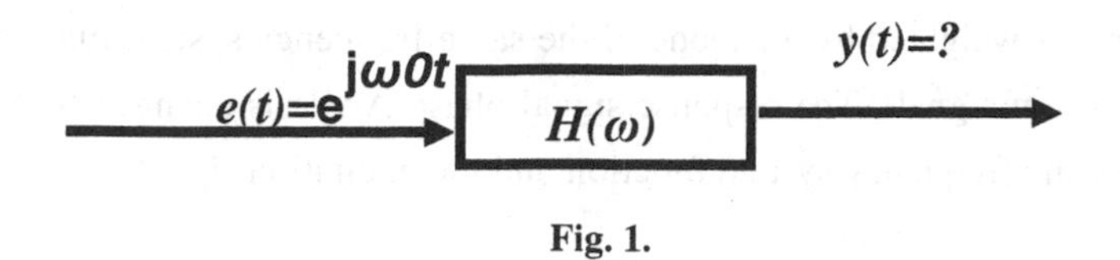

Fig. 1.

D. Jin and S. Lin (Eds.): Advances in ECWAC, Vol. 1, AISC 148, pp. 501–504.
springerlink.com © Springer-Verlag Berlin Heidelberg 2012

The system function: $H(\omega) = |H(\omega)|e^{j\varphi(\omega)}$

By the time domain convolution:

$$y(t) = h(t) * e^{j\omega_0 t}$$

$$y(t) = \int_{-\infty}^{\infty} h(\tau)e^{j\omega_0(t-\tau)}d\tau = \int_{-\infty}^{\infty} h(\tau)e^{-j\omega_0\tau}d\tau . e^{j\omega_0 t} \tag{1-1}$$

$$y(t) = H(\omega_0)e^{j\omega_0 t};$$

The response of LTI system with basic plural exponential is weighted by the system function (signal frequency ω_0).

2 The Response of LTI System Is Acted by Sine (Cosine)Signals

Because the domain of excitation signal $e(t)=A\cos\omega_0 t$ is $(-\infty, \infty)$ .And $t=-\infty$,the system state can always think of 0 . So the response can be thought to be the zero-state response.

Euler formula: $$A\cos\omega_0 t = A\frac{e^{j\omega_0 t} + e^{-j\omega_0 t}}{2}$$

by the linear characteristic of LTI system With formula (1-1),

$$y(t) = \frac{A}{2}\{|H(\omega_0)|e^{j[\omega_0 t+\varphi(\omega_0)]} + |H(-\omega_0)|e^{j[-\omega_0 t+\varphi(-\omega_0)]}\}$$

When the impulse response h (t) is real. $|H(\omega)|$,the model of $H(\omega)$ is even function ; $\varphi(\omega)$,the phase Angle of $H(\omega)$ is odd function.

Formula (2-1) can be simplified and consolidated:

$$\begin{aligned} y(t) &= \frac{A}{2}\{|H(\omega_0)|e^{j[\omega_0 t+\varphi(\omega_0)]} + |H(\omega_0)|e^{j[-\omega_0 t-\varphi(\omega_0)]}\} \\ &= \frac{A}{2}|H(\omega_0)|\{e^{j[\omega_0 t+\varphi(\omega_0)]} + e^{-j[\omega_0 t+\varphi(\omega_0)]}\} \\ &= A|H(\omega_0)|\cos[\omega_0 t + \varphi(\omega_0)] \end{aligned} \tag{2-2}$$

The response is weighted by the system function (signal frequency ω0). The mode of response signal is weighted by the mode of the same frequency system function to the mode of excitation signal . The response signal phase Angle is changed by the phase Angle of the same frequency system function and the excitation signal.

3 The Response of LTI System Is Acted by Causal Sine (Cosine)Signals

When excitation signal is $e(t) = A\cos\omega_0 t\,\varepsilon(t)$, what is the difference between response of e(t)and $e(t) = A\cos\omega_0 t$?

$e(t) = A\cos\omega_0 t\,\varepsilon(t)$ acts system h (t), get response y (t) ,

$$y(t) = e(t) * h(t) = A\cos\omega_0 t\varepsilon(t) * h(t)$$

In frequency domain, y(t)<-> Y(ω),

$$\begin{aligned} Y(\omega) &= E(\omega)H(\omega) \\ &= \frac{A}{2}\{[\delta(\omega+\omega_0)+\delta(\omega-\omega_0)] * [\frac{1}{j\omega}+\pi\delta(\omega)]\}H(\omega) \\ &= \frac{A}{2}[\pi\delta(\omega+\omega_0)+\pi\delta(\omega-\omega_0)]H(\omega) \\ &\quad + \frac{A}{2}[\frac{1}{j(\omega+\omega_0)}+\frac{1}{j(\omega-\omega_0)}]H(\omega) \qquad (3-1) \\ &= \frac{A}{2}[\pi\delta(\omega+\omega_0)|H(\omega_0)|e^{-j\varphi(\omega_0)}+\pi\delta(\omega-\omega_0)|H(\omega_0)|e^{j\varphi(\omega_0)}] \\ &\quad + \frac{A}{2}[\frac{1}{j(\omega+\omega_0)}+\frac{1}{j(\omega-\omega_0)}]H(\omega) \end{aligned}$$

Formula (3-1) will be transformed from frequency domain transform to the time domain ,The first part of Y(ω)is set to Y1(ω)

$$Y_1(\omega) = \frac{A}{2}[\pi\delta(\omega+\omega_0)|H(\omega_0)|e^{-j\varphi(\omega_0)}+\pi\delta(\omega-\omega_0)|H(\omega_0)|e^{j\varphi(\omega_0)}]$$

$$Y_2(\omega) = \frac{A}{2}[\frac{1}{j(\omega+\omega_0)}+\frac{1}{j(\omega-\omega_0)}]H(\omega)$$

so,

$$y_1(t) = \frac{A}{2}|H(\omega_0)|[\frac{1}{2}e^{-j\omega_0 t}e^{-j\varphi(\omega_0)}+\frac{1}{2}e^{j\omega_0 t}e^{j\varphi(\omega_0)}]$$

The second part of Y(ω)is set to $Y_2(\omega)$

$$y_1(t) = \frac{A}{2}|H(\omega_0)|\cos[\omega_0 t+\phi(\omega_0)]$$

$$y_2(t) = \frac{A}{2} e^{-j\omega_0 t} \int_{-\infty}^{t} h(\tau) d\tau + \frac{A}{2} e^{j\omega_0 t} \int_{-\infty}^{t} h(\tau) d\tau$$

$$y_2(t) = A \cos \omega_0 t \int_{-\infty}^{t} h(\tau) d\tau$$

When the system is stable system, then 't' is ∞, $\int_{-\infty}^{t} h(\tau) d\tau$ is zero, That is the system stability solution is yss(t)

$$y_{ss}(t) = \frac{A}{2} |H(\omega_0)| \cos[\omega_0 t + \varphi(\omega_0)] \quad (3\text{-}2)$$

To compare formula (3－2) with formula (2－2), we find it that the stable response from t=– ∞ and the stable response from t>0 ,all is weighted by system function.

4 Conclusion

The response of periodic signal is identified by system characteristics and excitation characteristics. The excitation characteristics should be paid attention For a sine signal.

References

1. Junli: Signals and Systems. Higher Education Press, Beijing (2000)
2. Oppenheim Signals and Systems. Science and Technology Press, Hangzhou (1991)
3. Guan, Z.: Signal and Linear Systems. Higher Education Press, Beijing (1992)

Construction of Chaotic System Based on Parameter Searching and Application in the Spread Spectrum Communication

Shaoyong Zhang

College of Energy and Electric, Hohai University, Nanjing, Jiangsu, China, 210098
zsy@hhu.edu.cn

Abstract. The dynamic characteristic of non-linear system is 吧controlled by the parameter in the chaotic system, and the state of dynamic system will change with the variation of the parameter. Based on the complexity analysis of the parameter variation in chaotic system, parameter variation sphere which causes chaos can be searched in collapse and non-linear map, the procedure from bifurcation to chaos can be directly observed by the bifurcation diagram, and then chaotic system can be constructed. Large quantity, arbitrary periods, simple production, and good quality are the requests for new chaotic sequences in the research of chaos spreading sequence communication, and constructing equation through parameter searching can realize this target in some extent. In this paper, parameter searching is used to construct a sort of chaotic equations, which is valuable to anti-disturbance in chaos spreading sequence communication and secrecy of communication.

Keywords: parameter searching, construction of chaotic equations, chaos spreading sequences, non-linear map, self/cross correlation.

1 Introduction

The study of discrete chaotic equation is mainly concentrated in Logistic、Kent、Chebyshev or the improved dynamic characteristics[1-2]. From the complexity analysis of chaotic equation with the variation of the parameter, dynamic characteristics of nonlinear system is controlled by the control constraint parameter equation.

Variation of control equation constraints parameters based on chaotic occurrence mechanism is analyzed in this paper, the construction of chaotic system is explored. Parameters search method is a new method of constructing chaotic system[3]. Chaotic spread spectrum communication field has a approximation statistical properties to gaussian white noise. The statistical properties is is key to the current research in the chaotic spread spectrum communication, The present study uses Logistic or its improved nonlinear mapping method generating chaotic spread spectrum sequence, including a peacekeeping high dimensional nonlinear mapping, and the statistical characteristics of computer simulation and theory analysis are analyzed, relevant

D. Jin and S. Lin (Eds.): Advances in ECWAC, Vol. 1, AISC 148, pp. 505–509.
springerlink.com © Springer-Verlag Berlin Heidelberg 2012

literature[4] further puts forward a kind of chaotic spread spectrum sequence generator, given a method of producing spread sequence based on the chaos of a class of chaotic maps. How to look for further new chaotic spread spectrum sequence of numerous, cycle arbitrary, producing simple and good performance is still chaotic spread spectrum sequence effort. This paper realizing the building of the chaotic system based on computer simulation research.MATLAB SIMULINK simulation verified performance of ds-cdma system.

2 Construction and Proof of One Dimension Chas System

2.1 Try Constructing Nonlinear Mapping

$$f_{\mu}(x)=\mu x\prod_{i=1}^{n}(i-x),n\in N \tag{1}$$

take n = 3, parameters search method construct search process of chaos phenomena. Search steps are as follows :

1) Ask the fixed point of derivatives value, that is Solving the following equation

$$f_{\mu}(x)=\mu x(1-x)(2-x)(3-x) \tag{2}$$

Its root $x=p_{\mu}$ is the fixed point.

2) Ask derivatives value of equation

$$f_{\mu}'(x)|x=p_{\mu} \tag{3}$$

3) According to the value judgment of fixed point attract (or exclusion) characteristics

If $\left|f_{\mu}'(x)|x=p_{\mu}\right|<1$,the fixed point will be attracting fixed point; if $\left|f_{\mu}'(x)|x=p_{\mu}\right|>1$,the fixed point is the exclusion point; if $\left|f_{\mu}'(x)|x=p_{\mu}\right|=1$ not satisfy hyperbolic, there will be appear cutting branching phenomenon.

4) Select the adjacent to the two or two repeller attractor interval

When the μ values is 0.9 and 0.98 respectively, Adjusting the parameter μ value and select the mapping interval, randomly giving initial value, programming iteration, observing the occurred chaotic phenomenon.

5) According to the search result, draw a branching diagram.

In the parameter values for parameters near the given range, preparation of branched diagram drawing program for drawing a branching diagram.

Have a branching diagram can be visually observed system chaos control parameter variation range.

Thus, construction of one-dimensional chaotic system ends basically, but the chaos system is a complex system, Complexity analysis of the chaotic system with parameter change,there will be need for intermittent chaotic situation analysis of chaotic system, this paper has made a thorough research, this thesis will no longer make specific analysis, please refer to literature [1-4], from the above construction steps and results can show that the chaotic system can be realized through parameter search construction.

2.2 Constructing Proofs of One Dimensional Chaotic System

Might as well take the nonlinear mapping $f(x) = \mu \exp(\cos x)$, in accordance with the parameter search steps, constructing parameter changes of system chaos, then rendering system bifurcation diagram. In the simulation study, when the system control parameter value changes, observing in the iterative process, producing bifurcations, along with the bifurcation number changing system chaos and period doubling bifurcation to chaos pathways, chaos phenomenon will occur after intermittent chaos, chaos, chaos is not consistent, how to prove the system chaos characteristic and measurement of chaos and to what extent, this paper selects Lyapunov index method. On the sensitivity of the initial value are all manifestations of chaos system essential properties, the properties of quantification, i.e. measuring system chaos to what extent is Lyapunov numbers, this value is defined as $\ln \lambda = \lim_{t \to \infty} \frac{1}{t} \ln J_t$, in which

$$J_t = \left| \frac{dF(x_t)}{dt} \right| \left| \frac{dF(x_{t-1})}{dt} \right| \ldots\ldots \left| \frac{dF(x_1)}{dt} \right| \tag{4}$$

Lyapunov numbers is greater than 0, then the system will occur chaos.

Through the above analysis, for the one-dimensional folding, aperiodic non-linear mapping can be constructed by search parameters of chaotic system, constructing the nonlinear mapping, because the mapping from parameter values of the control, the parameter values of the search, can construct chaotic systems. For multidimensional system parameters of vector, empathy can achieve multidimensional chaotic system ; for the differential equations describing the nonlinear chaotic equation construction, makes the following analysis

Balance points of Differential equation $\dot{x} = f(x, \mu)$ is the solution of differential equation $\dot{x} = f(x, \mu) = 0$, the stability of balance points can be judged by Lyapunov second method, that is to build the energy equation $V > 0$, then solve the derivation of V, if $\dot{V} < 0$,it is the stable balance point. To establish the corresponding relation : equilibrium point is corresponding to the fixed point ; stable equilibrium point (UEP) corresponds to the attractor (repellor), according to the similar method , chaos equation is built based on the search for the parameter μ values.

3 The Application of Chaotic Sequences in Spread Spectrum Communication

Spread spectrum communication technology has many excellent properties, signal is extended to a wide frequency band and transmitted, signal bandwidth of system RF is much wider than the original signal, spread-spectrum communication has very strong anti-jamming performance, multiple access communications, security and multipath, the spread spectrum communication application domain expands unceasingly. Various channel interference of adjacent channel interference is existed by using the traditional communication modulation and frequency division multiplex situation inevitably, it has good confidentiality, strong interference resistance and other advantages, the receiving machine can accurately synchronous conditions, the desired spreading sequence can suppress multiple access interference and multipath interference completely. Therefore, to seek a desired spreading sequence becomes the emphasis of research At the same time, due to the chaotic spread spectrum code group number is infinite, in the CDMA code division multiple access technology in the appropriate selected for spread spectrum address code, so the numerical chaotic sequences can be well applied to spread spectrum communication.

At present, in the field of Engineering, MATLAB has become a kind of important technology tool, the SIMULINK is used for a dynamic system modeling, simulation and analysis software package, dynamic model of chaotic spread spectrum DS-CDMA system with the SIMULINK communication block is shown in Figure 1, and its performance is analyzed.

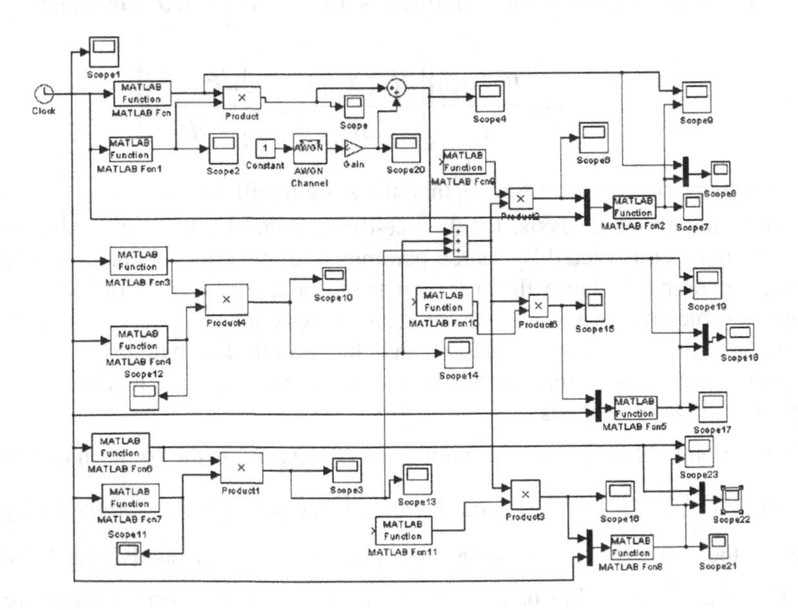

Fig. 1. DS-CDMA system model

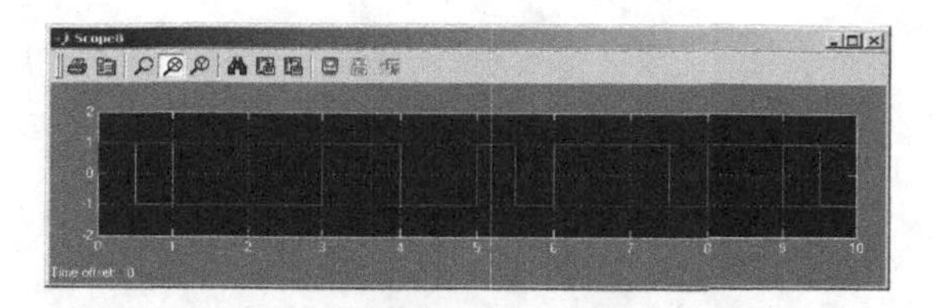

Fig. 2. Input/output waveform comparison

Simulation results is observed in the figure 2, simulation results show that spread spectrum signal has transmission error phenomenon, the main reason is that the truncation of chaotic spreading sequences generated truncated sidelobe and performance deterioration of the chaotic spread spectrum sequences, cross-correlation properties of spread spectrum code groups are destroyed. Module judgment device has no the the best comparison decision, the received signal may have misjudged, without the use of coding and error correcting code technology. It has no optimization of the chaotic spread spectrum code. This need to be further studied and discussed.

4 Summary

This paper realize the chaotic system through the parameter search method, chaotic system is sensitive to initial value, tiny deviations will be into different tracks, chaotic sequence has good self / cross-correlation properties, chaotic sequences generate spread spectrum sequences by digital chaotic, adapting the requirements of chaotic spread spectrum communication to chaotic spread spectrum sequence. This can be well applied to spread spectrum communication as the spreading and despreading codes, spread spectrum communications have a very good reference value.

References

1. Di, H., Chen, H., et al.: Chaotic maps and its application of a kind of folding times unlimited. Journal of Shanghai Traffic University 35 (January 2001)
2. Zhao, Y., Zhang, C., Wu, C.: Spread spectrum communication from digital chaotic sequences. Journals of Information Engineering University, The First (3) (September 2000)
3. He, S., Zhou, S.: The research of a class of chaotic maps spread sequence. In: The 26th Electronic and Information vol. 1(49/3) (February 2004)
4. Zhao, G., Fang, J.: Chaotic equation with the complexity of the parameter variation analysis. China Atomic Energy Science Research Annual (2004)

An Improvement of ACL Match Based on Hash Structure

Weihua Hu, Jing Xu, and Jia Lv

Institute of Computer Graphic Image, Hangzhou Dianzi University, Hangzhou, China

Abstract. This paper particularly analyzes the present message matching process in ACL system and proposes an improved method based on Hash structure according to the characteristic and the advantages and disadvantages of the message matching algorithm. With using SmartBits 6000B equipment, the method has been tested by a series of experiments on network devices. The experiment result turned out that the method improved the ACL system's ability of matching data well.

Keywords: ACL, rules, match, Hash.

1 Introduction

Access Control List (ACL) is an order-list on the network-device which is used to control packets which flow through the network devices. ACL can achieve the aim to prevent unauthorized users, control access and guarantee security of internal network by the way of security policies. ACL technology has a very wide range of application, for example, Firewall can rely on ACL to divide and control data flow. With the improvement of internet security's importance, how to optimize ACL's function and performance has been the problem to be solved for network device providers. Existing ACL system generally achieves data filtering by static rule-list and sequential matching mode. But as rule sets increasing and network traffic exploding, the performance of ACL system will decrease sharply, which will have a bad effect on the performance of network device. Under this background, our paper analyzes the arithmetic process of message matching in ACL, proposes an improved method based on Hash structure, and proves its effectiveness of the improved method by experiments.

2 Message Treating Scheme in ACL

If a message enters one port of a network device, operating system will provide a section of memory to save the message, and then referring to the condition of applying ACL group, decides whether to call function in ACL module which will match the message(this paper discusses the condition that the port has applied the ACL group). Finally ACL module will do the internal processing and return the result to other service modules which have applied ACL. The environment of ACL module in network device is shown in figure 1:

D. Jin and S. Lin (Eds.): Advances in ECWAC, Vol. 1, AISC 148, pp. 511–516.
springerlink.com © Springer-Verlag Berlin Heidelberg 2012

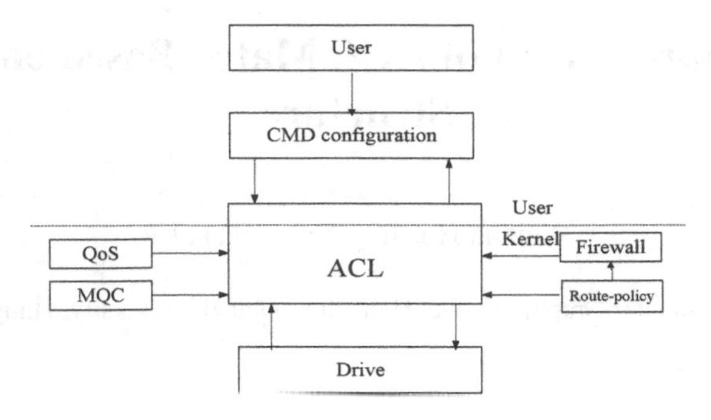

Fig. 1. ACL module's environment

CMD configuration translates user commands to machine ones. Drive modules delivers messages to ACL module. Qos, Firewall and so on consider whether to call ACL according to their services.

2.1 Matching Process of ACL

ACL module's main responsibility includes maintenance of ACL data, message matching process and ACL application. Message matching process in ACL is achieved in kernel. Matching work can fall into two parts: message analyzing and data matching. ACL matching process is shown in figure 2:

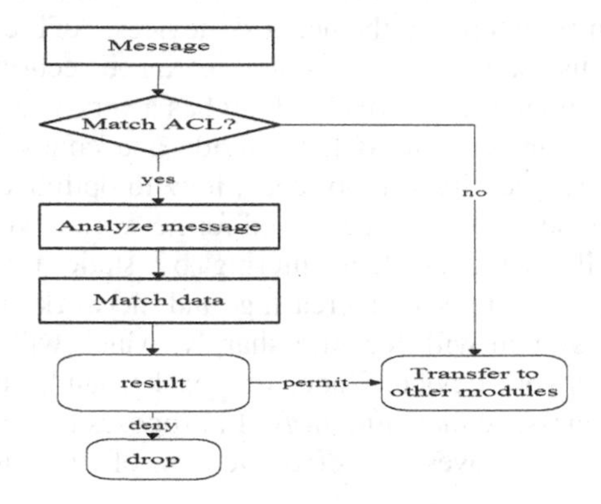

Fig. 2. ACL matching process

2.2 Process of Message Analyzing

According to the current demand characteristic of ACL module and IP message format, message information is saved in data structure as below:

```
typedef struct tagIP_packet
{
 inaddr_s srcIPAddr;
 unsigned char ucPID;
```

```
 ......
 unsigned short usFragType;
}ACL_Packet_S
```

When the drive module accomplishes delivering messages to ACL module, ACL module enables reading pointer to point to the head address which storages content of message, analyzes other necessary information depending on IP message head format, and saves the information to structure variable which specially saves message information(here we make it as pstPacket).

2.3 Process of Data Matching

After accomplishing message analysis, ACL module looks for the rule-list under this ACL group and sequentially matches information in pstPacket against rule-list. It won't stop until hitting a target. On the condition that all the rules aren't hit, it will return Rule_notfound. According to this theory, we can take the senior ACL for example. Matching algorithm is described as below:

```
Match (pstPacket, rule-list)
{
while (rule-list is not empty)
{
 if(need to match type of protocol)
  if(pstPacket.pro!=rule-list[x].pro)
      continue;
 if (need to match source addr )
  if(pstPacket.addr!=rule-list[x].addr)
      continue;
 ......
return Rule[x]-action;
}
return Rule-notfound;
}
```

3 An Improved Method Based on Hash Structure

Given the previous analysis, we can conclude: whenever ACL module receives a message, ACL needs to analyze this message once and match it several times. When rule number of ACL group is very huge(even hundreds and thousands) and the received message doesn't match any rule or just match the last one, ACL module will have to do matching work hundreds and thousands times. Hence, it can be learned that the efficiency of message matching function nearly decides the efficiency of the whole algorithm [1]. According to disadvantages of serial matching method, this paper proposes an improved method based on Hash structure.

Based on ACL rule's characters, Hash function named Hashkey can be designed as: take, fold and do xor operation of IP address, port number, protocol number and so on, then modulo the hash length [2]. Hash function is shown as below:

```
Hashkey(addr,port,......)
{
 int addrcp,key;
 addrcp=(addr&~(~0<<16)) ((addr>>16)&~(~0<<16);
 key = addrcp  port;
 ......
 return (key%LENGTH);
}
```

Because there is a limitation to length distribution of hash table, it's unavoidable to come up with hash conflict problem. If two ACL rules point to the same address after hash function operation, the two rules can be formed to a structure of linked list, which can prevent happening of conflict problem and avoid the second hash locating. The main algorithm is shown as below:

```
AvoidConfliction(Hashkey(rule.addr, rule.port))
{
 if(H[Hashkey(rule.addr,rule.port,...) ] = NULL)
   H[Hashkey(rule.addr,rule.port),...... ] = &rule;
 else   p=H[Hashkey(rule.addr,rule.port,......)];
 while(p->next!=NULL) {p=p->next;}
 p->next = &rule;
}
```

After receiving a message, ACL will call Hashkey function to look for the address of the related rule in hash table and do the further matching. Matching algorithm is shown as below:

```
Match (pstPacket)
{
 index=Hashkey(pstPacket.addr, pstPacket.port);
 if(Hashtable[index].addr ==
 pstPacket.addr && Hashtable[index].port ==
 pstPacket.port)
  {
   return Hashtable[index];
  }
 else
   return Rule-notfound;
}
```

Time complexity of hash searching is approaching to O (1), but time complexity of sequential searching is O (N) [3]. Obviously the rule-matching method based on hash structure is more efficient than the serial matching method, which highly improves ACL system's throughput and filtering property, especially as rule number grows up.

4 Experimental Design and Result Analysis

We performed tests of the improved method by using Smartbits 6000B. Smartbits network performance analysis system is a dedicated network testing system which is introduced by Spirent.

4.1 Process of Data Matching

The experiment's network environment of firewall application is shown in figure 3:

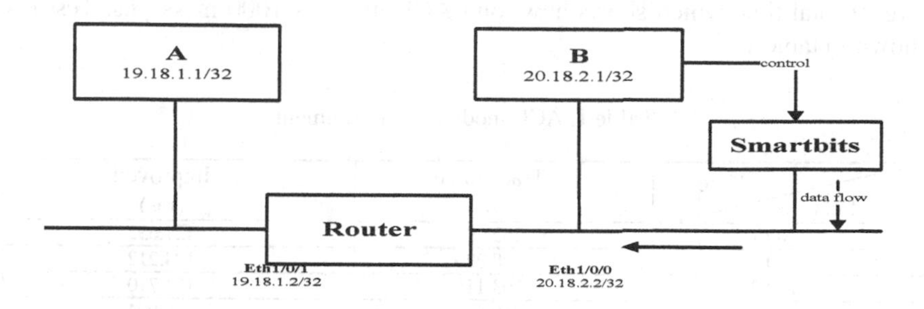

Fig. 3. Experiment network

In figure 3, A is a PC whose address is 19.18.1.1/32; B is a PC whose address is 20.18.2.1/32 and has installed SmartFlow 5.60 applied software; Smartbits test device is used for sending data flow to router, and data flow is controlled by the applied software on B; Router is a router with ACL module.

4.2 Configuration of ACL

According to network environment, we can configure ACL on port Eth 1/0/0 of Router, and enable firewall apply the senior ACL to divide and control data which flows through the port [4]. At first, we can set 500 ACL rules which don't match the messages, and then sets the ACL rule which matches messages. The configuration is shown as below:

```
<Router> system-view
# enable firewall
[Router] firewall enable
# configure senior ACL
[Router]acl number 3000 name acl3000
[Router-acl-adv-3000]rule 1 per icmp soource 1.2.2.1 32
……
[Router-acl-adv-3000]rule 501 deny tcp source 20.18.2.1
32 des 19.18.1.1 32 soource-port eq 11 destination-port
gt 50 logging
[Router-acl-adv-3000]quit
```

```
#port Ethernet1/0/0 enable firewall
[Router] interface Ethernet 1/0/0
[Router-Ethernet1/0/0] firewall packet-filter acl 3000
inbound
[Router-Ethernet1/0/0] quit
```

4.3 Test Result

In order to prove the improved method's effect on the ACL matching time, we install both the traditional ACL and the improved one on the Router, and make 5 group record of total time which shows how long ACL matches 1000 messages. Test result is shown in table 1:

Table 1. ACL module's environment

Time / Sets	Traditional (ms)	Improved (ms)
NO.1	202.001	121.632
NO.2	208.220	124.212
NO.3	203.115	121.710
NO.4	210.556	126.004
NO.5	207.040	121.101

5 Conclusion

This paper studies on ACL message matching process and proposes an improved method based on Hash structure. On the basis of accomplishing ACL, we designed network environment for firewall application of ACL and executed network tests alone with Smartbits 6000B device performing data test. Both theoretical analysis and experimental result have demonstrated the efficiency of the improved method based on hash structure.

References

1. Grout, V., McGinn, J., Davies, J.: Real-Time Optimization of Access Control Lists for Efficient Internet Packet Filtering. Journal of Heuristics 13(5), 435–454 (2007)
2. Liang, J., Zhao, Q., Chen, X.: Research of Firewall Matching Arithmetic Based on Rules. Computer Engineering and Applications (2005)
3. Knuth, D.E.: The Art of Computer Programming. Tsinghua University Press, Beijing (2002)
4. Lu, J.: The Use of ACL Technology to Enhance Network Security. Computer Security (2011)

New Theta Function Wave Solutions to the EW Equation Using Symbolic Computation

Xiaoxia Yang[1] and Junmin Wang[2]

[1] School of Mathematical Sciences, Pingdingshan University, Pingdingshan, China
wjm1261@sina.com
[2] Department of Mathematics and Information Science,
Henan University of Economics and Law, Zhenzhou, China
wjm1261@sina.com

Abstract. In this paper, theta functions and the auxiliary equation method is used to seek exact solutions for the the EW equation. As a result, some new theta function solutions are successfully obtained with the aid of symbolic computation for the first time.

Keywords: the auxiliary equation method, the EW equation, theta function, symbolic computation.

1 Introduction

The nonlinear phenomena exist in all the fields including either the scientific work or engineering fields, such as fluid mechanics, plasma physics, optical fibers, biology, solid state physics, chemical kinematics, chemical physics, and so on. It is well known that many non-linear partial differential equations (NLPDEs) are widely used to describe these complex phenomena. Research on solutions of NLEEs is popular, these exact solutions when they exist can help one to well understand the mechanism of the complicated physical phenomena and dynamical processes modelled by these NLPDEs. So, the powerful and efficient methods to find analytic solutions and numerical solutions of nonlinear equations have drawn a lot of interest by a diverse group of scientists. There have been many research on exact solutions to NLPDE, such as the famous Inverse scattering method, Backlund transformation, Darboux transformation, Hirota bilinear method, Painleve method and so on[2-6]. In recent years, direct search for exact solutions to NLPDE has become more and more attractive partly due to the availability of computer symbolic systems like Maple or Mathematica which allows us to perform some complicated and tedious algebraic calculation on computer, and helps us to find new exact solutions to NLPDE, such as Homogeneous balance method[7], tanh-function method [8], sine-cosine method[9], Jacobi elliptic functions method[10], F-expansion[11,12] and so on.

The well-known Korteweg and de Vries equation:

$$u_t + uu_x + u_{xxx} = 0 \tag{1}$$

D. Jin and S. Lin (Eds.): Advances in ECWAC, Vol. 1, AISC 148, pp. 517–522.
springerlink.com © Springer-Verlag Berlin Heidelberg 2012

is a nonlinear partial differential equation (NLPDE) that models the time-dependent motion of shallow water waves in one space dimension. Morrison et al. [1] proposed the one -dimensional PDE:

$$u_t + uu_x - \mu u_{xxt} = 0 \tag{2}$$

as an equally valid and accurate model for the same wave phenomena simulated by the KdV equation. This PDE is called the equal width (EW) equation because the solutions for solitary waves with a permanent form and speed, for a given value of the parameter μ are waves with an equal width or wavelength for all wave amplitudes.

In this paper, we apply the auxiliary equation method[13] to seek exact solutions the EW equation (2) by taking full advantages of the elliptic equation:

$$(F'(x))^2 = b_4 F^4(x) + b_2 F^2(x) + b_0 \tag{3}$$

and get some traveling wave solutions in terms of theta functions with the aid of symbolic computation for the first time.

This paper is arranged as follows. In section 2, we shall illustrate the auxiliary equation method and the properties of theta functions; In section 3, we apply the auxiliary equation method and a new solution of the elliptic equation (3) to seek exact solutions of the EW equation; Some conclusions are given in section 4.

2 The Method of Solution

We simply describe the auxiliary equation method as follows.

Consider a given nonlinear partial differential equation(NLPDE) with independent variables

$$x = (x_1; x_2; \cdots; x_l; t)$$

and dependent variable u :

$$P(u; u_t; u_{x_1}; u_{tt}; \cdots) = 0. \tag{4}$$

Generally speaking, the left-hand side of (4) is a polynomial in u and its various partial derivatives.

We seek its travelling wave solution in the formal solution

$$u(\xi) = a_0 + \sum_{i=1}^{N} a_i F^n(\xi) \tag{5}$$

by taking

$$u(x_1; x_2; \cdots; x_l; t) = u(\xi),$$

$$\xi = k_1 x_1 + k_2 x_2 + \cdots + k_l x_l - \lambda t \tag{6}$$

where $k_1, k_2, \cdots k_l, \lambda$ are constants to be determined, and $a_i (i = 1, 2, \cdots N)$ are also constants to be determined, $F(\xi)$ satisfies Elliptic equation (2). Inserting (5) into (4) yields an ODE for $u(\xi)$:

$$P(u; u'; u''; \cdots) = 0 \tag{7}$$

where Integer N can be determined.

For the elliptic equation (2), the following fact is needed to realize the aim of this paper.

Proposition: If we take

$b_4 = b_0 = \vartheta_2^2(0)\vartheta_3^2(0)$ and $b_2 = -(\vartheta_2^2(0) + \vartheta_3^2(0))$, then $F(z) = \vartheta_1(z)/\vartheta_4(z)$ satisfies the elliptic equation (3), where theta functions are defined as following

$$\vartheta[\begin{pmatrix} \varepsilon \\ \varepsilon' \end{pmatrix}](z|\tau) = \sum_{n=-\infty}^{\infty} \exp\{\pi i \tau (n + \frac{\varepsilon}{2})^2 + 2(n + \frac{\varepsilon}{2})(z + \frac{\varepsilon'}{2})\}$$

$$\vartheta_i(z) \triangleq \vartheta_i(z|\tau) = \vartheta[\varepsilon_i](z|\tau),$$

$$\varepsilon_1 = \begin{bmatrix} 1 \\ 1 \end{bmatrix}, \varepsilon_2 = \begin{bmatrix} 1 \\ 0 \end{bmatrix}, \varepsilon_3 = \begin{bmatrix} 0 \\ 0 \end{bmatrix}, \varepsilon_4 = \begin{bmatrix} 0 \\ 1 \end{bmatrix}. i = 1, 2, 3, 4.$$

To determine $u(\xi)$ explicitly, one may take the following steps:

Step 1: Determine N by considering the homogeneous balance between the governing nonlinear term(s) and highest order derivatives of $u(\xi)$ in (7).

Step 2: Substituting (5) into (7), and using (2), and then the left-hand side of (7) can be converted into a finite series in $F^k(\xi)(k = 0, 1, \cdots, M)$.

Step 3: Equating each coefficient of $F^k(\xi)$ to zero yields a system of algebraic equations for $a_i\ (i = 0, 1, \cdots, N)$.

Step 4: Solving the system of algebraic equations, with the aid of Mathematica or Maple, a_i, k_i, λ can be expressed by A, B, C (or the coefficients of ODE (7).

Step 5: Substituting these results into (6), we can obtain the general form of travelling wave solutions to (3).

Step 6: From propositon, we can give a series of theta function solutions to (3).

3 Exact Solutions of the EW Equation

In this section, we will make use of the auxiliary equation method and symbolic computation to find the exact solutions to the EW equation.

We assume that (2) has travelling wave solution in the form

$$u(x,t) = U(\xi),\ \xi = \rho x + \omega t \tag{8}$$

Substituting (8) into (2), then (2) is transformed into the following form:

$$\omega u' + \rho u u' - \rho^2 \omega u''' = 0 \qquad (9).$$

According to step 1 in section 2, by balancing u''' and uu' in (9), we obtain $n = 2$, and suppose that (9) has the following solutions:

$$U(\xi) = a_0 + a_1 F(\xi) + a_2 F^2(\xi), \qquad (10)$$

then

$$U'(\xi) = a_1 F'(\xi) + 2a_2 F(\xi) F'(\xi) \qquad (11)$$

$$U'''(\xi) = a_1 F''(\xi) + 2a_2 F(\xi) F''(\xi) + 2a_2 (F'(\xi))^2 \qquad (12)$$

Substituting (10)- (12)along with (3) into (9) yields a polynomial equation in $F(\xi)$. Setting their coefficients to zero yields a set of algebraic equations for unknown parameters a_0, a_1, a_2, ω.

$$\omega a_1 b_0 + \rho a_0 a_1 b_0 - \rho^2 \omega a_1 b_0 b_2 = 0$$

$$\rho a_1^2 b_0 + 2\omega a_2 b_0 + 2\rho a_0 a_2 b_0 - 8\rho^2 \omega a_2 b_2 b_0 = 0$$

$$3\rho a_2 a_1 b_0 + \omega a_1 b_2 + \rho a_0 a_1 b_2 - \rho^2 \omega a_1 b_2^2$$
$$-6\rho^2 \omega a_1 b_0 b_4 = 0$$

$$2\rho a_2^2 b_0 + \rho a_1^2 b_2 + 2\omega a_2 b_2 + 2\rho a_0 a_2 b_2$$
$$-8\rho^2 \omega a_2 b_2^2 - 24\rho^2 \omega a_2 b_0 b_4 = 0$$

$$3\rho a_1 a_2 b_2 + \omega a_1 b_4 + \rho a_1 a_0 b_4 - 7\rho^2 \omega a_1 b_2 b_4 = 0$$

$$2\rho a_2^2 b_2 + \rho a_1^2 b_4 + 2\omega a_2 b_4 + 2\rho a_0 a_2 b_4$$
$$-32\rho^2 \omega a_2 b_2 b_4 = 0$$

$$\rho a_1 a_2 b_4 - 2\rho^2 \omega a_1 b_4^2 = 0$$

$$\rho a_2^2 b_4 - 12\rho^2 \omega a_2 b_4^2 = 0$$

Solving these equations by using symbolic software--Mathematica, we can get the following solutions:

$$a_0 = \frac{a_2(4\rho^2 b_2 - 1)}{12\rho^2 b_4}, a_1 = 0, \omega = \frac{a_2}{12\rho b_4}. \qquad (13)$$

The travelling wave solution of the EW equation are given by

$$U(\xi)=\frac{a_2(4\rho^2b_2-1)}{12\rho^2b_4}+a_2F^2(\xi)$$

Where $\xi=\rho x+\omega t$, b_0,b_2,b_4,a_2 are arbitrary constants, and ω are given in (13).

From the proposition , if we choose $b_4=b_0=\vartheta_2^2(0)\vartheta_3^2(0)$ and $b_2=-(\vartheta_2^2(0)+\vartheta_3^2(0))$, we can get the solution to the EW equation in terms of theta functions:

$$u(x,t)=\frac{a_2(4\rho^2b_2-1)}{12\rho^2b_4}+a_2\frac{\vartheta_1^2(\xi)}{\vartheta_4^2(\xi)},$$

where where $\xi=\rho x+\omega t$, b_0,b_2,b_4,a_2 are arbitrary constants, and ω are given in (13).

To grasp the characteristcs of solutions of (2), we dipict the figure of the solution $u(x,t)$ by using the mathematica, its properties and profiles are displayed in figure a under chosen parameters: $a_2=0.01,\rho=-0.1,\tau=0.5$

From figure 1, it is easy to see that the solution $u(x,t)$ is doubly-periodic wave solutions.

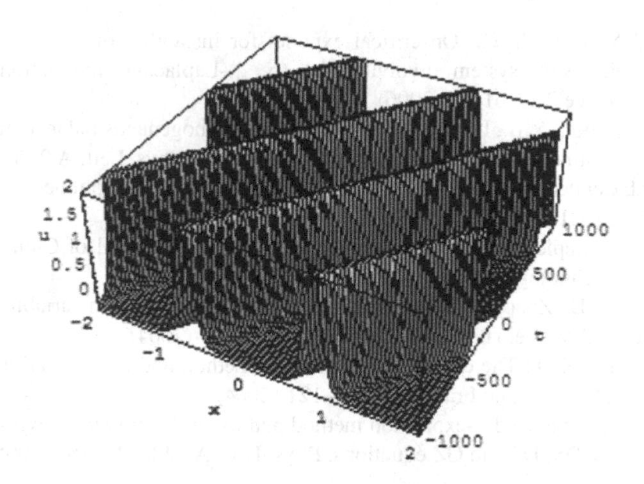

Fig. 1. Perspective view of the wave $u(x,t)$

4 Conclusions

In this paper, we have studied the EW equation. By using auxiliary equation method, some traveling wave solutions in terms of theta functions are successfully obtained with the aid of symbolic computation for the first time, they should be meaningful to explain some physics phenomena. It is shown that the auxiliary equation method is a very effective and powerful mathematical tool for solving nonlinear evolution equations in mathematics and physics. Moreover, with the aid of computer symbolic systems (Mathematica or Maple), the method can be conveniently operated .

Acknowledgments. This work has been supported by the Specific Public Service Sectors of National Forestry Research of China (No.201004044), the Basic and advanced technology projects of Henan Province (No.102300410262) and the National Science Foundation of Henan province (No.2010A110001).

References

1. Saka, B.: Algorithms for numerical solution of the modified equal width wave equation using collocation method. Math. Comput. Model (45), 1096–1117 (2007)
2. Abolowitz, M.J., Clarkson, P.A.: Solitons, nonlinear evolution equations and inverse scatting. Cambridge University Press, London (1991)
3. Ju, L.: On solution of the Dullin-Gottwald-Holm equation. International Journal of Nonlinear Science 1(1), 43–48 (2006)
4. Lu, D.C., Tian, L.X.: Backlund transformation and n-soliton-like solutions to the combined KdV-Burgers equation with variable coefficients. International Journal of Nonlinear Science 1(2), 3–10 (2006)
5. Wang, Y.C., Zhang, W.B.: Application of the adomian decmposition method to fully nonlinear Sine-Gordon equation. International Journal of Nonlinear Science 2(1), 29–38 (2006)
6. Afrouzi, G.A., Rousoli, H.: On critical exponet for instability of positive solutions to a Reaction- Diffussion system involving the (p, q)-Laplacian. International Journal of Nonlinear Science 2(1), 61–64 (2006)
7. Wang, M.L., Zhou, Y.B., Li, Z.B.: Application of a homogeneous balance method to exact solutions of nonlinear equations in mathematical physics. Phys. Lett. A 216, 67–75 (1996)
8. Fan, E.G.: Extended tanh-function method and its applications to nonlinear equations. Phys. Lett. A 277, 12–218 (2000)
9. Yan, C.Q.: A simple transformation for nonlinear waves. Journal of China University of Science and Technology 224, 7–84 (1996)
10. Fu, Z.T., Liu, S.D., Zhao, Q.: Exact solutions to KdV equations with variable coefficients or forcing. Applied Mathematics and Mechanics 25, 73–79 (2004)
11. Liu, J.B., Yang, K.Q.: The extended F-expansion method and exact solutions of nonlinear PDEs. Chaos, Solitons and Fractals 22, 11–121 (2004)
12. Yomba, E.: The extended F-expansion method and its application for solving the nonlinear wave, CKGZ, GDS, DS and GZ equations. Phys. Lett. A 340, 149–160 (2005)

The Upper Bound of Lifetime on Fixed Energy Budget in Wireless Sensor Networks

Lisheng Ren and Fang Wang

College of Information Science & Technology at the Agriculture University of Hebei, Lingyusi Street, BaoDing, Hebei, China
Rls60683@163.com, Wf9003@163.com

Abstract. Wireless sensor networks have a wide range of potential, Practical and useful applications[1]. Sensor nodes monitor some surrounding environmental phenomenon, process the data obtained and forward this data towards a base station located on the periphery of the sensor network. Base station(s) collect the data from the sensor nodes and transmit this data to some remote control station. Sensor network models considered by most researchers have a single static base station located on the periphery of the sensor network [2], [3], [4], [5]. Most sensors are equipped with non-rechargeable batteries[6]. While the network as a whole is required to provide fine resolution monitoring for an extended period of time, the upper bound of lifetime is an important problem. In this work, we develop closed-form expressions of the upper bound of lifetime in energy-efficient wireless sensor networks.

Keywords: Network, Wireless, Sensor, Lifetime.

1 Introduction

Recent advances in VLSI, microprocessor and wireless communication technologies have enabled the design and deployment of large-scale sensor networks, where thousands, or even tens of thousands of small sensors are distributed over a vast field to obtain fine-grained, high-precision sensing data [7],[8],[9]. To wireless sensor networks, especially the large scale WSN, it is important to calculate thc upper bound of lifetime with N nodes. Because of the complexity of WSN, a lot of factors can affect it. There are a lot of approach to describe the problem, e.g. [10] present a novel approach based on fuzzy logic systems to analyze the lifetime of a wireless sensor network, [11] gives a formulation and solution to the cost-constrained lifetime-aware battery allocation problem for sensor networks with arbitrary topologies and heterogeneous power distributions, [12] talk about a hexagon tessellation sensor network model with role assignment scheme and estimate the reliability and lifetime distribution. Otherwise, this problem is compounded by some other factors, e.g. network structure, detection model, sensoring area, failure model, time synchronization, sensor deployment strategy, and network connectivity. [13],[14],[15],[16],[17]. There has also been some work on the asymptotic energy-constrained capacity of wireless sensor networks [18]. In this work, we derive some inequalities to express the upper bound of lifetime of energy-efficient wireless sensor networks and discuss the approach to increase lifetime.

D. Jin and S. Lin (Eds.): Advances in ECWAC, Vol. 1, AISC 148, pp. 523–529.
springerlink.com © Springer-Verlag Berlin Heidelberg 2012

2 Assumptions

The following are the key assumptions in our work:

• N nodes are immobile and deployed with constant density in a two-dimensional area. The constant density implies that if the network size is increased, the deployment area grows proportionally.

• The radio radius of a node is R for all nodes.

• There are m atomic events that are sensed in the environment and the sensor network is deployed for a fixed time duration T0. The distribution of events is assumcd to bc uniform in the deployment area. It implies that if area is increased, m grows proportionally. It also implies that if lifetime is increased, m grows proportionally.

• We assume that every node has known the location of their owns and the sinks' (e.g., due to triangle locating algorithm or using GPS locating algorithm[19], [20]). It implies that the energy cost for searching and routing can be ignored.

• We assume that the data received by the sinks is in proportion to events m.

• The total energy cost is assumed to be proportional to the total number of transmissions. This is reasonable particularly for sleep-cycled sensor networks where radio idle times are kept to a minimum.

• We assume that the links over which transmissions take place are lossless (e.g., using blacklisting) and present no interference due to concurrent transmissions (e.g., due to low traffic conditions or due to the use of a scheduled MAC protocol).[21]

• The energy at each node is a constant amounts En. One nodes would consume a constant amounts e while transporting the data produced by one atomic events.

• The energy e is in proportion to R2. We get $e=\beta R^2$ (β is constant)

• Si denotes the circular regions whose radius is R and the sink i is located at the circular regions.

3 Problem Formulation and Implementation Details

We consider two scenerios of the number of sinks: one sink and multi-sinks.

First we consider a area with N nodes deployed with a uniform random distribution. There is only one sink and it is assumed that the network is sufficiently dense so that all nodes within a distance kR of the sink can be reached in k hops.

Figure 1 illustrates a sample network for one sink. The sink is denoted by an "+" while the nodes are denoted by a dot. Say the sink and the circular regions whose radius is R and the sink is located at the circular regions are wholly surrounded with the network area. Now we get the number of these sensor nodes

$$Z = \pi R^2 d \tag{1}$$

These sensor nodes' total energy E is

$$E = E_n Z \tag{2}$$

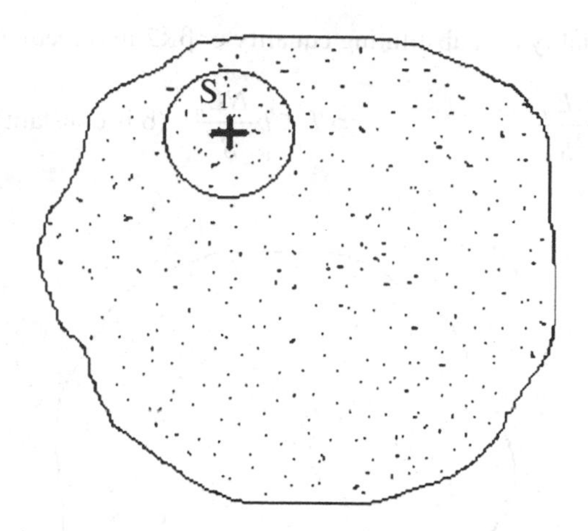

Fig. 1. One sink wholly inside the network

The nodes that are one hop away from the sink need to forward messages originating from many other nodes, in addition to delivering their own messages. Be in ideal conditions and adopt appropriate optimization [21],[22],before these sensor nodes deplete their energy and the network become inoperational., the data produced by m events can be received by the sink. We get

$$E \geq me \tag{3}$$

Since m is in proportion to area S and lifetime T where the sensor network is deployed, we get the equality

$$m = \mu ST \quad (\mu \text{is constant}) \tag{4}$$

Since N nodes are immobile and deployed with constant density in a two-dimensional area, we have that

$$d = \frac{N}{S} \tag{5}$$

Now we can derive the inequality (6) substituting equality (1),(2), (4) and (5) into inequality (3) as follows;

$$\pi R^2 \frac{N}{S} E_n \geq \mu STe \tag{6}$$

From inequality (6),say

$$T \leq \frac{\pi R^2 N E_n}{\mu e S^2} \tag{7}$$

we derive the inequality (8) substituting equality e=βR2 into inequality (7) as follows;

$$T \le \frac{\pi R^2 N E_n}{\mu \beta R^2 S^2} \quad \Rightarrow T \le b\frac{NE_n}{S^2} \text{ (b is constant)} \tag{8}$$

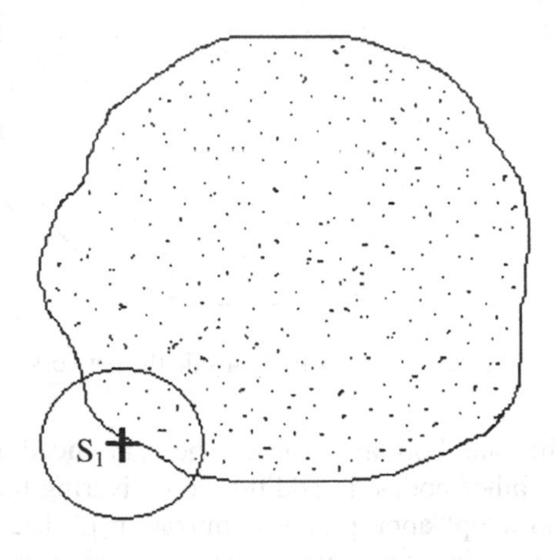

Fig. 2. One sink not wholly inside the network

Figure 2 illustrates another sample network for one sink. Say S are not wholly surrounding by the network area. Now we get the number of these sensor nodes

$$Z' = (\xi \times \pi R^2)d = \xi \cdot Z \quad (\xi \le 1, \xi \text{ is constant}) \tag{9}$$

These sensor nodes' total energy is

$$E' = E_n Z' \tag{10}$$

It is similar to the first scenario illustrated on Figure 1. We get

$$T' \le b\frac{\xi N E_n}{S^2} = \xi T \tag{11}$$

Figure 3 illustrates a sample network for Multiple sink. consider the scenario that all Si are wholly surrounded with the network area and each Si are wholly separated from each other. The symbol c denotes the number of sinks. It is similar to the foregoing statement.

$$T'' \le b\frac{cNE_n}{S^2} = cT \tag{12}$$

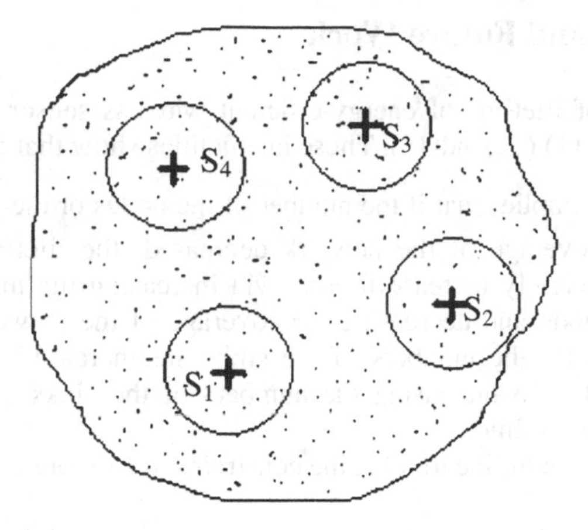

Fig. 3. Multi-sink wholly inside the network and Si is separated from each other

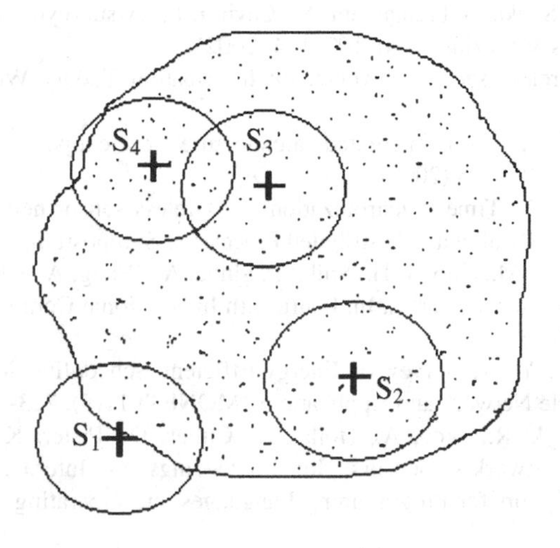

Fig. 4. Multi-sink not wholly Inside the network and Si is not wholly separated from each other

Figure 4 illustrates another sample network for Multiple sink. consider the scenario that each Si are not wholly separated from each other or all Si are not wholly surrounded whith the network area. The number of the nodes including in all Si would decrease, so the number of the node in whole network would decrease too. It can be express follows;

$$T''' \leq \lambda T \quad (\lambda \text{ is constant and } \lambda<1) \tag{13}$$

4 Conclusions and Future Work

The upper bound of lifetime of energy-efficient wireless sensor networks.are show with inequality (8),(11),(12)and(13). Those inequalities show that S is in proportion to N, $1/s^2$ and En. It implies that if the number of the nodes or the energy of node are increased, or the coverage of the network decreased ,the lifetime of the network grows. We can efficiently increase lifetime via increasing the number of the nodes and the energy of node and decreasing the coverage of the network. Inequality (12) and (13) show that if the numbers of the sinks are increased, the lifetime grows nearly proportionally. So increasing the numbers of the sinks is also one efficient approach to increase lifetime.

In the future, we would like to relax the constrains to fit more common conditions.

References

1. Akyildiz, W.S., Sankarasubramaniam, Y., Cayirci, E.: A surveyon sensor networks. IEEE Communications Magazine 40(8), 102–114 (2002)
2. Pottie, G.J.: Wireless sensor networks. In: Information Theory Workshop, pp. 139–140 (1998)
3. Agre, J., Clare, L.: An integrated architecture for cooperative sensing networks. Computer 33(5), 106–108 (2000)
4. Elson, J., Estrin, D.: Time synchronization for wireless sensor networks. In: Proceedings 15th International Parallel and Distributed Processing Symposium, pp. 1965–1970 (2001)
5. Min, R., Bhardwaj, M., Cho, S.-H., Shih, E., Sinha, A., Wang, A., Chandrakasan, A.: Low-power wireless sensor network.s. In: Fourteenth International Conference on VLSI Design, pp. 205–210 (2001)
6. Wang, L., Xiao, Y.: A Survey of Energy-Efficient Scheduling Mechanisms in Sensor Networks. Mobile Network and Applications (MONET) 11(5), 723–740 (2006)
7. Hill, J., Szewczyk, R., Woo, A., Hollar, S., Culler, D., Pister, K.: System architecture directions for networked sensors. In: Proceedings of International Conference on Architectural Support for Programming Languages and Operating Systems, ASPLOS-IX (2000)
8. Intanagonwiwat, C., Govindan, R., Estrin, D.: Directed diffusion: A scalable and robust communication paradigm for sensor networks. In: Proceedings of ACM International Conference on Mobile Computing and Networking, MOBICOM 2000 (2000)
9. Pottie, G., Kaiser, W.: Wireless integrated network sensors. Communications of the ACM 43(5), 51–58 (2000)
10. Shu, H., Liang, Q., Gao, J.: Wireless Sensor Network Lifetime Analysis Using Interval Type-2 Fuzzy Logic Systems. IEEE Computational Intelligence Society, 416–427 (April 2008)
11. Long, H., Liu, Y., Wang, Y., Dick, R.P., Yang, H.: Battery Allocation for Wireless Sensor Network Lifetime mMaximization Under Cost Constraints. In: Proceedings of the 2009 International Conference on Computer-Aided Design, pp. 705–712 (November 2009)
12. Jin, Y.-L., Lin, H.-J., Zhang, Z.-M., Zhang, Z., Zhang, X.-Y.: Estimating the Reliability and Lifetime of Wireless Sensor Network. In: 4th International Conference on Proceedings of Wireless Communications, Networking and Mobile Computing, WiCOM 2008, pp. 1–4 (October 2008)

13. Bhardwaj, M., Chandrakasan, A.P.: Bounding the Lifetime of Sensor Networks Via Optimal Role Assignments. In: Proceedings of IEEE INFOCOM (June 2002)
14. Chang, J.-H., Tassiulas, L.: Maximum Lifetime Routing In Wireless Sensor Networks. IEEE/ACM Transactions on Networking 12(4), 609–619 (2004)
15. Xiong, S., Li, J., Yu, L.: Maximize the Lifetime of a Data-gathering Wireless Sensor Network. In: 6th Annual IEEE Communications Society Conference on Sensor, Mesh and Ad Hoc Communications and Networks, SECON 2009, pp. 1–9 (June 2009)
16. Chu, M., Haussecker, H., Zhao, F.: Scalable information-driven sensor querying and routing for ad hoc heterogeneous sensor networks. International Journal of High Performance Computing Applications 16(3), 293–313 (2002)
17. Guang, S.Y., Ming, B.X.: Node Placement Strategies to Prolong Lifetime in One-Dimensional Linear Wireless Sensor Network. Applied Mechanics and Materials 55-57, 1705–1710 (2011)
18. Hu, Z., Li, B.: Fundamental Performance Limits of Wireless Sensor Networks. In: Yang, X., Yi, P. (eds.) Ad Hoc and Sensor Networks,, Nova Science Publishers (2004)
19. Savarese, C., Rabaey, J.: Locationing in distributed Ad-hoc wireless sensor network. In: Proceedings of the IEEE International Conference on Acoustics, Speech and Signal Processing, ICASSP (2001), `http://bwrc.eecs.berkeley.edu/Publications/2001/Locatng_distrb_ad-hoc_wrlss_snsr_ntwks/icassp2001_final.pdf`
20. Bulusu, N., Heidemann, J., Estrin, D.: GPS-less Low Cost Outdoor Localization for Very Small Devices. IEEE Personal Communications Magazine 7(5), 28–34 (2000)
21. Ahn, J., Krishnamachari, B.: Fundamental scaling laws for energy-efficient storage and querying in wireless sensor networks. In: MobiHoc 2006, pp. 334–343 (2006)
22. Ahn, J., Krishnamachari, B.: Derivations of the Expected Energy Costs of Search and Replication in Wireless Sensor Networks. USC Computer Engineering Technical Report CENG-2006-3 (April 2006)

An Automatic Initialization Method of Reference Model in Target Tracking

Junxiang Gao[1] and Jingtao Xu[2]

[1] School of Science, Huazhong Agricultural University, Wuhan 430070, China
[2] School of Information and Communication Engineering,
Beijing University of Posts and Telecommunications, 100876, China
gao200@gmail.com

Abstract. The initialization of reference model in target tracking is usually performed manually; and there are still defects in current automatic method. To resolve these problems, this paper proposes an automatically and accurately initialization method. Firstly, the targets number and position are computed using connected components labeling algorithm. Consequently, minimal circumscribed rectangles of the targets are obtained according to radius of gyration tensor method. Finally, the reference models are estimated adopting Monte Carlo simulation. Experimental results show that the reference models can be accurately computed in different scenes; also they can reflect the visual features of the targets.

Keywords: machine vision, target tracking, intelligent video surveillance.

1 Introduction

Moving object tracking is an important component in many applications, such as video conferencing, distance learning, smart rooms, and video surveillance. Quite a few tracking algorithms have been proposed including the gradient descent method, the mean shift method, Kalman filter framework and particle filter framework. In these algorithms, a target model should be firstly defined in a reference image and then searched for in subsequent frames using a function that evaluates the similarity between the model and a candidate [1-3]. In current research, the construction of target model is generally performed manually, i.e., drawing a rectangle or an ellipse enclosing the target in a reference frame as the model [4-6]. Although automatic initialization methods are used in a few literatures, prior target knowledge is usually needed and thus the universality is reduced. For example, the approach proposed in [7] is only applicable for face tracking due to the use of color information. The projection method in [8, 9] is suitable for simple scene only; however the number, position or size of the targets may not be correctly computed if the projections of moving objects are overlapped.

To resolve these problems, we propose an automatic and accurate construction method of target model. Connected components labeling algorithm and radius of gyration tensor method are respectively used to compute the minimal circumscribed rectangles of the targets. The detailed steps are described in the following sections.

D. Jin and S. Lin (Eds.): Advances in ECWAC, Vol. 1, AISC 148, pp. 531–537.
springerlink.com © Springer-Verlag Berlin Heidelberg 2012

2 The Analysis of Projection Method

In practice, the procedure of moving objects detection has been finished before target tracking, thus we suppose that the mask of moving objects is binary image $M(x, y)$.

$$M(x, y) = \begin{cases} 1 & (x, y) \in \{\text{moving object pixels}\} \\ 0 & \text{otherwise} \end{cases} \tag{1}$$

where (x, y) represent the position of moving object pixel. Image $M(x, y)$ can be obtained by any algorithm of moving object detection, for example, optical flow method, frame difference method, background subtraction method, and so on.

For a binary image $M(x, y)$, the vertical projection of every column is defined by the pixel amount where $M(x, y)=1$, and the horizontal projection of every row is defined by the pixel amount where $M(x, y)=1$. In projection method, the vertical projection of $M(x, y)$ is firstly computed to get the amount, widths and horizontal positions of the targets. Consequently, the heights of the targets are determined by horizontal projection of binary image $M(x, y)$. The results of these two steps are shown in Fig. 1.

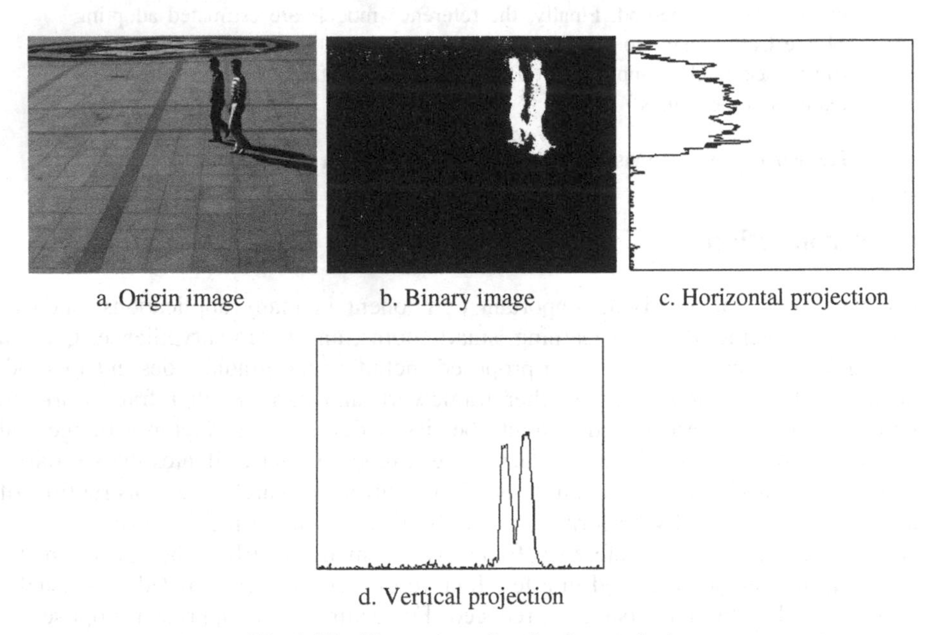

a. Origin image b. Binary image c. Horizontal projection

d. Vertical projection

Fig. 1. The illustration of projection method

The drawback of projection method can be briefed as follows. For one thing, if there is more than one target in the scene and their projection is overlapped, the amount, positions and sizes of these targets can not be accurately computed according to the projection information. For another, the target models of projection method are upright circumscribed rectangles; therefore many background pixels will be included in the target models if the targets are not upright in the image.

To conquer the problems above mentioned, two measures are taken in this paper. In the beginning, the uncertainties caused by overlapped projection are inherent defect of the project method. Thereby another solution, called connected components labeling algorithm, is adopted to get the amount, positions and sizes of the targets in this paper. Consequently, both upright circumscribed rectangle and minimal circumscribed rectangle can enclose all the target pixels. The projection method using the former, however, the latter can describe the target more accurately because of less background pixels in it. So we choose the latter to be the target model.

3 Automatic Initialization Method of Target Model

3.1 Connected Components Labeling Algorithm

Suppose a pixel set $M(x_i, y_i)=v$ is a subset of $M(x, y)$, where i=0, 1, 2,…, n, v=0 or v=1. If any two pixels can be connected by value v, the pixel set $M(x_i, y_i)$ is called a connected component. A labeled image *LB* can be obtained according to the algorithm detailed in literature [10]. The pixel values in labeled image *LB* are the labels of connected components at the same position, and the labels are integers to index connected components. A binary image with five connected components is shown in Fig. 2-a, and the labeled image is shown in Fig. 2-b.

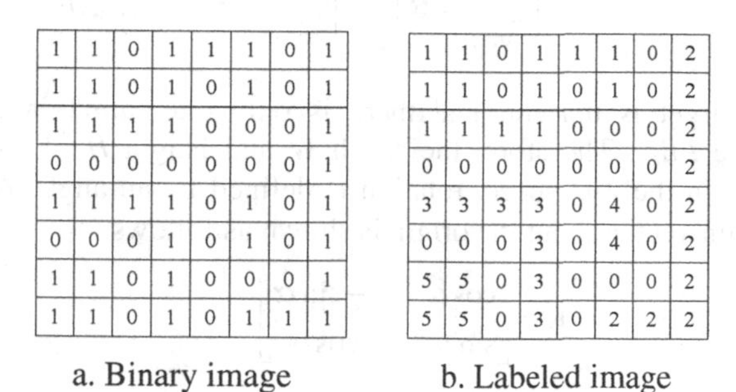

1	1	0	1	1	1	0	1
1	1	0	1	0	1	0	1
1	1	1	1	0	0	0	1
0	0	0	0	0	0	0	1
1	1	1	1	0	1	0	1
0	0	0	1	0	1	0	1
1	1	0	1	0	0	0	1
1	1	0	1	0	1	1	1

a. Binary image

1	1	0	1	1	1	0	2
1	1	0	1	0	1	0	2
1	1	1	1	0	0	0	2
0	0	0	0	0	0	0	2
3	3	3	3	0	4	0	2
0	0	0	3	0	4	0	2
5	5	0	3	0	0	0	2
5	5	0	3	0	2	2	2

b. Labeled image

Fig. 2. Binary image and labeled image with five connected components

3.2 Minimal Circumscribed Rectangles of the Targets

If the target is not upright in the image, it is difficult to calculate the parameters of their minimal circumscribed rectangles. Therefore, it is necessary to move and rotate the target pixels to situate their center at the origin and let the axes of the targets point to the same direction with coordinate axes, while keeping the shape of the targets fixed. As a result, the upright circumscribed rectangles and minimal circumscribed rectangles are superposed, and it is easy to compute the position and size of the target. A detailed description of the procedures is given as follows.

Suppose LB_k (k=1, 2, …, K) is a connected component image including one target, the coordinates of n target pixels are (p_i, q_i), where i=0, 1, 2,…, n, then the means of coordinates (p_i, q_i) of n target pixels, represented by (m, c) are easy to be computed. Consequently, the relative coordinates (u, v) are computed according to (2).

$$(u \quad v) = (p \quad q) - (m \quad c) \tag{2}$$

An n-by-2 matrix **X** can be constructed using (u, v) as follows

$$\mathbf{X} = \begin{bmatrix} u_1 & v_1 \\ u_2 & v_2 \\ \vdots & \vdots \\ u_n & v_n \end{bmatrix} \tag{3}$$

Suppose 2-by-2 matrix **V** is the tensor of gyration radius of the target, i.e., $\mathbf{V}=(\mathbf{X}^T\mathbf{X})/n$, then a matrix **U** can be constructed using the two eigenvectors of matrix **V**. Rotation matrix **R** is determined according to (4).

$$\mathbf{R} = \begin{bmatrix} -1 & 1 \\ -1 & 1 \end{bmatrix} \bullet \mathbf{U} \tag{4}$$

The rotated coordinates (x_i, y_i) are computed as the following equation.

$$\begin{pmatrix} x_i \\ y_i \end{pmatrix} = \mathbf{R} \begin{pmatrix} u_i \\ v_i \end{pmatrix} \qquad i = 1, 2, \ldots, n \tag{5}$$

After that, the target is upright, and there is only one target in each connected component image LB_k . Therefore, the width W and height H of the target can be easily computed. In the 2-D space, rotation is defined by an angle α . The rotation matrix of a column vector about the origin is shown as follows.

$$\mathbf{R} = \begin{bmatrix} \cos\alpha & -\sin\alpha \\ \sin\alpha & \cos\alpha \end{bmatrix} \tag{6}$$

So α is determined by (7).

$$\alpha = \arcsin(\mathbf{R}[2,1]) \tag{7}$$

As a result, we can get the minimal circumscribed rectangle of the target in connected component image LB_k represented by a 5-dimension vector $(m, c, W, H, -\alpha)^T$, where (m, c) represent the center of rectangle, (W, H) represent the width and height, $-\alpha$ is the rotation angle. The target model we need is the minimal circumscribed rectangle above mentioned.

4 Experimental Results

We test the proposed method on different sequences from public data set. Three of them are extracted to demonstrate the performance of the algorithm. The detailed information of these sequences is listed in Table 1.

Table 1. Detailed information of the sequences

title	targets	Data set
MOV045	One lean person	http://210.44.184.112/shadowInv
MOV031	Three balls	http://210.44.184.112/shadowInv
Test	Two persons	http://peipa.essex.ac.uk/ipa/pix/pets/

The results of the three sequences processed by three algorithms are shown in Fig.3-Fig.5. In every figure, the left image represent manual target model, while the mid one and the right one are results of projection method and the presented method, respectively. The target models are indicated by a white rectangle.

In Fig. 3, the target is lean in the image, which causes a large number of background pixels are included in target model in Fig. 3-b. However, the amount of misclassified pixels is sharply decreased in Fig. 3-c. The reason for this decrease is that the rotation of the rectangle can adapt lean status of the target. In Fig. 4, three balls can be recognized by vertical projection, yet horizontal projections of the balls are partly overlapped. The overlapped projection cause the height of the targets markedly increased in Fig. 4-b. Although the projections of the targets are overlapped, there is no occlusion among the balls in the image. Accordingly, it is not difficult to segment the targets. As a result, the proposed algorithm is not affected by overlapped projection, and it can accurately compute the target model in Fig. 4-c. In Fig. 5, both the vertical projection and the horizontal projection are overlapped, which results in the projection method mistakes two persons for one target. Similar to Fig. 4-c, the overlapped projections have no effect on the presented method, and the target model can be successfully computed in Fig. 5-c.

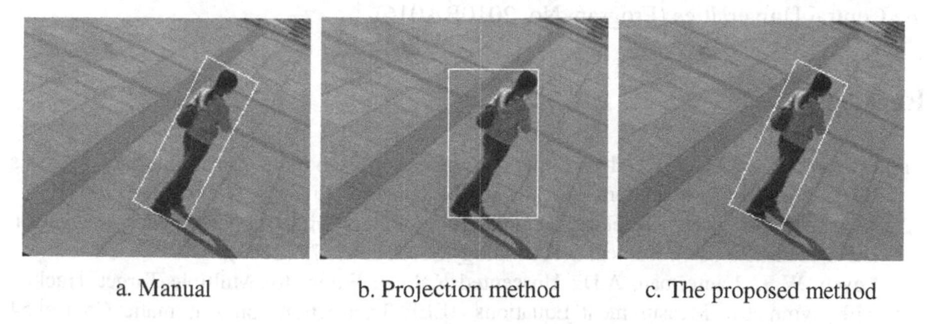

a. Manual b. Projection method c. The proposed method

Fig. 3. Target model of sequence Mov045 computed by three methods

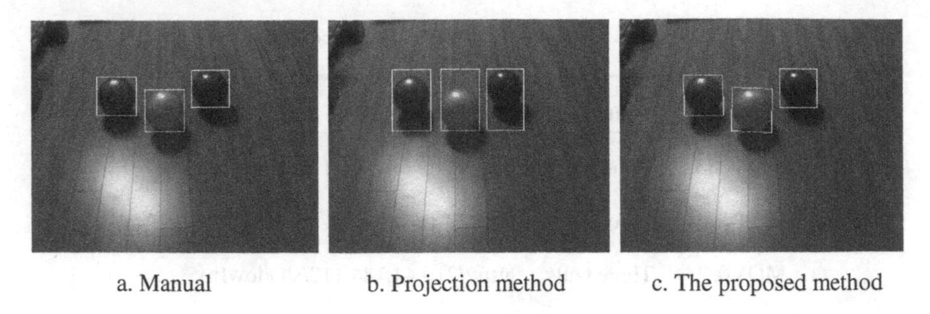

a. Manual b. Projection method c. The proposed method

Fig. 4. Target model of sequence Mov031 computed by three methods

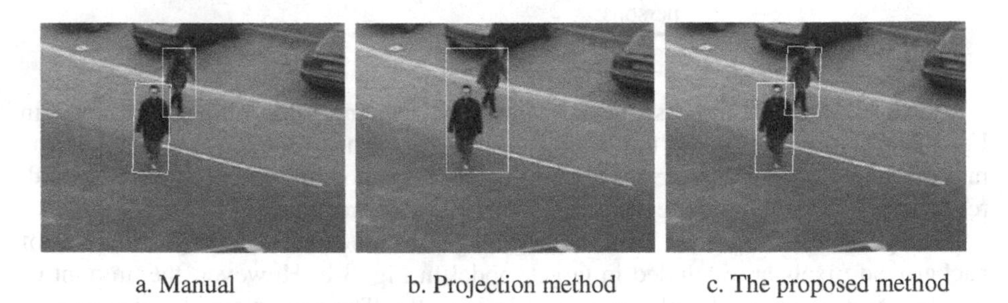

a. Manual b. Projection method c. The proposed method

Fig. 5. Target model of sequence Test computed by three methods

5 Conclusion

In this paper, we propose an automatic and accurate method for initialization of target model in object tracking. Connected components labeling algorithm and radius of gyration tensor method are respectively used to compute the minimal circumscribed rectangles of the targets. Experimental results show that the presented method can construct target model effectively, and it outperforms projection method obviously.

Acknowledgments. This work is supported by the Fundamental Research Funds for the Central Universities (Program No. 2010BA016).

References

1. Comaniciu, D., Meer, P.: Mean Shift: A Robust Approach toward Feature Space Analysis. IEEE Transactions on Pattern Analysis and Machine Intelligence 24, 603–619 (2002)
2. Kabaoglu, N.: Target Tracking Using Particle Filters with Support Vector Regression. IEEE Transactions on Vehicular Technology 58, 2569–2573 (2009)
3. Leven, W.F., Lanterman, A.D.: Unscented Kalman Filters for Multiple Target Tracking with Symmetric Measurement Equations. IEEE Transactions on Automatic Control 54, 370–375 (2009)

4. Birchfield, S.T., Sriram, R.: Spatiograms Versus Histograms for Region-Based Tracking. In: Proceedings of International Conference on Computer Vision and Pattern Recognition, pp. 1158–1163 (2005)
5. Kyriakides, I., Papandreou-Suppappola: Sequential Monte Carlo Methods for Tracking Multiple Targets with Deterministic and Stochastic Constraints. Signal Processing 56, 937–948 (2008)
6. Maggio, E., Smerladi, F., Cavallaro, A.: Adaptive Multifeature Tracking in a Particle Filtering Framework. Circuits and Systems for Video Technology 17, 1348–1359 (2007)
7. Pernkopf, F.: Tracking of Multiple Targets Using Online Learning for Reference Model Adaptation. IEEE Transactions on Systems, Man, and Cybernetics, Part B: Cybernetics 38, 1465–1475 (2008)
8. Wang, J.T.: Research on Object Detection, Tracking and Behavior Recognition in Video Sequences. Nanjing University of Science and Technology (2008)
9. Wu, C.D., Guo, L.F., Zhang, Y.Z., et al.: Method for Touching Objects in Multi-Vehicle Tracking. Journal of Northeastern University(Natural Science) 29, 1065–1068 (2008)
10. Shapiro, L.G., Stockman, G.C.: Computer Vision. Prentice-Hall (2001)

4. Birchfield, S.T., Rangarajan, S.: [illegible] Versus Histograms for Region-Based Tracking. In: Proceedings of International Conference on Computer Vision and Pattern Recognition, pp. [illegible] (2005)
5. [illegible]: Sequential Monte Carlo Methods for Tracking Multiple Targets with Deterministic and Stochastic Constraints. Signal Processing 56, [illegible] (2008)
6. Maggio, E., Smeraldi, F., [illegible]: Adaptive Multifeature Tracking in a Particle Filtering Framework. [illegible] Circuits and Systems for Video Technology 17, 1348–[illegible] (2007)
7. [illegible]
8. [illegible]
9. [illegible] Multi-Vehicle Tracking [illegible] (2008)
10. Shapiro, L.G., Stockman, G.C.: Computer Vision. Prentice Hall (2001)

An Integration Method of Small Batch Data Transmission between Network DNC and PDM

Guangrong Yan, Fei Wang, and Tao Ding

School of Mechanical Engineering & Automation, Beihang University, Beijing, P.R. China

Abstract. The application of Network DNC(Distributed Numerical Control) and PDM(Product Data Management) software in enterprise has improved the efficiency of the manufacturing process. However, lacking of uniform planning for information systems, the product data in DNC and PDM system can not be shared and reused. This paper proposed an integration method for small batch product data transmission between Network DNC and PDM system. Control-component interface integration method was adopted for the integration scheme. By means of port querying of the key products' attribute, integration topology, product BOM(Bill of Material) relationship and logical process of data importing were be involved. Finally this paper verifies the feasibility and effectiveness by using CAXA network DNC software platform

Keywords: Network DNC, Product Data Management, Control-component Interface Integration, Bill of Material.

1 Introduction

DNC is a grading system to distribute and manage the data between management-end and workshop manufacturing-end, it is commonly used in modern digital workshop as the production management mode. It connects the manufacturing equipment with the upper layer computers and realizes the data exchanging between them. There are many models that the Network DNC functions included: data communication, production state collection and management, statistical analysis, production scheduling and management and tool management. PDM is the technology that manages product-related information and process, it is based on software and its functions include document management, workflow management, task coordination, etc. PDM, centered on product, is a data management system running through the whole life of product and it is also a plat for data integration and unified management. With the development of information technology, DNC and PDM system have been universally applied in the enterprise. However, the information system is purchased individually and implemented asynchronous, the information system is lack of master plan, so the storage structure and exchange standard between DNC and PDM are different very much, the data between these systems cannot be shared and reused, manufacturing and management is disjointed in enterprise. This paper presents an integration method applicable to small batch product data transmission between DNC and PDM, develops a control-component interface individually, realizes integration by means of port querying of the key product attribute and conditional extraction. The

D. Jin and S. Lin (Eds.): Advances in ECWAC, Vol. 1, AISC 148, pp. 539–545.
springerlink.com © Springer-Verlag Berlin Heidelberg 2012

method is convenient to operate and has high integration level, it is suitable for small batch product data importing or updating. At last, this paper verifies the feasibility and effectiveness of the method by using CAXA network DNC platform.

2 Integrated Solution

At present, there are two kinds of information system integration method, one is point-to-point integration and the other is distributed technology package integration. Point-to-point method achieves integration of two systems, it is simple implementation and low-cost but limited functionality. Point-to-point integration method includes data transmission integration, control component interface integration and seamless functional integration according to data structure and interface standard [1]. Distributed technology package integration builds a unified standard integration platform using distribution technique other systems are integrated to this platform to achieve the integration of multiple systems. However distributed integration is realized its' complicated and high-cost, and the development period of it is very long, so it's not suitable for SME information integration.

Considering from the data structure, interface standard and other factors, the integration of network DNC and PDM researches in this paper adopts point-to-point integration. Fig. 1 gives the integration architecture diagram. Control component interface integration need to be developed by special interface tool, component interface adopted intermediate table or file to store, exchange data with two sides of system conversion by different standards. However the method is hard to implement, it is easy to operate and high universalizable.

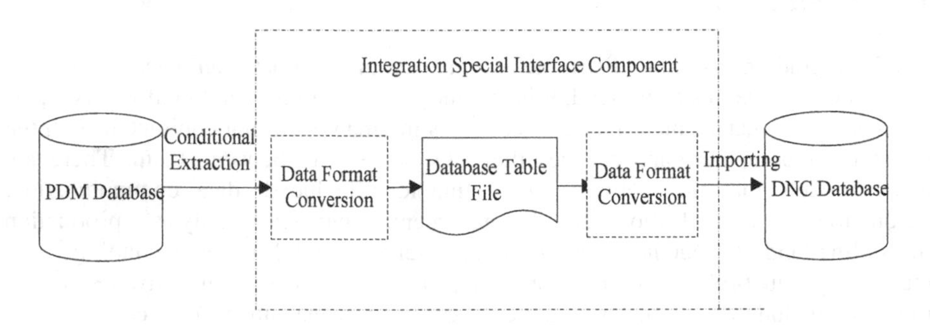

Fig. 1. Integration architecture diagram of DNC and PDM

Integration control component has its own interactive interface, it can access PDM database by setting data source parameters and extract information by means of port querying of the key product attribute and conditional selection. The product attribute and BOM relationship extracted displayed in the port interface in the form of tree-structure, user select required product structure and import into DNC system, at the same time data information is stored into DNC database after a format conversion process. The product structure imported is in form of temporary product tree structure, manager can hitch it to the specified location using shear and paste.

3 Business Logic Processing

The integration of Network DNC and PDM adopted three-tier topology structure. The key point of integration processing is to deal with the logical relationship. And the data to transmit among systems includes product having assembly relation attribute, sub-product information and design BOM relationship. It is managed by single-layer mode when extracting BOM relation in DNC system, that means it only need to extract its direct sub-product when user inputs a product. The imported product and part must be release, some logical operations such as update and delete should be considered when the same product or BOM be imported multiple times. In the following section, we will discuss the implementation of integration in details from three aspects of integration topology, BOM relationship handling and data importing logic.

Integration Topology. The integration of DNC and PDM adopts three tier sever architecture. As Fig 2 shown, the product data is be managed by the processing department which is based on PDM sever, the workshop equipment are managed by Network DNC through workshop sever. The integration method suitable to small batch product data transmission which the paper researched is the process shown by ① in Fig. 2, it is mainly used when the product structure tree of DNC and PDM is inconsistent or product data in PDM system is update. Now the network DNC system extracts the latest product and BOM data to achieve synchronization among systems.

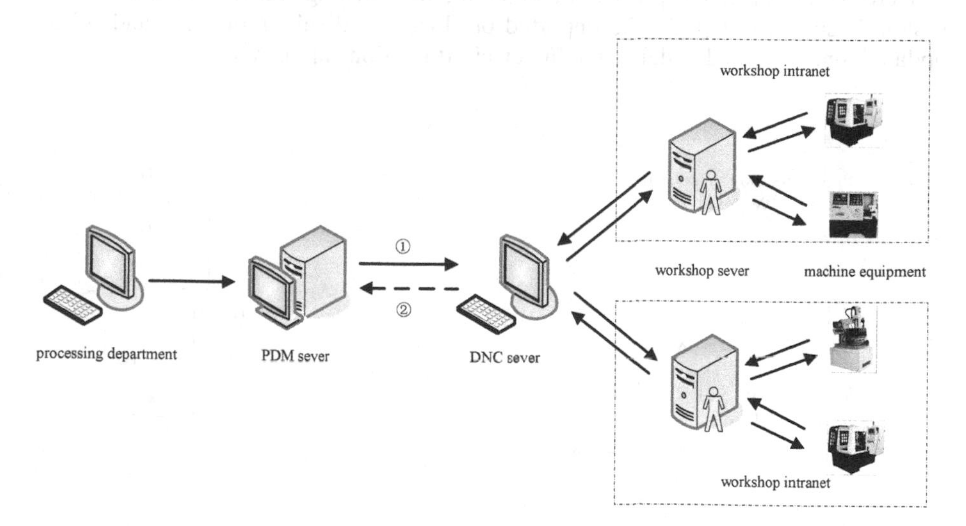

Fig. 2. Integration Topology Structure

The Product information and manufacturing data in network DNC are delivered to workshop servers and NC machines by the function of assignment. The file data modified in workshop can return to DNC system. The assignment folder and the receive folder are independent in order to prevent coverage. The manager can upload files to PDM sever manually after verified correct, as the number ② shown in Fig. 2, thus we realize the data consistent and synchronization of the whole system.

BOM Relationship Handling. In the PDM database there are two kinds of tables, field attribute table and BOM relationship table. Both of them are used to store product attribute and BOM structure. Interface tool obtain complete product attribute and BOM relationship by means of traversal querying the two kinds of tables, conditional to compare the product key attribute in order to get satisfying constraint result set, then select required product structure from the result set and imported it into the network DNC system. Firstly, need to judge whether the product is already exist when importing product and BOM structure into network DNC system, the different treatment of product BOM structure according to the query result, specific operation process is shown in the following.

1、If the importing product BOM structure has been already exist in DNC system, update the product structure, specially include following operation.

① Delete the existing product relationship, include following two steps.

a. Query all sub-products of the original product, if no sub-product has father-child relationship with other product, delete the product, sub-products and BOM relationship.

b. If some sub-products have father-child relationship with other products, only delete the relationship of the sub-product and the father-product.

② Query the version status of the sub-product of the new product, import the release sub-product into DNC system and build BOM relationship with the father product. Sub-product not release can not be operated.

Here an example, the process of updating the existing BOM structure in DNC system is shown in Fig. 3. The imported product A and sub-product A1 and A2 are updated, but only need to delete the father-child relationship of A and A3.

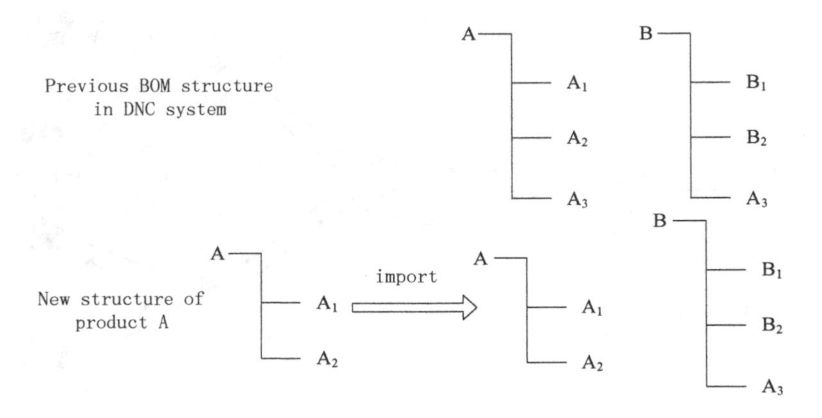

Fig. 3. Import existent product structure

2、If the imported product BOM structure has not been exist in DNC system, then import the product structure directly, specially include following operations.

① Query if sub-products of the importing product have father-child relationship with other products, update attributes of all sub-products which have relationship.

② Query the status of sub-products of the imported product, import release sub-products into DNC system and build BOM relationship with the importing product.

The process of importing a new product BOM structure is shown as Fig. 4. Sub-product A3 of product B needs to update after importing.

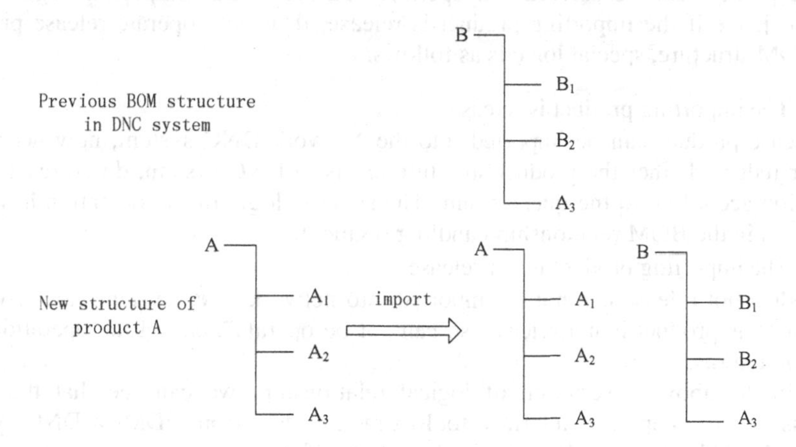

Fig. 4. Import new product structure

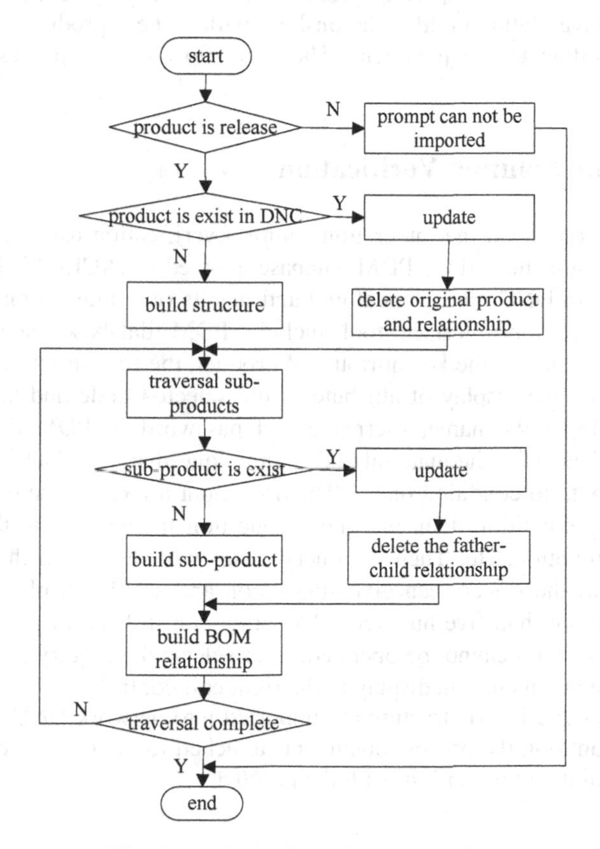

Fig. 5. The logic flow of importing data

Data Importing Logic. The premise of the extracting product BOM structure from PDM database to network DNC is a version status of the product, it means that only release product can be derived and operated. Therefore data importing logic firstly need to judge if the importing product is release, then only operate release product and BOM structure, special logic is as follows.

1、The importing product is release

Release product can be imported into the Network DNC system; now need to a further judge whether the product structure exists in DNC system, do corresponding operation according to the query result. The specific logic of the operation has been discussed in the BOM relationship handling segment.

2、The importing product is not release

Product not release can not be imported into network DNC system, now give the prompt "this product is not release, so can not be operate", cancel the operation and function returns.

From the above description of logical relationship, we can see that the logic process of the component interface tool extracting data from PDM to DNC system can be divided into three key steps: (1) judging if the product and sub-product are release, (2) judging if the importing product exist in DNC system and (3) querying sub-products have father-child relationship with other products. Do different operations according to the judgment. The logic flow of this process is shown in Fig. 5.

4 Integration Solution Verification

This paper has completed the integration solution verification using CAXA network DNC software platform, and the PDM database adopted ORACLE 9i. The integration tool is triggered in DNC management end and has its own interaction interface. The main functions of the interface tool include PDM database parameter setting, querying and selection to the key attribute of product, the tree-structure display of the product result set, the display of attribute of the selected node and the status of the interface tool. Input the name, username and password of PDM data source and connect to the database. The material code can unique mark product, so it is selected as the key attribute to condition query. The user input a fixed value of material code and select query condition, then the query condition includes more than, less than, equals to, subsumption, etc. Then the query result will display in the tree-structure preview window, the system can give suggestion and set the display number if the query result is more than five hundreds. The gray icon in front of part represents the part is not release, so it cannot be operated. User selects the inquiry product node and its material code and name can display in the right edit control.

When product and BOM structure are imported into network DNC system making use of integration tool, the process documents attached to the product enter into DNC database and maintain relationship with the product.

5 Conclusion

With the widely using of Network DNC and PDM in enterprise, there are many problems appear at the same time. For example, the data among systems cannot be shared and reused and the integrity of enterprise is destroyed.

This paper proposed an integration method suitable for small batch product data transmission between the Network DNC and PDM. The component tool and interactive interface are developed separately. Achieve integration by means of condition querying the key attribute of product and extracting. This paper designed integration scheme and researched the integration topology, BOM relationship handling and data importing logic in details. Finally the paper proved the feasibility and effectiveness of this integration method using CAXA network DNC software platform.

Acknowledgment. The author thanks China's 863 program *Key Technologies of i-Plane for Large Aircraft Development* for the sponsorship of this paper. The project number is 2009AA043302.

References

1. Ren, R.: Application of ERP and DNC Integration Technology. CAD/CAM and Manufacturing Information (5), 36–38 (2009)
2. Chen, X.: Study and Application for PDM. Electro-Mechanical Engineering (6), 2–4 (2001)
3. META Grup. The Future of Data Intergration Technologies. A META Group Whitepaper (6), 1–3 (2004)
4. Li, Z.: Research and Development of Network DNC system based on PDM. Hefei University of Technology Master Dissertation, Hefei, pp. 3–6 (2008)
5. Yan, W.: Study on Key Techniques of DNC System Oriented Networked Manufacturing. Dalian University of Technology, Dalian, pp. 8–14 (2003)
6. Wang, S., Liu, F.: The Development of DNC Integration Technology. Chinese Mechanical Engineering 9(2), 54–56 (1998)
7. Zhang, F., Liu, Q., Lei, B., Dan, D.N.C.: Integrated DNC: A Case Study. International Journal of Production Research 39(17), 2–4 (2001)
8. Li, Y., Wang, Z., Zheng, J.: Discussion on Integrated DNC Reconstruction in Enterprises. Journal of HeFei University of Technology 23(2), 332–335 (2002)

A Text Document Clustering Method Based on Topical Concept

Yi Ding and Xian Fu

The College of Computer Science and Technology
Hubei Normal University, Huangshi, China
{teacher.dingyi,teacher.fu}@yahoo.com.cn

Abstract. Nowadays, document clustering technology has been extensively used in text mining, information retrieval systems and etc. The conventional document clustering methods rely on the classical vector-space model using the key words as the feature. However, these methods ignore the semantic relations among the keywords, do not really address the special problems of document clustering: high dimensionality of the data, and high computation complexity. To solve these problems, based on topic concept clustering, this paper proposes a method for Chinese document clustering. In this paper, we introduce a novel topical document clustering method called Document Features Indexing Clustering (DFIC), which can identify topics accurately and cluster documents according to these topics. In DFIC, "topic elements" are defined and extracted for indexing base clusters. Additionally, document features are investigated and exploited. Experimental results show that DFIC can gain a higher precision (92.76%) than some widely used traditional clustering methods.

Keywords: document clustering, clusters indexing, topical concept.

1 Introduction

With the abundance of text documents available through the Web or corporate document management systems, the dynamic partitioning of document sets into previously unseen categories ranks high on the priority list for many applications like business intelligence systems. However, current text clustering approaches tend to neglect several major aspects that greatly limit their practical applicability. In this paper, a new clustering method is presented, which is named Document Features Indexing Clustering (DFIC) method. The main question in a topical concept clustering method is how to describe a topic. We believe that a topic, which consists of a series of closely related events[1], is represented by a set of "topic elements", such as participants, locations, dates, properties, and activities. According to the above principle, we set up "topic elements indexes", whereby documents on the same topic can be indexed and form base clusters.

D. Jin and S. Lin (Eds.): Advances in ECWAC, Vol. 1, AISC 148, pp. 547–552.
springerlink.com © Springer-Verlag Berlin Heidelberg 2012

2 Related Work

All clustering approaches based on frequencies of terms/concepts and similarities of data points suffer from the same mathematical properties of the underlying spaces. These properties imply that even when "good" clusters with relatively small mean squared errors can be built, these clusters do not exhibit significant structural information as their data points are not really more similar to each other than to many other data points. Therefore, we derive the high level requirement for text clustering approaches that they either rely on much more background knowledge or that they cluster in subspaces of the input space. In order to improving this problem, the most relevant works are that of Zamir et.al. [2][3]. They proposed a phrase-based document clustering approach based on Suffix Tree Clustering (STC) [4][5]. The method basically involves the use of a tree structure to represent share suffix, which generate base clusters of documents and then combined into final clusters based on connect-component graph algorithm.

3 Document Features Indexing Clustering

3.1 Clusters Indexing

As introduced before, STC method has some obvious advantages. Actually, these good qualities should all be ascribed to the smart way of forming base clusters by creating an inverted index of phrases for the documents. For the sake of convenience, we call this kind of way indexing base clusters hereafter. The precise meaning of indexing base clusters can be formulated as below: Let $F=\{F_1,F_2,\ldots,F_n\}$ be a document collection and $I=\{I_1,I_2,\ldots,I_n\}$ be a set of chosen indexes. Document F_i will be placed in the cluster indexed by I_j if and only if the number of times that I_j occurs in F_i exceeds a predetermined threshold T.

In clusters indexing, no halting constraint is needed. That's deserving commendation because it is impossible to determine the optimal number of clusters beforehand with the scales of documents and the numbers of actual topics varying a lot in practice. Besides, a single document may be on two or more topics simultaneity. Thus, it seems reasonable to allow a document to appear in more than one cluster. This need is met by indexing base clusters as a document can be indexed by several indexes. Furthermore, the clusters can be described by indexing base clusters since each base cluster created in this way has an index which represents the cluster's content.

3.2 Exploiting Document Features

We take advantage of document features in forming indexes. In natural language processing, a document is generally represented as a vector of words [4]. The weights of words are usually calculated using statistical techniques (such as "tf.idf"). Nevertheless, the document features of words themselves should also be given enough attention. Take POS (Part-of-Speech) for example, intuitively, words with different POSes should have different contributions to characterizing a document [5].

Usually, nouns and verbs are the most indicative. Adjectives and adverbs are less valuable. Function words have little or no influence and should be excluded as stop words. In addition, NEs are more discriminative than normal words and should be assigned higher weight. In DFIC, we form the indexes by utilizing NEs coupled with important nouns and verbs.

3.3 Main Steps

Step 1: Document is submitted to sequential preprocessing modules including word segmentation, POS tagging, and NE recognition. In this step, stop words are removed. Here, a stop list containing punctuations, common used words and some news specific words is maintained.

Step 2: In the stage of document representation, Vector Space Model (VSM) is employed. The vector terms here contain only NEs, nouns, and verbs. The tf.idf is used for weighing the vector terms.

Step 3: Forming indexes and creating base clusters: An index used in DFIC consists of two parts: an NE-part and a keyword-part:

Definition 1: Let F be a document, $A=\{a_1,a_2,\ldots,a_n\}$ be a set of NEs occurring at least twice in F, $B=\{b_1,b_2,\ldots,b_n\}$ be a set of keywords (nouns and verbs) in F whose tf.idf weights exceed a preset threshold T. $\forall a\in A$ and $\forall b\in B$, the two-tuple of (x,y) is defined as one of D's indexes.

If the size of A and B are x and y, then document F has x×y chances to be indexed by any of its indexes. This makes it possible that a single document can be indexed on different topics and put into several base clusters. With the well designed indexes combining NEs and keywords, DFIC constructs base clusters by merging documents that share common indexes.

Step 4: Combining base clusters into clusters: The base clusters formed in the last section overlap a lot. Hence we combine base clusters to reduce duplication and form more complete clusters. Let c_i,c_j be two base clusters. If their distance is less than a preset threshold T_r, then they will be combined. In order to measure the distance between two base clusters, the centroids of them have to be calculated. The distance measure used in the combination algorithm is the cosine measure:

$$\cos\left(c_i,c_j\right)=c_i\cdot c_j/\left\|c_i\right\|\left\|c_j\right\| \tag{1}$$

4 Experiments

4.1 Data and Metrics

The method is evaluated using a collection of 2271 computer sciences documents collected from the web. From these documents, we have manually identified 300 topics, whose maximum size is 28 documents and minimum is 4. In this paper, the precision and recall are computed respectively in evaluation.

Given a particular topic Ti of size y_i and a particular cluster C_j of size y_j, suppose y_{ij} documents in the cluster C_j belong to Ti, then the precision of this topic and cluster is defined to be:

$$precision\left(T_i,C_j\right)=y_{ij}/y_j \tag{2}$$

The precision of T_i is the maximum precision value attained at any cluster in the cluster set C:

$$precision\left(T_i\right)=\max_{c_j\in c} precision\left(T_i,C_j\right) \tag{3}$$

The overall precision is computed by taking the weighted average of the individual precision:

$$precision=\sum_{i=1}^{N_T}\frac{y_i}{Y}precision\left(T_i\right) \tag{4}$$

where N is the total number of documents and NT is the number of topics. Similarly, the recall of the entire clustering results can be defined as

$$recall=\sum_{i=1}^{N_T}\frac{y_i}{Y}recall(T_i) \tag{5}$$

Where $recall(T_i)=\max_{c_j\in c} precision\left(T_i,C_j\right)$ and $recall\left(T_i,C_j\right)=y_{ij}/y_i$.

4.2 Comparisons with Other Methods

We conduct two sets of experiments. In the first set, the DFIC is compared with K-means Clustering (KMC), STC and Coagulation Hierarchical Clustering (CHC). The stopping criterion for CHC and KMC is set to 300, which is the factual number of topics. First of all, the precision and recall of the above four methods are computed and compared (Fig.1, Table.1). As expected, the DFIC method scores highest. We believe that this positive result is mainly due to DFIC's well designed indexes which can identify topics more accurately.

4.3 Comparisons of Different Parts

In the second set of experiments, we evaluate the contributions of DFIC's different parts. Firstly, we try to find out whether the indexes involving both NEs and keywords work better than using only NEs or keywords. Table.2 compares three runs of DFICs which use different kinds of indexes. It is obvious that the DFIC indexed by keywords plus NEs performs much better than that indexed by keywords in both precision and recall. This indicates that it is not enough to describe topics using keywords alone. We can also see that the run using NEs alone achieves the highest recall but the lowest precision. This is because, in the experimental data, more than one topic may be related to the same NE entity. Thus a single NE may index several topics. This results in many "large" base clusters which should be responsible for the high recall and low precision.

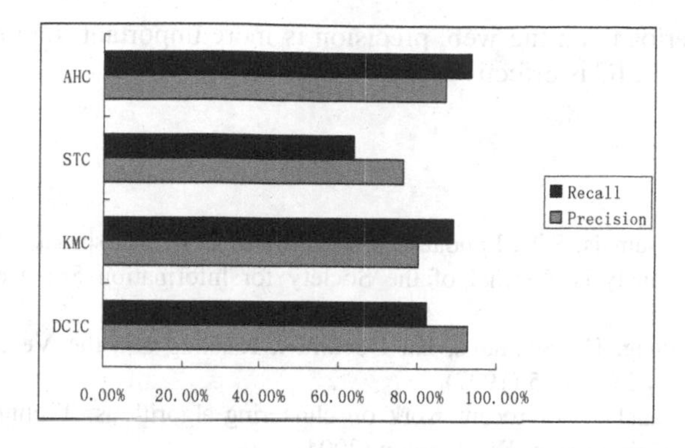

Fig. 1. Precision and Recall of four clustering methods

Table 1. Precision and Recall of four clustering methods

	DFIC	KMC	STC	CHC
Precision	92.76%	80.14%	76.21%	87.04%
Recall	82.17%	88.97%	63.71%	93.45%

Table 2. Comparison of DFICs using different kinds of indexes for creating base clusters

Different kinds of indexes	Precision(%)	Recall(%)
keywords + NEs	92.76	82.17
keywords only	87.66	77.82
NEs only	77.43	85.39

Table 3. Comparison of DFICs having base clusters combined or not

Base clusters combined or not	Precision(%)	Recall(%)
combined	92.76	82.17
uncombined	95.34	50.47

"partitioned" into several base clusters, and the combination is necessary for producing larger, more complete clusters. We can conclude that combining base clusters can improve recall significantly.

5 Conclusion

In this paper, we propose a novel clustering method DFIC. We conduct a number of evaluations that compare DFIC with some traditional clustering methods, including CHC, KMC and STC. Results show that the DFIC method can achieve a higher precision while maintaining an acceptable level of recall. Considering the "information overload" on the web, precision is more important than recall. Thus we can conclude that DFIC is effective.

References

1. Deerwester, S., Dumais, S.T., Landauer, T.K., Furnas, G.W., Harshman, R.A.: Indexing by latent semantic analysis. Journal of the Society for Information Science 41(6), 391–407 (2002)
2. Lee, D.-L., Chuang, H., Seamons, K.: Document Ranking and the Vector-Space Model. IEEE Software 14(2), 67–75 (1997)
3. Fasulo, D.: An analysis of recent work on clustering algorithms. Technical Report UW-CSE-01-03-02, University of Washington (2004)
4. Zamir, O., Etzioni, O.: Web Document Clustering:A Feasibility Demonstration. In: Proceedings of the 21st International ACM SIGIR Conference on Research and Development in Information Retrieval, pp. 46–54 (1998)
5. Gusfield, D.: Algorithms on Strings,Trees and Sequences: Computer Science and Computational Biology. Cambridge University Press, Cambridge (1997)

Wireless Sensor Network Based on ZigBee in Aquaculture

Xingqiao Liu and Liqiang Cheng

School of Electrical and Information Engineering, Jiangsu University, 212013, Zhenjiang, China
xqliu@ujs.edu.cn, 1162203868@qq.com

Abstract. Monitoring the living environment of fish accurately and quickly plays a critical role in the aquaculture industry. The aquaculture monitoring environment has the characteristics of multi-measuring points, long measuring time, and high complexity measuring conditions. In order to solve the problems of the high cost and wiring difficulty during monitoring, a new type of wireless sensor network system for aquaculture based on ZigBee is designed. This system achieved the goals of collect, transmit and display multi-parameters, such as dissolved oxygen (DO) and temperature. By using wireless sensor network, transmissions of the data between sensor detecting nodes and coordinate nodes are fast and accurate. The system also enforces real-time remote monitoring. Because of its simple structure, fast data transmission, easy functional extension, high self-organization and self-recovery ability, the system is suitable for industrialized aquaculture.

Keywords: ZigBee, Aquaculture, Temperature, Dissolved Oxygen.

1 Introduction

In the modern aquaculture farm, the water condition affects the health of fish, therefore directly influences the farming yield. It is very essential to upgrade the automation and artificial intelligence level in monitoring the parameters of the water, such as temperature and dissolved oxygen [1].

Traditionally, sensors are placed in the monitoring areas and connected through wire style to transmit analog signals. These analog signals are converted into digital signals, stored, analyzed and demonstrated on the computer [2]. Nevertheless, this method is facing many difficulties. If more sensors are needed, the massive wires will not only increase the cost of the system, but also cause difficulties in installing and maintaining. What is wore, on some occasions, such as rotating parts and oil tank, sensors are not allowed to wire.

With the development of wireless sensor network technology, this paper designs a new type of multi-parameter wireless monitoring system based on ZigBee in order to further enhance and boost reliability, flexibility and the anti-interference of the long-distance online supervisory system for aquaculture [3]. The system adopts wireless transceiver chip cc2430 produced by Chipcon Corporation and wireless communication technology to establish wireless network. It can automatically build

D. Jin and S. Lin (Eds.): Advances in ECWAC, Vol. 1, AISC 148, pp. 553–558.
springerlink.com © Springer-Verlag Berlin Heidelberg 2012

networks without any physical wire, therefore has a lower cost. It also has the ability of self-organization and self-recovery, as well as real-time monitoring.

2 Experimental Details

The design uses the chip cc2430 promoted by Chipcon Corporation, which conforms to IEEE802.15.4 standard, and combines the excellent performance of the leading cc2420 RF transceiver with an industry-standard enhanced 8051 MCU [4].

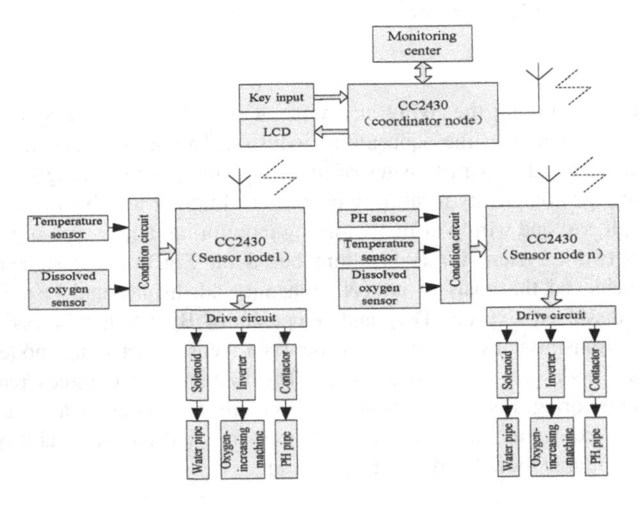

Fig. 1. The wireless sensor network architecture

As shown in Fig.1, this wireless sensor network is mainly composed of coordinators and sensor nodes. The system also has a monitoring center. The coordinator is the bridge between the sensor nodes and the monitoring center, so all the gathered data is transmitted to the monitoring center through the coordinator, and the control command is sent to the sensor node through the coordinator.

2.1 Dissolved Oxygen Measurement

While the temperature can be measured by the resistance of a thermal sensitive resistor, the dissolved oxygen measurement is more complicated.

We use the *thin-film electrode pole spectrum measurement*. The circuit is shown in Fig. 2. The negative pole of the electrode is a 4 mm gold piece, and the anode which is the reference electrode is the silver piece. The electrolyte is filled between the two-poles, the end of which is covered by the thin-film. When the polarizing voltage imposed on two-pole is about 0.7V, the oxygen seeping from the film returns to original state on the gold negative pole. Because of the oxidation-reduction reaction, the current generated by electronic transformation is in proportion to the oxygen partial pressure in the sample. When the water is lack of oxygen, there is no current in the sensor. When it has enough oxygen, the dissolved oxygen concentration information is delivered to condition circuitin form of current [5].

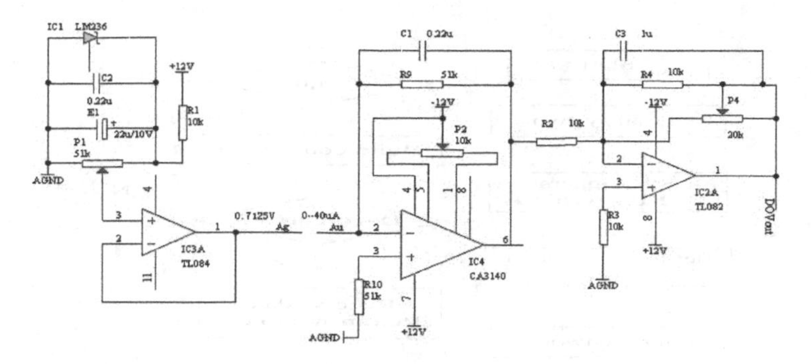

Fig. 2. Circuit for dissolved oxygen measurement

2.2 Software Design

The flow chart of coordinator is shown in Fig. 3. Under the electricity initialization condition, triggering the key event determines whether the devicc is a coordinator. Then, the coordinator starts a ZigBee network, and then enters into the network monitor waiting status. After receiving the joined request from the child nodes, it assigns network address and transmits confirmation message to child nodes to establish the binding. After that, the coordinator waits data request. Once received the data message transmitted from the sensor nodes, it carries on the analysis of the data packet. After confirming the information is data message, it demonstrates the data to the monitoring center.

The flow chart of sensor node is shown in Fig. 4. Under the electricity initialization condition, triggering the key event determines whether the device is a terminal node. It scans the channels to join an appropriate network. After successfully joining the network, it transmits the network address to coordinator. If receiving the information of gathering data, the procedure starts to enter the application layer, calling task processing function zb_HandleOsalEvent(), and triggering the corresponding event function, such as MY_REPORT_TEMP_EVT event, MY_REPORT_PH_EVT event. It starts the A/D sampling to gather each parameter value. CC2430 has 8~14-bit ADC with up to eight inputs, firstly chooses the input channel for gathering, and then sets the related ports and the disposition registers. The gathered data is in storage in the ADCH and the ADCL register, waiting to be transmitted to the coordinator.

Part of A/D sampling program for temperature is given as follows:

```
uint8 myApp_ReadTemperature (void)
{uint16 value;
ADCIF = 0; // Clear ADC interrupt flag
ADCCON1=0x3f;
ADCCON3 =0x32; //choose channel AIN2
while ( !ADCIF ); // Wait for the conversion to finish
value = ADCL;
value |= ((uint16) ADCH) << 8; //Get the result
}
```

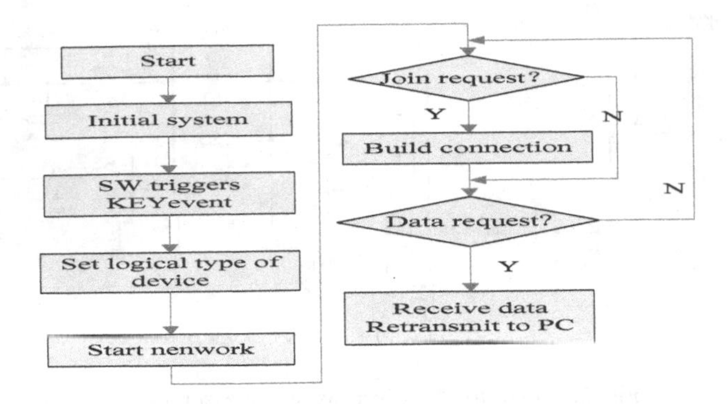

Fig. 3. Flow chart of coordinator

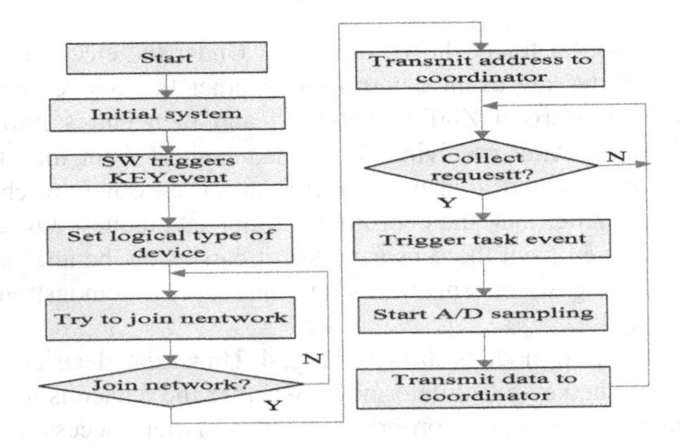

Fig. 4. Flow chart of the terminal node

3 Experimental Result

The wireless sensor network was tested by comparing actual values and measured values. The experimental data is shown in Table 1. Compared with actual value, the relative error rate of measurement values of sensor nodes is less than 1%. The error is limited to the range of requirement by the process of stable amplitude, filtering and temperature calibration.

The system selects the wireless method in the data communication process, and overcomes the shortcomings of wired communication. The error of communication approaches to zero, which shows that the network communication is stable and satisfies the design requirement. The variation of dissolved oxygen and temperature per day is shown in Table 2.

From the table, it can be seen that the various parameter variation is quite stable, the temperature fluctuation is within the scope of +0.5, the amount of dissolved oxygen variation stabilizes within +0.4mg/L.

Table 1. Comparison between observed values and standard values of parameters

Parameter	Actual Values	Measured Value	Relative Error Rate
Dissolved Oxygen	8.1	7.90	-2.5%
	8.0	7.88	-1.5%
	7.8	8.01	+2.7%
	7.0	7.21	+3%
Temperature	22.0	22.15	0.6%
	25.2	25.32	0.5%
	28.1	28.19	0.3%
	30.3	30.45	0.49%

Table 2. Daily change of temperature and dissolved oxygen

parameter	time											
	0	2:00	4:00	6:00	8:00	10:00	12:00	14:00	16:00	18:00	20:00	22:00
Tempreture (℃)	25.2	24.9	24.6	24.4	24.8	25.0	25.3	25.4	25.5	25.3	25.2	25.1
Dissolved oxygen ($mg{\cdot}L^{-1}$)	7.2	7.3	6.9	6.8	6.6	6.7	7.0	7.1	7.3	7.2	6.9	7.1

4 Conclusions

Experiment proofs that the system can be used to monitor and control temperature and dissolved oxygen in aquiculture. This paper applies ZigBee technology in the intelligence aquaculture system, improving the performance of monitoring system and realizing the multi-parameter collection, wireless transmission as well as display without region or time limit.

Acknowledgments. The subject is funded by agricultural science and technology program of Changzhou, Jiangsu Province, and the priority Academic program development of Jiangsu higher education institutions (No.6-2011). The authors would like to thanks sponsors for giving a chance to publish the paper.

References

1. Zhu, W.J., Ran, G.J.: Design of automatic supervise and control system on many environmental factors of aquaculture. Fresh Water Fishery 31(1), 60–61 (2001)
2. Haron, N.S., Mahamad, M.K.B., Aziz, I.A., Mehat, M.: International Symposium on ITSim 2008, vol. 4, pp. 1–7 (2008)

3. Morello, R., De Capua, C., Meduri, A.: Instrumentation and Measurement Technology Conference (I2MTC), pp.1150–1154 (2010)
4. Dissanayake, S.D., Karunasekara, P.P.C., Lakmanaarachchi, D.D., Rathnayaka, A., Samarasinghe, A.T.L.: 4th International Conference on Information and Automation for Sustainability, ICIAFS, pp. 257–260 (2008)
5. Liu, X.Q., Liu, Y., Gong, X.X., Zhang, H.L.: Measuring Technology and Mechatronics Automation, 85–88 (2009)

Legal Protection of Network Virtual Property

Hui Chang[1], Yang Tang, and Zhao-Jing Ma

[1] Law Department, School of Liberal Arts and Law, Tianjin University,
Tianjin 300072, China
tjdxchfx@163.com

Abstract. The rapid development of Internet has brought the rise of online games as well as legal disputes about it. Based on the legal definition of network virtual property, this paper elaborates the ownership of network virtual property in detail and proposes useful advice about protecting virtual property from legislative and judicial perspectives.

Keywords: Network virtual property, Ownership, Legal Protection.

1 Introduction

It is reported by China Internet Network Information Center (CNNIC) that by June 2010, the number of Chinese netizens has surpassed 400 million to reach 420 million. Compared to that by the end of 2009 it has increased 36 million, and the popularization rate rises to 31.8%. Mobile phone netizens have increased 43.34 million within the last half year and reached 277 million [1]. According to the latest statistics issued by *Analysys International*, by 2010, Chinese mobile game players have risen to 130 million and the market furthered to 3.3 billion; the number of mobile online game players also crossed 20 million.

Given the statistics above, it is apparent as a result of the rapid development of information technology, online game has become a huge industry. Due to the different understanding of its legal attributes and the lack of legislation, sometimes network virtual property lacks timely and effective protection. The author believes that the key to solve this problem includes a unified legal definition of virtual property and its ownership, and to improve the protection mechanism from legislative and judicial perspectives.

2 Legal Definition of Network Virtual Property

2.1 Definition of Network Virtual Property

What is network virtual property? Scholars do not have a unified and clear definition at present. Concluded from the late studies, the definition about network virtual property should be categorized in broad and narrow senses.

The broad sense of network virtual property emphasizes virtuality, which covers any property as long as it is digitalized and immaterial. In the narrow sense however, network virtual property specifically refers to the game resources that dependent on

D. Jin and S. Lin (Eds.): Advances in ECWAC, Vol. 1, AISC 148, pp. 559–564.
springerlink.com © Springer-Verlag Berlin Heidelberg 2012

cyberspace and belonging of game players. Therefore, some scholars believe that network virtual property, refers to any virtual objects that have use value and exchange value in a virtual community of online game and belong to the game player, like game account and equipment, currency, pets, books, etc.[2]. The author agrees to adopt the narrow sense of network virtual property, and discusses the legal issues concerning it in this thesis.

2.2 Characteristics of Network Virtual Property

In order to address the issue of virtual property through legislation, first of all we need to define it legally and to illustrate its connotation and denotation, so that we draw the boundaries between virtual property in legal sense and what we say of virtual property or virtual trading items in daily life. Hereby we shall give a legal definition to virtual property from the perspective of feature analysis.

(1) Virtuality of Virtual Property
Virtual property is dependent on virtual cyberspace, thus virtual property is first and foremost virtual, as the fundamental distinction from traditional property.

(2) Valuableness of Virtual Property
Property is made from human physical and mental labor, characterized by scarcity, use value and exchange value. Virtual property should also have such characteristics to become a kind of legal property.

(3) Reality of Virtual Property
If virtual property only exists and functions in virtual space and is unable to connect with the real world, it should be excluded from legal property.

(4) Legality of Virtual Property
Any right in law should base on its legality of existence. The way of obtaining and the object of virtual property should meet the requirement of legality [3].

(5) Limitedness of Virtual Property
Virtual property exists on the basis of cyberspace. As a service product, it has certain term of service, according to the game provider's state of operation and market demand [4].

3 Ownership of Network Virtual Property

Currently there are two views among scholars, one holds that the ownership pf network virtual property should belong to the service operators (also the developers, similarly hereinafter); the other believes that the ownership goes to its actual owner, not the service provider. We would like to take the second opinion. Thereinafter we would discuss the first point of view taking the virtual property in online games for example.

According to the first viewpoint, first, the Internet service operator has the original real right of the hardware, software and virtual property in the online game, based on its act of production which commands investment funds, technology and personnel, but the user only obtains the use right of virtual property through contract with the network service operator. Second, the service operator acquired the original real right by developing online games, thus has the exclusive control of game; Third, when the game has been developed, before it goes to market and the operator signs contract with users, the operator owns complete disposition of the game, as well as the virtual property of it. Since the service operator has the acquisition by production, exclusive possession and complete disposition, it enjoys the ownership of network virtual property [5]. However, the service operator is not entitled of the original real right of virtual property. The network service operator creates the game, which provides a basic template for all sorts of possible virtual property in the game, rather than design all the possible virtual property one by one. Players enter the game and take a series of operations which, either completing certain settings or according to certain odds, obtaining virtual property. As long as the network operator cannot get the original real right of virtual property in games, it doesn't have complete disposition to virtual property in games, or enjoy the complete real right of it either.

Another viewpoint is that virtual property in online games is obtained by rules that the game operator sets, so the ownership belongs to the game developers, administration authority belongs to the service operator, players only get the use right [6]. After the player enters the game, he/she plays according to the rules and earns virtual property. Therefore, the virtual property obtained by the player is actually the labor income generated on basis of the established procedures and rules. Besides, some holds that virtual property does not have "real right", since it only has use value or exchange value in certain virtual space, and once out of the particular virtual space it becomes worthless [7], which we couldn't agree. It is impossible for us to casually possess, use, profit or dispose virtual property without permissions from its actual owner, for reason that it could not be recognized and utilized off virtual cyberspace. Just the same as some object that only embodies value in a certain field, out of its particular field it may lose its value or get greatly depreciated, but the absolute ownership of its owner is still widely approved.

Some people believe that in the use of virtual property, the service operator is the greatest obligor who ensures players' right to their virtual property. Players under contract request the operator to realize the expected functions of virtual property, while the service operator has obligation to take actions. Without the cooperation of the operator, players cannot control any virtual property, thus players have right of claim instead of right of domain, which makes the main subjects of right and obligation of virtual property particular [8]. The relationship of rights and obligations concerning virtual property is not a debt relationship between netizens and network service providers. The contract between users and service operators is used to explain that users get access to the Internet, receive service for activities and pay for the service to network operators; it is not a contract with network operators focusing on virtual property.

To sum up, users of network service gain virtual property through proper behavior consistent with the rules, thus the ownership of the virtual property should fall to its real owner and the network operator is responsible for preserving it. This kind of right and obligation relationship comes into being the moment the user gets access to the network service.

4 Legal Protection of Network Virtual Property

In order to promote the healthy and orderly development of networks and to protect the legitimate rights and interests of online game players, it is necessary that we perfect the legal protection from legislative and judicial perspectives.

4.1 Perspective of Legislation

(1) Establish virtual property-specific laws to protect network virtual property
In laws and regulations that have been promulgated and implemented such as *Decision on Safeguarding Internet Security* and *Regulations on Protection of Computer Information System Security*, protection about network virtual property is still a blank. The legal characteristics and requirements, real right of virtual property, obligations of the operator and solutions to disputes have not been mentioned in established laws like *General Civil Law*. In addition, problems brought up by private servers, external hangings, regulations of virtual exchange markets, management of online game form contracts, require a fundamental law about virtual property to solve problems thoroughly.

At the same time we draft for specific law, we should pay attention to the consistency with established laws. In the domain of our current domestic civil legislation, the different properties of real right and credit has been made clear in *Real Right Law*, but a clear distinction between tangible and intangible property is not yet made in our civil law system. There is no provision about what intangible property is in China. Therefore legislative supplement should be carried on as soon as possible. An independent legislative regulation about virtual property would help clarify people's understanding about this concept. It is better if we take it as legitimate market behavior with legal regulations than let it develop in underground market, so that we regulate exchanges of virtual property to a great extent and lower the possibility of crimes of invading people's lawful property and interests.

(2) Include network virtual property to the protection range of existing laws
Due to the limitation of Management of Computer Information Network and Internet Security Insurance, sometimes it is hard to address disputes besides virtual property theft. Thus civil laws become the only reliance to solve the problems completely, which makes it urgent to widen the explanation of "other legitimate property" in Provision No.75 of *General Civil Law*, establish relevant judicial explanation and take virtual property into the domain of "other legitimate property".

Moreover, virtual property could be included to the protective range of *Real Right Law* for it fits for the characteristics of "object" in civil law. First, virtual property is usually in forms of accounts, currency, equipment, user grade, etc. resources that users have direct command of. As long as users could control, manage and use virtual property to satisfy their needs, it is tangible and controllable. Secondly, through exchange with physical property, virtual property could realize its currency value and bring profit to the user, which has already become a popular social phenomenon. Last, specificity is critical for controllability and negotiability, so virtual property is featured with specificity. Now that virtual property has characteristics of "object", confirmation of virtual property rights helps ensure the security of virtual property trade and social stability. As a result, to encompass network virtual property into protection of *Real Right Law* appears plausible and urgent.

4.2 Perspective of Judiciary

From perspective of judiciary, it concerns how to select appropriate legal system, how to calculate the actual loss of violated virtual property, how to set effective regulations against issues in the production, transaction and loss as well as problems about how to delimit network crimes.

(1) Distinguish between infringement subjects, allocate responsibilities with fair
To make distinction between infringement subjects of virtual property would allocate the responsibilities more reasonably and help victims get timely judicial aid [8]. Direct infringement subject should apply to the principle of non-fault liability, as long as the owner provides evidence of particular injurious behavior against the invader, no matter the invader actually is in fault or not, corresponding responsibilities fall upon him/her. As for indirect infringement subject, it's better to apply presumed-default liability principle.

There are two situations that virtual property could be infringed: one is because of game operators themselves, who terminate operations and lead to loss of virtual property, the other is due to invasion of the third party. We shall discuss the situations respectively.

In the first case, it is due to operational termination or other self reasons that virtual property expires. As online games have termination, virtual property is also characterized with time limit. Before players start the game, it should be expected that the game has expiration. If the operator stops operating for objective reasons such as incompetent management, players have no right to claim compensation. But before the operator determines to end the game, according to principle of good faith, it has the duty of advance information. Otherwise, for the players' loss it should take liability for breach of contract. Of course, if the operator terminates operation by intentional bankruptcy, it has to not only return players' virtual property but also take on compensation liabilities.

Secondly, it is the third party's infringement that results to loss of network virtual property. The infringement by the third party is usually through illegal means getting game accounts and passwords, making the lawful owners unable to possess their virtual property as a result, such as theft of virtual property. Since the operator has obligation to keep and reserve virtual property, when it is invaded by a third party, players have right to claim the operator's liability for breach of contract or ask the third party for tort liability. If the online game operator has no contract with players about protective technical standards and measures, infringement by the third party is regarded as interference with contract; players could only claim tort liability of the third party. However, if the operator promised to preserve virtual property for players, even if the third party steals the property with technically advanced measures, the operator should take liability for breach of contract.

(2) Establish a reasonable evaluation system
The particular feature of the virtual property, which different from other kinds of properties, makes it special in the process of evaluation. But in actual judicial practices,

the courts cannot make effective evaluation of the virtual property, that they can only roll back the record to protect the interests of game players. For this reason, it is urgent to establish a reasonable evaluation system [9]. We believe the following methods will be helpful.

First, account the true value of virtual property by calculating the social necessary labor time for production.

Second, the true value of virtual property can be determined by applying the market price of virtual equipment.

Third, conduct in-depth pricing based on the value that the online game operator provides. It is against the facts if the price set by the game operator is taken as the price of the virtual property.

5 Conclusion

Based on the legal definition of network virtual property, we have elaborated the ownership of network virtual property in detail and proposed useful advice about protecting virtual property from legislative and judicial perspectives.

References

1. China Internet Network Information Center, `http://www.cnnic.cn/index/0E/00/11/index.htm`
2. Jiang, M.X.: Analysis of legal protection of network virtual property. Jianghuai Tribune. 3, 140–142 (2010)
3. Chen, X.Q., Ge, B.Q.: Legal property of network virtual property. Zhejiang Academic Journal 5, 144–148 (2004)
4. Yang, L.X., Wang, Z.H.: On real right attributes of network virtual property and related basic rules. Journal of The National Procurators College 12, 1–13 (2004)
5. Yong, P.: Research on Cyber-security Law. Zhongguo ren min gong an da xue chu ban she, Beijing (2008)
6. Bai, Y.J.: Legal protection of virtual properties on the internet. New Heights 29, 116–118 (2010)
7. Zhang, L.: Civil law protection of network virtual property. Legal System and Society 10, 94–95 (2010)
8. Li, X.L., Chen, Y.: Value measurement and legal protection of network virtual property. Consume Guide 8, 96–98 (2010)
9. Wang, X.X.: Multidisciplinary analysis of network virtual property and its civil law protection system construction. Gansu Social Sciences 3, 37–40 (2009)

Discussion on the Education of the Most Sought-after Skill in Web Development: Framework Knowledge

Jianhong Sun[1], Qun Cai[2], and Yingjiang Li[1]

[1] Engineering College of Honghe University
[2] Science College of Honghe University, Yunnan Mengzi, 661100, China
sparkhonghe@gmail.com

Abstract. In recent years, more and more students those have just graduated from colleges and universities are hard to find a job for their expertise skills are inadequacy for the potential positions demand. As a guideline for these students of computer related majors preparing knowledge and expertise skills for getting a good position in web development, we discuss the education of the most sought-after skill, framework knowledge in this paper. From the required foundation knowledge to the latest new concepts related framework all are described. This work is one results of our research in reflection in education of which objective aim to solve the employment difficulties of students.

Keywords: framework knowledge, Web development, employment, education.

1 Introduction

In the last few years, more and more students those have just graduated from colleges and universities have to face tremendous employment pressure. In another hand, with the high-speed development of China economy, the high-level talented persons with expertise still keep in extreme shortage situation in many industries. To this end, we have to reflect on the aspects of education in which something may be wrong. Face to this issue, the education reform voice is becoming very strong in China, specially the computer education. The computer technology is one of the fastest development disciplines. Knowledge and technique updating is more important than other disciplines. The society asks for the computer talents no longer request them just know how to use common office software and install operating system. Only the high-end talents with excellent professional skills are always welcome by the society.

It is not difficult to find what the requirements of many positions related computer are. S. Somasegar, senior vice president of Microsoft's Software Developer Division published a report entitled "Key Software Development Trends [1]" on his blog on Feb 23, 2010. And he said these trends also are focus of Microsoft Corporation investment. Most of key software development trends in S.somasegar list concern web development, such as "The Web as a Platform", "Cloud Computing" and "Distributed Development". Therefore, there is no doubt that web development is a hot spot. Furthermore, in [2], "Framework Knowledge" has been recommended as the most sought-after skills in web development by the famous website, "Nettuts+" which

D. Jin and S. Lin (Eds.): Advances in ECWAC, Vol. 1, AISC 148, pp. 565–570.
springerlink.com © Springer-Verlag Berlin Heidelberg 2012

aimed at web developers and designers offering tutorials and articles on technologies, skills and techniques to improve how we design and build websites.

In order to improve the effectiveness of education, direct students preparing knowledge and expertise skills for potential position in the future. In this paper, we focus on the education of Framework Knowledge; explore how to master the development skills based on web framework.

2 Background

2.1 The Students Employment Situation of Computer Science and Technology Major

From survey results of one third-party education data evaluation and consulting institution, MYCOS, the number of unemployed graduates (have graduated for up half years) of Computer Science and Technology have already on the top 10 list from 2007 to 2010. As shown in Tab. 1, it is the latest years, 2010th unemployed graduates' statistical results of China. From the results, it is not difficult to find that employment situation of the students of Computer Science and Technology major is fairly grim.

Table 1. 2010th Unemployed Graduates of China (Top 10 Majors)

Majors	Number of unemployment(10 thousand)
English	0.94
Law	0.94
International Economy and Trade	0.93
Computer Science and Technology	0.91
Accounting	0.73
BBA	0.69
Finance	0.63
Electronic and Information Engineering	0.55
Marketing	0.52
Mechanical Engineering and Automation	0.46

* Note: Data come from MYCOS (My China Occupational Skills): http://www.mycos.com.cn/

2.2 About Honghe University, a Case Study

Honghe University (HU) is a developing university which from a normal specialized postsecondary college upgrade to a university since 2003. It is one of 1090 universities of China (* Data come from 2010 China Statistical Yearbook). HU serves a diverse population of full- and part-time, national and international students. In 2009~2010 academic year, more than ten thousand students enrolled in 12 undergraduate colleges with 36 undergraduate specialties [3]. In HU, from 2007 to

2010, the enrolled information is shown as Table 2. The table shows that the number of enrolled students of computer major is decreasing as the total number of enrolled students is increasing. This situation attributed to the graduates of computer major undergoes more pressures than other majors. According to the survey of Department of Computer Science & Technology of HU, the 2010th graduates of computer major only two students got a job after have graduated for three months.

Table 2. The Enrolled Information of HU from 2007 to 2010

Items \ Years	2007	2008	2009	2010
Total number of enrolled students	2175	2947	2987	3021
Enrolled students of computer major	82	98	98	61

2.3 The Main Factor Affecting the Employment

Cultivating applied talents for society is the goal of HU and for this HU's administration is promoting instructional technology as a crucial part of higher education for faculty and students. According to survey on the diploma projects of graduates in recent years, the survey results show that most students are unable to master the professional skills that the students are supposed to arm. Most of students choose developing web application system as their diploma projects, but most of them develop only an unsophisticated website. From an application point of view, their works did not achieve the purpose of any application, only a very few exceptions. So it goes without saying that the disappoint results of the employment statistics of graduates is not difficult to imagine. Faced to this severe situation, currently, how to improve the competitive forces of students is becoming a crisis of the Department of Computer Science & Technology of HU.

3 The Most Sought-after Skills in Web Development

Glen Stansberry, a web developer and blogger presented 10 most sought-after skills in web development with 155 comments on website, "Nettuts+" on Oct 20th, 2008. Even now, they still have important reference value. Among of them, the framework knowledge appears to be more important. Framework knowledge is widely used in software development, not limited to web development. If the students are proficient in the skill, they will be in a great position to find work and no matter how fierce competition. In next section, we introduce some concerning foundation knowledge and architectures of framework knowledge as an education guideline.

4 The Main Technologies of Framework

To learn framework knowledge, it needs to understand some technologies and architectures of framework first.

4.1 Required Foundation Knowledge

The students are expected to be familiar with the following basic client-side technologies: *HTML (and/or) XHTML/XML and related knowledge: Extensible Stylesheet Language Transformations (XSLT), XPath and Document Object Model (DOM); JavaScript; Cascading style sheets (CSS).*

The following knowledge also needs to learn, they are very useful. They are: *Database knowledge; Web services (REST and SOAP); TCP/IP; Ajax.*

The students will also need to be familiar with at least one of the following server-side technologies: *PHP; ASP.NET; Perl; Java; Perl; Ruby.*

4.2 Architectures of Framework

Model–view–controller (MVC) is a standard architectural pattern which separates an application into three main components: the model, the view, and the controller. Reference [5], MVC was first described in 1979 by Trygve Reenskaug and working on Smalltalk at Xerox PARC. In the influential paper [6][7], the original implementation is described in depth. And currently it has become an important concept in software design. The concept of MVC is shown as Fig.1.

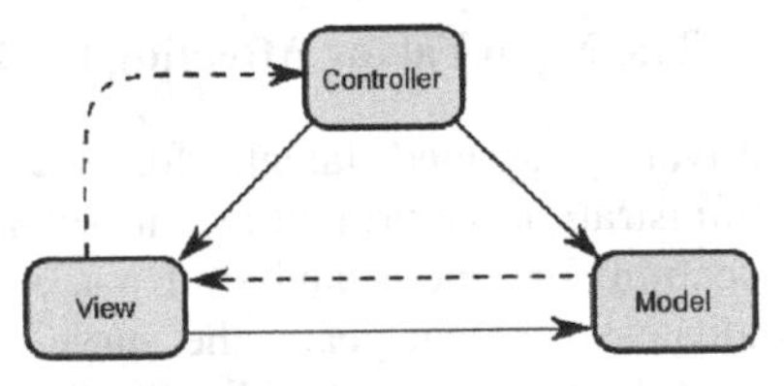

Fig. 1. Model-view-controller concept

In Fig.1, the solid line represents a direct association, the dashed an indirect association. The related knowledge is described as following:

Controller: Responsible for handles the business-logic of the application, it should use the Models to access the database along with user-input to construct the information or execute the action the user wishes for.

View: Views deal with everything graphical. They request data from their model, and display the data.

Model: responsible for database access and executing queries. In event-driven systems, the model notifies observers (usually views) when the information changes so that they can react [8].

1) Advantages of MVC [9]

Multiple views types can share a same model. A Web application usually provides multiple user interfaces. For example, the Web site should provide Internet interface and WAP interface based on same model if Web site is supposed to offer visiting service for users via both mobile phone and browser. In the MVC design pattern, the model response to user requests and return response data, the view is responsible for formatting the data and present them to users, business logic and presentation layer separation, the same model can be reused in different views, so greatly improving the code reusability.

Controller self-contains some high independent cohesive objects which remain relatively independent with models and views, so we can easily change the application's data layer and business rules. For example, migrating data from MySQL to Oracle, or change the data source from RDBMS to LDAP, we just need change the controller. With a good Controller, the View will display the data properly regardless it come from the database or LDAP server. As the three modules of MVC pattern independent of each other, changing one will not affect the other two. Therefore, according to this design, we can construct some good components with little mutual interference.

In addition, the controller increases application flexibility and configurability. Controller can be used to connect different models and views to complete the needs of users; can also be a powerful tool for construct applications. Given a number of reusable models and views, the controller can select the appropriate a model according to the users' needs, and then select the appropriate view of the results to display for the users.

2) Pull VS. Push MVC Architecture [10]

According to the push model, user actions should be interpreted by the Controller which will generate the data and push it on to the View, hence the "push". On the other hand, the pull model assumes that the user requires some kind of output (like a list items from the database). The View will access the Controller in order to get the data it needs in order to display the user the kind of output he requested. This is much like the View is "pulling" data from the Controller.

4.3 Some Common Sub-frameworks of Frameworks

A complete application framework usually composed by some common sub-frameworks, they are:

- Testing Framework(s)
- DB migration frameworks
- Security Frameworks
- Template Frameworks
- Caching Frameworks
- Form Validation Frameworks.

5 Summary

Learning frameworks is one of the best ways to increase the students' skill set and the potential positions can be as a developer work for a company or a freelancer work by own business. Some popular frameworks are worth of spend time on it for getting a good pay job, such as *Spring Framework, Rails Framework, Django and CakePHP*. The framework concerning much foundation knowledge, it is not easy to familiar the framework, but taking some time to master a new skill can get a huge advantage in finding work. Website owners pay big bucks for development that allows them to make money, because they see it not as a regrettable expense, but as an investment in future profit [4]. This is the driving force of market demand.

Acknowledgments. Thanks for pecuniary aid of Honghe University Bilingual Course Construction Fund (SYKC0802) and Honghe University Discipline Construction Fund (081203).

References

1. Somasegar, S.: Key Software Development Trends, http://blogs.msdn.com/b/somasegar/
2. Stansberry, G.: 10 Most Sought-after Skills in Web Development, http://net.tutsplus.com
3. Sun, J., Zhu, Y., Fu, J., Xu, H.: The Impact of Computer Based Education on Learning Method. In: 2010 International Conference on Education and Sport Education, ESE 2010, vol. 1, pp. 76–79 (2010)
4. Stansberry, G.: 10 Most Sought-after Skills in Web Development, http://net.tutsplus.com
5. Reenskaug, T.: Thing-Model-View-Editor an Example from a planning system. Which Proposed in a note on DynaBook Requirement and then Working on Smalltalk at Xerox PARC (1978-1979)
6. Burbeck, S.: Applications Programming in Smalltalk-80: How to use Model–View–Controller. This Paper Originally Described the MVC Framework as it Existed in Smalltalk-80 V2.0. and updated in 1992 (1992)
7. Krasner, G., Pope, S.: A Cookbook for Using the Model-View-Controller user Interface Paradigm in Smalltalk_80. Journal of Object Oriented Programming, JOOP (1988)
8. Model–view–controller, Wikipedia, http://en.wikipedia.org/wiki/Model/controller/
9. MVC, Wikipedia, http://zh.wikipedia.org/wiki/MVC/ (in Chinese)
10. Rutenberg, G.: Pull vs. Push MVC Architecture, http://www.guyrutenberg.com/2008/04/26/pull-vs-push-mvc-architecture/

Study on Full Text Retrieval of Spatial Data Based on Geocoding

Fan Gao[1], Feng Gao[1], and Bingliang Cui[2]

[1] Hainan Bureau of Surveying and Mapping Haikou China
[2] South China Normal University Guangzhou China

Abstract. This paper introduced the concept and developing situation of full-text retrieval technology, analyzed the search engine mechanism of full-text retrieval technology in this paper. Based on full-text retrieval technology and geocoding technology, a new retrieval method of spatial data is proposed in this paper. Finally, we developed a full-text retrieval system for spatial data to validate the correctness and reliability.

Keywords: Full text retrieval, Geocoding, Spatial data.

1 Introduction

With the development of computers and GIS, there is more and more electronic information in GIS using computer storage device as the carrier. The information can be divided into two categories: structured data and unstructured data. Structured data is such as mapping sector production data, the various elements of data; unstructured data is the number of text data, image data, video and other multimedia. According to statistics, the amount of information across unstructured data occupies more than 80%. For structured data, using an RDBMS (relational database management system) technology is the best way to manage [1] [2]. However, due to structural reasons, underlying RDBMS itself makes it a lot of non-end management inherent structure of the data somewhat, especially in these massive unstructured data query slower. Therefore, it is a problem worthy of study how to efficiently store and query unstructured data, in which the full-text searches technology to become hot research scholars at home and abroad [3] [4] [5].

The electronic information as the carrier of computer storage device implicitly contains a large number of spatial information, such as place names. GIS information on the management of spatial objects is usually composed of two parts, one part is the space object location information, and the other part is the space object attribute information. With the development of information technology, the urgent need to manage the object space position information and documents that describe the object information, the full-text retrieval system does not currently have this GIS spatial analysis. How to effectively use spatial information to locate the information entity directly enhance the retrieval efficiency. making the computer information system can also have the full-text retrieval capabilities and GIS capabilities, therefore, spatial information retrieval is an important issue faced. This article aims to introduce the

D. Jin and S. Lin (Eds.): Advances in ECWAC, Vol. 1, AISC 148, pp. 571–575.
springerlink.com © Springer-Verlag Berlin Heidelberg 2012

full-text retrieval of GIS technology, spatial data and attribute information of the full-text retrieval results map to the relative map.

2 Full-Text Retrieval of Spatial Data

2.1 Full Text Retrieval

Full-text retrieval technology is a new information retrieval technology produced at the late 1950s. It provides efficient data management tools and powerful data query tools, which can help people to collate and manage a lot of documentation quickly. Full-text retrieval provides access to full text (referring to the original records) space, any text characters and strings can be used as the entry point to retrieve full-text retrieval based on the original record of the search words, words between the operation targeted a specific location , no indexing of the literature. There is no standard reference word. Therefore, the full-text retrieval is a thesaurus which doesn't need to rely on direct retrieval method using the word freedom. For its ease of use and practicality, full-text retrieval system has become synonymous with a new generation of management information systems. The full-text retrieval engine as the core technology has become the mainstream technology of the internet age.

2.2 Spatial Attribute Data Retrieval

Spatial data retrieval stage of the most urgent problem to be solved is to achieve the object of written material in space (i.e. attributes) to retrieve. Traditional GIS attribute information for the relational database management in use of management, the property is defined as much as possible before cutting into the smallest unit of information that is required by the relational database paradigm of the first atomic value of the minimum requirements, and in fact, many properties of spatial objects is difficult with a simple segmentation property field after the full expression, such as the names of property resources, history, etc. It can only use the document form. It is weak that traditional GIS spatial objects for describing a document, especially document the specific management. GIS for the spatial attribute data indexing and retrieval is based on the paradigm of relational database theory, requires that each field must be the smallest atomic attribute values, once the property does not meet the requirements of the first paradigm, to retrieve an attribute field in the least significant units, such as retrieving the name attribute field name, you can traverse the entire table.

2.3 Spatial Location Information Retrieval

To make geographic information system technology and full-text retrieval technology integration, spatial data and documents must be established between an intermediate data table associated with table. As a result, when GIS in spatial data table space to do a particular query, the basis of the spatial data subset, through the association table, we can get tables in the document data corresponding document data subset to achieve through the spatial query function query document features. To be full-text of the original literature implicitly contains a large number of spatial information, such as place names, etc. If the information can be transformed into the space available for GIS analysis using spatial data, GIS technology is about to introduce full-text search, will greatly improve the spatial data search capabilities. Geocoding technology is the non-spatial data and spatial data linked to the most important and most practical way, it is a space-based positioning technology coding, geographic objects of reference in determining the rules by giving certain unique identification code and the establishment of geographic objects (names indicated by the geographical entity) and the mapping coordinate system, which will location information into a GIS can be used for the geographical coordinates [7] [8] [9]. The use of geocoding technology in geospatial data resources to determine the reference range in the location of the establishment of spatial information and the link between non-spatial information to achieve in a variety of address space (such as administrative, census areas, streets, etc.) within the information integration.

In the geocoding process (shown in Figure 2), two types of data need to match the data, one is the only entity that contains geographic location information, and no associated map location information (i.e. spatial coordinates) of the address data (such as street address, zip code, administrative divisions, etc.); the other is the map already contains the relevant location information (spatial coordinates) of the geo-referenced data (including street map data, zip code map data, map data and other administrative divisions). These data sets or databases in the address matching process play the role of spatial reference, which the completion of a match, to former gives geospatial coordinates.

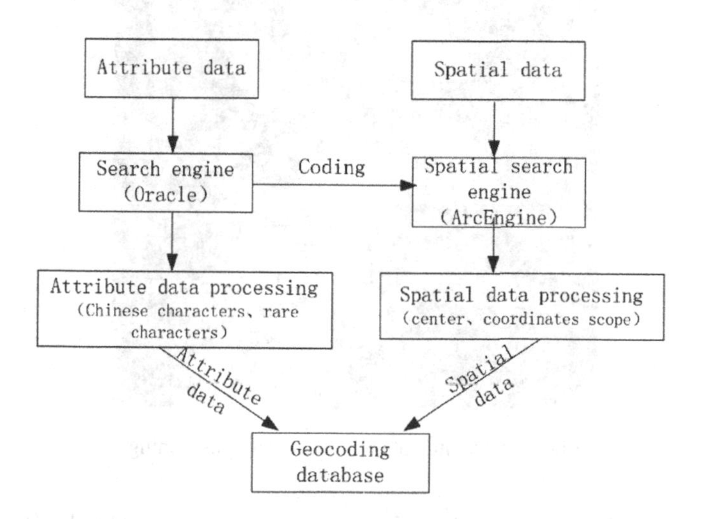

Fig. 2. Geocoding process

3 Implementation of Spatial Data Retrieval

In this paper, Visual C # 2008 for the development platform, database use Oracle 10g, full-text retrieval engine using Oracle Text, spatial data query, display, analysis using ESRI's ArcEngine 9.3 space engine, designed and implemented a GIS system in the full-text retrieval module, and with simulated data for this method of experimental verification, system implementation and process modules shown in Figure 3.

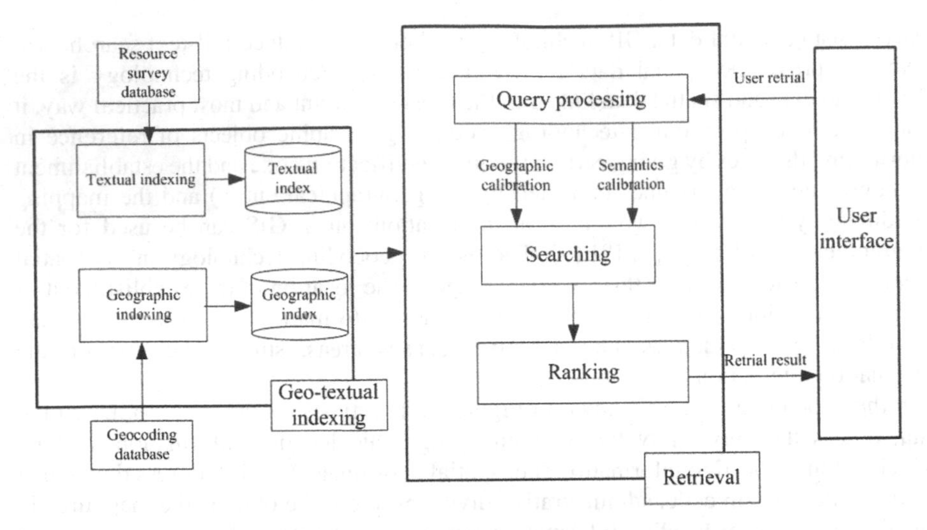

Fig. 3. System modules and system flow chart

Fig. 4. Attribute data to spatial data positioning

Oracle Text full-text retrieval engine as a component of Oracle 10g, provides powerful text search functions. Oracle Text uses inverted index (Inverted Index), this method is the use of word segmentation algorithm will split the content into a series of key documents in the index record of all the keywords and the keywords of all documents there. In this way, all the text in the document indexing, retrieval is actually a text search index. Since the index itself is a series of text with a pointer to the original document, so the search is fast, accurate positioning, experimental results also verify this. Experimental system to achieve a number of resources is (property data) geocoding feature to create a geocoding database, to achieve the interface shown in

Figure 4. Experiment to achieve through geocoding attribute data and spatial interaction data retrieval, and search results are arranged according to certain rules displayed in the user interface, click on search results can be achieved on a map location (Figure 4).

4 Conclusion

This paper analyzes the full-text retrieval and geocoding technology, based on the combine the two, presents a GIS-based spatial data encoding full-text retrieval method. Its core idea is the first attribute information for spatial data in the full-text search, and then through the geocoding technology, the result is mapped to the map. It overcomes the GIS data in the unstructured data retrieval problems, but also an effective solution to the current full-text retrieval results can not be completed to the physical location of information problems, to improve the retrieval efficiency, enables the system to also have the ability to full-text retrieval and GIS. However, the method also on the full-text retrieval algorithms to be improved, while geocoding optimality to be verified, the function prototype needs further improvement.

Acknowledgment. This research was supported by the "Science and technology innovation project of Hainan Bureau of Surveying; Innovation project of the construction of digital geo-spatial framework in Hainan Island; Mapping" and "Guangdong Province of talent introduction Fund (No. 2009-26)".

References

1. Hendler, J.: Agents and the Semantic Web. IEEE Intelligent Systems (March 4, 2001)
2. Lawrence, S., Giles, C.L.: Searching the world wide web. Science, 280 (4) (1998)
3. Feng, L., Chen, J.-H.: Research on database full text searching. Journal of Shanghai Normal University(Natural Sciences) 39(2), 151–157 (2010)
4. Yan-jun, U., Zhang, H.-J., Hao, W.-N.: Design and Realization of DB Full Text Retrieval System Based on P2P. Computer Technology and Development 17(9), 28–34 (2007)
5. Lang, X., Wang, S.: Research and Development of Full Text Search Engine Based on Lucene. Computer Engineering 32(4), 95–99 (2006)
6. Jiang, Z., Li, Q.: Research on the Applications of Geocoding. Geography and Geo-Information Science (02), 133–134 (2004)
7. Tang, X., Chen, X., Zhang, X.: Research on Toponym Resolution in Chinese Text. Geomatics and Information Science of Wuhan University 35(8), 930–935 (2010)

Analysis of Performance Optimization Principles and Models in Web

Xin Wang

School of Computer and Communication Engineering
Weifang University, Shandong 261061, China
www268@126.com

Abstract. The high speed development in Web, causes the optimized question to be getting more and more prominent. the first principle of Web Performance Optimization is to understand, to know that income will have to pay, and return is diminishing; Simultaneously the probability will decrease Web the performance. Web Technical models to improve the performance are: sharing costs, high-speed caching, profiles. Based on this study, given the crucial Web performance optimization recommendations, which improve the performance of Web usage, accelerate the efficient use of Internet has an important significance.

Keywords: optimization, principle, model, Web performance.

1 Introduction

Now, Web has grown from a novelty into an essential tool to disseminate information. However, with the continuous development of Internet, Web performance problems more seriously, this is more a wealth of information and modern services making. Therefore, we must optimize the performance of the Web. Performance optimization in the Web, there are some general principles applicable to a variety of specific solutions and unified pattern.

2 The Principle of Optimal Performance Web

2.1 Has Obtained, Must Have Loses

Only after doing research to see if I can improve the performance of a system. But you also risk the risk that: Put a lot of time wasted on research, the last only a can not do anything to improve the performance of the conclusions, especially in the tight financial constraints and a greater risk of the case. You must carefully analyze the difficulty of the problem and possible size of the harvest. If the gains can be large, then it should be determined to find a solution to improve performance, or not worth the time.

D. Jin and S. Lin (Eds.): Advances in ECWAC, Vol. 1, AISC 148, pp. 577–581.
springerlink.com © Springer-Verlag Berlin Heidelberg 2012

2.2 The Understanding Is the Base Element

Enters a house when darkness, will have the possibility to hit hurts itself. If turned on the light to be many easily. The lamp has given you the information, enables you to optimize the path. Optimize Web performance is the same reason. Consideration of issues in the mind the more clearly, the problem easier to solve. By now, instructed your light was along with the workstation, the software, and the router reference manual. Therefore must first do the matter is reads the reference manual and understands it.

When the reference manual has reduces your pain the key, they appear are rare. By changing the setting and measuring performance is feasible, but to figure out why the difference between different settings, and their relationship to other settings and the relationship between subsystems is even more important. This knowledge can be found in the reference manual. Somewhere in the increase in performance may cause performance degradation elsewhere. If you do not know what to pay, then you will not know whether it is worthwhile to modify the system.

2.3 Has Obtained Must Pay

Although the key of optimal performance is to do more with less, but any increase in performance, as little effort to understand and solve this problem, will have to be paid. In some cases, you have to buy new hardware, re-planning system architecture, may also reduce system portability, maintainability, security, reliability, or increase development time. You can let the system bus running day and night to improve server performance, but such systems are also more likely to collapse. To improve performance, you can remove the firewall and any encryption device, but you will also be exposed, more vulnerable to external attack. In fact the optimization of known usage patterns will inevitably hurt some other different usage patterns.

2.4 The Repayment Is Decreasing Progressively

The size of the investment can be seen that when the hardware has been optimized to complete the task. When it comes to optimizing the system, it is usually easy to find the most simple problem and fix it. With the optimization of the system to increase the performance becomes more difficult and more dependent on the particular configuration and usage patterns. When does not match the return on investment and optimize over. Some clues to explain the optimization process has ended, at least so in the following cases:

(1)Users no longer aware of any performance improvement.

(2)Changed the good programming style for the performance, making the code can not be transplanted, and difficult to maintain.

(3)When you consider programs written in assembly language.

(4)The total cost per page view than a person employed in accordance with customer's request to call them and fax costs.

(5)When you're bored time.

3 Web Improve Performance Model

The performance improvement may according to the model grouping, this way be more advantageous than each one concrete proposal to the work.

3.1 Cost-Sharing

For the purpose of economy, performance improvements usually involve how the cost sharing between multiple transaction processing:

(1)HTTP allows a single TCP connection for downloading multiple files. This feature is known as the "persistent connection". Therefore, a TCP connection establishment and cancellation of more than one file involved, not only has a relationship with a file.

(2)Design image map is another. Not to send multiple small images, but to send a large image. That if the original image is clickable, So by having this big picture into a clickable image of the way, you can also get the same functionality.

(3)Java.jar document with this similar, the java.class document package, like this may download them through a TCP connection, but is not establishes an independent TCP connection to each kind of document.

3.2 Caching Technology

The idea of caching technology is simple: those frequently accessed data at hand. Only when some data than other data access is more frequent in practice, the high speed cache technology only then has an effect.

(1)By the following approach to the storage space for more good performance: Run the most popular input program input offline data to the CGI program, and will cache all the results together. Then, users can quickly access to static HTML without having to generate dynamic HTML.

(2)Increase the memory can reduce the server to find content on the disk needs, thereby reducing the access time.

(3)Web proxy server caches some of the most popular Web pages, so that organizations can reduce the load on Internet access, while also reducing the time to access these pages.

3.3 Profile

Profile can be used to discover the reality of some use patterns, use it to find bottlenecks in your code, you can also optimize the usage patterns. We can follow the following "Amdahl's recommendations" to quickly solve commonly encountered problems.

(1)Discovers from configuration files' code most often the code which visits, in order to maximum limit optimize these codes.

(2)Can be configured for the user, and web site settings will be closer in their place based on the information.

(3)Write down the user download time, and assuming about what they have access to throughput, to adjust the content to make them more suitable for the user's access type.

3.4 Using Known Information

Do not underestimate the value of information, even the most trivial information:

(1)You knew that the next visit is possibly on a HTML page's image, as the matter stands, theoretically speaking, the Web server may to the HTML page analysis, prefetch the image.

(2)Once has used some connection, will have the possibility also to use to connect. Therefore HTTP has the lasting connection.

(3)If the Web server can identify a particular user's usage patterns, you can optimize these models, you can prepare in advance of its contents, or the content needs to be dynamically generated.

3.5 Parallel Processing

Web services have many problems due to multiple entities at the same time solving the same problem:

(1)Netscape and some other browser can open multiple connections to the server, to issue multiple requests in parallel, and want the server to determine the order of one of the most effective services to these requests, rather than a random order so that customers request.

(2)The Java procedure benefits from the multithreading, when some threads are blocked, the multithreading allows other threads to continue to carry out. For example: A Java application procedure's user registers when needs to fill in some things on the screen, then another different thread may use this opportunity to download other kind of documents. Only if the order is indeed very important, otherwise do not carry on the serial processing easily.

(3)Symmetric multiprocessing hardware can map multiple threads to multiple CPU, and can execute code in parallel.

4 Summary

Of course, to make the Web the fastest way is to do nothing. That is, if you can remove a part of the system, then this part should be removed. One way is to observe whether there is some redundancy, if any, are removed. But there is a better way, that is not possible to consider the use of certain equipment. you may not need to use for a specific business Web. This Web performance issues was also gone.

References

1. Killelea, P.: Web Performance Tuning. Tsinghua University Press (2003)
2. Cao, D.: Network Design Fundamental. Hope Electronic Press, Beijing (2000)

3. Wu, G.: Key technology e-commerce. Economic Science Press (2002)
4. Mining, K.R.: E-commerce data:the good,the bad,and the ugly (2001)
5. Shahabi, C., Banaei-Kashani, F.: A framework for efficient and anonymous Web usage mining based on client-side tracking (2002)
6. Spiliopoulou, M., Mobasher, B., Berendt, B.: A framework for the evaluation of session reconstruction heuristics in Web-usage analysis (2) (2003)
7. Srivastava, J., Cooley, B., Deshpande, M.: Web usage mining:discovery and applications of usage patterns from Web data (2) (2000)
8. Park, J.S., Chen, M.S., Yu, R.S.: Using a Hash. Based Method with Transaction Trimming for Mining Association Rules. IEEE Trans. on Knowledge and Data Engineering 9(5), 813–825 (1997)
9. Michele Facca, F., Luca Lanzi, P.: Recent developments in Web usage mining research (2003)

Research on Saving Electric Energy of Practical Teaching in Colleges and Universities

Shuangguo Niu* and Xinfa Dong

Yellow River Conservancy Technical Institute, Kaifeng, China
niushuangguo@163.com, hhsydxf@163.com

Abstract. In the practical teaching, colleges and universities commonly use gold machine tools such as the engine lathe, the universal miller, the universal shaper, the surface grinding machine, etc. and adopt the three-phase AC asynchronous cage motor as JO_2 and Y series, which shares 60% of electromechanical equipments used in colleges and universities' practical teaching. According to the relevant principles and scientific research achievements, and combined with teaching practices, this paper mainly pay attention to the soft start of electromechanical equipments and reactive power compensation to save electric energy in conventional practical teaching . Meanwhile, this paper also discusses the soft start saving method of the synchronous motor and the asynchronous motor as well as the theory and method of how the motor drag the perceptual active load equipment to realize the reactive compensation power saving on the spot.

Keywords: colleges and universities, practical teaching, electromechanical equipments, motor, electric energy saving.

1 Introduction

The professional gold processing practical teachings in colleges and universities such as the "machinofacture and design", "numerical control technology" and "mechanical-electrical integration" and "mould processing" make use of some precision-machined tools as the CNC lathe, CNC milling machine, processing center, the cutting machine, and the carving machine etc. though the number of gold dragging motors is large, power is higher, all of them have been adopted power saving measures as soft start, frequency control of motor speed and reactive power compensation. Basic skill training teaching commonly use the gold machining tools as the engine lathe, the universal miller, the universal shaper, and the flat grinder etc. using the gold the drag motor of JO_2 and Y series three-phase cage type asynchronous motors, it produces bigger impact current (general for 4 ~ 7 times In power system) when switched on which influences the voltage fluctuations in the electrical power system. For the power of more than 10 KW three-phase cage type asynchronous motors, the traditional method of starting the induce voltage is electromotor stator winding Y/train by law, stator windings comes close to the step-down transformers to limit the starting current

* Shuangguo Niu (1957-). Research on electrical engineering in college.

D. Jin and S. Lin (Eds.): Advances in ECWAC, Vol. 1, AISC 148, pp. 583–588.
springerlink.com © Springer-Verlag Berlin Heidelberg 2012

of the electromotor. Although this can play a certain role of limiting the flow, it could not solve the problem radically. Promoting researches on how to save the electric energy in practical teaching commonly used gold machining tools as the soft start of the drag gold motor, the frequency control of motor speed, and the on-site reactive power compensation is one of the emergency problems to be solved in the development of the colleges and universities. This paper tries to combine with the teaching practice; mainly pay attention to the methods of saving electric energy in the practical teaching commonly used electromechanical equipments' soft start and the reactive power compensation in colleges and universities.

2 Motors of the Major Electromechanical Equipments in the Practical Teaching Adopt the Soft Starter

In the colleges and universities, the basic skill training teaching commonly use the CA6140A and CA630 engine lathe, the X62W universal miller, the K52 vertical milling machine, the YB32-100 A four pillar type universal hydraulic press, the shaper, and the flat grinder etc. the drag electro motors of those gold machine tools are motor JO2 and Y series three-phase cage type asynchronous motors. All of them did not adopt energy saving measures. Studies on how to find out the methods of saving energy in the electromotor soft start and reactive power compensation is the important problem to be solved in the present development of colleges and universities. Intelligent electromotor soft starter is a new motor control equipment which sets the soft start, the soft stop, the under loading and energy conservation and multifunction protection in one. Not only does it achieve no impact and smooth starting current in the whole process of starting but also adjust the parameters such as the current limit value of the starting process and the starting time etc. according to the characteristics of the motor load. This equipment has multifunctional protection to the electromotor and could fundamentally solve many disadvantages of the traditional methods in reduced voltage starting [1].

2.1 The Asynchronous Motor Soft Start-Up Methods

A. Electromotor Soft start-up process.

(1). Electromotor soft start. When use the ramp voltage to activate the SCR soft starter, at the beginning the soft starter should be made to output an initial voltage (initial voltage which could be adjusted between 80 ~ 280 V) enough to overcome the initial static friction torque between the mechanical equipments, then the drag equipment begins to turn, its starting current is Is. Under the microcomputer control, we continue to increase the output voltage to make the motor speed up. When the output voltage of the Soft starter is close to the rated voltage, the electromotor has reached the rated speed, the starting current Is reduced to the load current In. when the Starting time t1 ends, the soft starter will output the rated voltage, and send out the bypass signal which will make the contactor bypass closed, then output voltage of the soft starter stops, the electromotor shifts to the normal operation. The Soft start initial torque can also be adjusted by a given initial voltage and a start-up time to control the starting current between 2 ~ 4.5 times of the rated current·

(2). electromotor soft parking. Soft parking can eliminate the inertial impact brought by free parking. At the parking moment t2, press down the terminal parking instructions. The terminal voltage will slowly drop from Un, at the minute the voltage drops there will be a small current shock caused by the electromotor, and the electric current will decrease with the voltage until the motor stops.

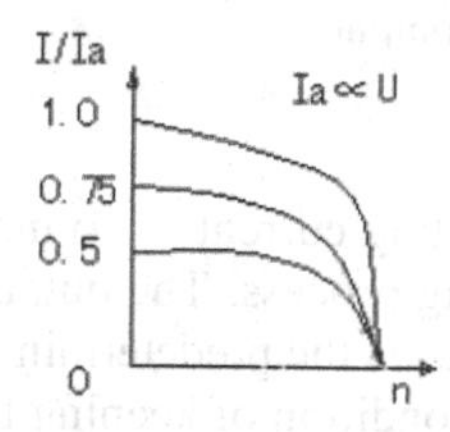

Fig. 1. When we lower the electromotor working voltage, the characteristic curve of the starting current

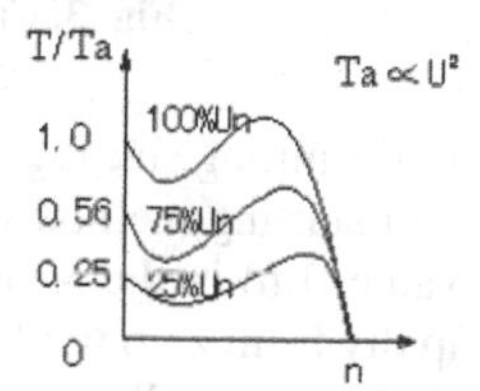

Fig. 2. When we lower the electromotor working voltage, the characteristic curve of the starting torque

B. The mechanical characteristics of electromotor soft start. The starting torque TSt of the three-phase ac asynchronous motor directly has something to do with the square of the voltage ($T_{St} \propto U_1^2$). Change the voltage on the stator winding, voltage, the Tst on the electromotor will be affected (see Figure 1). The working principle of electromotor soft start is by controlling the conduction angles of three opposite parallel thyristors which is in the series circuit between the power supply and the accused electromotor in order to make the terminal voltage increases from the setting value set in advance to the rated voltage.

C. Applicable occasions of electromotor soft start
Asynchronous motor can be used under any applications if there is no need of speed governing. It is suitable for various pump class load or fan class load. For motors long time working on light carriage, the soft starter have the effect of light load and energy saving.

2.2 Asynchronous Motor Soft Start

A. Start method of asynchronous motor soft start.

(1) Voltage doubles slope starting
As chart 3 shows, during the process of starting, the output torque increases with the voltage, and provide an initial starting voltage Us. According to the load Us is adjustable. When the Us be transferred to higher than the static friction torque load, the load can immediately begin to turn. At this minute, the output voltage starts to rise according to a certain slope coefficient, (the slope coefficient is adjustable) the motor will accelerate continuously. When the output voltage reach the speed voltage Ur, the motor is basically reaching the rated speed. The soft starter will automatically detect the speed voltage in the process of starting. When the motor reaches the rated speed, the output voltage will be made to reach the rated voltage.

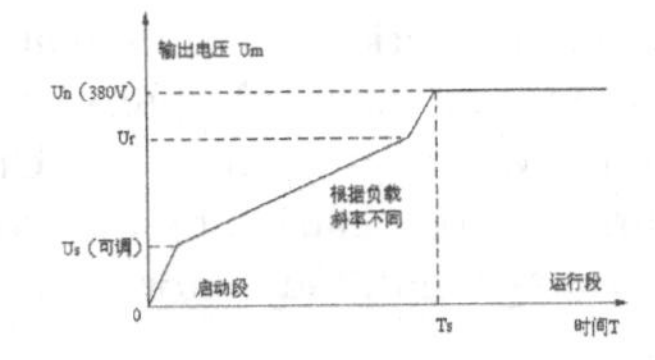

Fig. 3. Characteristics of soft start output

(2) The current-limiting starting
This is the soft starting method which will limit the starting current to no more than a certain set value (Im) in the asynchronous motor starting process. The output voltage increases rapidly from zero until the output current achieves the predetermined current limit Im, and then gradually rise the voltage under the condition of keeping the output current I < Im. When it reaches the rated voltage, make the motor revolving speed gradually rise up to the rated speed.

The advantage of this kind of starting is the current is small, and can be adjusted according to the need, which causes small effects of power grid. Its disadvantage is it is hard to know the starting pressure drop that we can't make full use of the space [2].

B. The operation characteristics of asynchronous motor soft start

(1) It can make the starting voltage smoothly rising to a constant slope. Low starting current; no impulse current on the power grid and can reduce the mechanical shock to the load.

(2) The rising rate of the starting voltage can be adjustable, the starting voltage can be adjusted continuously according to different load between 30% ~ 70% Un (Un is the rated voltage) to ensure the smoothness of the starting voltage.

(3) Starting time can be set according to different load.

(4) Starter also has functions as the SCR short circuit protection, phase lack protection, overheating protection, low-voltage protection.

3 Situ Reactive Compensation Section

The perceptual load is widely used in the electromechanical equipments. It causes a lot of reactive power and makes the grid power factor extremely low, which enormously affected the power grid and the efficient operation of power system security. Therefore, improve the power factor of power grid, basically balance the reactive power, decrease the on line loss and improve the quality of voltage is a very important work. The commonly used equipments such as conventional cars, mill, planer and modern machining center, CNC lathe, milling machine in the gold production practical teaching in colleges and universities all belong to the motor-drive and are part of the perceptual active load. Situ reactive compensation section can be used to save electricity.

3.1 Situ Reactive Compensation Power Saving Method

The reactive power consumed by Compensation on-spot electrical equipments (mainly motors) is directly install the capacitor beside the electric equipments. The capacitor bank set with the motor putting into or out of service simultaneously to make up part of the reactive power consumed by the motor be compensated on-spot. Thus reduces the reactive power delivered by transmission and distribution line which is above the outfit point and gains great drop benefit [3]. Dynamic reactive compensation circuit is mainly composed with the fixed compensation capacitor, high-power antiparallel SCR and the unsaturated reactor, as shown in Figure4.

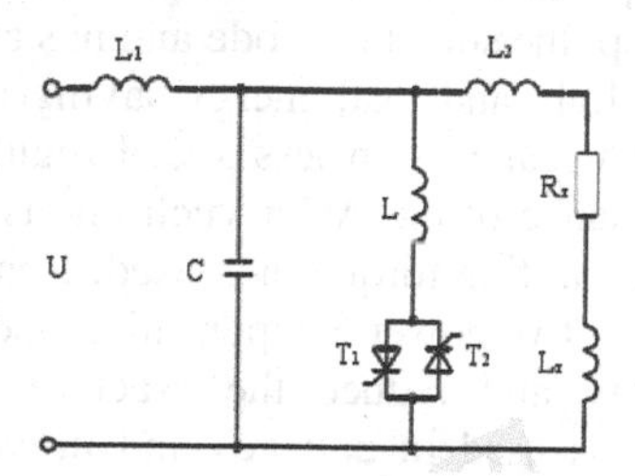

Fig. 4. The circuit principle diagram of dynamic reactive power compensation

Among them, L1 and L2 are inductances to retrain the high-frequency harmonics. Because the current flows through the reactor is under the silicon controlled phase shifting control, the susceptance of the parallel branch between capacitance and inductance can be continuously adjusted in the range from capacitive to perceptual, thus to adjust the reactive power supplied to the power grid continuously and complete the role power factor of the dynamic reactive power compensation played in the dynamical system. Among them, the equivalent susceptance of the parallel branch between capacitance and inductance is:

$$b = j(\omega C - \frac{2\pi - 2\alpha + \sin 2\alpha}{\pi \omega L})$$

Alpha for silicon controlled triggering Angle. When $\alpha=\pi/2$ and $1/\omega L>\omega C$, $b=-j[(1/\omega L)-\omega C]$, the circuit is perceptual. The parallel branch absorbs reactive power from the power grid; When $\alpha=\pi$, $b = j\,\omega\, C$, the circuit is capacitive, the parallel branch outputs reactive power to the power grid. Usually, parallel branch changes from the perceptual to the capacitive; it's correlated with the triggering angle. If power factor $\cos\varphi$ of the power grid is low, the control device output signal that the SCR triggering angle amplifies to make the parallel branch provide the power grid with more advanced reactive to compensate the demand of reactive power the perceptual load needed from the power grid and improve the load power factor. Because of the triggering Angle can be continuously changed, so the compensation is dynamic. After Capacitance C being placed in the circuit the phase difference that the current lags behind the voltage is being reduced so as to improve the power factor of the circuit. The active power derives from the power grid will increase, which explains the join of compensation devices will greatly reduce the energy loss of the system, and proceeds to achieve the energy saving.

Advantages of compensation on-spot is: when the proportion of large and medium-sized asynchronous motor is larger and use more time, the effect of reducing loss and saving electricity is remarkable for this kind of compensation mode. The balance will be basically achieved on the spot.

4 Conclusions

The electromechanical equipments and drive motor used in the practice teaching in colleges and universities adopt the soft start mode and has characteristics including the soft start, soft parking, soft light and load energy saving as well as multi-functional protection. The soft start can realize step less speed regulating of the electromotor, improve the starting performance of the AC asynchronous motor, and largely reduce the starting current, then the starting torque increased. Reactive power compensation commonly uses capacitive reactive power compensation and perceptual reactive power etc. to save electric energy and reduce the reactive load of the power grid. Electromechanical equipments used in colleges and universities' practical teaching adopt the soft start and situ reactive compensation methods to save electric energy can bring considerable power saving benefits. Colleges and universities should strengthen and promote the research of this aspect.

References

1. Niu, S.G.: Research on problems of saving electric energy in university laboratory. Journal of Yellow River Conservancy Technical Institute (4), 47–49 (2009)
2. Wei, D.F.: Electrical Machinery Technology, vol. 8, pp. 240–241. China water power press, Beijing (2004)
3. Liu, C.F.: The Application of Low Voltage Reactive Power Compensation and Performance Analysis. Electrical Technology Magazine (5), 32–35 (2002)

Prediction and Evaluation Methods of Mining Damage Based on Computer Simulation

Hong Ji[1,1] and Xueyi Yu[2,2]

[1] School of Geologic and Environment Engineering,
Xi'an University of Science and Technology, Xi'an, Shaanxi, 710054, China
modena@sohu.com
[2] School of Energy Engineering, Xi'an University of Science and Technology, Xi'an, Shaanxi, 710054, China
yuxy@xust.edu.cn

Abstract. Based on the mining and geological conditions of Xigu Village in Pu bai mine, the synthetic prediction software (Yhl-12) is applied to predict surface displacement and deformation of Xigu Village. Numerical simulation and expected function simulation are used to determine the mining method and related parameters. The basic laws and the particularity of surface subsidence caused by strip mining are analyzed. The harmonious strip mining method is proposed. It is more effective to implement economical and reasonable mining with the method.

Keywords: strip mining, surface subsidence, prediction, YHL-12, FLAC3D.

1 Introduction

The large areas of surface buildings, roads and public facilities are severely damaged because of surface subsidence and crack damage caused by coal mining. It seriously affects the mining sustainable economic development. The mining damage caused by coal mining under buildings is severe, thus it is important to predict the surface displacement and deformation. The study on these problems can improve the life quality, protect ecological environment and realize sustainable development of mining area.

2 Mining and Geological Conditions

Nanpan mine of Pubai is located in the southeast of mining area which is 4500 meters long and 775 ~ 1320 meters wide. Xigu Village is located in the eastern boundary of Nanpan mine. The coal area is about 100000 m2 which contains 2.4246 million tons of coal. Nanpan mining area just crosses the centre of Xigu Village which does greatly harm to the surface buildings. The main coal seam is No5, and the surface area is covered by thick loess layers which is generally 78 ~ 88 meters and the average

D. Jin and S. Lin (Eds.): Advances in ECWAC, Vol. 1, AISC 148, pp. 589–594.
springerlink.com © Springer-Verlag Berlin Heidelberg 2012

thickness of overburden is 90 ~ 165 meters.The basic conditions of surface buildings are mentioned as follows. Xigu Village is composed of 12 villages which have about 3847 people. There are 75 three-storey buildings, 227 two-storey buildings and 348 bungalows which are seriously impacted by the surface mining damage. The comparison diagram of buildings and coal pillars is shown as Fig.1.

3 Prediction Analysis of Surface Displacement and Deformation

3.1 Prediction and Evaluation Software (YLH-12)

YLH-12 is a synthetical prediction and evaluation software which involved preferred mining method, prediction of surface movement and deformation, classification and evaluation of surface building damage and parameter optimization. Surface mining subsidence is multi-dimensional influence function about space and time. Therefore YLH-12 can evaluate subsidence of surface building facilities generally.

3.2 Mathematical Prediction Model of Surface Displacement and Deformation

$$D(x,y,z)=\sum_{i=1}^{m}C_i\sum_{k=1}^{l}\int_{q}^{q_{k+1}} f_{j1}(R_k,z)f_{j2}(q)dq \quad (1)$$

where, x,y,z denote coordinate of calculated nodes; C_i denotes time inflection coefficient; m denotes calculated segment amount ; l denotes turning point of calculated segment; qk denotes angle between X axis and the line of calculated nodes p(x,y) and turning point k; qk=qk+1, ql+1=q1 and fj1, fj2 are operation functions. Radius of polar coordinate as

$$R_k(q,x,y)=\frac{(x_k-x)(y_{k+1}-y_k)-(y_k-y)(x_{k+1}-x_k)}{(y_{k+1}-y_k)\cos(q)-(x_{k+1}-x_k)\sin(q)} \quad (2)$$

where, x_k, y_k denote coordinate of turning point k. Based on the above principle, the mathematical model of surface dynamic subsidence in polar coordinate is as

$$W(x,y,t)=\frac{W_{\max}}{2\pi}\sum_{i=1}^{m}C_i\sum_{k=1}^{l}\int_{q_k}^{q_{k+1}}\left[1-e^{-\pi\frac{R_k^2}{r^2}}\right]dq \quad (3)$$

where, C_i is subsidence time coefficient of the i mining segment. According to Sulstowicz's hypothesis that the volume of subsidence basin is proportional to the volume of non-compaction gob, and start mining time t-->0, the interval time between start mining time and prediction time $t=t_i$, surface stable time t--> ∞. Thus, subsidence time coefficient is $C_i=(1-e^{-ct})$, where, c is mainly related to mining depth, mining rate and characters of overburden. Generally, the range of c is from 1.0 to 3.0; When mining depth is shallow and overburden is soft, the range of c is from 2.5 to 3.0; When mining depth is shallow and overburden is hard, the range of c is from 2.0 to 2.5; When mining depth is great and overburden is soft, the range of c is from 1.5 to 2.0; When mining depth is great and overburden is hard, the range of c is from 1.0 to 1.5; If repeated mining is implemented, the value of c is less than 1. General integral calculation model in polar coordinate is shown as Fig2.

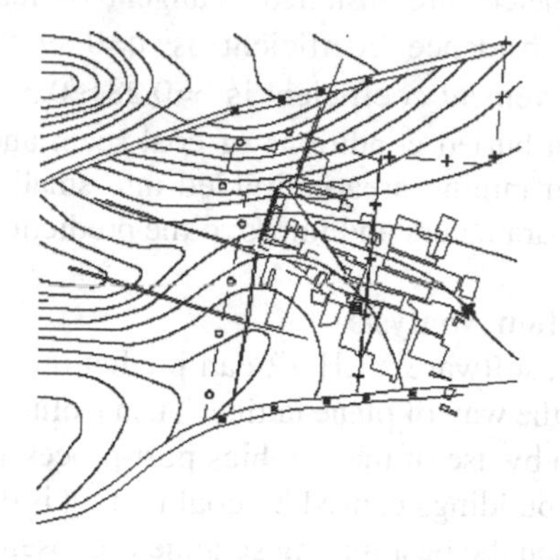

Fig. 1. The comparison diagram of buildings and coal pillars

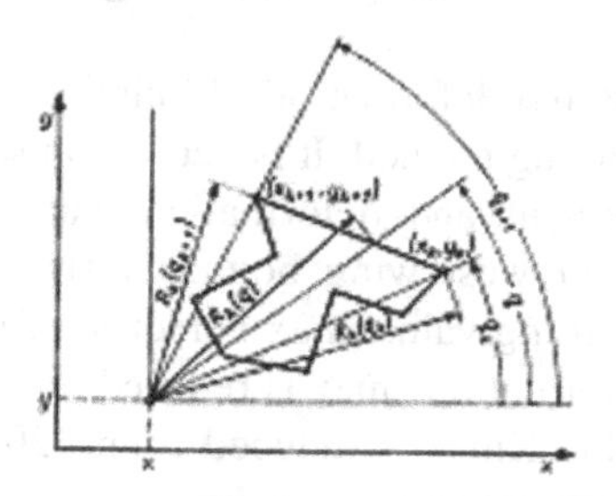

Fig. 2. General integral calculation model in polar coordinate

4 Mining Scheme and Interrelated Parameters

4.1 Mining Scheme

Based on distribution of mine openings, coal seam hosting conditions, distribution of surface buildings, principle of strip mining, mining width and leaving width of pillars, two mining schemes are proposed. Mining scheme 1 is denoted the hading and variable strip mining method; Mining scheme 2 is denoted the trend and variable strip mining method. In mining scheme 1, mining width is install as b = 18~32 meters, the first mining strip from boundary of mining area is advanced by 18 meters, the second mining strip is by 20 meters, the following strips gradually increase with 2 meters interval and the largest mining width is by 32 meters. leaving width of coal seam is installed as a=25 meters. In mining scheme 2, mining width of coal seam is b =25 meters and leaving width is a=25 meters.

4.2 Computer Simulation and Prediction

4.2.1 Prediction Parameters

According to characters of overburden in Nanpan mine and observation data of strip mining, prediction parameters are installed. Tangent of main influencing angle is $tag\beta=2.5$, $\beta=68.2°$; Subsidence coefficient is $\eta=0.89$; Inflexion excursion is $d=0.03H$; Horizontal movement coefficient is $b=0.28\sim0.32$; Maximum subsidence angle is $\varphi=86°$. Based on buried conditions of coal seam and characteristics of strip mining, the calculation of mining area is divided into small units in order to get the same geological mining parameters and improve the prediction accuracy.

4.2.2 Results of Prediction Analysis

Prediction and evaluation software YLH-12 can predict the surface displacement and deformation on gob with the way of plane lattice. Then isoline drawing of displacement and deformation is gotten by use of the graphics post-processing software. The mining impact degree of surface buildings caused by coal mining is determined by subsidence value of buildings, and then the best mining scheme is chosen. The isoline drawings of surface subsidence are shown in Fig. 3, 4. The mining impact degree and scope of surface buildings are shown in Table1. The following conclusions can be drawn with prediction analysis.

The surface displacement and deformation of buildings can be reduced effectively with the harmonious strip mining method. It is verified in scheme 1 that by comparing the differences of displacement and deformation located in east wing boundary (coordinate arrangement) and west wing boundary (non-coordinated arrangement) respectively, the maximum tilting value imax is reduced from 2.9mm/m to 1.0mm/m, the maximum horizontal deformation εmax is reduced from ±1.4mm/m to ±0.3mm/m, the maximum curvature value Kmax is reduced from ±20e-6/m to ±6e- 6/m and the maximum horizontal displacement Umax is reduced from 90mm to 40mm, as shown in Fig.3. Moreover, as illustrated in scheme 2, the maximum tilting value imax is reduced from 2.9mm/m to 2.0mm/m, the maximum horizontal deformation εmax is reduced from ±1.4mm/m to ±0.6mm/m, the maximum curvature value Kmax is reduced from ±20e-6/m to ±10e-6/m and the maximum horizontal displacement Umax is reduced from 90mm to 60mm, as shown in Fig.4.

When the surface is covered by collapsible loess, the horizontal displacement of mining surface is more than 1mm/m, then crakes will occur in the surface and the building foundations will be damaged. So the decrease discussed previously is very important for building protection located in east wing boundary.

The mining strip is gradually widened from east to west. This is conducive to adjust the width of mining strips to ensure the safety of the building according to extent of the damage of buildings on gob. The trend and variable strip mining method is proper for coal seam which angle is smaller than 15 degree. The hading and variable strip mining method is adapted for inclined coal seam, and is adapted more widely.

Table 1. The mining impact degree and scope of surface buildings

Index	Unit	Location	surface subsidence of buildings in scheme 1	surface subsidence of buildings in scheme 2
W	[mm]	East wing	-20～-100	-50～350
		West wing	-20～-142	-50～150
U	[mm]	East wing	0～100	0～155
		West wing	0～100	0～120
εmax	[mm/m]	East wing	0.2～1.50	0.25～1.50
		West wing	0.2～1.30	0.25～1.50
εmin	[mm/m]	East wing	0.25～1.0	-0.25～0.8
		West wing	0.25～0.50	0.0～0.60
Kmax	[e-6/m]	East wing	-0.50～-28.0	-5.0～-25.0
		West wing	-5.0～-26.0	-5.0～-30.0
Kmin	[e-6/m]	East wing	-5.0～-20.0	5.0～-15.0
		West wing	-5.0～-10.0	0.0～-12.0
i	[mm/m]	East wing	0.40～1.60	0.5～2.5
		West wing	0.40～1.80	0.0～2.2
P.S	Negative value of horizontal deformation means compress distortion, while positive value means tensile distortion; Negative value of curvature means convexity, while positive value means concavity.			

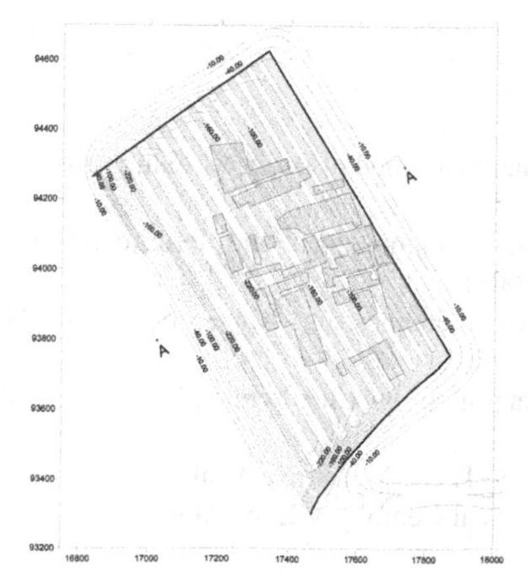

Fig. 3. The isoline drawing of surface subsidence for scheme 1

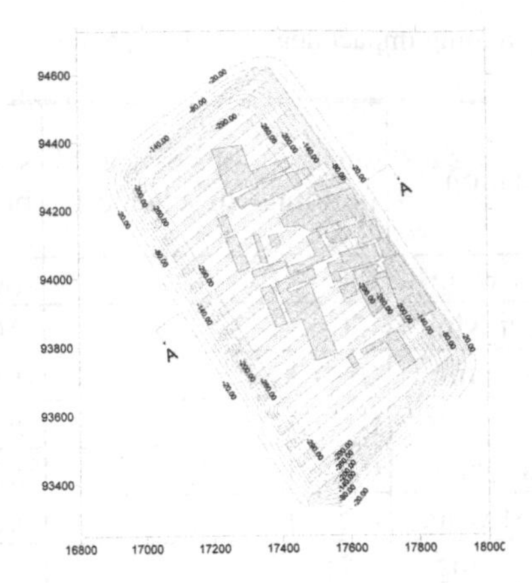

Fig. 4. The isoline drawing of surface subsidence for scheme 2

5 Summary

Surface displacement and deformation caused by coal mining is a special issue in the western mine area. Coal mining destructs overburden seriously and surface subsidence continues for a long time, so surface displacement and deformation damage has a certain particularity and complexity. This paper applies prediction software YHL-12 to expect surface displacement and deformation in Xigu village of Pubai. Based on the geological and mining conditions, the basic laws and characteristics of surface subsidence for strip mining method are obtained. Environmental protection and sustainable development methods are benefit from the strip mining method.

References

1. Yu, X.: Prediction Evaluation Method of surface mining subsidence. Mine Design 5(2), 8–9 (1997)
2. Huang, S., Yu, X.: Prediction onmovement and deformation of surface in steep coal seam. Shanxi Coal 1(2), 5–6 (2006)
3. Liubo, Han,Y.: Principles, Examples and Application of FLAC. China Communications Press (2005)
4. Zhang, J.: Analysis on stability of building Foundation in Goaf. Mine Surveying 9(3), 28–29 (2003)
5. Yu, X.-Y., Liu, Z., Niu, Z.-T., Li, W., et al.: Analysis on the structure stability of hard and massive overlying strata. Coal Geology & Exploration 35(5), 38–41 (2007)

Research and Application on Automatic Generation Technology of JavaScript Input Validation

Yongchang Ren[1], Jie Hu[1], Lisha Ning[2], and Tao Xing[3]

[1] College of Information Science and Technology, Bohai University,
Jinzhou, 121013, China
rycryc@sina.com, hujie26833@gmail.com
[2] Liaoning Agricultural Economy School,
Jinzhou, 121000, China
ninglisha@126.com
[3] College of Engineering, Graduate University of Chinese Academy of Sciences
Beijing, 100190, China
xingtao@mail.tsinghua.edu.cn

Abstract. Client authentication for Web applications require a lot of script writing, poor maintainability, this paper follows the system engineering principles and methods to study the data set to the page table, the application automatically generated by the automatic generation of JavaScript technology. Input validation in the way of the Web based on the comparison, first designs when the automatic production uses "Checker Dictionary" and "Control Settings Page" two table structures; Then carries on the automatic generating routine design, including program design process and the main validation function; Finally carries on the confirmation through the automatic production example. The results show that the automatically generated input validation of JavaScript, reduces the technical requirements for software developers to improve software development productivity, changed the traditional software development model.

Keywords: Web, JavaScript, input validation, automatic generation technology.

1 Introduction

Because the Web application is open, all of the browser on the network can use the application, so the application page by entering the data collected is very complex, not only contains the normal user the wrong input, but also possibly contains the malicious user the malicious input. A robust application system must prevent the illegal import of these applications, prevent these illegal inputs to enter the system, so as to ensure the system is not affected. Therefore, the input validations is a question which all Web application must be solved [1].

Usually two ways to verify the validity of input data: One way is to verify the server side, another way is to authenticate the client. Server-side validation is to submit data to the server, the server performs a form of validation, then returns the results to the client. The disadvantage of this approach is to go through the server each time authentication is not only a long response time, but also increase the burden on the server. The

D. Jin and S. Lin (Eds.): Advances in ECWAC, Vol. 1, AISC 148, pp. 595–600.
springerlink.com © Springer-Verlag Berlin Heidelberg 2012

client-side validation is a local call in the script directly to verify this approach, the response time is short, reducing the server-side computing [2]. At present can be found is a lot of data validation using the client source code on the Internet, but too many entries, you need a lot of script writing validation code, maintainability, seriously affecting the development efficiency [2]. JavaScript auto-generation technology is an effective way to solve these problems.

2 Comparison Of Web Input Test Methods

The Web input validation way is divided into client and server side, the server-side divides into the ordinary server and Ajax, presently take Struts2 as the example, each way comparison as shown in Table 1 [4].

Table 1. Comparison of Web Input Test Methods

Mode	Validation Framework	Ajax	Script
Location	Server	Server	Client
Whether to refresh the page	Yes	No	No
Test code	Java	Java	JavaScript
Implementation	(1) Definition of class Action : Action.java ; (2) Custom profile : Action-validation.xml ; (3) Action is defined in struts.xml ; (4) JSP page defines the Form。	It is basically the same to the " Validation Framework ", but need to do the following .modification : (1) Install dojo plugin ; (2) JSP using dojo plugin ; (3)Added the head tag in the head (introduce the dojo js library) ; (4) Modify the definition of Form。	(1) Set up : <script language="JavaScript"> ; (2)Set check properties ; (3) Written verification function (to the number of items for each check) ; (4) Return the check results。

3 Automatically Generate Table Structure Design Technology

Automatically generated test input supports JavaScript needs "Checker Dictionary" and "Control Settings Page" two tables. Between the tables through the "Checker China" field to be connected.

3.1 Checker Dictionary Table

Checker dictionary tables used to set all of the testing device, any validator controls, must be the dictionary of the validator. Checker dictionary table structure as shown in Table 2.

Table 2. Checker Dictionary Table Results

No	Chinese Name	Field Name	Type	Width	Null
0	ID	ID	varchar	14	0
1	Numeric code	SZ	varchar	10	0
2	Checker china	CheckerChina	varchar	50	0
3	Alphabet code	PY	varchar	50	0
4	Checker name	CheckName	varchar	50	0
5	Allows the use of sign	FlagRX	tinyint		0
6	Remark	Remark	varchar	100	1

3.2 Control Settings Page

Control Settings Page table store all the set up controls of the pages, as well as information on the control test, according to the table data to generate the JavaScript validation code. Control Settings Page Table Structure shown in Table 3.

Table 3. Control Settings Page Table Structure

No	Chinese Name	Field Name	Type	Width	Null
0	ID	ID	varchar	14	0
1	Chinese name	CName	varchar	50	0
2	Controls name	CtrlName	varchar	50	1
3	Field name	Name	varchar	50	1
4	Field type	Type	varchar	10	1
5	Jave type	JavaType	varchar	50	1
6	Null symbol	FlagKZ	Tinyint		1
11	Checher china	CheckerChina	varchar	50	1
12	Check the minimum	minValue	varchar	50	1
13	Check the maximum	maxValue	varchar	50	1
14	Integer places	intNum	tinyint		1
15	Decimal places	dotNum	tinyint		1
16	Expression	Expression	varchar	100	1
17	Check tips	CheckPrompt	varchar	100	1

4 Automatic Generation Technology Programming

Program design process shown in figure 1.

Programming generally divided into four steps:

(1) From the "Control Settings Page," the page-table to inquire this page's all the "validator" whether exist in the "checker dictionary" table, if there is, the next step; otherwise that there are unknown can not generate code validator, the build process to terminate;

(2) Check the generated JavaScript file name and path, if you meet the requirements to open the file, then the next step; Otherwise, the generation process to terminate;

(3) To cycle all of this page's "validator", according to "Check type" and execution related functions, then to generate JavaScript;

(4) Output the prompt and returns the check results.

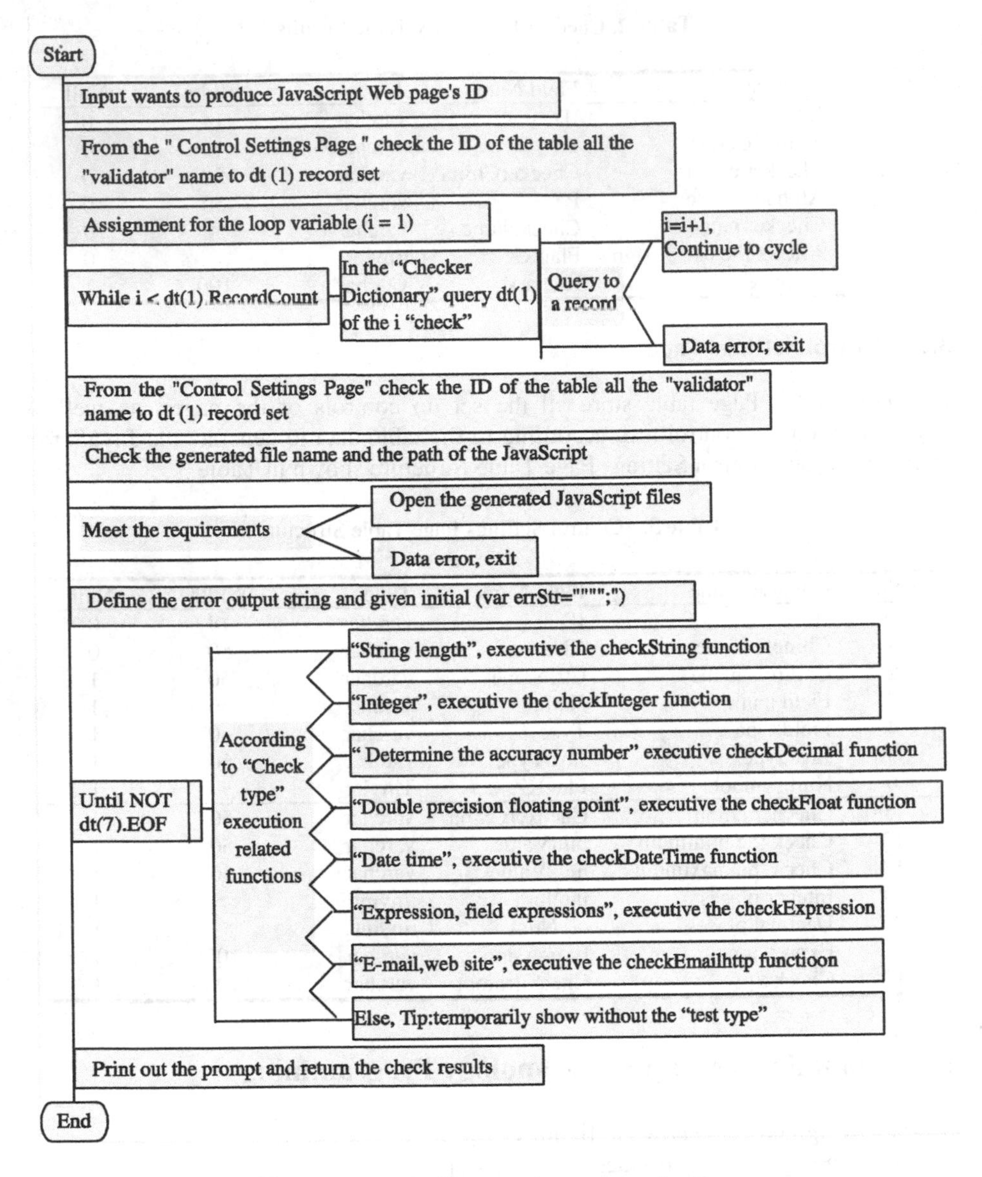

Fig. 1. Automatically generate programming flow

5 Automatic Generation Examples

The use of Web page maintenance for global variables, you need to enter the serial number, Chinese description, variable name, variable value, variable unit, initialization flags, notes and other information, through the automated generation technology to generate this JavaScript code is as follows:

```
<!—Input validation -->
function inputCheck(){
    //When definition is error, output string
    var errStr="";
    //Check all input values
    errStr+=checkInteger("Serial number",document.all.no.value,0,0,9999,"","");
    errStr+=checkString("Chinese
description",document.all.describe.value,0,1,50,"","");
    errStr+=checkString("Variable name",document.all.name.value,0,1,50,"","");
    errStr+=checkString("Variable value",document.all.value.value,0,1,50,"","");
    errStr+=checkString("Variable unit",document.all.unit.value,1,0,10,"","");
    errStr+=checkInteger("Initialization flags",document.all.flagCS.value,0,0,1,"","");
    errStr+=checkString("Notes",document.all.remark.value,1,0,200,"","");
    //Output the check tips and return the check results
    return outputResult(errStr);
}
```

6 Conclusion

The design idea of JavaScript input validation technology automatically generates is to follow the systems engineering principles and methods, according to the data settings in the data table, automatically generates JavaScript by the application. Developers do not need to master JavaScript and database technology, just to complete design of the data in the database table, can automatically generate JavaScript, reduces the technical requirements for software developers and improves software development productivity, then changes the traditional software development model. This study content for Web application developers in the high reference value.

Acknowledgment. This work is supported by, Education department of liaoning province key laboratory fund project (2008S002), General project education department of liaoning province (L2010011), Small medium-sized enterprise innovation fund of liaoning province(20104110), and Teaching reform project of Bohai University(2009).

References

1. Li, G.: Struts2 Definitive Guide. Publishing House of Electronic Industry (2008)
2. Kong, L., Guo, X., Song, G.: Development of Application Program of Enterprise Dynamic Website —— Real-time Checking Function of Input Data and Applying Method (Javascript). Tool Engineering 39(11), 51–54 (2005)
3. Wang, C.L., Geng, X.Y.: Research on Input Validation of the Struts2. Software Guide 8(5), 41–42 (2009)
4. Wu, M.S.: Authentication method of the Struts2. CSDN Blog (July 20, 2011), http://blog.csdn.net/a8567022/article/details/4011745
5. Baike, B.: Regular Expressions (July 20, 2011), http://baike.baidu.com/view/94238.htm

6. Zhang, K.: Design for Input Data Verification and Software Reliability. Computer Development & Applications 17(2), 17–18 (2004)
7. Yin, Z.Y., Zhao, H., Zhao, P.G., et al.: The Design of Cyclical Redundancy Check Arithmetic Based on Parallel Data Input. Computer Engineering and Applications 42(27), 1–5 (2006)

Research and Application on Representation Tools of Software Detailed Design

Yongchang Ren[1], Wei Cai[1], Lisha Ning[2], and Tao Xing[3]

[1] College of Information Science and Technology, Bohai University,
Jinzhou, 121013, China
rycryc@sina.com, caiweipk@126.com
[2] Liaoning Agricultural Economy School,
Jinzhou, 121000, China
ninglisha@126.com
[3] College of Engineering, Graduate University of Chinese Academy of Sciences
Beijing, 100190, China
xingtao@mail.tsinghua.edu.cn

Abstract. When detailed design stage, need decide realized arithmetic of each module, and accurately express these algorithms. Aims at characteristics of software detailed design have too much presentation, hard to select and master. This paper researches on primary representation tools. On the base of research sequence type, select type, first judge type loop, after judge type loop and many situations select type etc 5 type control structures of structural program designing, through the method of combine pictorial representation with textual description, emphasize research representation of these 5 type control structures use representation tools, the representation tools of research contain flow chart, box diagram, problem analysis diagram, and IPO diagram etc 7 types. The research contents of this paper are reference and guide of software detailed design, and it has important guiding significance for software designer.

Keywords: software design, detailed design, control structure, representation tools.

1 Introduction

Software design divides into general design and detailed design. General design confirms population structure of software, divide software to some function independent modules, and describe external characteristic of every module (function and interface), and definition module universal global data structure and the style of user interface. Detailed design is further specified for general design, confirm two bulk properties of module, it means describe implementation of every module (how to do) and define local data structure of module [1]. Generated detailed design document after pass the reexamine is the milestone of detailed design stage over, and it is according to use program design language on the encoding stage.

On the stage of detailed design, will decide realized arithmetic of every module, and accurately describe this arithmetic. In an ideal situation, arithmetic procedure description use natural language expression, like this it is easier to understand description for the person

D. Jin and S. Lin (Eds.): Advances in ECWAC, Vol. 1, AISC 148, pp. 601–606.
springerlink.com © Springer-Verlag Berlin Heidelberg 2012

who is unfamiliar with software, and needn't relearn. But natural language often has ambiguity on the grammar and semantically, so it often rely on context can describe problem clearly. So we must use the way which more strong binding force to describe part of the process. This paper research representation tools can content the require of most of the software detailed design, it offer guide for designer.

2 Basic Control Structure

The concept of "structural program designing" is called the third milestone in Software developing, it suggested by the Dutch computer scientists, Edsgar Wybe Dijkstra. C·Bohm and G·Jacopini prove that as long as there is sequence, select and loop three type control structures, it enough to express all program structures in program design language [2].

If code block of a program only through sequence, select and loop three type control structures to link, and every code block only has one entrance and one exit, then this program is structured.

For using wrote structured program, it must prescribe a limit to use control structure in detailed design representation tools. Through sequence, select and loop three type structures can write structured program, but select extend many situations select type, judge also can divide into first judge type loop (While type) and after judge type loop (until type). Five type essential control structures as follow:

- Sequence type: several continuous job steps successively arrange constitute;
- Select type: By short-cut process of some logic type, decide select one of two processes;
- First judge (While) type loop: When loop controlled condition holds, repeat executes specific process;
- After judge (Until) type loop: repeat executes some specific processes, until controlled condition holds;
- Many situations (Case) select type: list various process conditions, according to the short-cut process of control variable, choose one executive.

3 Representation Tools

For detailed design tools, will can provide unambiguous sex description for design, can point out control flow, dispose function and data organization etc implementation details, thus can direct translate the description for design into program code on the programming stage.

3.1 Flow Chart

Flow Chart is program chart, also known as program flow chart, it is pictorial representation using unified regulation standard symbol to describe program execution concrete steps, it is one kind describe tool which usage history longest and prevalent widest. From at the end of the 1940's to the middle of 1970's, flow chart is the main tool of software design all the time [3]. Basic control structure is shown in figure 1 shows.

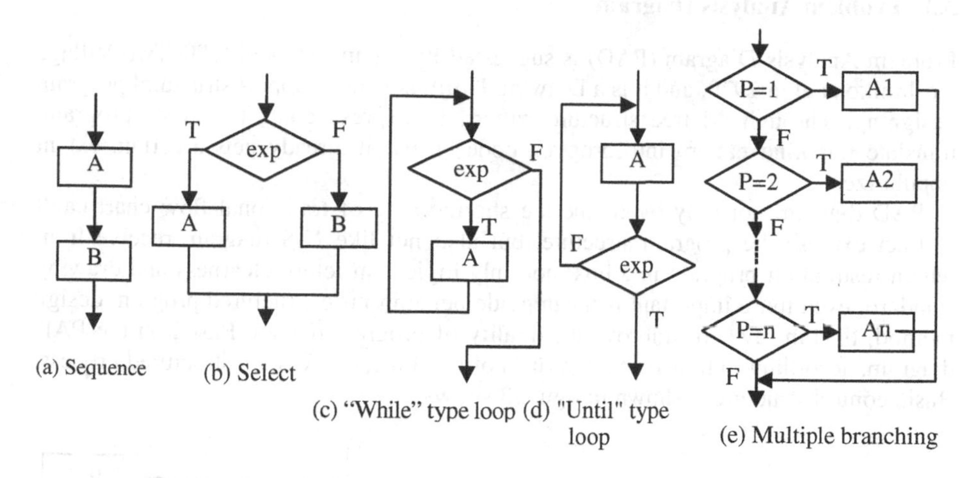

Fig. 1. Basic control structure of flow chart

3.2 Box Diagram

1983 years, American I.Nassi and B.Sheiderman together put forward one kind structuring flow chart which needn't GOTO statement and flow line, it call box diagram, also call N-S diagram. In N-S diagram, each processing step is expressed by a box, box can nest. Box only can enter from top, walk out from below, in addition to this no other entrances, so box diagram impose restriction on optional control transfer, guarantee good structure of program [4]. Basic control structure is shown in figure 2 shows.

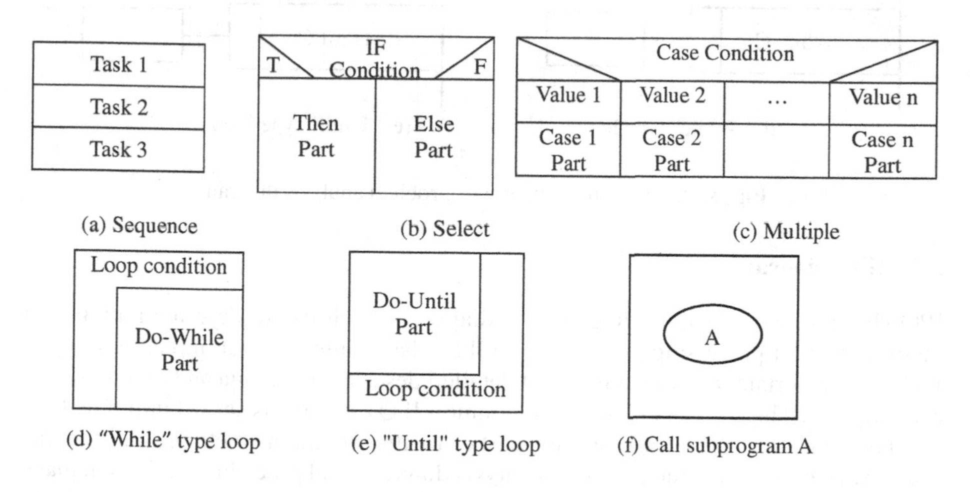

Fig. 2. Basic control structure of box diagram

3.3 Problem Analysis Diagram

Problem Analysis Diagram (PAD) is suggested by Japan's Hitachi LTD Two Village Yoshihiko et al in 1979, and it is a Drawing Utilities which support structural program designing. Through 2d tree structure pattern to express control flow of program, translate this kind pattern into program code easier, it already gets a certain extent popularize.

PAD diagram not only overcome the shortcoming of traditional flow chart can't distinct express the program structure, but also not like N-S diagram receive limit which restrain all program in a box, not only logical structure clearness and drawing standard, even more important is it can guide person to use structured program design method, thus in favor of improve the quality of program design. Based on the PAD diagram, according to transformation rules of machine, can write a structured process. Basic control structure is shown in figure 3 shows.

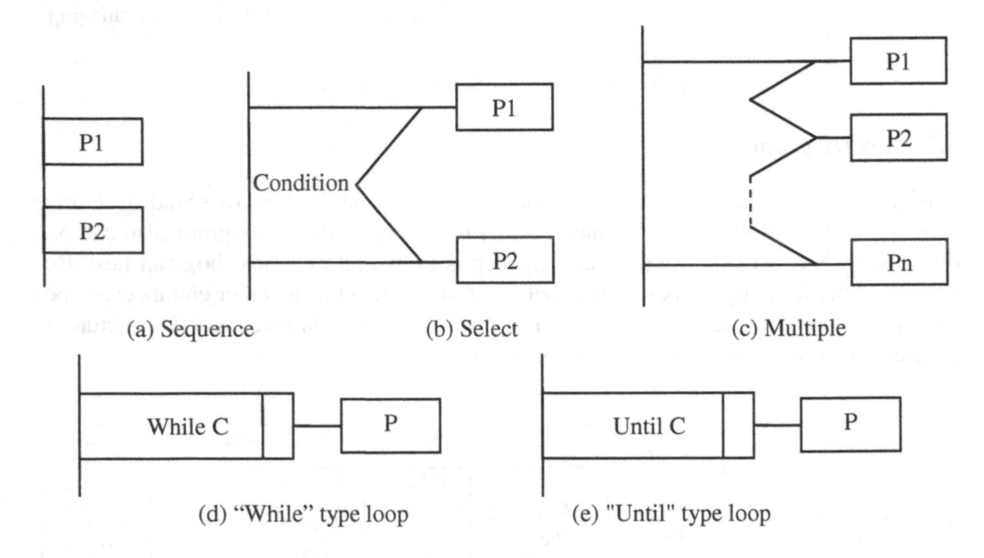

Fig. 3. Basic control structure of problem analysis diagram

3.4 IPO Diagram

IPO (Input Process Output) diagram is a diagram which use to describe a particular module internal processing procedure and the relationship between input and output. IPO is an important tool which fit HIPO detailed describe input data and output data of each module. The basic components of common IPO diagram as shown in figure 4.

The main part of IPO diagram is the part of treatment and instructions, this part can use flow chart, N-S diagram, problem analysis diagram, and procedure design language etc tools to describe, these methods each has different strengths and application scope, which method shall we choose in practical work, it need according to particular case and designer's habit, cardinal rule for choose is accurately and concisely describe the details of module execution.

In IPO diagram, input data and output data derived from data dictionary. Variable declaration is to point to the defined variables within the module, and it has nothing with other parts of system, defined, storage and used only by this module. Memo is to make necessary introductions for concerned problem of this module.

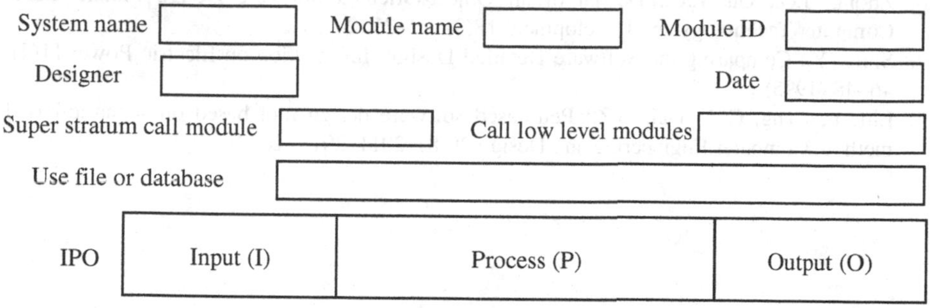

System name | Module name | Module ID
Designer | Date
Super stratum call module | Call low level modules
Use file or database

IPO	Input (I)	Process (P)	Output (O)

Fig. 4. Basic components of IPO picture

4 Conclusion

Software detailed design is core technology of software engineering, it has great effect for code generation, software testing, software maintenance and software quality etc, it is the important stage of ensure software reliability, maintainability and intelligibility, must give enough attention [7]. In team development process, there are different understand to design among member, use the means of expression is also different, such design usual use characters to describe, but text description also exist different meanings on natural language, it goes against the design effort proceed, better way is generate normative document to exchange through design tool [8]. Detailed design representation tools are key factor of ensure detailed design quality and efficiency, it is a research subject of great value. The research of this paper is reference and guide of software detailed design, and it has important guiding significance for software designer.

Acknowledgment. This work is supported by, Education department of liaoning province key laboratory fund project (2008S002), General project education department of liaoning province (L2010011), Small medium-sized enterprise innovation fund of liaoning province(20104110), and Teaching reform project of Bohai University(2009).

References

1. Ren, Y.C., Xing, T.: Software Development Process Management. Beijing Jiaotong University Press (2010)
2. Ren, Y.C., Li, C.J., et al.: Method and Management of Software Project Development. Tsinghua University Press (2011)

3. Zhang, H.F.: Introduction to Software Engineering, 4th edn. Tsinghua University Press (2003)
4. Zhang, H.F.: Software Engineering, 2nd edn. Posts & Telecom Press (2006)
5. Xu, Z.Q.: Design and Implementation of Software Design Pattern Library System. Master's degree of Huazhong University of Science and Technology (2009)
6. Zhong, J.Q., Gu, L.C.: Design of an Object-Oriented Software Design Pattern Base. Computer Technology and Development 18(9), 22–25 (2008)
7. Song, Y.: Comparing the Software Detailed Design. Information on Electric Power 11(4), 46–48 (1995)
8. Liu, J., Deng, C.Z., Dai, G.Z.: Pen-based software design tool based on scenario-based method. Computer Engineering and Design 30(8), 2011–2014 (2009)

Research on Support Vector Regression in the Stock Market Forecasting

Chun Cai, Qinghua Ma, and Shuqiang Lv

Dept of Information and Science,
College of Arts and Science, Beijing Union University,
Beijing 100191, China

Abstract. As the stock market has high noise, nonlinearity, uncertainty characteristics as well as traditional neural network forecasting method deficiencies of the problem, this paper presents a support vector regression (SVR) and financial time series methods to forecast future stock market; follows the value investment philosophy, reduces dimensional space by support vector machines the input vector mapped to a high dimensional space, and non-linear problem into a linear problem, establish the value of the equity investments of prediction models under the principle of structural risk minimization. For an empirical test of the model to get the stock opened at 0.02% error of 3.66%, closing error of 0.00% to 3.41%, the results showed that SVR prediction in the stock market index has certain advantages.

Keywords: support vector regression, time series, kernel function, parameter selection.

1 Introduction

As social economic development and gradually increase awareness of people to invest in the stock. But investing in stocks is associated with considerable risk, the higher the investment, the greater the possible risk. Currently, the method on the stock price trend forecast is done in two ways: fundamental analysis and technical analysis method [1]. Fundamental analysis, which as far as possible to identify all thc factors that affect stock price volatility, stock price established between these factors and the model to predict stock prices. This approach has a strong theoretical basis, but due to the existence of widespread speculation in financial markets , and therefore stock prices often make a serious departure from its fundamental value to the basic analysis has been recognized in practice is not high. A technical analysis and the other is from the stock market's historical data, charts, technical indicators for stock prices to forecast changes in the law. This method is based on the assumption: the securities market behavior has included the macro and micro economy, all the information; price is always some kind of movement in accordance with the trend movement; prices tend to repeat history operation mode [2].

Support Vector Machines is a new type of machine learning methods. Data mining is a new technology which is to solve the problem of machine learning new tools. Is a statistical method, it is 90 in the 20th century by the Vapnik proposed in

D. Jin and S. Lin (Eds.): Advances in ECWAC, Vol. 1, AISC 148, pp. 607–612.
springerlink.com © Springer-Verlag Berlin Heidelberg 2012

recent years in the theoretical research and algorithm have made a breakthrough. Its greatest feature is to change the traditional empirical risk minimization principle, put forward the principle of structural risk minimization, which has a global optimum and good generalization ability. Also support vector machines for nonlinear problems, the first non-linear problem into a high dimensional space of the linear problem, then use the kernel function to replace the high dimensional inner product space, which cleverly solved the complex computational problems, effectively overcome the "curse of dimensionality", "over learning" and the local minimum problem. By this method can successfully deal with classification and regression problems [3]. This paper uses support vector regression method to find historical stock data, law of the broader stock indicators to analyze and judge, to look forward to this method can predict the best time to buy stocks, helping the stock manipulator success in the investment.

2 Support Vector Regression Algorithm

Support vector machines are divided into support vector machines and support vector regression two. Originally designed for the classification, but in recent years, support vector regression algorithm in the research also showed an excellent performance. The concept of support vector regression: Let the training set are known, which is the input index, otherwise known as input, otherwise known as model, is the corresponding output. Regression problem is, for any given a new input, according to the type training set to infer that it corresponds to. Mathematical language to describe the problem as follows: Given training set, $T = \{(x_1, y_1), (x_2, y_2)\hbar\ , (x_l, y_l)\}$, Where $x_i \in R^n$, $y_i \in R$. Assume that the training set is based on a distribution of the selected $R^n \times R$ independent and identically distributed sample points, try to find a real function $\overline{f}(x)$, so as $y = \overline{f}(x)$ to infer the value of a corresponding input $x_i \in R^n$, and make the expected risk on the training set $y \in R$ .And on the training set makes the expected risk.

$$R(\overline{f}) = \int c(x, y, \overline{f}) dP(x, y) \qquad (1)$$

2.1 Alogorithm Steps

Support vector regression is not sensitive to loss of function using support vector regression. Algorithm is as follows:

(1) Given the training set: $T = \{(x_i, y_i), i = 1,2,\hbar\ , l\}$.

(2) Select the appropriate positive ε and C, as well as the appropriate kernel function [4].

(3) Construct and solve the optimization problem [5], the optimal solution

$$\bar{a}^{(*)} = (\bar{a}_1, \bar{a}_1^*, \hbar, \bar{a}_l, \bar{a}_l^*)^T \tag{2}$$

(4) Constructor function regression estimates:

$$\bar{f}(x) = \sum_{i=1}^{l} (\bar{a}_i^* - \bar{a}_i) K(x, x_i) + \bar{b} \tag{3}$$

Which $\bar{b}$ calculated by the following formula: in the range of $(0, C)$ choice $\bar{a}_j$ or $\bar{a}_k^*$, if elected to that $\bar{a}_j$, the

$$\bar{b} = y_j - \sum_{i=1}^{l} (\bar{a}_i^* - \bar{a}_i) K(x_i, x_j) + \varepsilon \tag{4}$$

If the election to that $\bar{a}^*$, the

$$\bar{b} = y_j - \sum_{i=1}^{l} (\bar{a}_i^* - \bar{a}_i)(x_i, x_j) - \varepsilon \tag{5}$$

2.2 Algorithm Kernel Function

Kernel function is the basis of nuclear techniques, and support vector machine kernel trick is an important part. SVM training process, the algorithm complexity is only determined by the number of samples, especially sample number of support vectors. However, in support vector training process, in particular, the linear non-separable case, the calculation of sample product will become very complicated. Kernel function by introducing non-linear separable data samples in high dimensional space into linearly separable, cleverly avoided the high-dimensional space, a huge amount of math problems, support vector machines can make the finite sample, the effective treatment high-dimensional problem.

Kernel function is defined as: Suppose there is a nonlinear mapping $\varphi : X \to \phi(X)$ from input space to feature space, feature transformation, you can define a kernel function as:

$$K(x, y) = (\varphi(x), \varphi(y)), x, y \in X \tag{6}$$

It can be seen within the kernel function is defined in the form of plot, this expression makes an implicit nonlinear mapping into operation. This means do not know the specific expression of the nonlinear mapping, and only the selected kernel function is enough. In this form the linear learning machine that can make the calculation process of machine learning is not dependent on the dimension of feature space, which can overcome the "curse of dimensionality." In other words, the calculation of the kernel function is to carry out the input space, SVM only a minority of the final decision function determined by SVM, computational complexity and dimension of feature space depends on the number of support vectors, not the sample space dimension, and ultimately it is only a sample of the input space, calculating the inner product, and not in the high-dimensional feature space is calculated. Kernel function used in this paper is the radial basis function, ie

$$K(x, x_i) = \exp\{-\frac{|x - x_i|^2}{\sigma^2}\} \tag{7}$$

The support vector machine function in the most widely used, the parameter values directly affect the performance of SVM learning. Kernel function parameter selection now widely used all over the cross-validation method, which although large amount of calculation, the calculation accuracy.

3 Data Experiment

180 Index on the Shanghai Stock Exchange SSE 30 Index on the original and renamed made adjustments. A shares in all markets to extract the most representative of the 180 sample stocks[6], the Shanghai 180 Index as a large scale, good liquidity, development and stability, to truly reflect the market value of the main part of the level.

This paper used on 180 of 1 July 2009 to 26 February 2010 data, in which 1 July 2009 to 31 December 2009 as a training set of data, January 5, 2010 to 2010 on February 26 as the test data set[7]. In predicting the open index, the max index of the day selected, the min index, open index, the date of the close index and trading volume as the input vector to predict the next day's open index, see Figure 1; in predicting the close index, see Figure 2, select the max index of the day, the min index, close index, trading volume predict the next day's open index, see Figure 3. the next day's close index, see Figure 4.

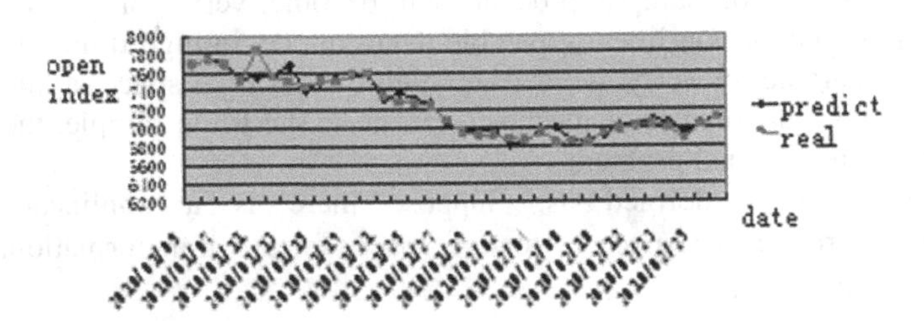

Fig. 1. 2010.1.5-2010.2.26 open index

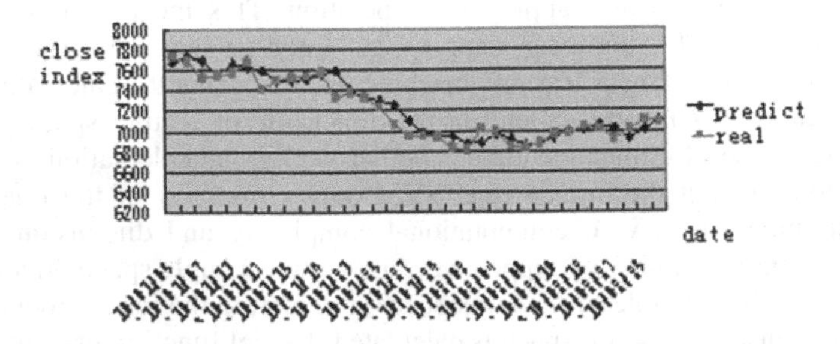

Fig. 2. 2010.1.5-2010.2.26 close index

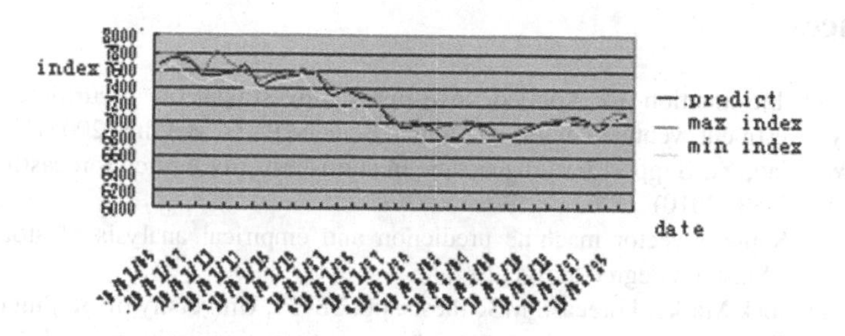

Fig. 3. The next day open index

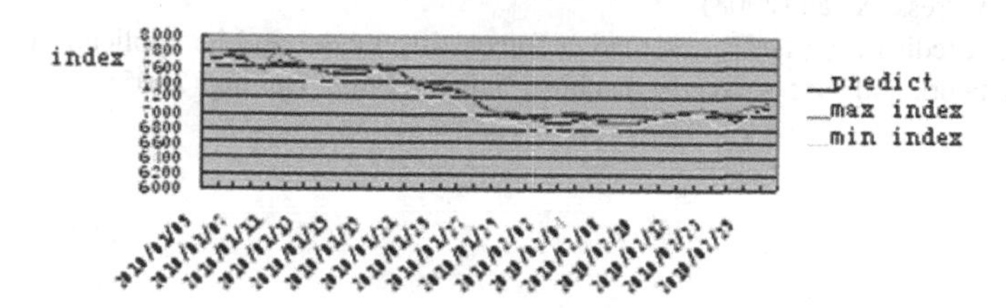

Fig. 4. The next day close index

Figure 4.3, Figure 4.4 shows the predictive value of daily essential in the daily max and min index.

Icons can be seen from the above figures, support vector machine applications in the stock prediction is quite good, using this method to forecast the stock manipulators of data can be used as a good reference document.

4 Summary

These tests showed that support vector machine approach to financial time series of short-term prediction is possible, although we can see the space visually financial time series is random, short-term unpredictable. And because SVM can map this space into a high dimensional vector space, in high-dimensional space, linear regression, and through machine learning to predict and achieve an acceptable prediction. Therefore, we once thought were financial time series are short-term understanding of the views should be updated and thinking more deeply. Thus, support vector machines to forecast the stock market is a very good prospect.

Acknowledgment. The paper is supported by Academic Human Resources (PHR201008292) and Beijing Education Committee Research Project (KM201111717004).

References

1. Wang, S.: Introduction to Applied Investment: investment in securities, investment, property investment, venture capital. Economic Science Press, Beijing (2004)
2. Zhou, W., Yao, Y.: Support vector machine in short-term stock price forecasting. Business Research, 85–88 (2010)
3. Zhang, L.: Support vector machine prediction and empirical analysis of stock. Qingdao University, a Master's degree thesis (2007)
4. Tao, X.S.: Stock Market Forecast. MSc thesis, pp. 38–39, University of Beijing (2005)
5. Yang, D., Tian, Y.: Support vector machines - theory, algorithms and Development. Science Press, Beijing (2009)
6. Bai, P., Zhang, X.: Support Vector Machine Theory and Engineering Application. Xidian University Press, Xi'an (2008)
7. Pang, S.: Credit rating and stock market forecasting model and its application - statistical, neural network and support vector machine. Science Press, Beijing (2005)

Research on Teaching Evaluation System Model Based on Trusted Network

Wen Qin[1], Hai-ying Li[1], and Wu Qin[2]

[1] College of Education, Hebei Normal University of Science & Technology
066004 Qinhuangdao, China
[2] Tianjin Railway Technical and Vocational College
300240 Tianjin, China
{qinwen003,haiyingli,qinwu001}@163.com

Abstract. Teaching evaluation is an important part in school teaching activities, which has good promoting action on improving teaching method and accelerates teaching level. In the viewpoint of trusted network, a novel evaluation model on teaching system was proposed. The model gives computation method of student credibility and integrates credibility into students-centered teaching evaluation. It also takes full account far-reaching and progressive characters of teaching, which not only making school students participate in the evaluation, but also allow graduate students to evaluate.

Keywords: trusted network, teaching evaluation, evaluation model, credibility.

1 Introduction

The rapid development of modern network made people increasingly dependent on the network. Individuals can chat with friends on the network. Many enterprises seek for business partners on the network by breaking boundaries of time and space. A lot of organizations perform various awards and enacting in accordance with result of network voting. Now more and more works depend on hand are increasingly being replaced by network information system. The subsequent result is that network credibility has attracted attention from experts [1-3]. As large amount of people or enterprises now conduct electrical business activities with network, users expect real and credible related materials by network when they want to perform network transaction with distance stranger cooperators. Therefore, the credibility under network environment includes not only security and credibility from service provider, but also authenticity and reliability of information under network environment. Teaching evaluation system is an important part in school teaching, the goal of which is to assess teaching effectiveness and teaching level of teachers. Now many systems has transformed from traditional manual operation to network operation and many schools have established their own network teaching evaluation system, which played a good role in promoting teaching in schools. Similar to common network system, the evaluation model of network teaching evaluation system plays a crucial role in results. But when the evaluation results of teachers are far away from what expected, the promotion of evaluation on teaching may result in delimitating side effects, which

D. Jin and S. Lin (Eds.): Advances in ECWAC, Vol. 1, AISC 148, pp. 613–617.
springerlink.com © Springer-Verlag Berlin Heidelberg 2012

deviate from the original intention of teaching assessment. Therefore, it is very significant to research on credible teaching evaluation system model.

Starting from the viewpoint of credible network, the paper presented a novel teaching evaluation system model. It establishes trust bond between two most important subjects in teaching with an opening network environment. The students are divided based on credibility. Thus, in case of evaluation of students on teachers, they integrate their credibility into assess data. The credibility of students is provided by classroom teachers, head teachers and familiar students. Main features of the model are as follows. Firstly, participants of teaching evaluation not only include students being educated, but also graduates that received education from this teach, which reflect the time extension effectiveness of education. Secondly, outstanding students on aspects of academic and others have more opportunities to participate in teaching evaluation in that these students undoubtedly have a high credibility. Thirdly, the teaching level is not raised in one day. Teaching evaluation reflects asymptotic character of teaching level and the result will not lead to biased assessment because of a mistake.

The paper is organized as follows: section 2 gives credible network model for teaching evaluation; section 3 proposes parameter setting method and section 4 concludes our work.

2 Network Model

Teaching evaluation is assessment on various comprehensive capabilities on teaching effectiveness of teachers. Seen from network structure, it should be a star network as shown in Fig. 1. Teachers are center of star nodes. In broadly speaking, each node connected to center include not only present lecture object, school students, but also include graduates now toward community, because the teaching effectiveness is a long-term effect on the students. Furthermore, the teaching levels will mot increase over one night, and will not instantly fall to nothing. Therefore, direct scoring from graduates is an essential factor to objectively assess on teaching effectiveness. In this way, teaching evaluation will be a constraint network system. It not only enables school students directly assess on teaching effectiveness, but also enables students that has listened to the teacher can re-assess on teaching effectiveness.

Currently, most of teaching evaluation system scores on teaching from many aspects as sound, combination of theory and practice, classroom atmosphere, knowledge depth, humor and others. Then the result will be arrived from summation of weighted aspects, which can be expressed as:

$$\begin{cases} f = \sum_{i=1}^{m} \lambda_i x_i \\ x_i = \dfrac{1}{n_i} \sum_{j=1}^{n_i} m_{ij} \end{cases} \tag{1}$$

Where, $\lambda_i (i = 1,2,\hbar\ ,m)$ is the weight of i-th scoring; x_i is average score given by evaluators and $m_{ij} (j = 1,2,\hbar\ ,n_i)$ is the i-th score from the j-th evaluator.

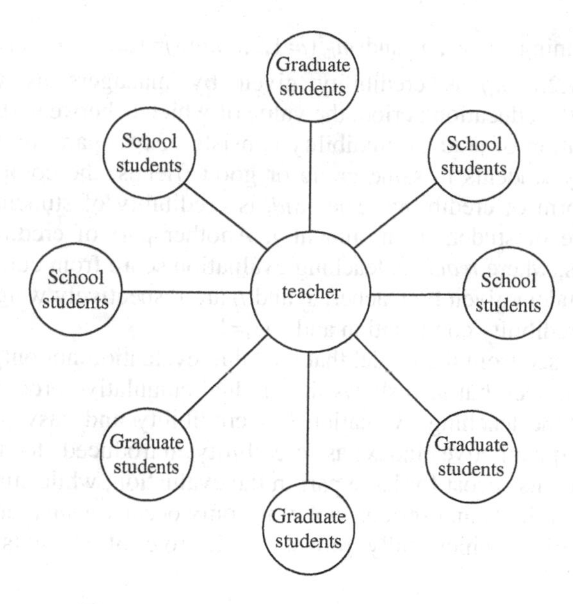

Fig. 1. Star teaching evaluation network

2.1 Evaluation Model

Although (1) is relatively straightforward, it is easy to be accepted by most teachers and students. The model has ignored some key factors in teaching. For example, teaching is a gradual process. Many teachers are recognized and highly valued by society, the result of many years engaged in teaching, but not depend on temporary power. So the model algorithm does not consider progressive nature of teaching effectiveness and teaching level. Secondly, as subject in the teaching, students have great difference. As students are in the specific age, we should take full account of various factors in the evaluation process, such as learning, thinking, behavior and growth conditions, which has not been reflected in the model. So, we present the following model:

$$\begin{cases} f = \sum_{i=1}^{m} \lambda_i x_j \\ x_i = \dfrac{\sum_{j=1}^{n_i} stud_j m_{ij}}{\sum_{j=1}^{n_i} stud_j} \\ stud_j = s_j \dfrac{\sum_{i,i \neq j} stud_i stud_{ij}}{\sum_{i,i \neq j} stud_i} + t_j \dfrac{\sum_i teach_i f_{ij}}{\sum_i teach_i} \end{cases} \tag{2}$$

Where, the meaning of λ_i, x_i and m_{ij} (i=1,2,...,m; j=1,2,..., n_i) are same as above. The $stud_j$ (j=1,2,...,n_i) is credibility given by managers as well as familiar schoolmates in the education period, the value of which is between 0 an 100.

The computation of student credibility consists of two parts of different weights. One is given by students in same grade or good friends, the computation of which also take the form of credibility. The $stud_i$ is credibility of students, while $stud_{ij}$ is evaluation score of student i on student j. Another part of credibility is given by relative teachers, where $teach_i$ is teaching evaluation score from current teachers; f_{ij} is credibility of student given by teacher; s_j and t_j are respectively weight of student and teacher in the credibility computation and $t_j+s_i=1$.

Thus, we can see from the model that teaching evaluation not only reflects the idea of students as subject, but also shows the gradual cumulative process of teaching. At the same time, the teaching evaluation has credibility and easy to be accepted by teachers. The quantitative index as credibility introduced to the model make outstanding students accost for large part in the evaluation, while students that has not correct learning attitude and serious violations only occupy a smaller proportion in the teaching evaluation, which fully reflects main role of students in the teaching assessment.

2.2 Model Evolution

The raising of teaching level is a gradual process. Therefore, each teaching evaluation will play supervisory role and teaching level is just improved in the ongoing teaching evaluation process. So the result should has previous cumulative. We give the following evaluation evolution model:

If this is the first time to participate in teaching evaluation, f_n=f; otherwise, $f_n = \lambda f + \lambda_{last} f_{last}$.

Where, f_n is the final evaluation result; f and f_{last} are results in the time and last time respectively; λ and λ_{last} are weights in this time and last time. $\lambda_{last} + \lambda = 1$.

3 Model Data Access and Parameters Setting

3.1 Value Range for Credible Evaluation

As a means to test teaching effect, teaching evaluation has been implemented in many schools. The evaluation result is reflected mostly with score. The score may be tenth or hundred-mark system. From the point of teaching, it is not appropriate to show negative point in the evaluation [4].

3.2 Source of Model Data

From the model we can know that the main difference between proposed algorithm and common one is access of credibility from evaluators. Generally speaking, the subject to evaluate teaching effect is student, while they show a greater difference in schools regardless of primary, secondary or university. The main reason is that growth environment, family education; life habits of each person are quite different.

So, a reasonable credulity on students is help to improve aspects of students. For primary and secondary students, it is not difficult of given credibility because teachers and headers are more familiar to students. The credibility can reflect attitude of students, political ideology and so on. As far as university is concerned, the credibility of students mainly accessed from events that can reflect study attitude and thinking as award-winning cases, disciplinary cases, inventions and others.

3.3 Model Parameter Selection

There are many weight parameters in the model. The settings of these parameters should be determined according to actual situation and long-term exploration. For these graduated students, their evaluation on teachers is based on personal experience after participate in practical working, so their assessment should account for some ratio. Because of factors as school conditions, regional differences, the weight setting can be adapted to local conditions and took reasonable choice.

4 Conclusion

Based on the idea of credible network, the paper presented a novel teaching evaluation system model. The model not only reflects progressive nature and far-reaching effects of educational activities, but also integrates credibility of evaluators. The future work is to develop teaching evaluation system based on the model and to test its reasonability in practical applications. Some parameters in the system should also be optimized, especially evolution parameters in the teaching process so that the teaching evaluation system has more credibility and legitimacy.

References

1. Lewicki, R., McAllister, D., Bies, R.: Trust and distrust: new relationships and Realities. Academy of Management Review 23, 438–458 (1998)
2. Kamvar, S.D., Schlosser, M.T., Garcia-Molina, H.: The engentrust algorithm for reputation management in P2P networks. In: Proceedings of the 12 th International Conference on World Wide Web, pp. 640–651 (2003)
3. Ding, L., Kolari, P., Ganjugunte, S.: Modeling and evaluating trust network inference. In: The 7 th Inernational Workshop on Trust in Agent Societies at AAMAS, pp. 21–32 (2004)
4. Guha, R., Kumar, R., Raghavan, P., Tomkins, A.: Propagation of trust and distrust, International World Wide Web Conference, pp. 403–412 (2004)

Author Index